Pitman Research Notes in Mathematics Series

Submission of proposals for consideration

Suggestions for publication, in the form of outlines and representative samples, are invited by the Editorial Board for assessment. Intending authors should approach one of the main editors or another member of the Editorial Board, citing the relevant AMS subject classifications. Alternatively, outlines may be sent directly to the publisher's offices. Refereeing is by members of the board and other mathematical authorities in the topic concerned, throughout the world.

Preparation of accepted manuscripts

On acceptance of a proposal, the publisher will supply full instructions for the preparation of manuscripts in a form suitable for direct photo-lithographic reproduction. Specially printed grid sheets can be provided and a contribution is offered by the publisher towards the cost of typing. Word processor output, subject to the publisher's approval, is also acceptable.

Illustrations should be prepared by the authors, ready for direct reproduction without further improvement. The use of hand-drawn symbols should be avoided wherever possible, in order to maintain maximum clarity of the text.

The publisher will be pleased to give any guidance necessary during the preparation of a typescript, and will be happy to answer any queries.

Important note

In order to avoid later retyping, intending authors are strongly urged not to begin final preparation of a typescript before receiving the publisher's guidelines. In this way it is hoped to preserve the uniform appearance of the series.

Longman Scientific & Technical
Longman House
Burnt Mill
Harlow, Essex, CM20 2JE
UK
(Telephone (0279) 426721)

Titles in this series. A full list is available from the publisher on request

Velleda Baldoni
Massimo A Picardello (Editors)

University of Rome, Italy

Representations of Lie groups and quantum groups

Copublished in the United States with
John Wiley & Sons, Inc., New York

Longman Scientific & Technical
Longman Group Limited
Longman House, Burnt Mill, Harlow
Essex CM20 2JE, England
and Associated companies throughout the world.

Copublished in the United States with
John Wiley & Sons Inc., 605 Third Avenue, New York, NY 10158

First published 1994

AMS Subject Classifications: 22E, 81R

ISSN 0269-3674

ISBN 0 582 23179 5

British Library Cataloguing in Publication Data

A catalogue record for this book is
available from the British Library

Library of Congress Cataloging-in-Publication Data

European School of Group Theory (1993 : Trento, Italy)
    Representations of Lie groups and quantum groups : proceedings of
the European School of Group Theory, 1993 / V. Baldoni and M.A.
Picardello (editors).
    p. cm. -- (Pitman research notes in mathematics series ;)
    1. Lie Groups--Congresses. 2. Representations of groups-
-Congresses. 3. Quantum groups--congresses. I. Baldoni, V.
II. Picardello, Massimo A., 1949- . III. Title. IV. Series.
QA387.E97 1993
512'.55--dc20   94-10255
                                                      CIP

Printed and bound in Great Britain
by Biddles Ltd, Guildford and King's Lynn

CONTENTS

PREFACE

These are the proceedings of the 1993 session of the European School of Group Theory. The European School of Group Theory provides high level courses for young researchers on recent developments in Group Theory. Each year the School is organized in a different european country. The first session was held in 1991 in Luminy, France. The second was in Enschede, The Netherlands, in 1992. The 1993 session took place in Villa Madruzzo, Trento, Italy, from July 19 to July 30. It was run jointly with the Congress *Advances in Representation Theory of Lie Groups and Quantum Groups*, which was intended to offer a high–level overview on recent advancements in current research, and to allow the participants to interact with a large number of leading experts.

The Congress was organized by Centro Internazionale di Ricerche Matematiche (CIRM). Both the School and the Congress were made possible by the generous support of CIRM, and of Italian National Research Council and Università di Roma "Tor Vergata", and by the skillful organizational work of the secretary of CIRM, Mr. Augusto Micheletti.

The scientific program consisted of five main courses, some additional advanced lectures by other specialists, many tutorial sessions and a seminar, in which several of the younger participants presented their own research work.

The main courses were as follows:

C. De Concini (Pisa, Italy): *Quantum Groups and Poisson Groups*

M. Duflo (Paris, France): *Characters of Lie Groups*

T.H. Koornwinder (Amsterdam, The Netherlands): *Quantum Groups and q-special Functions*

N. Wallach (San Diego, USA): C^∞-*vectors in Lie Group Representations*

A. Valette (Neuchâtel, Switzerland): *Old and new about Kazhdan's Property T*

Unfortunately, Duflo's lecture notes were not available at press time. We hope that they will be included in the proceedings of a future session of the School. With the kind permission of the author, we have replaced Duflo's notes with the lecture notes of the course on representations of p-adic groups held by Marko Tadić at the first session of the School, in 1991 in Luminy.

It is a pleasure to acknowledge the excellent work done by the lecturers, which made the School very stimulating. Notwithstanding the extremely heavy burden of lectures, conferences, seminars and tutorial sessions, the audience remained very high throughout the two weeks of activities. We wish to thank all the participants for their interest and enthusiasm, and even more for the unique environment, warm and friendly, that they were able to create.

Finally, the warmest thanks to Simonetta De Nicola for her heavy job of revising all the material of this book.

Velleda Baldoni
Massimo Picardello

QUANTUM GROUPS AND POISSON GROUPS

ILARIA DAMIANI AND CORRADO DE CONCINI

INTRODUCTION.

As stated in the title, these notes cover (with some exceptions) the material of the course given by the second named author with a very valuable collaboration of the first at the Summer school in Group theory held in Trento in the summer of 1993.

Since this new subject (the basic general definitions have been given in 1985 by Drinfeld and Jimbo) has been growing at a very fast rate, our aim in this notes is not to give any kind of comprehensive account of the theory, but rather to give a basic introduction explaining some of the aspects which we hope could motivate further reading.

With this view in mind we start in chapt. 1 with the easiest (and historically first) example of Quantum Group, namely $U_q(s\ell(2))$ and, after having given its definition, we cover its finite dimensional representation theory and we explicitly determine its R-matrix. Also we remark that by choosing a suitable *integral form* for $U_q(s\ell(2))$, this algebra can be viewed as a deformation of the coordinate ring of a solvable algebraic group equipped with a Poisson structure.

In chapt. 2 we first recall a few standard facts about root weights etc. and then we introduce, following Drinfeld and especially Jimbo, the quantized enveloping algebra $U_q(\mathfrak{g})$ for any semisimple Lie algebra $\mathfrak{g}$ as Hopf algebras over the field of rational functions $\mathbb{C}(q)$ defined by generators and relations which are q analogues of Chevalley generators and Serre relations. We review a remarkable action, introduced by Lusztig, of the generalized braid group associated to the root system of $\mathfrak{g}$ which allows us to introduce some kind of root vectors which provide a Poincaré Birkhoff Witt (PBW) basis of $U_q(\mathfrak{g})$. We then briefly review the finite dimensional representation theory for $U_q(\mathfrak{g})$. This chapter contains essentially no proofs but references are given.

In chapt. 3 we study the R-matrix for $U_q(\mathfrak{g})$. We first review Drinfeld duality following a paper of Tanisaki. Using this we introduce the R-matrix and then we give an explicit formula for it due to Levendorskii and Soibelman in terms of the PBW basis and quantum exponentials. On the way we study the behavior of the Braid group action under the comultiplication. In this paragraph rather detailed proofs are given, at least in the simply laced case.

Partially supported by 40% M.U.R.S.T. program

Chapt. 4 is almost essentially dedicated to the theory of Poisson algebraic groups. We first give their definition. Then we introduce the notion of a Manin triple and then, using this, we give a set of examples which will turn out to be central in the theory of quantum groups. The notion of Manin triple also allows us to essentially reduce the classification of symplectic leaves in a Poisson group to a classification of orbits for a dual group acting on a suitable space. In our set of examples we show how this is connected to the theory of conjugacy classes in a semisimple group with a flavor reminiscent of the Kirillov-Kostant Poisson structure for the coadjont representation of a Lie algebra. We then pass to introduce a suitable notion of quantum deformation of a Poisson group. We show how with this notion we can regard the Universal enveloping algebra of a Lie algebra as a quantization of the Kirillov-Kostant structure. Finally we introduce a suitable $\mathbb{C}[q, q^{-1}]$-subalgebra A of $U_q(\mathfrak{g})$ and show that it is indeed a quantum deformation of a Poisson group H which appears in our set of examples.

Chapt. 5 has to do with the specialization U_ε of the algebra A at a primitive ℓ-th root of 1 ε (with some restrictions on ℓ). We show that this algebra can be viewed as the algebra of global sections of a suitable bundle over H. By some results obtained in section 4 this allows us to connect the representation theory of U_ε to the geometry of symplectic leaves in H and in turn to the geometry of conjugacy classes in the semisimple adjoint group having $\mathfrak{g}$ as Lie algebra. We apply these ideas together with some facts from non commutative algebra to compute the degree of U_ε. In bypassing we study a suitable filtration on U_ε whose associated graded algebra is defined by extremely simple relations (we call these algebras quasipolynomial algebras).

Finally it is a great pleasure for us to thank the organizers Velleda Baldoni and Massimo Picardello of the Summer school and its participants for creating an extremely pleasant atmosphere during the school.

1. THE QUANTUM ALGEBRA OF $s\ell(2)$.

1.1. Let q be an indeterminate. We define $U_q(s\ell(2))$ as the Hopf algebra over $\mathbb{C}(q)$ generated by the elements E, F, $K^{\pm 1}$ with relations

$$KEK^{-1} = q^2 E$$
$$KFK^{-1} = q^{-2} F$$
$$[E, F] = \frac{K - K^{-1}}{q - q^{-1}} \tag{1.1.1}$$

with comultiplication Δ, antipode S, counit ε given by

$$\Delta(E) = E \otimes 1 + K \otimes E$$
$$\Delta(F) = F \otimes K^{-1} + 1 \otimes F \tag{1.1.2}$$
$$\Delta(K) = K \otimes K$$

$$S(E) = -K^{-1}E$$
$$S(F) = -FK \tag{1.1.3}$$
$$S(K) = K^{-1}$$

$$\varepsilon(E) = \varepsilon(F) = 0$$
$$\varepsilon(K) = 1 \tag{1.1.4}$$

We leave to the reader to verify that our definitions are compatible with the axioms of a Hopf algebra.

For $n \in \mathbb{Z}$ we let

$$[n]_q = (q^n - q^{-n})/(q - q^{-1}), \ [n]_q! = [1]_q[2]_q \ldots [n]_q,$$

$$\begin{bmatrix} n \\ j \end{bmatrix}_q = [n]_q[n-1]_q \ldots [n-j+1]_q/[j]_q! \text{ for } j \in \mathbb{N}, \ \begin{bmatrix} n \\ 0 \end{bmatrix}_q = 1.$$

We want to study the finite dimensional representations of $U_q(s\ell(2))$. We start with the irreducible representations

Theorem. *1) Let V be an irreducible representation of $U_q(s\ell(2))$. Then both E and F act nilpotently on V. Dim (Ker E)=dim (Ker F) = 1. For any vector $v \in$ Ker E, $Kv = \epsilon q^n v$ with $\epsilon = \pm 1$ and n a non negative integer (we call the pair (ϵ, n) the highest weight of V).*

2) Given a pair $(\epsilon, n) \in \{\pm 1\} \times (\mathbb{N} \cup \{0\})$ there exists, up to isomorphism, a unique irreducible representation $V_{\epsilon,n}$ of highest weight (ϵ, n). dim $V_{\epsilon,n} = n+1$ and $char(V_{\epsilon,n}) = Tr_V(K) = \epsilon[n+1]_q$.

Proof. Let V be an irreducible representation; since V is finite dimensional and K is invertible, there exists an eigenvector w of K with eigenvalue $\alpha \neq 0$. Consider the vectors $E^s w$ for $s = 1, 2, \ldots$. By the relations (1.1.1) one gets that $KE^s w = \alpha q^{2n} E^s w$, so if the vectors $E^s w$ were all not equal to 0, they would be linearly independent contradicting the finite dimensionality of V. It follows that there exists s such that $E^s w \neq 0$ while $E^{s+1} w = 0$. Set $v = E^s w$. Then $v \in$ Ker (E) and v is an eigenvector for K of eigenvalue $\beta = \alpha q^{2s}$.

Consider now the vectors $F^s v$. Reasoning exactly as above we see that there exists n such that $F^n v \neq 0$ while $F^{n+1} v = 0$. Consider the space V' spanned by the vectors $F^s v$. Clearly a basis for V' is given by $v, Fv, \ldots, F^n v$ which are eigenvector for K of eigenvalue $\beta, q^{-2}\beta, \ldots, q^{-2n}\beta$. We claim that $V = V'$. For this it suffices to see, by irreducibility, that V' is a subrepresentation V. Now V' is clearly stable under F and K, furthermore the relations (1.1.1) easily imply that

$$EF^s v = m_s F^{s-1} v \tag{1.1.5}$$

with $m_s = [s]_q \frac{q^{-s+1}\beta - q^{s-1}\beta^{-1}}{q - q^{-1}}$, for all $s = 0, 1, 2, \ldots$ so that $V = V'$. This also proves that both E and F act nilpotently on V. It remains to compute β. Let us

3

apply (1.1.5) for $s = n + 1$. Since $F^{n+1} = 0$, we obtain

$$0 = EF^{n+1} = [n+1]_q \frac{q^{-n}\beta - q^n\beta^{-1}}{q - q^{-1}} F^n$$

hence $\frac{q^{-n}\beta - q^n\beta^{-1}}{q - q^{-1}} = 0$ so that $\beta = \pm q^n$.

To prove part 2) it clearly suffices, from the above considerations, to exhibit an irreducible representation $V_{\epsilon,n}$ of highest weight (ϵ, n), then all the rest will follow immediately. For this we take the vector space $\mathbb{C}(q)^{n+1}$ with basis $v_0, \ldots, v_n$ and we define a $U_q(s\ell(2))$-module structure by

$$
\begin{aligned}
Kv_i &= \epsilon q^{n-2i} v_i \\
Ev_i &= \epsilon [i]_q v_{i-1} \\
Fv_i &= [n-i]_q v_{i+1}.
\end{aligned}
\tag{1.1.6}
$$

$\square$

1.2. We now want to see that, in general, finite dimensional representations of $U_q(s\ell(2))$ are completely reducible i.e. they are isomorphic to a direct sum of irreducible representations.

Let us start by considering the Casimir element

$$C = Kq + K^{-1}q^{-1} + (q - q^{-1})^2 FE.$$

It is easy to see by direct computation that C lies in the center of $U_q(s\ell(2))$. Furthermore C acts on $V_{\epsilon,n}$ by multiplication by the scalar $\epsilon(q^{n-1} + q^{1-n}) = c_{\epsilon.n}$.

We then have

Theorem. *Let V be a finite dimensional representation of $U_q(s\ell(2))$. Let $U \subset V$ be a subrepresentation. Then there exists a subrepresentation $W \subset V$ such that $V = U \oplus W$.*

Proof. In order to show our claim we need to see that, given an exact sequence

$$0 \to U \to V \xrightarrow{\pi} U' \to 0$$

of representations, there exists $\psi : U' \to U$ such that $\pi\psi = id_{U'}$ and $\psi(av) = a\psi(v)$ for all $a \in U_q(s\ell(2))$ and $v \in U'$. Indeed $W = \psi(U')$ will be the desired complement to U.

Our first step is to reduce to the case in which U' is the trivial representation (i.e. $U' = \mathbb{C}(q)$ and $a \in U_q(s\ell(2))$ acts by multiplication by $\varepsilon(a)$). To see this notice that since $U_q(s\ell(2))$ is a Hopf algebra, $U_q(s\ell(2))$ acts on $Hom(U', V) \cong V \otimes U'^*$. Consider the subspace $H = \{\phi \in Hom(U', V) | \pi\phi = \alpha_\phi id, \alpha_\phi \in \mathbb{C}(q)\}$. It is easy to see that H is a $U_q(s\ell(2))$ submodule and that the map $\phi \to \alpha_\phi$ is a homomorphism of $U_q(s\ell(2))$ modules $H \to \mathbb{C}(q)$. Suppose this splits. Then by taking the image of

1 under the splitting we get an element $\psi \in Hom(U', V)$ such that $\pi\psi = id_{U'}$ and $\psi(av) = a\psi(v)$ for all $a \in U_q(s\ell(2))$ and $v \in U'$, thus solving our original problem.

Suppose now we have an exact sequence

$$0 \to U \to V \to \mathbb{C}(q) \to 0.$$

It is easy to see that in order to show that it splits it suffices to show that for any subrepresentation $V' \subset U$ the exact sequence

$$0 \to U/V' \to V/V' \to \mathbb{C}(q) \to 0$$

splits. Thus we can further assume that U is irreducible.

To summarize, in order to prove our theorem we are reduced to show that any exact sequence

$$0 \to U \to V \to \mathbb{C}(q) \to 0$$

with U irreducible splits.

Suppose $U \cong V_{\epsilon,n}$ is not the trivial representation i.e. $(\epsilon, n) \neq (1,0)$. Then the operator $\frac{C - c_{\epsilon,n}}{c_{1,0} - c_{\epsilon,n}}$ on V clearly has U in its Kernel and induces the identity on $\mathbb{C}(q)$ thus giving the desired splitting. If also U is trivial we leave to the reader the immediate verification that V is the direct sum of two trivial representations. $\square$

Consider now the ring $\mathbb{Z}[q, q^{-1}, \eta]/(\eta^2 - 1)$. Notice that any $U_q(s\ell(2))$ module V is the direct sum $V_+ \oplus V_-$, where V_+ (resp. V_-) is the span of the eigenvectors for K with eigenvalue q^s (resp. $-q^s$) for some integer s. Both V_+ and V_- are $U_q(s\ell(2))$ submodules and we can define

$$char(V) = Tr_{V_+}(K) + \eta Tr_{V_-}(K) \in \mathbb{Z}[q, q^{-1}, \eta]/(\eta^2 - 1).$$

We have already remarked that $Tr_{V_{\epsilon,n}}(K) = \epsilon[n+1]_q$. From this it is immediate to see

Corollary. *Let V_1 and V_2 be two finite dimensional representations of $U_q(s\ell(2))$. Then 1) $V_1 \cong V_2$ iff $char(V_1) = char(V_2)$.*
2) $ch(V_1 \otimes V_2) = ch(V_1)ch(V_2)$. In particular $V_1 \otimes V_2 \cong V_2 \otimes V_1$.

$\blacksquare$

1.3. Both the results and proofs about the finite dimensional representations of $U_q(s\ell(2))$ are completely analogous to those for the Lie algebra $s\ell(2)$. Indeed we can consider $U_q(s\ell(2))$ as a deformation of the universal enveloping algebra $U(s\ell(2))$. To make this precise let us now give a basis of $U_q(s\ell(2))$ as a vector space.

Proposition. *The monomials $F^m K^s E^n$ with m, n non negative integers and $s \in \mathbb{Z}$, are a basis of $U_q(s\ell(2))$ over $\mathbb{C}(q)$.*

Proof. It is clear from the relations that the given elements linearly span $U_q(s\ell(2))$. We need to prove linear independence.

Suppose we have a linear combination

$$\sum_{m,p \geq 0} F^m P_{m,p}(K) E^p = 0$$

with $P_{m,p}(K) \in \mathbb{C}(q)[K, K^{-1}]$. Assume at least one of the Laurent polynomials $P_{m,p}(K)$ is non zero. Let $s = \min\{p | P_{m,p}(K) \neq 0$ for some $m\}$. Consider $V_{1,n}$ with $n \geq s$. Let $v_0, \dots, v_n$ be the basis of $V_{1,n}$ given in Theorem 1.1. Then

$$0 = \sum_{m,p \geq 0} F^m P_{m,p}(K) E^p v_s = \sum_{m \geq 0} \frac{[s]_q! [n]_q!}{[n-m]_q!} P_{m,s}(q^n) v_m$$

Thus $P_{m,s}(q^n) = 0$ for large enough n. This clearly implies $P_{m,s}(K) = 0$ giving a contradiction. $\square$

1.4. Set now $R = k[q, q^{-1}]$ and let U be the R subalgebra in $U_q(s\ell(2))$ generated by the elements E, F, K and $H = [E, F]$. Clearly one has $K - K^{-1} = (q - q^{-1})H$. Furthermore one has that $\Delta(U) \subset U \otimes U$, $S(U) = U$ and $\varepsilon(U) = R$ so that U is a Hopf algebra over R. One has

Proposition. *1) The algebra $U/(q-1)$ is isomorphic as an algebra to $U(s\ell(2)) \otimes k[\mathbb{Z}/2\mathbb{Z}]$.*
2) The algebra $U/(q-1, K-1)$ is isomorphic as a Hopf algebra to $U(s\ell(2))$.

Proof. 1) It follows easily from Proposition 1.3 that the monomials $F^m H^n K^\gamma E^p$ with $m, n, p \geq 0$ and $\gamma \in \{0, 1\}$ form an R basis for U and hence a k basis for $U/(q-1)$. Furthermore $U/(q-1)$ is generated by E, F, H and K (by abuse of notation we denote with the same letter an element in U and its class modulo $q - 1$) and they satisfy the relations

$$[E, F] = H \quad [H, F] = 2E$$
$$[H, F] = -2F \quad K^2 = 1.$$

This immediately implies the claim.

2) is an immediate consequence of 1) and the definition of the Hopf algebra structure. $\square$

We have seen how to look at $U_q(s\ell(2))$ as a Hopf algebra deformation of the enveloping algebra of $s\ell(2)$. On the other hand suppose in $U_q(s\ell(2))$ we take the R subalgebra A generated by the elements K, $\overline{E} = (q - q^{-1})E$ and $\overline{F} = (q - $

$q^{-1})F$. Then one easily sees that also A is a Hopf algebra over R but $A/(q-1)$ is a commutative Hopf algebra and indeed it coincides with the coordinate ring of the algebraic group

$$H = \left\{ \left(\begin{pmatrix} t^{-1} & 0 \\ y & t \end{pmatrix}, \begin{pmatrix} t & x \\ 0 & t^{-1} \end{pmatrix} \right) \right\} \Big/ \{\pm(I,I)\}$$

with $x, y \in k$, $t \in k - \{0\}$ and I is the identity matrix. Thus we can also think $U_q(s\ell(2))$ as a deformation of an algebraic group.

1.5. We end this section by making some comments on the fact that, given two representations V_1 and V_2 of $U_q(s\ell(2))$, $V_1 \otimes V_2 \cong V_2 \otimes V_1$. If $U_q(s\ell(2))$ had been a cocommutative Hopf algebra this would have been a trivial fact since the transposition $P(v_1 \otimes v_2) = v_2 \otimes v_1$ would have been a $U_q(s\ell(2))$-module isomorphism of $V_1 \otimes V_2$ with $V_2 \otimes V_1$. Since $U_q(s\ell(2))$ is not cocommutative P is in general not $U_q(s\ell(2))$ equivariant. Let us consider now the following formal sum

$$\overline{R} = \sum_{n \geq 0} \frac{q^{-\binom{n}{2}}(q^{-1}-q)^n}{[n]_q!} E^n \otimes F^n \tag{1.5.1}$$

Notice that since E and F act nilpotently in any finite dimensional representation $\overline{R}$ gives a well defined invertible operator on the tensor product of two such representations.

From now on all the identities have to be understood as identities for operators acting on the tensor of finite dimensional representations of $U_q(s\ell(2))$.

Proposition. $\overline{R}$ *commutes with* $K \otimes K$.

$$\overline{R}(E \otimes K^{-1} + 1 \otimes E)\overline{R}^{-1} = E \otimes K + 1 \otimes E$$
$$\overline{R}(F \otimes 1 + K \otimes F)\overline{R}^{-1} = F \otimes 1 + K^{-1} \otimes F. \tag{1.5.2}$$

Proof. The fact that $\overline{R}$ commutes with $K \otimes K$ is clear. Let us show the commutation relation between $\overline{R}$ and $\Delta(E)$. For this we need the following identity in $U_q(s\ell(2))$, namely for all $n \geq 1$

$$[E, F^n] = [n]_q F^{n-1} \frac{Kq^{-n+1} - K^{-1}q^{n-1}}{q - q^{-1}}. \tag{1.5.3}$$

To see this let us proceed by induction. For $n = 1$ the identity is one of the relations (1.1.1). Suppose we have verified it for n. Then

$$EF^{n+1} = F^n EF + [n]_q F^{n-1} \frac{Kq^{-n+1} + K^{-1}q^{n-1}}{q - q^{-1}} F$$

$$= F^{n+1}E + F^n \frac{K([n]_q q^{-n-1} + 1) - K^{-1}([n]_q q^{n+1} + 1)}{q - q^{-1}}$$

$$= F^{n+1}E + [n+1]_q F^n \frac{Kq^{-n} + K^{-1}q^n}{q - q^{-1}}.$$

7

Let us go back to the proof of our proposition. We have

$$\overline{R}(E \otimes K^{-1} + 1 \otimes E) - (E \otimes K + 1 \otimes E)\overline{R} = [\overline{R}, 1 \otimes E] + \overline{R}(E \otimes K^{-1}) - (E \otimes K)\overline{R}.$$

Using (1.5.3) we get

$$[\overline{R}, 1 \otimes E] = -\sum_{n \geq 0} \frac{q^{-\binom{n}{2}}(q^{-1} - q)^n}{[n]_q!} E^{n+1} \otimes (K - K^{-1}q^{-2n})F^n = (E \otimes K)\overline{R}$$

$$-\overline{R}(E \otimes K^{-1})$$

as desired. The proof of the remaining identity is completely analogous. $\square$

1.6. We want now to define a new invertible operator $\mathcal{S}$ (usually this operator is denoted by $q^{\frac{-H \otimes H}{2}}$) acting on the tensor product of any two finite dimensional representations V_1, V_2 of $U_q(s\ell(2))$. Since K is diagonalizable on finite dimensional representations with eigenvalues ϵq^m with $\epsilon \in \{-1, 1\}$ and $m \in \mathbb{Z}$ it suffices to define $\mathcal{S}$ on vectors of the form $v_1 \otimes v_2$ with the property that $Kv_i = \epsilon_i q^{m_i}$, $i = 1, 2$. We set

$$\mathcal{S}(v_1 \otimes v_2) = \epsilon_1^{\frac{m_2}{2}} \epsilon_2^{\frac{m_1}{2}} q^{-\frac{m_1 m_2}{2}} v_1 \otimes v_2$$

(in order to define $\mathcal{S}$ we need to add a square root of q to our field $\mathbb{C}(q)$ and choose a square root of -1). We then have

Proposition. $\mathcal{S}$ commutes with $K \otimes K$.

$$\mathcal{S}(E \otimes 1)\mathcal{S}^{-1} = E \otimes K^{-1} \quad \mathcal{S}(1 \otimes E)\mathcal{S}^{-1} = K^{-1} \otimes E$$
$$\mathcal{S}(F \otimes 1)\mathcal{S}^{-1} = F \otimes K \quad \mathcal{S}(1 \otimes F)\mathcal{S}^{-1} = K \otimes F.$$

Proof. The fact that $\mathcal{S}$ commutes with $K \otimes K$ is clear. As for the other relations let us prove the first since all the remaining ones are proved in the same fashion. For v_1, v_2 as above one has

$$\mathcal{S}(E \otimes 1)\mathcal{S}^{-1}(v_1 \otimes v_2) = \epsilon_1^{\frac{-m_2}{2}} \epsilon_2^{\frac{-m_1}{2}} q^{\frac{m_1 m_2}{2}} \mathcal{S}(Ev_1 \otimes v_2) = \epsilon_2 q^{-m_2} Ev_1 \otimes v_2$$
$$= E \otimes K^{-1}(v_1 \otimes v_2)$$

proving the claim. $\square$

Take now the operator $R = \overline{R}\mathcal{S}$. One gets

Theorem. Let $a \in U_q(s\ell(2))$. Then

$$R\Delta(a)R^{-1} = \Delta'(a).$$

where if $\Delta(a) = \sum a_i \otimes b_i$, $\Delta'(a) = \sum b_i \otimes a_i$

Proof. Propositions 1.5 and 1.6 immediately imply the claim for the elements E, F, K of $U_q(s\ell(2))$. The rest follows. $\square$

Remarks. *1) From the above theorem it is clear that if we denote by P the transposition, the operator PR give an $U_q(s\ell(2))$ equivariant isomorphism between $V_1 \otimes V_2$ and $V_2 \otimes V_1$.*

2) It is well known that R is a solution to the so called Quantum Yang-Baxter equation.

2. QUANTUM GROUPS.

2.1. The quantum groups which will be the object of our study arise as q–analogues of the universal enveloping algebras of semisimple Lie algebras. The way in which we will introduce them, which is also the one in which they have been discovered, is to generalize the classical presentation of semisimple Lie algebras by Chevalley generators and Serre's relations. In this approach the main starting ingredient, which one might consider as a genetic code for the theory, is the Cartan matrix or its Dynkin diagram and the associated root system. We thus recall first quickly some basic notions and notations of this theory.

Let $C := (a_{ij})$ be a $n \times n$–matrix with integer entries such that $(i, j = 1, \ldots, n)$:

$$a_{ii} = 2, \ a_{ij} \leq 0 \text{ if } i \neq j, \tag{2.1.1}$$

and there exists a vector $(d_1, \ldots, d_n)$ with relatively prime positive integral entries d_i such that

$$(d_i a_{ij}) \text{ is a symmetric positive definite matrix.} \tag{2.1.2}$$

This is the definition of a *Cartan matrix*.

To the Cartan matrix C there is associated a finite reduced root system R, its weight and root lattices P and Q, the Weyl group W, a set of positive roots R^+, the set of simple roots Π, the fundamental weights $\omega_1, \ldots, \omega_n$, etc. Let us recall for convenience the basic definitions.

Let P be a lattice over $\mathbb{Z}$ with basis $\omega_1, \ldots, \omega_n$; P is called the *weight lattice* and the elements ω_i the *fundamental weights*. Let $Q^\vee = \mathrm{Hom}_\mathbb{Z}(P, \mathbb{Z})$ be the dual lattice with dual basis $\alpha_1^\vee, \ldots, \alpha_n^\vee$ (called the *coroots*), i.e. $\langle \omega_i, \alpha_j^\vee \rangle = \delta_{ij}$. One introduces the following objects:

$$\text{Dominant integral weights} \quad P_+ = \sum_{i=1}^{n} \mathbb{Z}_+ \omega_i$$

$$\text{A special weight} \quad \rho = \sum_{i=1}^{n} \omega_i,$$

$$\text{The simple roots} \quad \alpha_j = \sum_{i=1}^{n} a_{ij} \omega_i \quad (j = 1, \ldots, n),$$

$$\text{Root lattice} \quad Q = \sum_{j=1}^{n} \mathbb{Z} \alpha_j \subset P,$$

$$\text{Positive root lattice} \quad Q_+ = \sum_{j=1}^{n} \mathbb{Z}_+ \alpha_j$$

Define the usual partial ordering on P (*the dominant order*) by $\lambda \geq \mu$ if $\lambda - \mu \in Q_+$. For $\beta = \sum_i k_i \alpha_i \in Q$ let $ht\beta = \sum_i k_i$.

Define reflection automorphisms s_i of P by $s_i(\omega_j) = \omega_j - \delta_{ij} \alpha_i \ (i, j = 1, \ldots, n)$. Then $s_i(\alpha_j) = \alpha_j - a_{ij} \alpha_i$. Let W (the Weyl group) be the subgroup of $GL(P)$ generated by $s_1, \ldots, s_n$. Recall that W is a *Coxeter group* on generators s_i ($i = 1, \ldots, n$) and defining relations

$$s_i^2 = 1 \quad \text{and} \quad (s_i s_j)^{m_{ij}} = 1 \quad \text{when } i \neq j,$$

where $m_{ij} = 2, 3, 4$ or 6 for $a_{ij} a_{ji} = 0, 1, 2$ or 3 respectively ($i \neq j$).

Together with the Weyl group it is useful to introduce the (generalized) braid group. It is an infinite group $\mathcal{B}$ generated by elements T_i, $i = 1, \ldots, n$ and the following *braid relations*: for $i \neq j$ we take the word of (even) length $(T_i T_j)^{m_{ij}}$, split it in half and impose that the first half be equal to the second written in reverse order.

Of course the Weyl group W is the quotient of $\mathcal{B}$ under the further relations $T_i^2 = 1$.

It will be convenient to use the following abbreviated notation:

$$T_{ij}^{(m)} = T_i T_j T_i \ldots (m \text{ factors}).$$

For example, the braid relations read:

$$T_{ij}^{(m_{ij})} = T_{ji}^{(m_{ji})} \quad \text{if } i \neq j. \tag{2.1.3}$$

Let

$$\Pi = \{\alpha_1, \ldots, \alpha_n\}, \ \Pi^\vee = \{\alpha_1^\vee, \ldots, \alpha_n^\vee\},$$
$$R = W\Pi, \ R^+ = R \cap Q_+, \ R^\vee = W\Pi^\vee.$$

R is the set of *roots*, $R^\vee$ the *coroots*, Π the *simple roots*. The map $\alpha_i \longmapsto \alpha_i^\vee$ extends uniquely to a bijective W–equivariant map $\alpha \longmapsto \alpha^\vee$ between R and $R^\vee$. The reflection s_α defined by $s_\alpha(\lambda) = \lambda - \langle \lambda, \alpha^\vee \rangle \alpha$ lies in W for each $\alpha \in R$, so that $s_{\alpha_i} = s_i$.

Define a bilinear pairing $P \times Q @ >>> \mathbb{Z}$ by $(\omega_i | \alpha_j) = \delta_{ij} d_j$. Then $(\alpha_i | \alpha_j) = d_i a_{ij}$, giving a symmetric $\mathbb{Z}$–valued W–invariant bilinear form on Q such that $(\alpha | \alpha) \in 2\mathbb{Z}$. We may identify $Q^\vee$ with a sublattice of the $\mathbb{Q}$-span of P (containing Q) using this form. Then:

$$\alpha_i^\vee = d_i^{-1} \alpha_i, \ \alpha^\vee = 2\alpha/(\alpha | \alpha). \tag{2.1.4}$$

2.2. Every element of W has a length $\ell(w)$ which can be defined as the length of a shortest expression of w as a product of s_i and equals the cardinality of $R_w := \{\beta \in R^+ | w(\beta) < 0\}$; such an expression is called a *reduced expression* of w. Recall that

$$\ell(ws_i) = \ell(w) + 1 \text{ if } w(\alpha_i) > 0 \text{ and } \ell(ws_i) = \ell(w) - 1 \text{ if } w(\alpha_i) < 0.$$

In particular, since W acts transitively on the bases there exists a unique element w_0 of longest length $N = |R^+|$ such that $w_0(R^+) = -R^+$. Of course $w_0 = w_0^{-1}$, $w_0(\Pi) = -\Pi$ and $w_0(P_+) = -P_+$. For $\alpha \in R^+$ (resp. $\lambda \in P_+$) we let $^t\alpha = -w_0(\alpha) \in R^+$ (resp. $^t\lambda = -w_0(\lambda)$). For a fundamental weight ω, the weight $^t\omega$ is also fundamental and since $\Pi = -w_0(\Pi)$ we denote by $j \mapsto \bar{j}$ the permutation of $1, 2, \ldots, n$ such that $\alpha_{\bar{j}} = -w_0(\alpha_j)$.

We have that $s_j w_0 = w_0 s_{\bar{j}}$. More precisely, writing

$$w_0 = s_j s_{i_1} s_{i_2} \ldots s_{i_{N-1}} = s_{i_1} s_{i_2} \ldots s_{i_{N-1}} s_{\bar{j}}$$

we deduce:

Lemma. $s_{i_1} s_{i_2} \ldots s_{i_{N-1}}(\alpha_{\bar{j}}) = \alpha_j$.

Proof. $s_{i_1} s_{i_2} \ldots s_{i_{N-1}}(\alpha_{\bar{j}}) = s_j w_0(\alpha_{\bar{j}}) = s_j(-\alpha_j)$. $\square$

If $w = ab \in W$ is such that $\ell(w) = \ell(a) + \ell(b)$ we will say that this is a *reduced decomposition*.

Given an element $w \in W$ we set $\bar{w} := w_0 w w_0^{-1}$ (so that $\bar{s}_j = s_{\bar{j}}$).

Fix a reduced expression

$$w_0 = s_{i_1} s_{i_2} \ldots s_{i_N}, \text{ where } N = |R^+|, \tag{2.2.1}$$

then we have the corresponding *convex* ordering of R^+:

$$\beta_1 = \alpha_{i_1}, \; \beta_2 = s_{i_1}(\alpha_{i_2}) \ldots, \; \beta_N = s_{i_1} \ldots s_{i_{N-1}}(\alpha_{i_N}).$$

(The name "convex" refers to the property that if $i < j$ and $\beta_i + \beta_j \in R^+$, then $\beta_i + \beta_j = \beta_k$ for some k between i and j.)

For $R_{w^{-1}}$ the notion of convex ordering needs one further condition. If $\beta_i, \beta_k \in R_{w^{-1}}$, $\beta_k = \beta_i + \gamma$ with γ a positive root not in $R_{w^{-1}}$, then $i < k$.

A major tool of the computations to follow is a result of Matsumoto [M]:

Theorem. *If $s_{i_1} s_{i_2} \ldots s_{i_m}$ is a reduced expression of an element $w \in W$, then the element $T_w := T_{i_1} T_{i_2} \ldots T_{i_m}$ in $\mathcal{B}$ depends only on w (and not on its reduced expression).* $\qquad\square$

In other words we can pass from a reduced expression of w to another using only the braid relations.

In particular we have defined a canonical section $w \to T_w$ of W in $\mathcal{B}$; of course this section is not multiplicative but $T_a T_b = T_{ab}$ if $\ell(ab) = \ell(a) + \ell(b)$.

2.3. There are several objects, which have recently appeared and may deserve the name of quantum group. Here we introduce as quantum groups a simple variation of the construction of Drinfeld and Jimbo of the quantized enveloping algebra associated to a given Cartan matrix. The *dual* of such Hopf algebras could also be justly considered as interesting quantum groups (cf. for example [S], [LS]).

As we have already explained at the beginning of this chapter we follow the line of describing a q−analogue of the presentation of a semisimple Lie algebra by generators and relations. The reader should look at the standard presentation where the Chevalley generators are denoted by e_i, f_i, h_i and compare the usual Serre's relations with the ones we will present soon, he will find some mysterious differences. The explanation of these is really unclear, the discovery was made for $SL(2)$ first of all by Kulish-Reshetikhin who were trying to quantize the sin-Gordon equation.

One defines the *simply connected quantum group* U_P associated to the matrix (a_{ij}) as an algebra over the field $\mathbb{C}(q)$ on generators E_i, F_i ($i = 1, \ldots, n$), K_α ($\alpha \in P$) subject to the following relations

$$
\begin{aligned}
&K_\alpha K_\beta = K_{\alpha+\beta}, \; K_0 = 1, \\
&\sigma_\alpha(E_i) = q^{(\alpha|\alpha_i)} E_i, \; \sigma_\alpha(F_i) = q^{-(\alpha|\alpha_i)} F_i, \\
&[E_i, F_j] = \delta_{ij} \frac{K_{\alpha_i} - K_{-\alpha_i}}{q^{d_i} - q^{-d_i}}, \\
&(adE_i)^{1-a_{ij}} E_j = 0, \; (adF_i)^{1-a_{ij}} F_j = 0 \; (i \neq j),
\end{aligned}
\tag{2.3.1}
$$

where if $\Delta(x) = \sum x_j \otimes y_j$, $ad(x)(y) = \sum x_j y S(y_j)$. One should think of E_i, F_i as q−analogues of the e_i, f_i while the K_{α_i} are q−analogues of $q^{d_i \alpha_i}$. It has been shown that U_P has a Hopf algebra structure with comultiplication Δ, antipode S and counit ϵ defined by:

$$\Delta E_i = E_i \otimes 1 + K_{\alpha_i} \otimes E_i, \ \Delta F_i = F_i \otimes K_{-\alpha_i} + 1 \otimes F_i, \ \Delta K_\alpha = K_\alpha \otimes K_\alpha,$$

$$S E_i = -K_{-\alpha_i} E_i, \ S F_i = -F_i K_i, \ S K_\alpha = K_{-\alpha},$$

$$\epsilon E_i = 0, \ \epsilon F_i = 0, \ \epsilon K_\alpha = 1.$$

$$(2.3.2)$$

In order to verify this statement one has first of all to verify that the maps are well defined, i.e. compatible with the relations, and then the various Hopf algebras identities. To make these verifications is rather tedious (especially when $a_{ij} = -3$) but straightforward. It is usual to write $K_i := K_{\alpha_i}$. The quantum group $U_q(\mathfrak{g})$ of Drinfeld–Jimbo is the subalgebra of U_P over $\mathbb{C}(q)$ generated by the E_i, F_i, K_i, K_i^{-1} ($i = 1, \dots, n$). More generally, for any lattice M between P and Q one may consider the intermediate quantum group U_M generated by the E_i, F_i ($i = 1, \dots, n$) and the K_β with $\beta \in M$. In this notes we shall treat the case of $U_q(\mathfrak{g})$ although, regarding some question such as the determination of the center, it is more convenient to consider the simply connected case.

We denote by U^+, U^- and U^0 the $\mathbb{C}(q)$–subalgebra of $U_q(\mathfrak{g})$ generated by the E_i, the F_i and the K_β respectively. By $U^{\geq 0}$ (resp. $U^{\leq 0}$) the algebras generated by U^+ and U^0 (resp. U^- and U^0). Before going into deeper facts about the structure of these algebras we want to make a remark. We wish to analyze its 1-dimensional representations.

In other words we need to substitute to the given generators numbers satisfying the same relations.

From $[E_i, F_i] = \frac{K_{\alpha_i} - K_{-\alpha_i}}{q^{d_i} - q^{-d_i}}$ we deduce $K_{\alpha_i} - K_{-\alpha_i} = 0$ or $K_{\alpha_i} = \pm 1$.

From the commutation relations between the K_i and E_i, F_i we deduce $E_i = F_i = 0$. Thus we have 2^n 1-dimensional representations given assigning the values ± 1 to the K_i, we identify these representations with the characters of $P/2P$. Given any representation M and a character δ, since $U_q(\mathfrak{g})$ is a Hopf algebra we can perform the tensor product between these representations which will be denoted by M_δ.

2.4. We review the method which has been developed to construct quantum analogues of Poincaré Birkhoff Witt bases.

Lusztig ([L2]) has defined an action of the braid group $\mathcal{B}_W$ (associated to W), whose canonical generators one denotes by T_i, as a group of automorphisms of the algebra $U_q(\mathfrak{g})$ by the formulas:

$$T_i K_\alpha = K_{s_i(\alpha)}, \ T_i E_i = -F_i K_{\alpha_i},$$

$$T_i E_j = (-ad E_i)^{-a_{i,j}} (E_j), \qquad (2.4.1)$$

$$T_i \kappa = \kappa T_i,$$

where κ is a conjugate–linear anti–automorphism of $U_q(\mathfrak{g})$, viewed as an algebra over $\mathbb{C}(q)$ with conjugation given by $\kappa q = q^{-1}$, defined by:

$$\kappa E_i = F_i, \ \kappa F_i = E_i, \ \kappa K_\alpha = K_{-\alpha}, \ \kappa q = q^{-1}. \tag{2.4.2}$$

Again one might verify directly that the maps and their inverses are well defined and satisfy the braid relations.

Notice that these automorphisms do not respect the comultiplication or the antipode so they are not Hopf algebra automorphisms.

One of the main ingredients consists in using the braid group to construct analogues of root vectors associated to non simple roots.

2.5. Let us now take a reduced expression $w_0 = s_{i_1} \cdots s_{i_N}$ for the
longest element w_0 in the Weyl group. Set $\beta_t = s_{i_1} \cdots s_{i_{t-1}}(\alpha_{i_t})$ and get a total ordering on the set $\{\beta_1, \dots, \beta_N\}$ of positive roots (the convex ordering defined in 2.2).

We define the elements $E_{\beta_1}, \dots, E_{\beta_N}$ by $E_{\beta_t} = T_{i_1} \cdots T_{i_{t-1}}(E_{\alpha_{i_t}})$. Similarly the elements $F_{\beta_1}, \dots, F_{\beta_N}$ by $F_{\beta_t} = T_{i_1} \cdots T_{i_{t-1}}(F_{\alpha_{i_t}})$. These elements depend heavily on the choice of the reduced expression.

Theorem. ([L],[LS]). 1) $E_{\beta_t} \in U^+ \ \forall t = 1, \dots, N$ (resp. $F_{\beta_t} \in U^- \ \forall t = 1, \dots, N$.)

2) The monomials $E_{\beta_1}^{k_1} \cdots E_{\beta_N}^{k_N}$ (resp. $F_{\beta_N}^{k_N} \cdots F_{\beta_1}^{k_1}$) are a $\mathbb{C}(q)$ basis of U^+ (resp. U^-).

3) The monomials

$$F_{\beta_N}^{h_N} \cdots F_{\beta_1}^{h_1} K_\lambda E_{\beta_1}^{k_1} \cdots E_{\beta_N}^{k_N} \tag{2.5.1}$$

with $\lambda \in Q$, are a $\mathbb{C}(q)$ basis of $U_q(\mathfrak{g})$.

4) [LS] For $i < j$ one has:

$$E_{\beta_i} E_{\beta_j} - q^{(\beta_j|\beta_i)} E_{\beta_j} E_{\beta_i} = \sum_{k \in \mathbb{Z}_+^N} c_k E^k, \tag{2.5.2}$$

where $c_k \in \mathbb{Q}[q, q^{-1}]$ and $c_k \neq 0$ only when $k = (k_1, \dots, k_N)$ is such that $k_s = 0$ for $s \leq i$ and $s \geq j$ and $E^k = E_{\beta_1}^{k_1} \cdots E_{\beta_N}^{k_N}$.

$$F_{\beta_i} F_{\beta_j} - q^{-(\beta_j|\beta_i)} F_{\beta_j} F_{\beta_i} = \sum_{k \in \mathbb{Z}_+^N} c_k F^k, \tag{2.5.3}$$

where $c_k \in \mathbb{Q}[q, q^{-1}]$ and $c_k \neq 0$ only when $k = (k_N, \dots, k_1)$ is such that $k_s = 0$ for $s \leq i$ and $s \geq j$ and $F^k = F_{\beta_N}^{k_N} \cdots F_{\beta_1}^{k_1}$.

5) [R2] We have the tensor product decomposition (as vector spaces over the field $\mathbb{C}(q)$)

$$U_q(\mathfrak{g}) = U^- \otimes U^0 \otimes U^+.$$

$\square$

2.6. We want now to state a few facts on the representation theory of the algebras $U_q(\mathfrak{g})$, which, as in the case of $U_q(s\ell(2))$, is completely parallel to the representation theory of $\mathfrak{g}$.

Let us start with the finite dimensional irreducible $U_q(\mathfrak{g})$-modules. Fix a dominant weight $\lambda \in P_+$ and let $\varepsilon \in C_2^n$ ($C_2 = \{\pm 1\}$). Then one has

Theorem [L1]. *1) There exists a unique (up to isomorphism) irreducible $U_q(\mathfrak{g})$-module $V_{\varepsilon,\lambda}$ characterized by the existence of a non zero vector $v_{\varepsilon,\lambda} \in V_{\varepsilon,\lambda}$ (unique up to non zero multiples) such that*

$$K_i v_{\varepsilon,\lambda} = \varepsilon_i q^{\langle \lambda, \alpha_i^\vee \rangle} v_{\varepsilon,\lambda} \quad E_i v_{\varepsilon,\lambda} = 0.$$

2) Every finite dimensional irreducible $U_q(\mathfrak{g})$-module is isomorphic to $V_{\varepsilon,\lambda}$ for some pair (ε, λ).

Furthermore $V_{\varepsilon,\lambda}$ is a sum of weight spaces for the K_i's and the function $ch(V_{\varepsilon,\lambda})$: $K_i \longrightarrow Tr_{V_{\varepsilon,\lambda}}(K_i)$ is given by ε times the Weyl character formula.

2) Every finite dimensional irreducible $U_q(\mathfrak{g})$-module is isomorphic to $V_{\varepsilon,\lambda}$ for some pair (ε, λ).

For a general finite dimensional $U_q(\mathfrak{g})$-module one has

Proposition. *Any finite dimensional $U_q(\mathfrak{g})$-module V is completely reducible. Hence it is isomorphic to a direct sum of irreducible submodules.*

2.7. At this point as in section 1 we can define the character of a finite dimensional module V, first decomposing V as the direct sum $V = \oplus V_\varepsilon$ where V_ε is the sum of the irreducible components of type ε. One can the define the character of V as the function with values in $\mathbb{Z}[q, q^{-1}] \otimes \mathbb{Z}[C_2^n]$ as

$$char(V)(K_i) = \sum_\varepsilon \varepsilon Tr_{V_\varepsilon}(K_i).$$

A straightforward application of Weyl character formula then shows

Corollary. *Let V_1 and V_2 be two finite dimensional representations of $U_q(\mathfrak{g})$. Then*
1) $V_1 \cong V_2$ iff $char(V_1) = char(V_2)$.
2) $ch(V_1 \otimes V_2) = ch(V_1)ch(V_2)$. In particular $V_1 \otimes V_2 \cong V_2 \otimes V_1$.

In the following section we shall see how the isomorphism $V_1 \otimes V_2 \cong V_2 \otimes V_1$ can be explicitly realized by the so called R-matrix.

3. THE R-MATRIX.

In section one we have exhibited a certain formal operator R acting on tensor product of finite dimensional $U_q(s\ell(2))$-modules which exchanges the comultiplication with the opposite comultiplication. We call this operator the R-matrix. In this section we shall firstly briefly recall the existence of a similar operator for $U_q(\mathfrak{g})$ in the general case and show how to explicitly write down such a R-matrix. This formula has been obtained in [LS2] but, as far as we know, no proof of it has appeared in printed form. For the sake of simplicity of the exposition we shall restrict to the case in which the Dynkin diagram is simply laced. Let us recall a few facts about Drinfeld duality (for a clear treatment see [T]). One constructs a pairing $(\cdot|\cdot)$ between $U^{\geq 0}$ and $U^{\leq 0}$ and shows how to find an R-matrix using the canonical elements of this pairing.

3.1. We recall the definition of the pairing between $U^{\geq 0}$ and $U^{\leq 0}$.

Proposition. *There exists a unique pairing* $(\cdot|\cdot) : U^{\geq 0} \times U^{\leq 0} \to \mathbb{C}(q)$ *such that*

$$(x_1 x_2 | y) = (x_2 \otimes x_1 | \Delta(y))$$
$$(x | y_1 y_2) = (\Delta(x) | y_1 \otimes y_2)$$
$$(K_\lambda | K_\mu) = q^{-(\lambda,\mu)}$$
$$(K_\lambda | F_i) = 0$$
$$(E_i | K_\lambda) = 0$$
$$(E_i | F_j) = \delta_{ij} \frac{1}{q^{-1} - q}$$

This pairing is such that $(\cdot|\cdot)|_{U^{\geq 0}_{\beta} \times U^{\leq 0}_{-\gamma}} = 0$ *if* $\beta \neq \gamma$ *and* $(\cdot|\cdot)|_{U^{\geq 0}_{\beta} \times U^{\leq 0}_{-\beta}}$ *is non degenerate.* $\square$

3.2.

Proposition. *Let* $C_\beta \in U^+_\beta \otimes U^-_{-\beta}$ *be the canonical element of the pairing* $(\cdot|\cdot)$ *restricted to* $U^+_\beta \times U^-_{-\beta}$; *then if we put* $R \doteq (\sum_{\beta \in Q^+} C_\beta) q^{-\sum_i \tilde{H}_i \otimes \tilde{H}_i}$ *where* $(\tilde{H}_i, \tilde{H}_j) = \delta_{ij}$ *we have that* R *is an* R*-matrix for* U. *This means in particular that, for any element* $a \in U_q(\mathfrak{g})$ $R\Delta(a)R^{-1} = \Delta'(a)$, Δ' *denoting the opposite comultiplication.* $\square$

For the proof of the two preceding propositions see [T].

3.3. In the study of the R-matrix of the quantum group of a simple Lie algebra with simply laced Dynkin diagram, the R-matrix of $sl(2)$ plays a fundamental role. It is then convenient to exhibit it (in fact, not surprisingly we shall find the R-matrix already found in **1.5**).

Proposition. *The R-matrix of* $sl(2)$ *is*

$$R_{sl(2)} = \sum_{n \geq 0} \frac{(q^{-1} - q)^n q^{-\frac{n(n-1)}{2}}}{[n]_q!} E^n \otimes F^n q^{-\frac{H \otimes H}{2}}$$

where $q^H = K$.

Proof. We know that

$$R_{sl(2)} = \sum_{n \geq 0} C_n q^{-\frac{H \otimes H}{2}}$$

where, if $(E^n|F^n) = \frac{1}{a_n} \left(\in \mathbb{C}(q) \right)$, $C_n \doteq a_n E^n \otimes F^n$.

Let's calculate the a_n's.

Obviously we have that $a_0 = \frac{1}{(1|1)} = 1$, $a_1 = \frac{1}{(E|F)} = (q^{-1} - q)$; suppose $n > 1$; then

$$(E^n|F^n) = (\Delta E^n|F^{n-1} \otimes F) = \left(\sum_{r=1}^{n} E^{r-1} K E^{n-r} \otimes E \middle| F^{n-1} \otimes F \right) =$$

$$= q^{n-1}[n]_q (E^{n-1}|F^{n-1})(E|F) = \frac{[n]_q! q^{\frac{n(n-1)}{2}}}{(q^{-1} - q)^n},$$

so that $a_n = \dfrac{(q^{-1} - q)^n q^{-\frac{n(n-1)}{2}}}{[n]_q!}$. $\qquad\qquad\square$

Remark. Let us define another q-analogue of n by setting $(n)_q \doteq \dfrac{q^{2n} - 1}{q^2 - 1}$; then

$$R_{sl(2)} = \sum_{n \geq 0} \frac{((q^{-1} - q)E \otimes F)^n}{(n)_q!} q^{-\frac{H \otimes H}{2}} = exp_q((q^{-1} - q)E \otimes F) q^{-\frac{H \otimes H}{2}}.$$

In what follows we shall assume U to be the quantum algebra of a simple Lie algebra whose associated Dynkin diagram is simply laced; U is, as usually, Q-graded and its generators are E_i, F_i, $K_i^{\pm 1}$ $(i = 1, ..., n)$, where K_i can be seen as q^{H_i} (the generators of the Lie algebra are e_i, f_i, H_i).

3.4. We shall now introduce, for each $i = 1, ..., n$ ($n =$ rank of U) the ith *R-matrix* of U, which can be seen as a partial *R*-matrix, useful in order to calculate the *R*-matrix of U.

Lemma. $\forall\, x_\alpha \in U_\alpha$ and $\forall\, x_\beta \in U_\beta$ we have

$$q^{\frac{H_i \otimes H_i}{2}} (x_\alpha \otimes x_\beta) q^{-\frac{H_i \otimes H_i}{2}} = q^{\frac{\alpha(H_i)\beta(H_i)}{2}} (x_\alpha \otimes x_\beta) \left(K_i^{\frac{\beta(H_i)}{2}} \otimes K_i^{\frac{\alpha(H_i)}{2}} \right)$$

and analogously

$$q^{-\frac{H_i \otimes H_i}{2}} (x_\alpha \otimes x_\beta) q^{\frac{H_i \otimes H_i}{2}} = q^{-\frac{\alpha(H_i)\beta(H_i)}{2}} (x_\alpha \otimes x_\beta) \left(K_i^{-\frac{\beta(H_i)}{2}} \otimes K_i^{-\frac{\alpha(H_i)}{2}} \right).$$

17

Proof. Let h be such that $q = e^h$; then

$$q^{\frac{H_i \otimes H_i}{2}} (x_\alpha \otimes x_\beta) q^{-\frac{H_i \otimes H_i}{2}} = \sum_{r,s \geq 0} \frac{(-1)^s}{r!s!} \left(\frac{h}{2}\right)^{r+s} H_i^r x_\alpha H_i^s \otimes H_i^r x_\beta H_i^s =$$

$$= \sum_{t \geq 0} \frac{1}{t!} \left(\frac{h}{2}\right)^t \sum_{s=0}^{t} (-1)^s \binom{t}{s} x_\alpha (H_i + \alpha(H_i))^{t-s} H_i^s \otimes x_\beta (H_i + \beta(H_i))^{t-s} H_i^s =$$

$$= (x_\alpha \otimes x_\beta) \sum_{t \geq 0} \frac{1}{t!} \left(\frac{h}{2}\right)^t \left((H_i + \alpha(H_i)) \otimes (H_i + \beta(H_i)) - H_i \otimes H_i\right)^t =$$

$$= (x_\alpha \otimes x_\beta) \left(q^{\frac{\beta(H_i)H_i}{2}} \otimes 1\right) \left(1 \otimes q^{\frac{\alpha(H_i)H_i}{2}}\right) q^{\frac{\alpha(H_i)\beta(H_i)}{2}} =$$

$$= q^{\frac{\alpha(H_i)\beta(H_i)}{2}} (x_\alpha \otimes x_\beta) \left(K_i^{\frac{\beta(H_i)}{2}} \otimes K_i^{\frac{\alpha(H_i)}{2}}\right).$$

The claim is thus proved. $\square$

Definition. $\forall i$ let us define R_i to be the ith R-matrix of U, that is

$$R_i \doteq \exp_q\left((q^{-1} - q)E_i \otimes F_i\right) q^{-\frac{H_i \otimes H_i}{2}} \left(= \sum_{n \geq 0} \frac{\left((q^{-1} - q)E_i \otimes F_i\right)^n}{(n)_q!} q^{-\frac{H_i \otimes H_i}{2}}\right);$$

also, let's define $\tilde{R}_i$ to be

$$\tilde{R}_i \doteq q^{\frac{H_i \otimes H_i}{2}} R_i = \sum_{n \geq 0} \frac{(q^{-1} - q)^n q^{-\frac{n(5n-1)}{2}}}{[n]_q!} E_i^n K_i^{-n} \otimes F_i^n K_i^n.$$

Remark. R_i is invertible, and R_i^{-1} is given by

$$R_i^{-1} = q^{\frac{H_i \otimes H_i}{2}} \sum_{n \geq 0} \frac{(q - q^{-1})^n q^{\frac{n(n-1)}{2}}}{[n]_q!} E_i^n \otimes F_i^n =$$

$$= \sum_{n \geq 0} \frac{(q - q^{-1})^n q^{\frac{n(n-1)}{2} - 2n^2}}{[n]_q!} (E_i^n K_i^{-n} \otimes F_i^n K_i^n) q^{\frac{H_i \otimes H_i}{2}}.$$

Proof.

$$\sum_{n,m \geq 0} \frac{(q^{-1} - q)^n q^{-\frac{n(n-1)}{2}}}{[n]_q!} \frac{(q - q^{-1})^m q^{\frac{m(m-1)}{2}}}{[m]_q!} E_i^{n+m} \otimes F_i^{n+m} =$$

$$= \sum_{u \geq 0} \left(\sum_{n=0}^{u} \frac{(-1)^{u-n}(q^{-1} - q)^u q^{\frac{-n(n-1)+(u-n)(u-n-1)}{2}}}{[n]_q![u-n]_q!}\right) E_i^u \otimes F_i^u =$$

18

$$= \sum_{u \geq 0} \frac{(q^{-1} - q)^u}{[u]_q!} \left(\sum_{n=0}^{u} (-1)^{u-n} \begin{bmatrix} u \\ n \end{bmatrix}_q q^{\frac{u^2 - 2un - u + 2n}{2}} \right) E_i^u \otimes F_i^u =$$

$$= \sum_{u \geq 0} \frac{(q^{-1} - q)^u q^{\frac{u(u-1)}{2}}}{[u]_q!} \left(\sum_{n=0}^{u} (-1)^{u-n} \begin{bmatrix} u \\ n \end{bmatrix}_q q^{(1-u)n} \right) E_i^u \otimes F_i^u = 1 \otimes 1.$$

So R_i^{-1} is a right inverse of R_i. Analogous calculations show that R_i^{-1} is a left inverse of R_i, so that the claim is proved. $\qquad\qquad\square$

From the above remark it clearly follows that also $\tilde{R}_i$ is invertible. Its inverse is given by

$$\tilde{R}_i^{-1} = \sum_{n \geq 0} \frac{(q - q^{-1})^n q^{\frac{n(n-1)}{2} - 2n^2}}{[n]_q!} (E_i^n K_i^{-n} \otimes F_i^n K_i^n).$$

Proposition. $\forall x \in U$ we have the following relation between the comultiplication, the action of the braid group and the ith R-matrix:

$$\Delta T_i(x) = \tilde{R}_i^{-1} \cdot (T_i \otimes T_i)(\Delta(x)) \cdot \tilde{R}_i = R_i^{-1} \cdot q^{-\frac{H_i \otimes H_i}{2}} \cdot (T_i \otimes T_i)(\Delta(x)) \cdot q^{\frac{H_i \otimes H_i}{2}} \cdot R_i.$$

Proof. Since ΔT_i, $(T_i \otimes T_i)\Delta$ and the conjugation by $\tilde{R}_i$ are algebra homomorphisms, it is enough to prove the thesis on the generators, that is for $x \in \{K_\lambda, E_j, F_j\}$.
1) $x = K_\lambda$; then

$$(T_i \otimes T_i)(\Delta(x)) = (T_i \otimes T_i)(K_\lambda \otimes K_\lambda) = K_{s_i(\lambda)} \otimes K_{s_i(\lambda)};$$

since $\forall n \geq 0$ and $\forall \mu \in Q$ $E_i^n K_i^{-n} \otimes F_i^n K_i^n$ commutes with $K_\mu \otimes K_\mu$, we have that $\tilde{R}_i$ commutes with $K_{s_i(\lambda)} \otimes K_{s_i(\lambda)}$, so that $\tilde{R}_i^{-1} \cdot (T_i \otimes T_i)(\Delta(K_\lambda)) \cdot \tilde{R}_i = K_{s_i(\lambda)} \otimes K_{s_i(\lambda)} = \Delta(K_{s_i(\lambda)}) = \Delta T_i(K_\lambda)$;
2) $x = E_i, F_i$: it is a verification in $U(sl(2)_i) \cong U(sl(2))$ $\big(U(sl(2)_i)$ is the $\mathbb{C}(q)$-subalgebra of U generated by $E_i, F_i, K_i)$. We have that

$$q^{-\frac{H_i \otimes H_i}{2}} \cdot (T_i \otimes T_i)(\Delta(E_i)) \cdot q^{\frac{H_i \otimes H_i}{2}} =$$

$$= q^{-\frac{H_i \otimes H_i}{2}} \left(T_i(E_i) \otimes 1 + K_i^{-1} \otimes T_i(E_i) \right) q^{\frac{H_i \otimes H_i}{2}} =$$

$$= -(F_i K_i \otimes K_i + 1 \otimes F_i K_i) = -\Delta'(F_i K_i) = \Delta' T_i(E_i),$$

and, in the same way, that

$$q^{-\frac{H_i \otimes H_i}{2}} \cdot (T_i \otimes T_i)(\Delta(F_i)) \cdot q^{\frac{H_i \otimes H_i}{2}} = \Delta' T_i(F_i).$$

Thus we have to prove that $\Delta T_i(x) = R_i^{-1} \cdot \Delta' T_i(x) \cdot R_i$, which is an immediate consequence of the fact that $R_{sl(2)}\Delta(y)R_{sl(2)}^{-1} = \Delta'(y)$ $\forall y \in U(sl(2))$;

3) $x = E_j$, F_j with $a_{ij} = 0$: then $T_i(x) = x$, $(T_i \otimes T_i)(\Delta(x)) = \Delta(x)$; on the other hand it's clear that $\tilde{R}_i$ commutes with $\Delta(x)$, hence the thesis is obvious:

$$\Delta T_i(x) = \Delta(x) = \tilde{R}_i^{-1} \cdot \Delta(x) \cdot \tilde{R}_i = \tilde{R}_i^{-1} \cdot (T_i \otimes T_i)(\Delta(x)) \cdot \tilde{R}_i;$$

4) $x = E_j$, F_j with $a_{ij} = -1$; we want to prove that

$$\tilde{R}_i \cdot \Delta T_i(x) = (T_i \otimes T_i)(\Delta(x)) \cdot \tilde{R}_i;$$

to this aim let's recall that

$$\Delta T_i(E_j) = T_i(E_j) \otimes 1 + K_i K_j \otimes T_i(E_j) + (q^{-2} - 1)E_i K_j \otimes E_j,$$

$$\Delta T_i(F_j) = 1 \otimes T_i(F_j) + T_i(F_j) \otimes K_i^{-1} K_j^{-1} + (q - q^{-1})F_j \otimes F_i K_j^{-1},$$

$$T_i(E_j)E_i^n = q^{-n} E_i^n T_i(E_j), \quad T_i(F_j)F_i^n = q^{-n} F_i^n T_i(F_j),$$

$$[T_i(E_j), F_i^n] = q^{n-1}[n]_q F_i^{n-1} E_j K_i^{-1}, \quad [T_i(F_j), E_i^n] = -q^n[n]_q E_i^{n-1} F_j K_i.$$

Hence

$$(T_i \otimes T_i)(\Delta(E_j)) \cdot \tilde{R}_i =$$

$$= \big(T_i(E_j) \otimes 1 + K_i K_j \otimes T_i(E_j)\big) \sum_{n \geq 0} \frac{(q^{-1} - q)^n q^{-\frac{n(5n-1)}{2}}}{[n]_q!} E_i^n K_i^{-n} \otimes F_i^n K_i^n =$$

$$= \sum_{n \geq 0} \frac{(q^{-1} - q)^n q^{-\frac{n(5n-1)}{2}}}{[n]_q!} (E_i^n K_i^{-n} \otimes F_i^n K_i^n)\big(T_i(E_j) \otimes 1 + K_i K_j \otimes T_i(E_j)\big) +$$

$$\sum_{n \geq 0} \frac{(q^{-1} - q)^n q^{-\frac{n(5n-1)}{2}} q^{5n-4}}{[n]_q!} [n]_q E_i^{n-1} K_i^{1-n} E_i K_j \otimes F_i^{n-1} K_i^{n-1} E_j =$$

$$= \tilde{R}_i\big(T_i(E_j) \otimes 1 + K_i K_j \otimes T_i(E_j)\big) +$$

$$+ \sum_{n \geq 0} \frac{(q^{-1} - q)^{n+1} q^{-\frac{(n+1)(5n+4)}{2} + 5n+1} q^{5n-4}}{[n]_q!} [n]_q (E_i^n K_i^{-n} \otimes F_i^n K_i^n)(E_i K_j \otimes E_j) =$$

$$= \tilde{R}_i\big(T_i(E_j) \otimes 1 + K_i K_j \otimes T_i(E_j) + (q^{-2} - 1)E_i K_j \otimes E_j\big) = \tilde{R}_i \cdot \Delta(T_i(E_j)).$$

With analogous calculations we get that

$$(T_i \otimes T_i)(\Delta(F_j)) \cdot \tilde{R}_i = \tilde{R}_i \cdot \Delta(T_i(F_j)).$$

The thesis immediately follows.

$\square$

3.5. We shall now generalize the notion of partial R-matrices, introducing, for any element w of the Weyl group, R_w, which depends on a fixed reduced expression of w (recall that a reduced expression for w can be always completed to a reduced expression of the longest element in the Weyl group). This will enable us to study more precisely the relation between comultiplication and Braid group action.

Definition. *Let* $s_{i_1} \cdots s_{i_N} = w^0$ *be a reduced expression of the longest element of the Weyl group and* $\forall k$ *put* $w_k \doteq s_{i_1} \cdots s_{i_k}$; *define* $\tilde{R}_{w_k}$ *in the following way:*

$\tilde{R}_{w_1} \doteq \tilde{R}_{i_1}$, *and, if* $k > 1$, $\tilde{R}_{w_k} \doteq (T_{w_{k-1}} \otimes T_{w_{k-1}})(\tilde{R}_{i_k})\tilde{R}_{w_{k-1}}$.

Lemma. *Given a fixed reduced expression of* w^0 *we have that* $\forall k$ $\forall x \in U$

$$\Delta T_{w_k}(x) = \tilde{R}_{w_k}^{-1} \cdot (T_{w_k} \otimes T_{w_k})(\Delta(x)) \cdot \tilde{R}_{w_k}.$$

Proof. Use induction on k, the case $k = 1$ being obvious; let $k > 1$; then $T_{w_k} = T_{w_{k-1}}T_{i_k}$ so that

$$\Delta T_{w_k}(x) = \Delta T_{w_{k-1}}T_{i_k}(x) = \tilde{R}_{w_{k-1}}^{-1} \cdot (T_{w_{k-1}} \otimes T_{w_{k-1}})(\Delta T_{i_k}(x)) \cdot \tilde{R}_{w_{k-1}} =$$

$$= \tilde{R}_{w_{k-1}}^{-1} \cdot (T_{w_{k-1}} \otimes T_{w_{k-1}})(\tilde{R}_{i_k}^{-1} \cdot (T_{i_k} \otimes T_{i_k})(\Delta(x)) \cdot \tilde{R}_{i_k}) \cdot \tilde{R}_{w_{k-1}} =$$

$$= \tilde{R}_{w_k}^{-1} \cdot (T_{w_k} \otimes T_{w_k})(\Delta(x)) \cdot \tilde{R}_{w_k}.$$

□

Definition. *Let* $s_{i_1} \cdots s_{i_N} = w^0$ *be a reduced expression* (ι) *of* w^0 *and put* $w_k^\iota \doteq s_{i_1} \cdots s_{i_k}$, $\beta_k^\iota \doteq w_{k-1}^\iota(\alpha_{i_k})$. *We call* $U_{k,\iota}^{\geq 0}$ *and* $U_{k,\iota}^{\leq 0}$ *the linear subspaces of* $U^{\geq 0}$ *and* $U^{\leq 0}$ *linearly generated by* $\{E_{\beta_{k-1}^\iota}^{r_{k-1}} \cdot \ldots \cdot E_{\beta_1^\iota}^{r_1} K_\lambda | r_1, \ldots, r_{k-1} \in \mathbb{N}, \lambda \in Q\}$ *and* $\{E_{\beta_{k-1}^\iota}^{r_{k-1}} \cdot \ldots \cdot F_{\beta_1^\iota}^{r_1} K_\lambda | r_1, \ldots, r_{k-1} \in \mathbb{N}, \lambda \in Q\}$ *respectively (we omit the indication of the index* ι *when there are no possible misunderstandings).*

Remark. $U_{k,\iota}^{\geq 0}$ *and* $U_{k,\iota}^{\leq 0}$ *are subalgebras of* $U^{\geq 0}$ *and* $U^{\leq 0}$ *respectively.*

Proposition. $\tilde{R}_{w_k^\iota} = \sum_u x_u \otimes y_u$ *with* $x_u \in U_{k+1,\iota}^{\geq 0}$, $y_u \in U_{k+1,\iota}^{\leq 0}$.

Proof. Induction on k, the case $k = 1$ being obvious; let $k > 1$; then

$$\tilde{R}_{w_k^\iota} = (T_{w_{k-1}^\iota} \otimes T_{w_{k-1}^\iota})(\tilde{R}_{i_k})\tilde{R}_{w_{k-1}^\iota} = \tilde{R}_{\beta_k^\iota}\tilde{R}_{w_{k-1}^\iota}$$

where

$$\tilde{R}_{\beta_k^\iota} \doteq \sum_{n \geq 0} \frac{(q^{-1} - q)^n q^{-\frac{n(5n-1)}{2}}}{[n]_q!} E_{\beta_k^\iota}^n K_{\beta_k^\iota}^{-n} \otimes F_{\beta_k^\iota}^n K_{\beta_k^\iota}^n.$$

The thesis then follows from remark 3.5, noticing that $U_{k,\iota}^{\geq 0} \subset U_{k+1,\iota}^{\geq 0}$, $U_{k,\iota}^{\leq 0} \subset U_{k+1,\iota}^{\leq 0}$.

□

Corollary 1. $\forall k$ we have that

$$\Delta(E_{\beta_k}') - (E_{\beta_k} \otimes 1 + K_{\beta_k} \otimes E_{\beta_k}) \in U_{\overline{k}}^{\geq 0} \otimes U^{\geq 0},$$

$$\Delta(F_{\beta_k}) - (1 \otimes F_{\beta_k} + F_{\beta_k} \otimes K_{-\beta_k}) \in U^{\leq 0} \otimes U_{\overline{k}}^{\leq 0}.$$

Proof. We have that

$$\Delta(E_{\beta_k}) = \Delta T_{w_{k-1}}(E_{i_k}) = \tilde{R}_{w_{k-1}}^{-1} \cdot (T_{w_{k-1}} \otimes T_{w_{k-1}})(\Delta(E_{i_k})) \cdot \tilde{R}_{w_{k-1}} =$$

$$= \tilde{R}_{w_{k-1}}^{-1} \cdot (E_{\beta_k} \otimes 1 + K_{\beta_k} \otimes E_{\beta_k}) \tilde{R}_{w_{k-1}},$$

hence $\Delta(E_{\beta_k}) \in U_{\overline{k+1}}^{\geq 0} \otimes U^{\geq 0}$ and in the same way $\Delta(F_{\beta_k}) \in U^{\leq 0} \otimes U_{\overline{k+1}}^{\leq 0}$; on the other hand

$$\Delta(E_{\beta_k}) = E_{\beta_k} \otimes 1 + K_{\beta_k} \otimes E_{\beta_k} + \sum_{0 < \gamma < \beta_k} x_{\beta_k - \gamma} K_\gamma \otimes y_\gamma$$

with $x_\lambda, y_\lambda \in U_\lambda^+$ and

$$\Delta(F_{\beta_k}) = 1 \otimes F_{\beta_k} + F_{\beta_k} \otimes K_{-\beta_k} + \sum_{0 < \gamma < \beta_k} \tilde{x}_\gamma \otimes \tilde{y}_{\beta_k - \gamma} K_{-\gamma}$$

with $\tilde{x}_\lambda, \tilde{y}_\lambda \in U_{-\lambda}^-$, from which the thesis follows. $\qquad\square$

Corollary 2. $\forall k$, $\forall n \in \mathbb{N}$ we have that

$$\Delta(E_{\beta_k}^n) - (E_{\beta_k} \otimes 1 + K_{\beta_k} \otimes E_{\beta_k})^n = \sum_{r=0}^{n-1} E_{\beta_k}^r x_r \otimes y_r$$

where $\quad x_r = \sum_{\underline{c} \neq (0)} a_{\underline{c}} E_{\beta_{k-1}}^{c_{k-1}} \cdots E_{\beta_1}^{c_1} P_{\underline{c}}(K), \quad y_r \in U^+,$

$$\Delta(F_{\beta_k}^n) - (1 \otimes F_{\beta_k} + F_{\beta_k} \otimes K_{-\beta_k})^n = \sum_{r=0}^{n-1} \tilde{x}_r \otimes F_{\beta_k}^r \tilde{y}_r$$

where $\quad \tilde{x}_r \in U^-, \quad \tilde{y}_r = \sum_{\underline{c} \neq (0)} a_{\underline{c}} F_{\beta_{k-1}}^{c_{k-1}} \cdots F_{\beta_1}^{c_1} P_{\underline{c}}(K).$

Corollary 3.

$$\Delta(E_{\beta_k}^n) = q^{n-1}[n]_q E_{\beta_k}^{n-1} K_{\beta_k} \otimes E_{\beta_k} + \sum a_{\underline{r}, \lambda} E_{\beta_k}^{r_k} \cdots E_{\beta_1}^{r_1} K_\lambda \otimes y_{\underline{r}, \lambda},$$

where the sum is taken over the set $\{(\underline{r}, \lambda) | \underline{r} = (r_0, ..., r_k) \neq (0, ..., n-1), \lambda \in Q\}$; an analogous formula is true for $\Delta(F_{\beta_k}^n)$:

$$\Delta(F_{\beta_k}^n) = q^{n-1}[n]_q E_{\beta_k}^{n-1} K_{\beta_k} \otimes E_{\beta_k} + \sum a_{\underline{r}, \lambda} E_{\beta_k}^{r_k} \cdots E_{\beta_1}^{r_1} K_\lambda \otimes y_{\underline{r}, \lambda}.$$

3.6. Let us now calculate our pairing on the elements of the basis of type Poincaré-Birkhoff-Witt $E_{\beta_N}^{r_N} \cdots E_{\beta_1}^{r_1}$ and $F_{\beta_N}^{r'_N} \cdots F_{\beta_1}^{r'_1}$ where $r_1, ..., r_N, r'_1, ..., r'_N \in \mathbb{N}$.

Proposition. Let $\alpha = \beta_k$ for a suitable reduced expression of w^0; then

$$(E_\alpha | F_{\beta_N}^{r_N} \cdots F_{\beta_1}^{r_1}) \neq 0 \text{ or } (E_{\beta_N}^{r_N} \cdots E_{\beta_1}^{r_1} | F_\alpha) \neq 0 \Rightarrow \sum_{i=1}^{N} r_i = 1, \quad \sum_{i=1}^{N} r_i \beta_i = \alpha,$$

that is $r_i = \delta_{i,k}$ for some k.

Proof. We prove the thesis by induction on $\mathrm{ht}(\alpha)$, the case $\mathrm{ht}(\alpha) = 1$ being obvious. Let's consider $\mathrm{ht}(\alpha) > 1$ and suppose $(E_\alpha | F_{\beta_N}^{r_N} \cdots F_{\beta_1}^{r_1}) \neq 0$ or $(E_{\beta_N}^{r_N} \cdots E_{\beta_1}^{r_1} | F_\alpha) \neq 0$. Then necessarily we have that $\sum_{i=1}^{N} r_i \beta_i = \alpha$; suppose that $\sum_{i=1}^{N} r_i \neq 1$ and define $M \doteq \max\{i | r_i \neq 0\}$; then $M > k$: otherwise we would have $w_M^{-1}(\alpha) = s_{i_{M+1}} \cdots s_{i_{k-1}}(\alpha_{i_k}) \in R_+$. On the other.hand

$$w_M^{-1}(\alpha) = \sum_{j=1}^{M} r_j w_M^{-1}(\beta_j) = \sum_{j=1}^{M} r_j s_{i_M} \cdots s_{i_j}(\alpha_{i_j}) = -\sum_{j=1}^{M} r_j s_{i_M} \cdots s_{i_{j+1}}(\alpha_{i_j}) \in R_-,$$

giving a contradiction. Moreover $\mathrm{ht}(\beta_M) < \mathrm{ht}(\alpha)$, hence by applying the inductive hypothesis and corollary 3.5.1 we get

$$(E_\alpha | F_{\beta_M}^{r_M} \cdots F_{\beta_1}^{r_1}) = (\Delta E_\alpha | F_{\beta_M} \otimes F_{\beta_M}^{r_M - 1} F_{\beta_{M-1}}^{r_{M-1}} \cdots F_{\beta_1}^{r_1}) = 0$$

and

$$(E_{\beta_M}^{r_M} \cdots E_{\beta_1}^{r_1} | F_\alpha) = (E_{\beta_M}^{r_M - 1} E_{\beta_{M-1}}^{r_{M-1}} \cdots E_{\beta_1}^{r_1} \otimes E_{\beta_M} | \Delta F_\alpha) = 0.$$

$\square$

3.7. We now have to compute the value of our pairing, on vectors of the same weight. We start with two *root* vectors.

Lemma. $\forall \alpha \in R_+ \quad (E_\alpha | F_\alpha) = \dfrac{1}{q^{-1} - q}.$

Proof. We prove by induction on k that if ι is a reduced expression of $w^0 = s_{i_1} \cdots s_{i_N}$ then $(E_{\beta_k^\iota} | F_{\beta_k^\iota}) = \dfrac{1}{q^{-1} - q}.$

If $k = 1$ the thesis is obvious; let $k > 1$; then we have $\beta_k^\iota = w_{k-2}^\iota s_{i_{k-1}}(\alpha_{i_k})$. We can have either $w_{k-2}^\iota(\alpha_{i_k}) < 0$ or $w_{k-2}^\iota(\alpha_{i_k}) > 0$:

a) $w_{k-2}^\iota(\alpha_{i_k}) < 0$: then $l(w_{k-2}^\iota s_{i_k}) < l(w_{k-2}^\iota)$ and this implies that $w_{k-2}^\iota = w s_{i_k}$ with $l(w) = k - 3$ and, since $w(\alpha_{i_{k-1}}) = w s_{i_k} s_{i_{k-1}}(\alpha_{i_k}) = \beta_k^\iota > 0$, $\exists$ a reduced expression ι' of $w^0 = s_{i_1'} \cdots s_{i_N'}$ such that $s_{i_1'} \cdots s_{i_{k-3}'} = w$, $i_{k-2}' = i_{k-1}$.

Let x be E or F; then

$$x_{\beta_k^\iota} = T_{w_{k-2}^\iota} T_{i_{k-1}}(x_{i_k}) = T_w T_{i_k} T_{i_{k-1}}(x_{i_k}) = T_w(x_{i_{k-1}}) = x_{\beta_{k-2}'}.$$

Hence $(E_{\beta_k^\iota}|F_{\beta_k^\iota}) = (E_{\beta_{k-2}^{\iota'}}|F_{\beta_{k-2}^{\iota'}}) = \dfrac{1}{q^{-1}-q}$.

b) $w_{k-2}^\iota(\alpha_{i_k}) > 0$: let's put $w \doteq w_{k-2}^\iota$, $i \doteq i_{k-1}$, $j \doteq i_k$; we notice that $w_k^\iota(\alpha_i) = w s_i s_j(\alpha_i) = w(\alpha_j) > 0$, thus we can suppose $i_{k+1} = i$; moreover $w(\alpha_j) > 0$ implies that $\exists$ a reduced expression ι' of $w^0 = s_{i_1''} \cdots s_{i_N''}$ such that $i_r' = i_r \ \forall r \le k-2$ and $i_{k-1}' = j$. Then we have $E_{\beta_k}^\iota = T_w T_i(E_j)$ and

$$F_{\beta_k}^\iota = T_w T_i(F_j) = q T_w(F_i) T_w(F_j) - T_w(F_j) T_w(F_i) =$$

$$= q T_w(F_i) T_w T_i T_j(F_i) - T_w T_i T_j(F_i) T_w(F_i) = q F_{\beta_{k-1}}^\iota F_{\beta_{k+1}}^\iota - F_{\beta_{k+1}}^\iota F_{\beta_{k-1}}^\iota.$$

Hence

$$(E_{\beta_k^\iota}|F_{\beta_k^\iota}) = (\Delta E_{\beta_k^\iota}|q F_{\beta_{k-1}^\iota} \otimes F_{\beta_{k+1}^\iota} - F_{\beta_{k+1}^\iota} \otimes F_{\beta_{k-1}^\iota}) =$$

$$= (\Delta E_{\beta_k^\iota}|q F_{\beta_{k-1}^\iota} \otimes F_{\beta_{k+1}^\iota}) = (\Delta T_w T_i(E_j)|q F_{\beta_{k-1}^\iota} \otimes F_{\beta_{k+1}^\iota}) =$$

$$= (\tilde{R}_w^{-1} \cdot (T_w \otimes T_w)(\Delta T_i(E_j)) \cdot \tilde{R}_w | q F_{\beta_{k-1}^\iota} \otimes F_{\beta_{k+1}^\iota}) =$$

$$= \big(\tilde{R}_w^{-1}(E_{\beta_k^\iota} \otimes 1 + K_{\beta_k^\iota} \otimes E_{\beta_k^\iota} + (q^{-2}-1)E_{\beta_{k-1}^\iota} K_{\beta_{k-1}^{\iota'}} \otimes E_{\beta_{k-1}^{\iota'}})\tilde{R}_w | q F_{\beta_{k-1}^\iota} \otimes F_{\beta_{k+1}^\iota}\big).$$

Let us recall that $\tilde{R}_w^{-1}(E_{\beta_k^\iota} \otimes 1)\tilde{R}_w = \sum x_s \otimes y_s$ with $y_s \in U_{k-1,\iota}^{\le 0}$ and $\tilde{R}_w^{-1}(K_{\beta_k^\iota} \otimes E_{\beta_k^\iota})\tilde{R}_w = \sum \tilde{x}_s \otimes \tilde{y}_s$ with $\tilde{x}_s \in U_{k-1,\iota}^{\ge 0}$. Noticing that $F_{\beta_{k+1}^\iota} = F_{\beta_{k-1}^{\iota'}}$ we get that

$$(E_{\beta_k^\iota}|F_{\beta_k^\iota}) = (q^{-2}-1)(\tilde{R}_w^{-1}(E_{\beta_{k-1}^\iota} K_{\beta_{k-1}^{\iota'}} \otimes E_{\beta_{k-1}^{\iota'}})\tilde{R}_w | q F_{\beta_{k-1}^\iota} \otimes F_{\beta_{k-1}^{\iota'}});$$

but the only term which gives a non-zero contribution is, because of homogeneity, $(q^{-2}-1)E_{\beta_{k-1}^\iota} K_{\beta_{k-1}^{\iota'}} \otimes E_{\beta_{k-1}^{\iota'}}$, so that

$$(E_{\beta_k^\iota}|F_{\beta_k^\iota}) \doteq (q^{-1}-q)(E_{\beta_{k-1}^\iota}|F_{\beta_{k-1}^\iota})(E_{\beta_{k-1}^{\iota'}}|F_{\beta_{k-1}^{\iota'}}) = \dfrac{1}{q^{-1}-q}.$$

The claim is thus completely proved. $\qquad\square$

We are now ready to deal with the general case. We have

Proposition.

$$(E_{\beta_N}^{r_N} \cdots E_{\beta_1}^{r_1} | F_{\beta_N}^{r_N'} \cdots F_{\beta_1}^{r_1'}) = \prod_{i=1}^{N} \delta_{r_i, r_i'} \dfrac{[r_i]_q! q^{\frac{r_i(r_i-1)}{2}}}{(q^{-1}-q)^{r_i}}.$$

Proof. Let us define $M \doteq \max\{i|r_i \ne 0\}$, $M' \doteq \max\{i|r_i' \ne 0\}$. There are three possibilities:

a) $M < M'$: then

$$(E_{\beta_N}^{r_N} \cdots E_{\beta_1}^{r_1} | F_{\beta_N}^{r_N'} \cdots F_{\beta_1}^{r_1'}) = (E_{\beta_M}^{r_M} \cdots E_{\beta_1}^{r_1} | F_{\beta_{M'}}^{r_{M'}'} \cdots F_{\beta_1}^{r_1'}) =$$

$$= \left(\Delta(E_{\beta_M}^{r_M} \cdots E_{\beta_1}^{r_1}) | F_{\beta_{M'}} \otimes F_{\beta_{M'}}^{r'_{M'}-1} F_{\beta_{M'-1}}^{r'_{M'-1}} \cdots F_{\beta_1}^{r'_1}\right) = 0$$

because $\Delta(E_{\beta_M}^{r_M} \cdots E_{\beta_1}^{r_1}) \in U_{M+1}^{\geq 0} \otimes U^{\geq 0} \subset U_{M'}^{\geq 0} \otimes U^{\geq 0}$;

b) $M > M'$: then

$$(E_{\beta_N}^{r_N} \cdots E_{\beta_1}^{r_1} | F_{\beta_N}^{r'_N} \cdots F_{\beta_1}^{r'_1}) = (E_{\beta_M}^{r_M} \cdots E_{\beta_1}^{r_1} | F_{\beta_{M'}}^{r'_{M'}} \cdots F_{\beta_1}^{r'_1}) =$$

$$= \left(E_{\beta_M}^{r_M-1} E_{\beta_{M-1}}^{r_{M-1}} \cdots E_{\beta_1}^{r_1} \otimes E_{\beta_M} | \Delta(F_{\beta_{M'}}^{r'_{M'}} \cdots F_{\beta_1}^{r'_1})\right) = 0$$

because $\Delta(F_{\beta_{M'}}^{r'_{M'}} \cdots F_{\beta_1}^{r'_1}) \in U^{\leq 0} \otimes U_{M'+1}^{\leq 0} \subset U^{\leq 0} \otimes U_M^{\leq 0}$;

c) $M = M'$: we'll prove the thesis by induction on $R' \doteq \sum_{i=1}^{N} r'_i$: the cases $R' = 0, 1$ have been already proved; let $R' > 1$; there are two possible cases:

i) $\exists k < M$ such that $r'_k \neq 0$; then

$$(E_{\beta_N}^{r_N} \cdots E_{\beta_1}^{r_1} | F_{\beta_N}^{r'_N} \cdots F_{\beta_1}^{r'_1}) = (E_{\beta_M}^{r_M} \cdots E_{\beta_1}^{r_1} | F_{\beta_M}^{r'_M} \cdots F_{\beta_1}^{r'_1}) =$$

$$= \left(\Delta(E_{\beta_M}^{r_M} \cdots E_{\beta_1}^{r_1}) | F_{\beta_M}^{r'_M} \otimes F_{\beta_{M-1}}^{r'_{M-1}} \cdots F_{\beta_1}^{r'_1}\right) =$$

$$= \left(\Delta(E_{\beta_M}^{r_M}) \cdot (K_{r_1\beta_1+\ldots+r_{M-1}\beta_{M-1}} \otimes E_{\beta_{M-1}}^{r_{M-1}} \cdots E_{\beta_1}^{r_1}) | F_{\beta_M}^{r'_M} \otimes F_{\beta_{M-1}}^{r'_{M-1}} \cdots F_{\beta_1}^{r'_1}\right) =$$

$$= \left(E_{\beta_M}^{r_M} K_{r_1\beta_1+\ldots+r_{M-1}\beta_{M-1}} \otimes E_{\beta_{M-1}}^{r_{M-1}} \cdots E_{\beta_1}^{r_1} | F_{\beta_M}^{r'_M} \otimes F_{\beta_{M-1}}^{r'_{M-1}} \cdots F_{\beta_1}^{r'_1}\right) =$$

$$= \delta_{r_M, r'_M} \frac{[r_M]_q! q^{\frac{r_M(r_M-1)}{2}}}{(q^{-1}-q)^{r_M}} \prod_{i=1}^{M-1} \delta_{r_i, r'_i} \frac{[r_i]_q! q^{\frac{r_i(r_i-1)}{2}}}{(q^{-1}-q)^{r_i}} =$$

$$= \prod_{i=1}^{N} \delta_{r_i, r'_i} \frac{[r_i]_q! q^{\frac{r_i(r_i-1)}{2}}}{(q^{-1}-q)^{r_i}};$$

ii) $r'_i = 0 \quad \forall i < M$; then

$$(E_{\beta_N}^{r_N} \cdots E_{\beta_1}^{r_1} | F_{\beta_N}^{r'_N} \cdots F_{\beta_1}^{r'_1}) = (E_{\beta_M}^{r_M} \cdots E_{\beta_1}^{r_1} | F_{\beta_M}^{r'_M}) =$$

$$= \left(\Delta(E_{\beta_M}^{r_M} \cdots E_{\beta_1}^{r_1}) | F_{\beta_M}^{r'_M-1} \otimes F_{\beta_M}\right) =$$

$$= \left(\Delta(E_{\beta_M}^{r_M}) \cdot (K_{r_1\beta_1+\ldots+r_{M-1}\beta_{M-1}} \otimes E_{\beta_{M-1}}^{r_{M-1}} \cdots E_{\beta_1}^{r_1}) | F_{\beta_M}^{r'_M-1} \otimes F_{\beta_M}\right) =$$

$$= \left(\prod_{i=1}^{M-1} \delta_{r'_i, 0}\right) (\Delta(E_{\beta_M}^{r_M}) | F_{\beta_M}^{r'_M-1} \otimes F_{\beta_M}) =$$

$$= \left(\prod_{i=1}^{M-1} \delta_{r'_i, 0}\right) q^{r_M-1} [r_M]_q (E_{\beta_M}^{r_M-1} \otimes E_{\beta_M} | F_{\beta_M}^{r'_M-1} \otimes F_{\beta_M}) = \prod_{i=1}^{N} \delta_{r_i, r'_i} \frac{[r_i]_q! q^{\frac{r_i(r_i-1)}{2}}}{(q^{-1}-q)^{r_i}}.$$

$\square$

3.8. We can summarize the results obtained till now in the following description of the R-matrix of U:

Theorem. *The R-matrix for $U_q(\mathfrak{g})$ is given by*

$$R = \left(\prod_{\beta \in R_+} exp_q\left((q^{-1} - q)E_\beta \otimes F_\beta\right) \right) q^{-\sum_i \tilde{H}_i \otimes \tilde{H}_i} =$$

$$= exp_q\left((q^{-1} - q)E_{\beta_N} \otimes F_{\beta_N}\right) \cdots exp_q\left((q^{-1} - q)E_{\beta_1} \otimes F_{\beta_1}\right) q^{-\sum_i \tilde{H}_i \otimes \tilde{H}_i}.$$

Proof. The theorem follows immediately from the computations performed above. $\square$

4. POISSON STRUCTURES.

4.1. Let us recall the notion of a Poisson structure over a commutative k−algebra A:

Definition. *A Poisson structure on A is a Lie product*

$$\{-, -\} : A \otimes_k A \longrightarrow A$$

satisfying the (Leibniz rule):

$$\{f_1 f_2, f_3\} = f_1\{f_2, f_3\} + f_2\{f_1, f_3\}. \tag{4.1.1}$$

If A, B are two Poisson algebras we have a canonical Poisson structure on $A \otimes_k B$ for which the two factors Poisson-commute and which extends the two given structures on the factors. It is given clearly by:

$$\{a \otimes b, c \otimes d\} := \{a, c\} \otimes bd + ac \otimes \{b, d\}.$$

Note that (4.1.1) implies that the map:

$$g \to \{f, g\}$$

is a derivation.

In a more geometric language, if A is the coordinate ring of a smooth affine variety, f gives rise to a vector field X_f such that:

$$X_f(g) = \{f, g\}.$$

X_f is called a *Hamiltonian vector field* and f is its Hamiltonian function.

The assumptions imply that the map $f \to X_f$ is a Lie algebra homomorphism (where the Lie product of vector fields is the usual commutator as operators).

They also imply that, if P is a point and g a function which vanishes in P to order ≥ 2 (i.e. $g(P) = 0, dg(P) = 0$), then $\{f, g\}(P) = 0$. This means that a Poisson structure on the coordinate ring of a smooth algebraic variety X can be interpreted as a particular type of bivector u (a section of $\wedge^2 T(X)$). Given a point P and two cotangent vectors α, β we have:

$$u_P(\alpha, \beta) := \{f, g\}(P), \text{ where } df(P) = \alpha, \ dg(P) = \beta. \tag{4.1.2}$$

The Jacobi identity can be interpreted as a cocycle identity on u (see [LW]). In general this bivector gives a skew symmetric form on the cotangent space which may be degenerate. In fact it is clear by the definitions that, if N is the kernel of this form in the space $T_X^*(P)$, its orthogonal in the tangent space $T_X(P)$ is the span of the values in P of the Hamiltonian vector fields. Let us call $H(P)$ this subspace; of course it can (and it will) be a subspace of variable dimension.

One should remark first of all that the definition can be globalized to any algebraic variety and also that it can be formulated as well in the analytic or differentiable categories.

Going back to the spaces $H(P)$ we have the following:

Definition. *A connected submanifold Y of X is called a symplectic leaf if for each $P \in Y$ the space $H(P)$ is the tangent space to Y in P and if Y is maximal with this property. The theorem of Frobenius implies the existence of a decomposition of X in symplectic leaves.*

A special case is when u is always non degenerate and so, if the manifold is connected, we have a unique leaf. In this case we also say that X is a *symplectic manifold*. Thus the leaves have an induced structure of symplectic manifolds. Given a Hamiltonian vector field H_f, since it is tangent in any point to the leaf through that point, it follows that if we integrate it locally to a germ of a 1-parameter subgroup, this subgroup of local diffeomorphisms preserves the leaves.

4.2. Assume now that $A = k[H]$ is the coordinate ring of an affine algebraic group H. We know that $k[H]$ is a Hopf algebra with comultiplication

$$\Delta : k[H] \to k[H] \otimes k[H] = k[H \times H]$$

given by $(\Delta f)(h_1, h_2) = f(h_1 h_2), \forall h_1, h_2 \in H$;
antipode

$$S : k[H] \to k[H]$$

given by $(Sf)(h) = f(h^{-1}), \forall h \in H$;
counit

$$\epsilon : k[H] \to k$$

given by $\epsilon(f) = f(e)$, where $e \in H$ is the identity element.

Definition. *H is said to be a Poisson algebraic group if $k[H]$ has a Poisson structure compatible with the Hopf algebra structures, that is if $\forall f_1, f_2 \in k[H]$ we have*

$$\Delta\{f_1, f_2\} = \{\Delta f_1, \Delta f_2\}, \tag{4.2.1}$$

$$S(\{f_1, f_2\}) = \{Sf_2, Sf_1\} \tag{4.2.2}$$

$$\epsilon\{f_1, f_2\} = 0. \tag{4.2.3}$$

4.3. It is useful to translate the axioms of a Poisson group in a more explicit form. As we have remarked, a Poisson structure is a special type of bivector, i.e. a skew form on the tangent bundle. For a group H it is convenient of course to trivialize the tangent bundle. Putting $\mathfrak{h} = Lie(H) = T_1(H)$ one identifies $T_1(H)$ with $T_a(H)$ using the differential at 1 of the map $l_a : x \to ax$. With this trivialization we think of a bivector as a family of skew forms ψ_a on $\mathfrak{h}$ and we want to translate the Poisson group axioms in this language.

First remark how the basic formulas for the group translate.

Consider the multiplication map $\mu : H \times H \to H$ and its differential in a point (a, b); we use the identifications $T_{(a,b)}H \times H = \mathfrak{h} \oplus \mathfrak{h}$, $T_{ab}(H) = \mathfrak{h}$ and call $\mu_{a,b} : \mathfrak{h} \oplus \mathfrak{h} \to \mathfrak{h}$ the resulting map, and $\mu^*_{a,b} : \mathfrak{h}^* \to \mathfrak{h}^* \oplus \mathfrak{h}^*$ its dual map. Since $(ab)^{-1}(axby) = b^{-1}xby$ we deduce that:

$$\mu_{a,b}(u, v) = b^{-1}u + v, \text{ resp. } \mu^*_{a,b}(\phi) = (b\phi, \phi) \tag{4.3.1}$$

where $b^{-1}u$, $b\phi$ refer respectively to adjoint and coadjoint action.

Similarly consider the inverse map $i : x \to x^{-1}$; its differential induces the family of maps:

$$u \to -au, \quad \phi \to -a^{-1}\phi.$$

Now let us translate the Poisson axioms in terms of the forms.

The compatibility with multiplication means that the inclusion maps $\mu^*_{a,b}$ are compatible with the corresponding forms ψ_{ab}, $\psi_a \oplus \psi_b$; from the formula (4.3.1) this means

$$\psi_{ab} = b^{-1}\psi_a + \psi_b, \tag{4.3.2}$$

where we have used the action of H on skew forms on $\mathfrak{h}$ induced by the coadjoint action.

The compatibility with the inverse means that the inverse map takes the Poisson form to minus itself and hence that:

$$\psi_{a^{-1}} = -a\psi_a. \tag{4.3.3}$$

It is clear that from (4.3.2) it follows that $\psi_1 = 0$ and also (4.3.3).

Take now the differential at 1 of the map $\psi : H \to \wedge^2\mathfrak{h}$; it defines a coproduct $\gamma : \mathfrak{h} \to \wedge^2\mathfrak{h}$ and dually an antisymmetric product on $\mathfrak{h}^*$.

One can easily verify, from the Jacobi identity for the Poisson bracket, that this is a Lie algebra product.

In fact one can easily deduce the same Lie algebra structure as follows: the maximal ideal M of 1 is a Lie ideal as well as M^2 (under Poisson bracket), hence

$h^* = T_H^*(1)$ is a Lie algebra. This Lie structure is the same as the one previously described.

Let us change for a while notations. Set $L := Lie(H) = T_H(1)$. We have a Lie algebra structure on L and on L^*, we have moreover an action (the coadjoint one) of L on L^* since L is a Lie algebra, and similarly an action of L^* on L. Define thus a multiplication on $L \oplus L^*$ setting:

$$[a + \varphi, b + \psi] := [a, b] + a.\psi - \psi.a + \varphi.b - b.\varphi + [\varphi, \psi] \qquad (4.3.4)$$

where the first and last term refer to the Lie algebra structures on L, L^* respectively, while the middle terms refer to the two coadjoint actions; we have then:

Theorem. *The given multiplication on $L \oplus L^*$ gives it a Lie algebra structure.*
The quadratic form $\langle (a, \varphi) | (b, \psi) \rangle := \psi(a) + \varphi(b)$ is associative (invariant) with respect to the Lie product.

We leave the proof of this theorem which can be verified from the various identities.

Notice that L, L^* are maximal isotropic with respect to the given form. It is then natural to define:

Definition. *A triple of Lie algebras $(\mathfrak{g}, \mathfrak{h}, \mathfrak{k})$ is called a Manin triple if it satisfies:*
i) $\mathfrak{h}, \mathfrak{k} \subset \mathfrak{g}$ as Lie subalgebras, $\mathfrak{g} = \mathfrak{h} \oplus \mathfrak{k}$ as vector spaces.
ii) There is a non degenerate invariant symmetric bilinear form (,) on $\mathfrak{g}$ with respect to which both $\mathfrak{h}$ and $\mathfrak{k}$ are maximal isotropic subspaces.

We aim to show that, as a Lie algebra is the infinitesimal datum of a Lie group, so is a Manin triple for a Poisson Lie group.

We go back to the identity (4.3.2) applied to the 1-parameter subgroups $a = exp(tu)$ and $b = exp(su)$. Set

$$A_u(t) := \psi_{exp(tu)},$$

recall that by definition $\gamma(u) = A_u'(0)$ and differentiate getting:

$$A_u'(t) = -u\,A(t) + \gamma(u).$$

This equation shows that the Manin triple determines completely the Poisson structure.

Let G be an algebraic group, $H, K \subset G$ two closed subgroups. Set $\mathfrak{g} = LieG$ (resp. $\mathfrak{h} = LieH$, $\mathfrak{k} = LieK$).

Definition. *The triple (G, H, K) is called an algebraic Manin triple if the triple of Lie algebras $(\mathfrak{g}, \mathfrak{h}, \mathfrak{k})$ is a Manin triple.*

Let us see how we can associate to an algebraic Manin triple, a structure of Poisson algebraic group on H (and symmetrically on K).

Let us now denote by $\pi : \mathfrak{g} \to \mathfrak{h}$ the projection with kernel $\mathfrak{k}$.

First we notice that the definition immediately implies that there is a natural isomorphism between $\mathfrak{h}^*$ and $\mathfrak{k}$. Under this isomorphism the coadjoint action of $\mathfrak{h}^*$ on $\mathfrak{k}$ is clearly read from formula (4.3.4):

$$a.\phi = (1 - \pi)[a, \phi].$$

For any $h \in H$, we set $\pi^h = adh^{-1} \circ \pi \circ adh$. Notice that also π^h maps $\mathfrak{g}$ to $\mathfrak{h}$. We can now define the bilinear form $\psi_h :=< \ , \ >_h$ on $\mathfrak{k} \cong \mathfrak{h}^*$ by

$$< x, y >_h := (\pi^h x, y), \forall x, y \in \mathfrak{k}. \tag{4.3.5}$$

Since the bilinear form (,) is invariant and the two subalgebras are maximal isotropic, we get

$$< x, y >_h = (\pi^h x, y) = (x, adh^{-1} \circ (1 - \pi) \circ adhy) =$$
$$- (x, adh^{-1} \circ \pi \circ adhy) = - < y, x >_h .$$

Thus $< \ , \ >_h$ is antisymmetric $\forall h \in H$.

As before consider the tangent and cotangent bundles on H as trivialized as $H \times \mathfrak{h}$, $H \times \mathfrak{h}^* = H \times \mathfrak{k}$ respectively. Their direct sum is the trivial bundle $H \times \mathfrak{g}$ in which we consider the flat connection making $\mathfrak{g}$ a space of horizontal sections for which the canonical scalar product is constant.

In this bundle we consider the constant projection π of $\mathfrak{g}$ to $\mathfrak{h}$ with kernel $\mathfrak{h}^*$ and the variable projection Π whose value in a point h is π^h. Notice that the restriction of π^h to $\mathfrak{h}^*$ is exactly the skew form ψ_h defined at the beginning of the section by (4.3.1).

Thus in compact form working in this larger bundle we have the formula for the Poisson bracket of two functions $f_1, f_2 \in k[H]$:

$$\{f_1, f_2\} = (\Pi df_1, df_2) \tag{4.3.6}$$

Proposition. *The bracket $\{ \ , \ \}$ defined above, provides H with the structure of a Poisson algebraic group with given infinitesimal Manin triple.*

Proof. From the definition we only have to verify the Jacobi identity and the compatibility with the Hopf structure.

A simple direct computation in the Lie algebra of sections of the bundle $H \times \mathfrak{g}$ shows that

$$d\{f, g\} = (1 - \pi)[(1 - \Pi)df, (1 - \Pi)dg]. \tag{4.3.7}$$

In fact, given a vector field X, we have:

$$(d\{f, g\}, X) = ([\Pi, ad(X)]df, dg) =$$
$$= (\Pi[X, df], dg) - ([X, \Pi df], dg) =$$
$$= ([X, df], (1 - \Pi)dg) - (X, [\Pi df, dg]) =$$
$$= (X, [df, (1 - \Pi)dg] - [\Pi df, dg]) =$$
$$= (X, (1 - \pi)[(1 - \Pi)df, (1 - \Pi)dg]).$$

From this we immediately get that

$$d\{f_1, \{f_2, f_3\}\} = (1 - \pi)[(1 - \Pi)df_1, [(1 - \Pi)df_2, (1 - \Pi)df_1]],$$

so that the Jacobi identity for $\mathfrak{g}$ clearly implies that

$$\{f_1, \{f_2, f_3\}\} + \{f_2, \{f_3, f_1\}\} + \{f_3, \{f_1, f_2\}\}$$

is a constant function $\forall f_1, f_2, f_3 \in k[H]$. Since clearly $\{f_1, f_2\}(e) = 0, \forall f_1, f_2 \in k[H]$, one gets that

$$\{f_1, \{f_2, f_3\}\} + \{f_2, \{f_3, f_1\}\} + \{f_3, \{f_1, f_2\}\} = 0.$$

It remains to prove that ψ_h (which is the restriction of π^h to $\mathfrak{h}^*$) satisfies the relation (4.3.2). We work in the bundle $H \times \mathfrak{g}$.

$$Ad(b)^{-1}\Pi_a(1 - \pi)Ad(b)(1 - \pi) + \psi_b =$$
$$Ad(b)^{-1}\Pi_a Ad(b)(1 - \pi) - Ad(b)^{-1}\pi Ad(b)(1 - \pi) + \psi_b = \psi_{ab}.$$

$\square$

There is an alternative description of the Poisson structure which can be described as follows:

Given an affine algebraic manifold X, denote by $\mathbb{C}[X]$ (resp. Vect X, resp. $\mathcal{D}X$) the space of regular functions (resp. vector fields, resp. differential 1–forms) on X. To define a Poisson bracket $\{,\}$ on $\mathbb{C}[X]$ is equivalent to defining a homomorphism of $\mathbb{C}[X]$–modules $\tau : \mathcal{D}X \to$ Vect X (satisfying the appropriate conditions), so that $\{f, g\} = < \tau(df), dg >$.

In our case $X = H$. We define a map $\tau : \mathcal{D}H \to$ Vect H as follows. We identify (Lie $H)^*$ with the space of left–invariant 1–forms on H. On the other hand, consider the unramified map $\tilde{\pi} : H \to K/G$. The right action of K on K/G gives an embedding Lie $K \subset$ Vect K/G. This gives a linear map $\tau_0 :$ (Lie $H)^* \to$ Vect K/G. Since $\tilde{\pi}$ is unramified we have a map $\tilde{\pi}^* :$ Vect $K/G \to$ Vect H. Then τ is the homomorphism of $\mathbb{C}[H]$–modules defined by the linear map $\tilde{\pi}^* \circ \tau_0$. One can easily verify that this definition agrees with the one we previously described.

Let us briefly discuss Poisson subgroups and quotients:

Consider a quotient Poisson group S of H, that is S is a quotient group of H and the ring $\mathbb{C}[S] \subset \mathbb{C}[H]$ is a Poisson subalgebra. Let U be the kernel of the quotient homomorphism $\varphi : H@ >>> S$, let $\mathfrak{s} = Lie\ S$, $\mathfrak{u} = Lie\ U$ and $d\varphi : \mathfrak{h}@ >>> \mathfrak{s}$ the Lie algebra quotient map. Then $\mathfrak{u}$ is an ideal in $\mathfrak{h}$ and we identify $\mathfrak{s}^*$ with a subspace of $\mathfrak{h}^* = \mathfrak{k}$ by taking $\mathfrak{u}^\perp \subset \mathfrak{g}$ under the invariant form and intersecting it with $\mathfrak{k}$. Then for $p \in S$ the linear map $\overline{\gamma}_p : \mathfrak{s}^*@ >>> \mathfrak{s}$ giving rise to the Poisson structure is given by:

$$\overline{\gamma}_p = (d\varphi) \cdot (\gamma_{\tilde{p}}|_{\mathfrak{s}^*})$$

where $\tilde{p} \in H$ is any representative of p ($\overline{\gamma}_p$ is independent of the choice of $\tilde{p}$).

The construction of the Manin triple corresponding to the Poisson manifold S is obtained from the following simple fact:

Lemma. *Let* $(\mathfrak{g}, \mathfrak{h}, \mathfrak{k})$ *be a Manin triple of Lie algebras, and let* $\mathfrak{u} \subset \mathfrak{h}$ *be an ideal such that* $\mathfrak{u}^\perp$ *(in* $\mathfrak{g}$*) intersected with* $\mathfrak{k}$ *is a subalgebra of the Lie algebra* $\mathfrak{k}$*. Then*

    (a) $\mathfrak{u}^\perp$ *is a subalgebra of* $\mathfrak{g}$ *and* $\mathfrak{u}$ *is an ideal of* $\mathfrak{u}^\perp$*.*

    (b) $(\mathfrak{u}^\perp/\mathfrak{u}, \mathfrak{h}/\mathfrak{u}, \mathfrak{k} \cap \mathfrak{u}^\perp)$ *is a Manin triple, where the bilinear form on* $\mathfrak{u}^\perp$ *is induced by that on* $\mathfrak{g}$*.*

Proof. Straightforward. $\square$

Notice that in this construction we have at the same time a quotient Poisson group of H and a Poisson subgroup of K: we are in a way in a *self dual* picture.

4.4. The construction that we have just described allows us to give various examples of Poisson algebraic groups.

Examples.

(1) The first such example is given by the so called Kostant-Kirillov structure on the coadjoint representation of a group K. In this case $H = \mathfrak{k}^*$ with the trivial Lie algebra structure and G is the semidirect product $K \times \mathfrak{k}^*$, K acting on $\mathfrak{k}^*$ via the coadjoint action. So, as a vector space, $\mathfrak{g} = \mathfrak{k} \oplus \mathfrak{k}^*$ and has a canonical symmetric bilinear form. We leave to the reader the easy verification that the triple (G, H, K) is an algebraic Manin triple and that the corresponding Poisson structure on H is the usual Kostant-Kirillov structure i.e. $\forall f_1, f_2 \in k[H], h \in H$,

$$\{f_1, f_2\}(h) = [df_1(h), df_2(h)](h).$$

It is also clear that, dually, the Poisson structure one gets on K is trivial.

(2) The second class of examples are the ones we shall be mostly interested in in these notes. Let K be a semisimple group. Fix a maximal torus $T \in K$ and a Borel subgroup $B^+ \supset T$, let B^- be the unique Borel subgroup such that $T = B \cap B^-$. Set $G = K \times K$. Denote by $\mu_\pm : B^\pm \to T$ the canonical projection homomorphisms, and consider the homomorphism $\phi : B^- \times B^+ \to T$ defined by $\phi(b_-, b+) = \mu_-(b_-)\mu_+(b_+)$. Then $H = \operatorname{Ker} \phi$, while K is embedded in G as the diagonal subgroup. Thus we have a triple (G, H, K) and in order to see that it is an algebraic Manin triple we need to define a non degenerate symmetric invariant bilinear form on $\mathfrak{g}$ satisfying the various properties in the definition. We first rescale the Killing form so that on the Cartan subalgebra it induces the normalized form giving square length 2 to the short roots. Then take the difference of the scaled Killing form on the second and first factor of $\mathfrak{g} = \mathfrak{k} \oplus \mathfrak{k}$, i.e. the unique form on $\mathfrak{g}$ coinciding with minus the scaled Killing form (resp. the scaled Killing form) on $\mathfrak{k} \oplus \{0\}$ (resp. $\{0\} \oplus \mathfrak{k}$), and such that $\mathfrak{k} \oplus \{0\}$ and $\{0\} \oplus \mathfrak{k}$) are mutually orthogonal. It is then trivial to verify all the required properties.

This example of course appears in the Beliavin Drinfeld [BD] classification of classical r-matrices.

4.5. A very interesting feature of algebraic Manin triples is the fact that the so called symplectic leaves in H are open sets in algebraic subvarieties and can be quite explicitly described.

Let us now consider an algebraic Manin triple (G, H, K). Let $p : H \to K\backslash G$ denote the restriction to H of the quotient map from G to $K\backslash G$. Then

Proposition. *The symplectic leaves in H coincide with the connected components of the preimages under p of the K orbits in $K\backslash G$.*

Proof. First notice that the map p is étale onto its image so that it induces an isomorphism between the tangent space TH_h and the tangent space $TK\backslash G_{p(h)}$ for each $h \in H$. Furthermore the right action of K on $K\backslash G$ induces a map of $\mathfrak{k}$ to $TK\backslash G_{p(h)}$ whose image clearly coincides with the tangent space to the K orbit through $p(h)$. Hence, composing with the above isomorphism we obtain a map $\phi : \mathfrak{k} \to TH_h$ whose image coincides with the tangent space to our candidate to be the symplectic leaf through h. If we now identify TH_h with $\mathfrak{h}$ using left translation we finally obtain a map $\phi_h : \mathfrak{k} \to \mathfrak{h}$ and a simple computation, which we leave to the reader, shows that

$$\phi_h = \pi^h. \tag{4.5.1}$$

Thus the vector fields $\{f, -\}, f \in k[H]$, span at h the tangent space to $p^{-1}(p(h)K)$ and our claim follows. $\qquad\square$

The above proposition allows us to determine the symplectic leaves in each of our examples. In the case of the Kostant-Kirillov structure on $\mathfrak{k}^*$ they are of course the coadjoint orbits. In the second example identify the right cosets of $K \times K$ modulo its diagonal subgroup K_Δ with K itself in the obvious way. Then the K_Δ orbits under right action are identified with the conjugacy classes in K and the symplectic leaves are the connected components of the preimages of conjugacy classes in K under the map

$$p : H \to K$$

defined by $p(b_1, b_2) = b_1^{-1} b_2$.

4.6. Let R be a commutative algebra and $h \in R$ an element. Suppose we have an R algebra A with the property that h is a non zero divisor in A and $A/(h)$ is commutative. Given an element $a \in A$ we shall denote by $\bar{a}$ its image modulo h.

Given $x, y \in A/(h)$, choose a and b such that $x = \bar{a}$ and $y = \bar{b}$. Since $A/(h)$ is commutative and h is a non zero divisor in A, the element $[a, b]/h \in A$ is well defined. Set:

$$\{x, y\} = \overline{[a, b]/h}, \tag{4.6.1}$$

It is immediate to see that $\{x, y\}$ is independent of the choice of the representatives a and b and in this way one defines a Poisson bracket on $A/(h)$.

Borrowing the language from Physics one says:

Definition. *A is a quantization or a quantum deformation of the Poisson algebra $A/(h)$.*

In practice one may be given a Poisson algebra and ask the problem of giving a suitable quantization of it. This problem does not a priori have a canonical solution, so that the procedures of quantization are quite variable.

If we assume that A is a Hopf algebra over R then $A/(h)$ is also a Hopf algebra and we easily see that its Hopf and Poisson structures are compatible, i.e. $A/(h)$ is a Poisson Hopf algebra.

Thus given a Poisson algebraic group H we may conversely ask for the existence of quantum deformations of H as Poisson group.

Definition. *A Hopf algebra A over R is called a quantum deformation of $k[H]$ if*
i) A is flat over R.
ii) there is an isomorphism $j : A/(h) \to k[H]$ of Hopf algebras such that

$$j(\{x,y\} = \{j(x), j(y)\},$$

the Poisson bracket on the commutative algebra $A/(h)$ being the one defined above.

Notice that one can give a more local definition of a quantum deformation using as base algebra the algebra $k[[h]]$ instead of the algebra R. In this case Reshetikin [Re], extending work of Drinfeld, has recently shown that any Lie bialgebra admits a quantum deformation. We do not know whether one can obtain similar results in this more global situation.

We end this section showing that in our first example, the Kirillov-Kostant structure on $\mathfrak{k}^*$, a quantum deformation is indeed the universal enveloping algebra $U(\mathfrak{k})$. Indeed if one sets A equal to the R algebra generated by $\mathfrak{k}$ with relations

$$xy - yx = (h)[x,y], \forall x, y \in \mathfrak{k},$$

with comultiplication
$$\Delta(x) = x \otimes 1 + 1 \otimes x, \forall x \in \mathfrak{k},$$

antipode
$$S(x) = -x, \forall x \in \mathfrak{k},$$

counit
$$\epsilon(x) = 0, \forall x \in \mathfrak{k},$$

one easily sees that A is a quantum deformation of the coordinate ring of $\mathfrak{k}^*$ with the Kirillov-Kostant structure.

4.7. We want to enrich our description of Poisson structures.

We start as in **4.6** with a commutative algebra R, an element $h \in R$, and an R-algebra A with the property that h is a non zero divisor in A, but we do not assume anymore that $A/(h)$ is commutative. Consider an element $a \in A$ such that its image $\bar{a}$ modulo h is in the center of $A/(h)$.

Let $y \in A/(h)$, $y = \bar{b}$ be given. Since $\bar{a}$ is in the center $\mathcal{Z}$ of $A/(h)$ and h is a non zero divisor in A, the element $[a, b]/h \in A$ is well defined. Set:

$$D_a[y] = \overline{[a,b]/h}. \tag{4.7.1}$$

It is immediate to see that:

Theorem.

i) D_a is a derivation.

ii) If a' is a different representative of $\bar{a}$ then $D_a - D_{a'}$ is an inner derivation.

iii) If φ is an automorphism of A (inducing one on $A/(h)$) we have:

$$\varphi \circ D_a \circ \varphi^{-1} = D_{\varphi(a)}.$$

In particular we get:

Corollary. i) The center $\mathcal{Z}$ of $A/(h)$ has a natural Poisson structure, given by:

$$\{x, y\} := D_a(y), \text{ where } x = \bar{a}.$$

ii) The group of automorphisms of A induces a group of Poisson automorphisms of $\mathcal{Z}$.

It may be worth to remark that we have several Lie algebras of derivations:

a) $[D_a, D_b] = D_{[a,b]/h}$, thus the derivations $D_a, \{a \in A | \underline{a} \in \mathcal{Z}\}$, form a Lie algebra $\mathcal{L}$.

b) The Poisson structure induces a Lie algebra of Derivations $\mathcal{L}'$ of $\mathcal{Z}$ and we have an exact sequence:

$$0 \to \mathcal{L}^0 \to \mathcal{L} \to \mathcal{L}' \to 0$$

where $\mathcal{L}^0$ is the algebra of inner derivations of $A/(h)$.

If A is a Hopf algebra and furthermore we assume that also $\Delta(\bar{a})$ is in the center of $A/(h) \otimes A/(h)$, then we also have $D_{\Delta(a)}(\Delta(y)) = \Delta(D_a(y))$. We want to deduce a useful corollary:

Assume we are in the previous setting (A is a Hopf algebra) and that we are given a Hopf subalgebra S of the center $\mathcal{Z}$ of $A/(h)$. Let T be the minimal subalgebra of $\mathcal{Z}$ containing S and closed under Poisson bracket, then:

Proposition. T is a Hopf subalgebra of $\mathcal{Z}$ and it is compatible with the Poisson structure.

Proof. From the previous analysis let $U := \{t \in T | \Delta(t) \in T \otimes T\}$. Clearly $S \subset U$ and U is a subalgebra. But now we have that U is also closed under Poisson bracket since, from the last remarks, we have for $x, y \in U$ that $\Delta(\{x, y\}) = \{\Delta(x), \Delta(y)\} \in T \otimes T$. $\qquad\square$

4.8. Before we revert to our examples we want to make a general remark which applies to the setting we have just discussed of an algebra $A/(h)$ which is a free module over a Poisson subalgebra Z^0 of its center, so that we have the derivations extending the Poisson structure.

Let us use the differentiable language rather than that of algebras. Assume that we have a manifold M and a vector bundle V of algebras with 1 (i.e. 1 and the multiplication map are smooth sections). We identify the functions on M with the sections of V which are multiples of 1. Let D be a derivation of V, i.e. a derivation of the algebra of sections which maps the algebra of functions on M into itself, and let X be the corresponding vector field on M.

Proposition. *For each point $p \in M$ there exists a neighborhood U_p of p and a map φ_t defined for $|t|$ sufficiently small on $V|U_p$ which is a morphism of vector bundles covering the germ of the 1–parameter group generated by X and is also an isomorphism of algebras.*

Proof. The hypotheses on D imply that it is a vector field on V, linear on the fibers, hence we have the existence of a local lift of the 1–parameter group as a morphism of vector bundles. The condition of being a derivation implies that the lift preserves the multiplication section i.e. it is a morphism of algebras. $\qquad\square$

We will have to consider a variation of the previous proposition:

Suppose M is a Poisson manifold. Assume furthermore that the Poisson structure lifts to V, i.e. each (local) function f induces a derivation on sections (as in **4.7**) extending the given Poisson bracket. This means that we have a lift of the Hamiltonian vector fields as in the previous proposition and we can deduce:

Corollary. *Under the previous hypotheses, the fibers of V over the points of a given symplectic leaf of M are all isomorphic as algebras.*

Proof. The proposition implies that in a neighborhood of a point in a leaf the algebras are isomorphic, but since the notion of isomorphism is transitive this implies the claim. $\qquad\square$

4.9. We want now to explain that the algebra $U_q(\mathfrak{g})$ that we have seen as a q analogue of the usual enveloping algebra is indeed in a suitable sense a deformation of the function algebra on the Poisson group H we have described in **4.4**. Take $R = k[q, q^{-1}], h = q - 1$, we want now to define a remarkable R-subalgebra $\mathcal{U}$ of $U_q(\mathfrak{g})$ which will actually be a quantum deformation of the Poisson algebraic group of H.

We use again the action of the generalized Braid group $\mathcal{B}$ introduced in (2.4.1).

We now define $\mathcal{U}$ to be the smallest $\mathcal{B}$ stable R-subalgebra of $U_q(\mathfrak{g})$ containing the elements:

$$\overline{E}_i := (q^{(\alpha_i, \alpha_i)/2} - q^{-(\alpha_i, \alpha_i)/2})E_i, \quad \overline{F}_i := (q^{(\alpha_i, \alpha_i)/2} - q^{-(\alpha_i, \alpha_i)/2})F_i$$

for $i = 1, \dots, r$ and the K_λ's.

Notice that, if we had chosen these new elements as generators of the quantum group U, we would have the same defining relations except for

$$[\overline{E}_i, \overline{F}_j] = \delta_{ij}(q^{d_i} - q^{-d_i})(K_{\alpha_i} - K_{-\alpha_i}).$$

Take now the root vectors:

$$\overline{E}_{\beta_t} = T_{i_1} \cdots T_{i_{t-1}}(\overline{E}_{\alpha_{i_t}}), \quad \overline{F}_{\beta_t} = T_{i_1} \cdots T_{i_{t-1}}(\overline{F}_{\alpha_{i_t}})$$

From the very definition of $\mathcal{U}$ the elements $\overline{E}_{\beta_t}$ and $\overline{F}_{\beta_t}$ all lie in $\mathcal{U}$ and the PBW monomials in them and the K_i's and their inverses considered in Theorem 2.5 are clearly linearly independent over R. Indeed one can show that they are a R-basis of $\mathcal{U}$, so that $\mathcal{U}$ is a free R module.

Let now H be the Poisson algebraic group associated to the algebraic Manin triple of example 2 in section **4.4**, then:

Theorem. *The algebra $\mathcal{U}$ is a quantum deformation of $k[H]$.*

The proof of this result extremely involved due to the fact that the objects at hand are given by generators and relation can be found in [DP]. It is hoped that a better understanding of the nature of the algebra $U_q(\mathfrak{g})$ could lead to a more conceptual proof of this statement.

5. THE QUANTUM GROUP AT ROOTS OF ONE.

5.1. The quantum group U_ε, where ε is a primitive ℓ^{th} root of 1, can be defined starting from the $R(= k[q, \bar{q}^{-1}])$-algebra $\mathcal{U}$ (introduced in **4.9** and generated by the elements $\overline{E}_i$, $\overline{F}_i$, K_λ and their $\mathcal{B}$ translates) and specializing q at ε, with ε a primitive ℓ^{th} root of 1. We shall limit ourselves to the case when ℓ is odd (and for G_2 also relatively prime to 3).

The idea that we shall follow is to connect the Poisson geometry of the group H considered at the end of last section (and of which U_1 is the function algebra), with the structure of representations of U_ε using the ideas explained in **4.7** and **4.8**.

Before doing this we need to recall a few facts of general nature (we shall recall the essential facts needed, referring to [DP] and the references given there for more precise results). Let C be an algebra over an algebraic closed field k which has no zero divisor. Assume that C is a finite module over its center Z and that Z is a finitely generated algebra. Let $Q(Z)$ denote the quotient field of Z. Then the algebra $D = C \otimes_Z Q(Z)$ is a division algebra. In particular one has $dim_{Q(Z)} D = d^2$, with d a positive integer which is called the degree of C. We denote by Spec C the set of equivalence classes of irreducible representations of C. We then have

Theorem. *1) Every irreducible representation of C is finite dimensional.*

2) Each irreducible representation of C has dimension at most d.

3) The canonical map Spec $C @> \chi >>$ Spec Z is surjective.

4) The set $\Omega_C = \{a \in$ Spec $Z|\ \chi^{-1}(a)$ consists of a single representation of dimension d is a non empty Zariski open set.

5.2. We want to work out an easy example of the above situation.

Take our primitive ℓ-th root of one ε.

Given an $n \times n$ skew–symmetric matrix $H = (h_{ij})$ over $\mathbb{Z}$, we construct the *twisted polynomial algebra* $\mathbb{C}_H[x_1, \ldots, x_n]$. This is the algebra on generators $x_1, \ldots, x_n$ and the following defining relations:

$$x_i x_j = \varepsilon^{h_{ij}} x_j x_i \quad (i, j = 1, \ldots, n). \tag{5.2.1}$$

It can be viewed as an iterated twisted polynomial algebra with respect to any ordering of the indeterminates x_i. Similarly, we can define the twisted Laurent polynomial algebra $\mathbb{C}_H[x_1, x_1^{-1}, \ldots, x_n, x_n^{-1}]$. Both algebras have no zero divisors.

To study its spectrum we start with a simple general lemma.

Lemma. *In any irreducible $\mathbb{C}_H[x_1, \ldots, x_n]$–module each element x_i is either 0 or invertible.*

Proof. It is clear that $\mathrm{Im}(x)$ and $\mathrm{Ker}\,(x)$ are submodules of M. $\qquad\qquad\square$

Given $a = (a_1, \ldots, a_n) \in \mathbb{Z}^n$, we shall write $x^a = x_1^{a_1} \ldots x_n^{a_n}$. The monomials x^a with $a \in \mathbb{Z}_+^n$ (resp. $a \in \mathbb{Z}^n$) are a basis for $\mathbb{C}_H[x_1, \ldots, x_n]$ (resp. $\mathbb{C}_H[x_1, x_1^{-1}, \ldots, x_n, x_n^{-1}]$).

We consider now the matrix H as a matrix of a homomorphism $H : \mathbb{Z}^n @>>> (\mathbb{Z}/\ell\mathbb{Z})^n$, and we denote by K the kernel of H and by h the cardinality of the image of H.

Proposition. *1) The elements x^a with $a \in K \cap \mathbb{Z}_+^n$ (resp. $a \in K$) form a basis of the center of $\mathbb{C}_H[x_1, \ldots, x_n]$ (resp. $\mathbb{C}_H[x_1, x_1^{-1}, \ldots, x_n, x_n^{-1}]$).*

2) Let $a^{(1)}, \ldots, a^{(h)}$ be a set of representatives of $\mathbb{Z}^n$ mod K. Then the monomials $x^{a^{(1)}}, \ldots, x^{a^{(h)}}$ form a basis of the algebra $\mathbb{C}_H[x_1, x_1^{-1}, \ldots, x_n, x_n^{-1}]$ over its center.

3) degree $\mathbb{C}_H[x_1, \ldots, x_n]$ = degree $\mathbb{C}_H[x_1, x_1^{-1}, \ldots, x_n, x_n^{-1}] = \sqrt{h}$.

Proof. Define a skewsymmetric bilinear form on $\mathbb{Z}^n$ by letting for $a = (a_1, \ldots, a_n)$, $b = (b_1, \ldots, b_n) \in \mathbb{Z}^n$: $\langle a|b \rangle = \sum_{i,j=1}^n h_{ij} a_i b_j$. Then we have

$$x^a x^b = \varepsilon^{\langle a|b \rangle} x^b x^a. \tag{5.2.2}$$

Since clearly a linear combination of the monomials x^a lies in the center if and only if each monomial appearing with non zero coefficient does, (5.2.2) implies 1).

2) follows from 1) and the fact that

$$x^a x^b = \varepsilon^{c(a,b)} x^{a+b}, \text{ where } c(a,b) = \sum_{i>j} h_{ij} a_i b_j.$$

3) follows from 2) and Theorem 5.1. $\qquad\qquad\square$

5.3. We now want to show how the algebra U_ε can be *degenerated* to a twisted polynomial algebra.

Using the explicit basis for $\mathcal{U}$ given in **4.9** (associated to a reduced expression of the longest element in W and hence to a convex ordering on R^+, see **2.2**), it is then not hard to see that the monomials

$$M_{k,r,\alpha} := \overline{F}_{\beta_N}^{h_N} \cdots \overline{F}_{\beta_1}^{h_1} K_\lambda \overline{E}_{\beta_1}^{k_1} \cdots \overline{E}_{\beta_N}^{k_N} \tag{5.3.1}$$

where $k = (k_1, \ldots, k_N)$, $r = (r_1, \ldots, r_N) \in \mathbb{Z}_+^N$ and $\alpha \in Q$ are a basis of U_ε. Given such a monomial define its total height by

$$d_0(M_{k,r,\alpha}) = \sum_i (k_i + r_i)\text{ht }\beta_i,$$

and its total degree by

$$d(M_{k,r,\alpha}) = (k_N, k_{N-1}, \ldots, k_1, r_1, \ldots, r_N, d_0(M_{k,r,\alpha})) \in \mathbb{Z}_+^{2N+1}.$$

We shall view $\mathbb{Z}_+^{2N+1}$ as a totally ordered semigroup with the lexicographical order $<$ such that $u_1 < u_2 < \ldots < u_{2N+1}$ where $u_i = (\delta_{i,1}, \ldots, \delta_{i,2N+1})$.

Introduce a $\mathbb{Z}_+^{2N+1}$-filtration of the algebra U by letting U_s ($s \in \mathbb{Z}_+^{2N+1}$) be the span of the monomials $M_{k,r,\alpha}$ such that $d(M_{k,r,\alpha}) \leq s$. Theorem 2.5.4 implies:

Proposition. *The associated graded algebra* Gr U_ε *of the* $\mathbb{Z}_+^{2N+1}$-*filtered algebra* U *is an algebra with generators* E_α ($\alpha \in R$) *and* K_β ($\beta \in P$) *subject to the following relations:*

$$K_\alpha K_\beta = K_{\alpha+\beta}, \ K_0 = 1; \tag{5.3.2}$$

$$K_\alpha E_\beta = q^{(\alpha|\beta)} E_\beta K_\alpha; \tag{5.3.3}$$

$$E_\alpha E_{-\beta} = E_{-\beta} E_\alpha \ \text{if } \alpha, \beta \in R^+; \tag{5.3.4}$$

$$E_\alpha E_\beta = q^{(\alpha|\beta)} E_\beta E_\alpha, \ E_{-\alpha} E_{-\beta} = q^{(\alpha|\beta)} E_{-\beta} E_{-\alpha} \tag{5.3.5}$$

if $\alpha, \beta \in R^+$ *and* $\alpha > \beta$ *in our convex ordering of* R^+.

Remarks. 1) Considering the degree by total height d_0, we obtain a $\mathbb{Z}_+$-filtration of U_ε; let $U^{(1)} = \overline{U_\varepsilon}$ be the associated graded algebra. Letting $d_1(M_{k,r,\alpha}) = r_N$, we obtain a $\mathbb{Z}_+$-filtration of $U^{(1)}$; let $U^{(2)} = \overline{U}^{(1)}$ be the associated graded algebra. Letting $d_2(M_{k,r,\alpha}) = r_{N-1}$, we similarly obtain $U^{(3)} = \overline{U}^{(2)}$, etc. It is clear that at the last step we get the algebra Gr U_ε defined by (5.3.2-5):

$$U^{(2N+1)} \simeq \text{Gr } U_\varepsilon. \tag{5.3.6}$$

2) The algebra Gr U_ε is a twisted polynomial algebra on generators $E_{\beta_1}, \ldots, E_{\beta_N}$, $E_{-\beta_N}, \ldots E_{-\beta_1}, K_1, \ldots, K_n$, with the elements $K_1, \ldots, K_n$ inverted.

A first simple application of this method is:

Theorem. *The algebra U_ε has no zero divisors.*

Proof. Since a twisted polynomial algebra has no zero divisor we get that Gr U_ε has no zero divisors. This immediately implies, looking at highest monomials, that also U_ε has no zero divisors. $\qquad\square$

5.4. We are now in the position to compute the degree of Gr U_ε. We obtain

Proposition. *The degree of Gr U_ε equals ℓ^N*

Proof. Consider the matrix

$$H = \begin{pmatrix} A & 0 & B \\ 0 & A & -B \\ -^t B & {}^t B & 0 \end{pmatrix},$$

where $(a_{ij}) = A$ is the skew symmetric matrix with $a_{ij} = (\beta_i|\beta_j)$ if $i < j$ and if $\alpha_1, \ldots, \alpha_n$ are the simple roots, $B = ((\alpha_i|\beta_j))_{1 \le i \le n, 1 \le j \le N}$, as a map

$$H : \mathbb{Z}^{2N+n} \to (\mathbb{Z}/(\ell))^{2N+n}.$$

By Remark 5.3.2 and Proposition 5.2.3 we know that the degree of Gr U_ε equals the square root of the cardinality of the image of H. By elementary rows and columns operations one can reduce the matrix H to the matrix

$$H' = \begin{pmatrix} 0 & A & 0 \\ A & 0 & B \\ 0 & -^t B & 0 \end{pmatrix}.$$

From this one immediately sees that the degree of Gr U_ε equals the cardinality of the image of the matrix $H_0 = (\,A \quad 0 \quad B\,)$ considered as a map $H_0 : \mathbb{Z}^{N+n} \to \mathbb{Z}^N$. Some elementary combinatorial considerations which we leave to the reader (see [DKP3]) show that H_0 is surjective. Since the algebra Gr U_ε is, up to inverting some elements, the quasipolynomial algebra associated to H, the claim follows from Proposition 5.2. $\qquad\square$

5.5. We now go back to the algebra U_ε and we introduce a particular subalgebra of the center of U_ε which will play a major role in its study. From the relations we can easily deduce that the elements $K_\alpha^\ell = K_{\ell\alpha}$, $\overline{E}_i^\ell$ and $\overline{F}_i^\ell$ are central. We apply now the Braid group $\mathcal{B}$ to these elements and obtain:

Definition. *The smallest subalgebra of U_ε containing the elements $K_\alpha^\ell, \overline{E}_i^\ell, \overline{F}_i^\ell$ and stable under $\mathcal{B}$ will be denoted by Z_0.*

By construction this algebra is contained in the center of U_ε. We want to describe now its main properties. We have already remarked that we have a Poisson bracket on the center of U_ε. We now have

Proposition (cf. [DP]. *1) Let* $x, y \in Z_0$; *then*

$$\{x, y\} \in Z_0$$

so that Z_0 *is a Poisson algebra.*
   2) Let $x \in Z_0$, *then*

$$\Delta(x) \in Z_0 \otimes Z_0, \quad S(x) \in Z_0,$$

so that Z_0 *is a Hopf subalgebra of* U_ε.
   3) The Poisson bracket and Hopf algebra structure on Z_0 *are compatible, so that* Z_0 *is the coordinate ring of a Poisson algebraic group.*

   .At this point the reader will pose himself the question of which Poisson group we obtain. The following theorem which is proved along the same lines as Theorem 4.9 (see [DKP]) gives a complete answer to this question.

Theorem. *There exists a unique* $\mathcal{B}$ *equivariant isomorphism*

$$\mathcal{F} : U_1 \rightarrow Z_0$$

of Poisson Hopf algebras such that

$$\mathcal{F}(\overline{E}_i) = \overline{E}_i^\ell \quad \mathcal{F}(\overline{F}_i) = \overline{F}_i^\ell \quad \mathcal{F}(K_i) = K_i^\ell.$$

The homomorphism $\mathcal{F}$ *will be called the Frobenius morphism.*

   It follows that Z_0 is the algebra of regular functions on the group H considered in section 4.

5.6. Using then the explicit basis for $\mathcal{U}$ given in **4.9**, it is not hard to see

Proposition. *i) The monomials*

$$\overline{F}_{\beta_N}^{\ell h_N} \cdots \overline{F}_{\beta_1}^{\ell h_1} K_{\ell\lambda} \overline{E}_{\beta_1}^{\ell k_1} \cdots \overline{E}_{\beta_N}^{\ell k_N} \tag{5.6.1}$$

are a basis of Z_0.
   The monomials

$$\overline{F}_{\beta_N}^{h_N} \cdots \overline{F}_{\beta_1}^{h_1} K_\lambda \overline{E}_{\beta_1}^{k_1} \cdots \overline{E}_{\beta_N}^{k_N} \tag{5.6.2}$$

with $0 \leq k_i, h_i < \ell$ *and* λ *running in a set of representatives of* $P/\ell P$ *are a basis of* U_ε *over* Z_0.

   The above proposition implies that U_ε, being a finite module over Z_0, is also a finite module over its center Z, so, since by Theorem 5.3 it has no zero divisors, it satisfies the hypotheses of Theorem 5.1. Notice that in contrast with the case of a generic q, this in particular implies that the irreducible representations of U_ε depend on *dim* ($\mathfrak{g}$) continuous parameters (this resembles the situation with Lie algebras in positive characteristic [KW] and [WK]).

We are now going to compute the degree of U_ε. We start with an estimate from below.

5.7. Let C be a filtered algebra, set $\overline{C}$ equal to the associated graded algebra and let $C[t]$ (resp. $C[t, t^{-1}]$) denote the ring of polynomials (resp. Laurent polynomials) over C. The *Rees algebra* $\mathcal{R}(C)$ of C is the following subalgebra of $C[t]$:

$$\mathcal{R}(C) = \sum_{j \in \mathbb{Z}_+} C_j t^j.$$

The following properties of the algebra $\mathcal{R}(C)$ are obvious:

Lemma. *1) If C has no zero divisors, then the same is true for $\mathcal{R}(C)$.*

2) If $\overline{C}$ is generated by homogeneous elements $\underline{a}_1, \underline{a}_2, \dots$ of degree $r_1, r_2, \dots$, then $\mathcal{R}(C)$ is generated by the elements $t, t^{r_1} a_1, t^{r_2} a_2, \dots$ where the a_i lift the $\underline{a}_i$.

3) $C[t, t^{-1}] = \mathcal{R}(C)[t^{-1}].$

4) $\mathcal{R}(C)/(t) \simeq \overline{C}.$

Using this we get

Proposition. *The degree of U_ε is at least ℓ^N.*

Proof. Consider the algebras $U_\varepsilon = U^{(0)}, U^{(1)}, \dots, U^{(2N+1)}$, considered in Remark 5.3.1. It is immediate to verify that since both U_ε and $U^{(2N+1)}$ satisfy the hypotheses of Theorem 5.1 so do all the $U^{(i)}$'s. We know that the degree of $U^{(2N)}$ is ℓ^N. So in order to prove our claim it suffices to show that for each $i = 0, \dots, 2N$ $\deg(U^{(i)}) \geq \deg(U^{(i+1)})$.

By definition $U^{(i+1)}$ is the associated graded algebra for a suitable gradation of $U^{(i)}$. Then taking the Rees algebra $\mathcal{R}(U^{(i)})$ it follows from part 1) and 2) in the Lemma that $\mathcal{R}(U^{(i)})$ satisfies the hypotheses of Theorem 5.1. From part 3 that $\deg(U^{(i)}) = \deg\mathcal{R}(U^{(i)})$ and from part 4 for that $\deg\mathcal{R}(U^{(i)}) \geq \deg(U^{(i+1)})$ proving the claim. $\qquad\square$

5.8. We are now going to give a construction of some U_ε-modules. Consider inside the algebra U_ε the subalgebra $\mathcal{B}_\varepsilon$ generated by the $\overline{E}_i$ and K_i and also the subalgebra S obtained by adding to $\mathcal{B}_\varepsilon$ the elements $\overline{F}^\ell_\alpha \alpha, \alpha \in R$. By the P.B.W. theorem these elements are algebraically independent over $\mathcal{B}_\varepsilon$ and S is the polynomial algebra $\mathcal{B}_\varepsilon[\overline{F}^\ell_\alpha], \alpha \in R^+$.

Consider a 1-dimensional representation $\mathbb{C}(\sigma)$ of S given by setting $\overline{E}_i := 0$, $\overline{F}^\ell_\alpha \alpha := b_\alpha, K_i := k_i$. We often identify σ with the point of coordinates (b_α, k_i). Set

$$V_\sigma := U_\varepsilon \otimes_S \mathbb{C}(\sigma) \tag{5.8.1}$$

Definition. V_σ is called the Baby Verma module of highest weight σ.

Set $v_\sigma := 1 \otimes 1 \in V_\sigma$.

Proposition. *1) The ℓ^N elements $F_{\beta_N}^{h_N} \cdots F_{\beta_1}^{h_1} v_\sigma$, $0 \le h_i < \ell$ are a basis of V_σ.*
2) $\overline{E}_i v_\sigma = 0$.
3) The elements of Z_0 act as scalars on V_σ, in fact as

$$\overline{F}_\alpha^\ell = b_\alpha, z_i = k_i^\ell, x_\alpha = 0.$$

Proof. 2) and 3) follow from the definition and 1) from the P.B.W. basis. $\square$

By Proposition 5.6 U_ε is a free module over Z_0 so following **4.7** and **4.8** we have that, considering U_ε as a sheaf of algebras over Z_0, the isomorphism type of the fibers is constant on symplectic leaves. Recall now (see **4.5**) that the symplectic leaves in H are just the connected components of the preimages of conjugacy classes in G under the map $p : H \to G$ defined by $p(b_1, b_2) = b_1^{-1} b_2$. The equations $\overline{E}_\alpha^\ell = 0$ define the subgroup $S \subset H$ consisting of pairs $\{b_1, b_2) \in H \mid t \in T\}$. Since $S = p^{-1}(B_-)$ and B_- meets every conjugacy class in G we deduce that S intersects every symplectic leaf in H.

It follows that the degree of U_ε is the dimension of a generic irreducible representation having as central character on Z_0 a point in S. For each such point we have an irreducible module which is a quotient of the Baby Verma module, thus we get ℓ^N as upper bound for the degree. Putting together this with Proposition 5.7 we have

Theorem. *The degree of U_ε equals ℓ^N.*

The analysis of the structure of U_ε and of its irreducible modules can be continued to a large extent. We refer the reader to [DP] for further results and references. Let us just finish by remarking that the knowledge of irreducible modules is rather satisfactory only in the case that they lie over a point in a symplectic leaf mapping to a regular class (for example in this case one knows that the representation has dimension exactly ℓ^N). For smaller conjugacy classes very little is known and it seems a very interesting problem to investigate these matters.

REFERENCES

[BD]   A. Belavin, V. Drinfeld, *On the solutions of the classical Yang-Baxter equations for simple Lie algebras*, Funk. Anal. Pri. **16** 3 (1982), 1-29.

[B]   N.Bourbaki, *Groupes et Algèbres de Lie*, Paris, 1980.

[DK]   C. De Concini, V.G. Kac, *Representations of quantum groups at roots of* 1, Progress in Math., vol. 92, Birhäuser, 1990, pp. 471–506.

[DK2]   C. De Concini, V.G. Kac, *Representations of quantum groups at roots of* 1: *reduction to the exceptional case*, in Infinite Analysis Part A, Adv. Series in Math. Phys. **16** (1992), 141–150.

[DKP]   C. De Concini, V.G. Kac, C. Procesi, *Quantum coadjoint action*, Journal of AMS **5** (1992), 151–190.

[DKP2]   C. De Concini, V.G. Kac, C. Procesi, *Some remarkable degenerations of quantum groups*, Comm. Math. Phys (1993).

[DKP3]   C.De Concini, V.G. Kac and C. Procesi, *Some quantum analogues of solvable groups*, preprint (1992).

[DP]   C.De Concini and C. Procesi, *Quantum Groups* (to appear in SLN in Math.).

[D1]   V.G. Drinfeld, *Hopf algebras and the quantum Yang-Baxter equation*, Soviet. Math. Dokl. **32**, 1 (1985), 254-258.

[D2]   V.G. Drinfeld, *Quantum groups*, Proc. ICM Berkeley **1** (1986), 789–820.

[J1]   M. Jimbo, *A q-difference analogue of $U(q)$ and the Yang-Baxter equation*, Lett. Math. Physf **10** (1985), 63–69.

[J2]   M. Jimbo, *A q-analogue of $U(gl(N+1))$, Hecke algebras and the Yang-Baxter equation*, Lett. Math. Phys. **11** (1986), 247–252.

[KW]   V.G. Kac, B. Yu. Weisfeiler, *Coadjoint action of a semi-simple algebraic group and the center of the enveloping algebra in characteristic p*, Indag. Math. **38** (1976), 136–151.

[KR]   Kirillov, A., Reshetikhin, N., *q-Weyl groups and R-matrices*, Comm. Math. Phys. **134** (1990), 421-431.

[LS]   S.Z. Levendorskii, Ya. S. Soibelman, *Algebras of functions on compact quantum groups, Schubert cells and quantum tori*, Comm. Math. Physics **139** (1991), 141–170.

[LS2]   S.Z. Levendorskii and Ya.S. Soibelman, *Quantum Weyl group and multiplicative formula for the R-matrix of a simple Lie algebra*, Funct. Analysis and its Appl, 2 **25** (1991), 143–145.

[LS3]   S.Z. Levendorskii and Ya.S. Soibelman, *Some applications of quantum Weyl group I*, J. Geom. Physics, 2 **7** (1990), 241–254.

[LW]   J.-H. Lu, A Weinstein, *Poisson Lie groups, dressing transformations and Bruhat decompositions*, J. Diff. Geom. **31** (1990), 501–526.

[L1]   G. Lusztig, *Quantum deformations of certain simple modules over enveloping algebras*, Adv. in Math. **70** (1988), 237-249..

[L2]   G. Lusztig, *Quantum groups at roots of* 1, Geom. Ded. **35** (1990), 89–114.

[L3]   G. Lusztig, *Finite dimensional Hopf algebras arising from quantum groups*, J. Amer. Math. Soc **3** (1990), 257-296.

[L4]   G. Lusztig, *Canonical bases arising from quantized enveloping algebras*, J. Amer. Math. Soc **3** (1990), 447-498..

[M]   Matsumoto, *Generateurs et relations de groupes de Weyl generalisés*, C.R. Acad. Sci. Paris **258** (1964), 3419–3422.

[Re]   N. Reshetikhin, *Quantization of Lie bialgebras*, Duke Math. Journal **7** (1992), 143-151.

[R1]   M. Rosso, *Finite dimensional representations of the quantum analogue of the enveloping algebra of a semisimple Lie algebra*, Comm. Math. Phys. **117** (1988), 581-593..

[R2]   M. Rosso, *Analogues de la forme de Killing et du théorème d'Harish-Chandra pour les groupes quantiques*, Ann. Sci. Ec. Norm. Sup. **23** (1990), 445–467.

[STS]   M.A. Semenov-Tian-Shansky, *Dressing transformations and Poisson group actions*, vol. 21, Publ. RIMS, 1985.

[S]   Ya.S. Soibelman, *The algebra of functions on a compact quantum group, and its representations*, Leningrad Math. J., 1 **2** (1991), 161–178.

[T]   T. Tanisaki, *Killing forms, Harish–Chandra isomorphisms, and universal R–matrices for quantum algebras*, in Infinite Analysis Part A, Adv. Series in Math. Phys., vol. 16, 1992, pp. 941–962.

[WK]   B. Yu, Weisfeiler, V.G. Kac, *On irreducible representations of Lie p–algebras*, Funct. Anal. Appl. **5** (1971), 28–36.

Scuola Normale Superiore, Piazza dei Cavalieri 7, 56100 Pisa. Italia

e–mail: deconcin@ux1sns.sns.it

e–mail: damiani@vaxsns.sns.it

COMPACT QUANTUM GROUPS AND q-SPECIAL FUNCTIONS

TOM H. KOORNWINDER

0. INTRODUCTION

It is the purpose of this paper to give a tutorial introduction to Hopf algebras and general compact quantum groups on the one hand (sections 1 and 2), and to q-hypergeometric functions and related orthogonal polynomials on the other hand (sections 3 and 4). The two parts can be read quite independently from each other. Sections 1 and 2 were specially written for my course at the European School of Group Theory 1993, Trento, Italy. Sections 3 and 4 were also used there, but written earlier as part of the notes for an intensive course on special functions aimed at Dutch graduate students. My Trento course had a third part dealing with $SU_q(2)$ and related q-special functions. My papers [26] and [28] were used as course material for that part, but they are not included here. For the reader who is new to this subject, it is crucial that he complements reading of the present notes with the study of papers dealing with $SU_q(2)$ and other special quantum groups.

I now describe the various sections in some more detail. Section 1 presents the basic theory of Hopf algebras and of corepresentations. Here (and in §2) proofs are either given in full detail or they are sketched such that the reader can easily fill in the gaps. No attempt has been made to high mathematical sophistication or big generality.

Section 2 deals with compact quantum groups. The original plan was to give here an account of Woronowicz's celebrated theory [41] of compact matrix quantum groups, but this section grew out into an alternative approach to compact quantum groups, avoiding C^*-algebras in the definition and in the proofs, but formulating everything on the Hopf *-algebra level. The C^*-algebra completion now appears as a final observation instead of an essential part of the definition. Both in [41] and in §2 the Haar functional plays a crucial role. The approach of §2, developed joint with M. S. Dijkhuizen [15], [16], may be somewhat shorter and easier to grasp than the C^*-algebra approach. We believe that this approach is very well suited for application to the special compact matrix quantum groups most commonly studied nowadays. The section concludes with a comparison of various approaches to compact quantum groups which have appeared in the literature.

Section 3 gives an introduction to q-hypergeometric functions. The more elementary q-special functions like q-exponential and q-binomial series are treated in a rather self-contained way, but for the higher q-hypergeometric functions some identities are given without proof. The reader is referred, for instance, to the encyclopedic treatise by Gasper and Rahman [20]. Hopefully, this section succeeds to give the reader some feeling for the subject and some impression of general techniques and ideas.

Finally, section 4 gives an overview of the classical orthogonal polynomials, where "classical" now means "up to the level of Askey-Wilson polynomials" [10]. The section starts with the "very classical" situation of Jacobi, Laguerre and Hermite polynomials and next discusses the Askey tableau of classical orthogonal polynomials (still for $q = 1$). Then the example of big q-Jacobi polynomials is worked out in detail, as a demonstration how the main formulas in this area can be neatly derived. The section continues with the q-Hahn tableau and then gives a self-contained introduction to the Askey-Wilson polynomials. Both section 3 and 4 conclude with some exercises.

I thank Mathijs Dijkhuizen and René Swarttouw for commenting on preliminary versions of sections 1, 2 resp. 3, 4.

1. GENERALITIES ABOUT HOPF ALGEBRAS

Standard references about Hopf algebras are the books by Abe [1] and Sweedler [37], see also Hazewinkel [22, §37.1]. Below we will assume ground field $\mathbb{C}$. By the *tensor product* $V \otimes W$ of two linear spaces V and W we will always mean the *algebraic* tensor product. Thus the elements of $V \otimes W$ are finite sums of elements $v_i \otimes w_i$ ($v_i \in V$, $w_i \in W$).

1.1. Hopf algebras. The reader will be familiar with the concept of an *associative algebra with unit* (or shortly an *algebra*), i.e., a linear space $\mathcal{A}$ with a bilinear mapping $(a, b) \mapsto ab : \mathcal{A} \times \mathcal{A} \to \mathcal{A}$ and with a special nonzero element $1 \in \mathcal{A}$ such that

$$(ab)c = a(bc) \quad \text{and} \quad 1a = a = a1. \tag{1.1}$$

When we define the *multiplication* as the linear mapping $m : \mathcal{A} \otimes \mathcal{A} \to \mathcal{A}$ such that $m(a \otimes b) = ab$ and the *unit* as the linear mapping $\eta : \mathbb{C} \to \mathcal{A}$ such that $\eta(1) = 1_{\mathcal{A}}$, then we can rephrase (1.1) as

$$m \circ (m \otimes \mathrm{id}) = m \circ (\mathrm{id} \otimes m), \tag{1.2}$$

$$m \circ (\eta \otimes \mathrm{id}) = \mathrm{id} = m \circ (\mathrm{id} \otimes \eta). \tag{1.3}$$

Here the two sides of (1.2) are linear mappings from $\mathcal{A} \otimes \mathcal{A} \otimes \mathcal{A}$ to $\mathcal{A}$, while the three parts of (1.3) are linear mappings from $\mathcal{A}$ to $\mathcal{A}$. In (1.3) we identify $\mathbb{C} \otimes \mathcal{A}$ with $\mathcal{A}$ by identifying $c \otimes a$ with ca ($c \in \mathbb{C}$, $a \in \mathcal{A}$). Thus $(\eta \otimes \mathrm{id})(a) = (\eta \otimes \mathrm{id})(1 \otimes a) = 1_{\mathcal{A}} \otimes a$.

Definition 1.1. A *coassociative coalgebra with counit* (or shortly a *coalgebra*) is a linear space $\mathcal{A}$ with linear mappings $\Delta : \mathcal{A} \to \mathcal{A} \otimes \mathcal{A}$ (*comultiplication*) and $\varepsilon : \mathcal{A} \to \mathbb{C}$ (*counit*) (nonzero) such that

$$(\Delta \otimes \mathrm{id}) \circ \Delta = (\mathrm{id} \otimes \Delta) \circ \Delta \quad (coassociativity), \tag{1.4}$$

$$(\varepsilon \otimes \mathrm{id}) \circ \Delta = \mathrm{id} = (\mathrm{id} \otimes \varepsilon) \circ \Delta. \tag{1.5}$$

Here the two sides of (1.4) are linear mappings from $\mathcal{A}$ to $\mathcal{A} \otimes \mathcal{A} \otimes \mathcal{A}$, while the three parts of (1.5) are linear mappings from $\mathcal{A}$ to $\mathcal{A}$. Here we used again the identification of $\mathbb{C} \otimes \mathcal{A}$ or $\mathcal{A} \otimes \mathbb{C}$ with $\mathcal{A}$.

If $\mathcal{A}$ is an algebra with identity element 1 then $\mathcal{A} \otimes \mathcal{A}$ naturally becomes an algebra with identity element $1 \otimes 1$ if we define the product of $a \otimes b$ and $c \otimes d$ as $ac \otimes bd$.

When we speak of an algebra homomorphism then we will always mean that the homomorphism mapping also sends 1 to 1 (*unital* algebra homomorphism).

Definition 1.2. A *bialgebra* is a linear space $\mathcal{A}$ which has the structure of both an algebra and a coalgebra such that the mappings $\Delta \colon \mathcal{A} \to \mathcal{A} \otimes \mathcal{A}$ and $\varepsilon \colon \mathcal{A} \to \mathbb{C}$ are algebra homomorphisms.

Definition 1.3. A *Hopf algebra* is a bialgebra $\mathcal{A}$ together with a linear mapping $S \colon \mathcal{A} \to \mathcal{A}$ (*antipode*) such that

$$m \circ (S \otimes \mathrm{id}) \circ \Delta = \eta \circ \varepsilon = m \circ (\mathrm{id} \otimes S) \circ \Delta \tag{1.6}$$

(identities of linear mappings from $\mathcal{A}$ to $\mathcal{A}$), i.e.,

$$(m \circ (S \otimes \mathrm{id}) \circ \Delta)(a) = \varepsilon(a)\, 1 = (m \circ (\mathrm{id} \otimes S) \circ \Delta)(a), \quad a \in \mathcal{A}. \tag{1.7}$$

If V is a linear space then the *flip operator* $\sigma \colon V \otimes V \to V \otimes V$ will be the linear operator such that $\sigma(v_1 \otimes v_2) = v_2 \otimes v_1$ for $v_1, v_2 \in V$. Note that an algebra $\mathcal{A}$ is commutative iff $m \circ \sigma = m$. By analogy, we define a coalgebra $\mathcal{A}$ to be *cocommutative* if $\sigma \circ \Delta = \Delta$.

Example 1.4. Let G be a group and let $\mathcal{A} := \mathrm{Fun}(G)$ be the algebra (under pointwise multiplication) of all complex-valued functions on G. To some extent, the algebra $\mathrm{Fun}(G \times G)$ of all complex-valued functions on $G \times G$ can be viewed as the tensor product $\mathcal{A} \otimes \mathcal{A}$. Just write $(a \otimes b)(x, y) := a(x)\, b(y)$ if $a, b \in \mathcal{A}$ and $x, y \in G$. However, $\mathrm{Fun}(G \times G)$ is not the algebraic tensor product of $\mathrm{Fun}(G)$ with $\mathrm{Fun}(G)$ except if G is a finite group. Let us for the moment not worry about this. Since $(m(a \otimes b))(x) = (ab)(x) = a(x)\, b(x) = (a \otimes b)(x, x)$ if $a, b \in \mathrm{Fun}(G)$, we can write

$$(m(F))(x) = F(x, x), \quad F \in \mathrm{Fun}(G \times G), \; x \in G. \tag{1.8}$$

Now define the comultiplication, counit and antipode by

$$(\Delta(a))(x, y) := a(xy), \quad a \in \mathcal{A}, \; x, y \in G, \tag{1.9}$$
$$\varepsilon(a) := a(e), \quad a \in \mathcal{A}, \tag{1.10}$$
$$(S(a))(x) := a(x^{-1}), \quad a \in \mathcal{A}, \; x \in G. \tag{1.11}$$

The general philosophy here is that all properties of and information about the group can be stored in the algebra $\mathrm{Fun}(G)$. Thus group multiplication, group identity and group inverse are described on the level of $\mathrm{Fun}(G)$ by Δ, ε and S, respectively.

Evidently, Δ and ε are algebra homomorphisms and S is a linear mapping (in this example also an algebra homomorphism, while $S^2 = \mathrm{id}$). Furthermore, the group axioms yield the Hopf algebra axioms (1.4), (1.5) and (1.6). This can be seen for the first identity in (1.7) by observing from (1.9) and (1.11) that $(((S \otimes \mathrm{id}) \circ \Delta)(a))(x,y) = a(x^{-1}y)$ and next from (1.8) that the left hand side of (1.7) evaluated in $x \in G$ equals $a(x^{-1}x) = a(e)$. By (1.10) the evaluation of the middle part of (1.7) in x yields the same.

Cocommutativity $\sigma \circ \Delta = \Delta$ would be equivalent here to $a(xy) = a(yx)$ for all $a \in \mathrm{Fun}(G)$. Thus $\mathrm{Fun}(G)$ is cocommutative iff the group G is abelian.

Everything above holds rigorously with algebraic tensor products if G is a finite group. Now suppose that G is a subgroup of $SL(n,\mathbb{C})$ (the group of complex $n \times n$ matrices of determinant 1). Usually (but not necessarily), this subgroup G may be thought to be *algebraic*, i.e. closed in the Zariski topology. Let $\mathrm{Pol}(G)$ consist of all complex-valued functions depending on $x \in G$ which can be written as polynomials in the matrix elements x_{ij} of x. Let $\mathrm{Pol}(G \times G)$ consist of all complex-valued functions of $(x,y) \in G \times G$ which can be written as polynomials in the matrix elements x_{ij} and y_{ij} of x and y. Then $\mathrm{Pol}(G \times G)$ can be identified with $\mathrm{Pol}(G) \otimes \mathrm{Pol}(G)$. Now $\mathcal{A} := \mathrm{Pol}(G)$ becomes a Hopf algebra in the algebraic sense with the above defined operations.

Let t_{ij} be the element of $\mathcal{A} = \mathrm{Pol}(G)$ such that $t_{ij}(x) = x_{ij}$ ($x \in G$). Then the t_{ij} form a set of generators of the algebra $\mathcal{A}$. In fact, $\mathcal{A}$ is the quotient algebra obtained when the free abelian algebra with 1 generated by the t_{ij} is divided by the ideal of all elements in this free algebra which vanish on G. The Hopf algebra operations can now be specified by defining them for the generators. Thus, since

$$(\Delta(t_{ij}))(x,y) = t_{ij}(xy) = (xy)_{ij} = \sum_{k=1}^{n} x_{ik}\, y_{kj} = \sum_{k=1}^{n} t_{ik}(x)\, t_{kj}(y),$$

we have

$$\Delta(t_{ij}) = \sum_{k=1}^{n} t_{ik} \otimes t_{kj}, \tag{1.12}$$

and similarly

$$\varepsilon(t_{ij}) = \delta_{ij}, \quad S(t_{ij}) = T_{ji}, \tag{1.13}$$

where T_{ji} is the cofactor of the (ji)th entry in the matrix $(t_{kl})_{k,l=1,\ldots,n}$.

The following notation is often useful. Let $\mathcal{A}$ be a coalgebra. If $a \in \mathcal{A}$ then we can choose sets of elements $a_{(1)i}$ and $a_{(2)i}$ in $\mathcal{A}$ (i running over a finite set) such that $\Delta(a) = \sum_i a_{(1)i} \otimes a_{(2)i}$. We write this symbolically as

$$\Delta(a) = \sum_{(a)} a_{(1)} \otimes a_{(2)}, \quad a \in \mathcal{A}. \tag{1.14}$$

Similarly, we write

$$(\Delta \otimes \mathrm{id})(\Delta(a)) = \sum_{(a)} a_{(1)} \otimes a_{(2)} \otimes a_{(3)}, \quad a \in \mathcal{A}.$$

This notation is justified by the coassociativity (1.4).

Now we can express identities for Hopf algebras which involve comultiplication by this notation. For instance, (1.5) (applied to $a \in \mathcal{A}$) and (1.7) can be written as

$$\sum_{(a)} \varepsilon(a_{(1)}) a_{(2)} = a = \sum_{(a)} \varepsilon(a_{(2)}) a_{(1)}, \tag{1.15}$$

$$\sum_{(a)} S(a_{(1)}) a_{(2)} = \varepsilon(a) 1 = \sum_{(a)} a_{(1)} S(a_{(2)}). \tag{1.16}$$

Proposition 1.5. Let $\mathcal{A}$ be a bialgebra. If an antipode S exists such that $\mathcal{A}$ becomes a Hopf algebra then S is unique.

Proof. Define a *convolution product* $F * G$ of linear mappings F and G from the bialgebra $\mathcal{A}$ to itself by

$$(F * G)(a) := (m \circ (F \otimes G) \circ \Delta)(a) = \sum_{(a)} F(a_{(1)}) G(a_{(2)}), \quad a \in \mathcal{A}. \tag{1.17}$$

Coassociativity of Δ and associativity of m then show that, under this convolution operation, $\mathrm{End}(\mathcal{A})$ is an associative algebra. The properties of counit and unit yield $\eta \circ \varepsilon$ as an identity element for this algebra. The antipode property (1.6) can now be interpreted as

$$S * \mathrm{id} = \eta \circ \varepsilon = \mathrm{id} * S. \tag{1.18}$$

Thus, if S exists then it is the two-sided inverse of id in this convolution algebra, and therefore unique. □

Because of (1.18) we can write the antipode as $S = S * \mathrm{id} * S$. This can be used in order to derive further properties of S. In particular, the next proposition states that S is unital, counital, anti-multiplicative and anti-comultiplicative.

Proposition 1.6. Let $\mathcal{A}$ be a Hopf algebra. Then, for $a, b \in \mathcal{A}$,

$$S(1) = 1, \tag{1.19}$$

$$\varepsilon(S(a)) = \varepsilon(a), \tag{1.20}$$

$$S(ab) = S(b) S(a), \tag{1.21}$$

$$\Delta(S(a)) = ((S \otimes S) \circ \sigma \circ \Delta)(a) = \sum_{(a)} S(a_{(2)}) \otimes S(a_{(1)}). \tag{1.22}$$

Proof. Formula (1.19) follows by putting $a := 1$ in (1.7), while (1.20) follows from (1.15) and (1.16). For the proof of (1.21) write

$$
\begin{aligned}
S(b)\,S(a) &= \sum_{(a),(b)} S(b_{(1)})\,S(a_{(1)})\,(\varepsilon(a_{(2)}\,b_{(2)})\,1) \\
&= \sum_{(a),(b)} S(b_{(1)})\,S(a_{(1)})\,a_{(2)}\,b_{(2)}\,S(a_{(3)}b_{(3)}) \\
&= \sum_{(a),(b)} \varepsilon(a_{(1)})\,\varepsilon(b_{(1)})\,S(a_{(2)}b_{(2)}) = S(ab).
\end{aligned}
$$

In the first identity (1.15) was applied twice. In the second identity we used (1.16) with $a := a_{(2)}b_{(2)}$ and also the fact that Δ is an algebra homomorphism. In the third identity we used (1.16) twice. The fourth identity contains two further applications of (1.15).

The proof of (1.22) is given by

$$
\begin{aligned}
\sum_{(a)} S(a_{(2)}) \otimes S(a_{(1)}) &= \sum_{(a)} \varepsilon(a_{(3)})\,S(a_{(2)}) \otimes S(a_{(1)}) \\
&= \sum_{(a)} (S(a_{(2)}) \otimes S(a_{(1)}))\,(\varepsilon(a_{(3)})\,1 \otimes 1) \\
&= \sum_{(a)} (S(a_{(2)}) \otimes S(a_{(1)}))\,(a_{(3)} \otimes a_{(4)})\,\Delta(S(a_{(5)})) \\
&= \sum_{(a)} (S(a_{(2)})\,a_{(3)} \otimes S(a_{(1)})\,a_{(4)})\,\Delta(S(a_{(5)})) \\
&= \sum_{(a)} \varepsilon(a_{(2)})\,(1 \otimes S(a_{(1)})\,a_{(3)})\,\Delta(S(a_{(4)})) \\
&= \sum_{(a)} (1 \otimes S(a_{(1)})\,a_{(2)})\,\Delta(S(a_{(3)})) \\
&= \sum_{(a)} \varepsilon(a_{(1)})\,\Delta(S(a_{(2)})) = \Delta(S(a)).
\end{aligned}
$$

For the third identity above, apply Δ to the second identity in (1.16).

Example 1.7. Let $\mathfrak{g}$ be a complex Lie algebra and $\mathcal{A} := \mathcal{U}(\mathfrak{g})$ its universal enveloping algebra. $\mathcal{U}(\mathfrak{g})$ is defined as the quotient algebra $\mathcal{T}(\mathfrak{g})/J$, where $\mathcal{T}(\mathfrak{g})$ is the tensor algebra of $\mathfrak{g}$ and J is the ideal in $\mathcal{T}(\mathfrak{g})$ which is generated by the elements $XY - YX - [X,Y]$ $(X,Y \in \mathfrak{g})$. We define Δ, ε and S first on $\mathfrak{g}$:

$$
\Delta(X) := X \otimes 1 + 1 \otimes X, \quad \varepsilon(X) := 0, \quad S(X) := -X \qquad \text{for } X \in \mathfrak{g}. \tag{1.23}
$$

On $\mathbb{C}$ we declare the operators to be unital. Now we can check the coalgebra axioms and the antipode axiom already for the operators acting on $\mathbb{C} \oplus \mathfrak{g}$. In fact, the counit and antipode axiom force ε and S to be as in (1.23). Next we extend Δ and ε to $\mathcal{T}(\mathfrak{g})$ as algebra homomorphisms and S as anti-algebra homomorphism. Now check that $\Delta(J) \subset \mathcal{T}(\mathfrak{g}) \otimes J + J \otimes \mathcal{T}(\mathfrak{g})$, $\varepsilon(J) = 0$ and $S(J) \subset J$. This allows us to consider Δ, ε and S as operators on $\mathcal{U}(\mathfrak{g})$. Finally we have to check the Hopf algebra axioms on all of $\mathcal{U}(\mathfrak{g})$, by using that they are already satisfied on a subspace of generators. Note that $\mathcal{A}$ is cocommutative, but generally not commutative, unless $\mathfrak{g}$ is an abelian Lie algebra. Note that the antipode satisfies $S^2 = \mathrm{id}$.

Remark 1.8. A further motivation for the concept of comultiplication is the wish to construct tensor products of representations of algebras. Suppose $\mathcal{A}$ is an algebra. Let π_1 and π_2 be algebra representations of $\mathcal{A}$ on finite dimensional linear spaces V_1 and V_2 resp., i.e., algebra homomorphisms of $\mathcal{A}$ to the algebras of linear endomorphisms of V_1 and V_2 respectively. Then $a_1 \otimes a_2 \mapsto \pi_1(a_1) \otimes \pi_2(a_2) : \mathcal{A} \otimes \mathcal{A} \to \mathrm{End}(V_1 \otimes V_2)$ is a representation of $\mathcal{A} \otimes \mathcal{A}$ on $V_1 \otimes V_2$. In order to obtain from this representation a representation of $\mathcal{A}$ on $V_1 \otimes V_2$, we need an algebra homomorphism from $\mathcal{A}$ to $\mathcal{A} \otimes \mathcal{A}$. Without further structure on $\mathcal{A}$ there is no canonical method for this. However, if $\mathcal{A}$ is a Hopf algebra, the desired mapping is provided by the comultiplication Δ. Then we can define the tensor product representation $\pi_1 \otimes \pi_2$ of $\mathcal{A}$ as the composition $(\pi_1 \otimes \pi_2) \circ \Delta$. Associativity of this tensor product is precisely assured by the coassociativity (1.4) of Δ.

Other Hopf algebra axioms are also meaningful in this context. For instance, the counit ε gives a one-dimensional algebra representation of $\mathcal{A}$. Then the counit axiom (1.5) implies that, for each finite dimensional algebra representation π of $\mathcal{A}$, we have that $\pi \otimes \varepsilon = \pi = \varepsilon \otimes \pi$. Also, if π is an algebra representation of $\mathcal{A}$ on a finite dimensional linear space V and if V^* is the linear dual of V, then we can define an algebra representation π^* of $\mathcal{A}$ on V^* by

$$\langle \pi^*(a) v^*, v \rangle := \langle v^*, \pi(S(a)) v \rangle, \quad v \in V, \quad v^* \in V^*.$$

Recall that the tensor product of two representations π_1 and π_2 of a Lie algebra $\mathfrak{g}$ is provided by $((\pi_1 \otimes \pi_2)(X))(v_1 \otimes v_2) = (\pi_1(X) v_1) \otimes v_2 + v_1 \otimes (\pi_2(X) v_2)$. Compare this with the comultiplication for $\mathfrak{g}$ as given by (1.23).

One should be aware that the antipode in a Hopf algebra $\mathcal{A}$ is not necessarily invertible as a linear mapping from $\mathcal{A}$ to itself.

We may consider $\mathcal{A}$ also as a Hopf algebra with opposite multiplication $m \circ \sigma$ and opposite comultiplication $\sigma \circ \Delta$, and ε, η and S unchanged. Then S is a Hopf algebra homomorphism from $\mathcal{A}$ with the old Hopf algebra structure to $\mathcal{A}$ with the new one. However, if we change only one of the two operations of multiplication and comultiplication into its opposite then we still have a bialgebra, but this will be a Hopf algebra if and only if S is invertible. The new antipode will then be precisely

S^{-1}. Indeed, let $\mathcal{A}'$ be $\mathcal{A}$ equipped with opposite multiplication and suppose $\mathcal{A}'$ is a Hopf algebra with antipode S'. Now apply S to

$$\sum_{(a)} a_{(2)} S'(a_{(1)}) = \varepsilon(a)1 = \sum_{(a)} S'(a_{(2)}) a_{(1)}.$$

This yields that SS' is the two-sided inverse of S under convolution. Hence $SS' = \mathrm{id}$, and also $S'S = \mathrm{id}$ by reverting the roles of S and S'. Conversely, if S is invertible, then define $S' := S^{-1}$ and show that S' is an antipode for $\mathcal{A}'$.

Recall that an (associative) algebra $\mathcal{A}$ (with 1) is a *-*algebra* if there is a mapping $a \mapsto a^* \colon \mathcal{A} \to \mathcal{A}$ (an *involution*) such that $(a^*)^* = a$, $(a+b)^* = a^* + b^*$, $(\lambda a)^* = \overline{\lambda}\, a^*$, $(ab)^* = b^* a^*$, $1^* = 1$. If $\mathcal{A}$ is a *-algebra then $\mathcal{A} \otimes \mathcal{A}$ becomes a *-algebra with $(a \otimes b)^* := a^* \otimes b^*$. If $\mathcal{A}$ and $\mathcal{B}$ are *-algebras then $F \colon \mathcal{A} \to \mathcal{B}$ is called a *-*homomorphism* if F is an algebra homomorphism such that $F(a^*) = (F(a))^*$.

Definition 1.9. A Hopf *-algebra is a Hopf algebra $\mathcal{A}$ which, as an algebra, is also a *-algebra such that $\Delta \colon \mathcal{A} \to \mathcal{A} \otimes \mathcal{A}$ and $\varepsilon \colon \mathcal{A} \to \mathbb{C}$ are *-homomorphisms.

Originally, the property stated in the next Proposition was part of the definition of a Hopf *-algebra. I thank S. Zakrzewski for helpful correspondence about this.

Proposition 1.10. In a Hopf *-algebra $\mathcal{A}$ we have

$$S \circ * \circ S \circ * = \mathrm{id}. \tag{1.24}$$

In particular, S is invertible with inverse $* \circ S \circ *$.

Proof. Apply (1.6) to a^* and next apply the involution to all members of (1.6). This shows that $* \circ S \circ *$ is an antipode for $\mathcal{A}'$, i.e., for $\mathcal{A}$ with opposite multiplication. Hence $* \circ S \circ * = S^{-1}$. □

Example 1.11. Let G be a group and let $\mathcal{A} := \mathrm{Fun}(G)$ as in Example 1.4 For $f \in \mathrm{Fun}(G)$ define $f^*(x) := \overline{f(x)}$ $(x \in G)$. Similarly, for $F \in \mathrm{Fun}(G \times G)$ define $F^*(x, y) := \overline{F(x, y)}$ $(x, y \in G)$. Then $\mathrm{Fun}(G)$ and $\mathrm{Fun}(G \times G)$ are commutative *-algebras and $\Delta \colon \mathrm{Fun}(G) \to \mathrm{Fun}(G \times G)$ and $\varepsilon \colon \mathrm{Fun}(G) \to \mathbb{C}$ are *-homomorphisms. So, apart from the fact that Δ does not map $\mathrm{Fun}(G)$ into $\mathrm{Fun}(G) \otimes \mathrm{Fun}(G)$, we have $\mathrm{Fun}(G)$ as an example of a Hopf *-algebra.

Example 1.12. Let G be a complex Lie group given as a closed connected subgroup of $SL(n, \mathbb{C})$ and let G_0 be a real connected Lie group and a real form of G. Then every $a \in \mathrm{Pol}(G)$ is completely determined by its restriction to G_0. Suppose that, for each $a \in \mathrm{Pol}(G)$, there exists $a^* \in \mathrm{Pol}(G)$ such that $a^*(x) = \overline{a(x)}$ for $x \in G_0$. Then the Hopf algebra $\mathrm{Pol}(G)$ becomes a Hopf *-algebra with this mapping *. Conversely, if the Hopf algebra $\mathrm{Pol}(G)$ is a Hopf $*$-algebra then we may define a real form G_0 of G by $G_0 := \{x \in G \mid \overline{a(x)} = a^*(x) \quad \forall a \in \mathrm{Pol}(G)\}$.

Thus, on the level of polynomial function algebras, a real Lie group is described by the algebra of functions on its complexification together with an involution for this algebra.

In particular, let the t_{ij} (cf. Example 1.4) be the generators of the algebra $\mathcal{A} = \mathrm{Pol}(G)$. If we suppose that $G_0 \subset SL(n, \mathbb{R})$ then $t_{ij}^*(x) = \overline{t_{ij}(x)} = t_{ij}(x)$ for $x \in G_0$. Hence $t_{ij}^* = t_{ij}$. This gives the action of $*$ on the generators and next, by anti-linear homomorphic continuation, on all of $\mathcal{A}$. However, if $G_0 \subset SU(n)$ then $t_{ij}^*(x) = \overline{t_{ij}(x)} = t_{ji}(x^{-1}) = (S(t_{ji}))(x)$ if $x \in G_0$. Hence $t_{ij}^* = S(t_{ji})$. Again, the involution on all of $\mathcal{A}$ follows by anti-linear homomorphic continuation.

1.2. Duality for Hopf algebras. Let $\mathcal{A}$ be a Hopf algebra and let $\mathcal{A}^*$ be its algebraic linear dual, i.e., the space of all linear mappings $f \colon \mathcal{A} \to \mathbb{C}$. We will write $\langle f, a \rangle := f(a)$ ($f \in \mathcal{A}^*$, $a \in \mathcal{A}$). The algebraic tensor product $\mathcal{A}^* \otimes \mathcal{A}^*$ is a subspace of $(\mathcal{A} \otimes \mathcal{A})^*$ by the rule

$$\langle f \otimes g, a \otimes b \rangle = \langle f, a \rangle \langle g, b \rangle, \quad f, g \in \mathcal{A}^*, \; a, b \in \mathcal{A}.$$

Unless $\mathcal{A}$ is finite dimensional, this will be a proper subspace. By duality, the Hopf algebra operations on $\mathcal{A}$ can be transferred to $\mathcal{A}^*$. We define

$$\langle fg, a \rangle := \langle f \otimes g, \Delta(a) \rangle = \sum_{(a)} \langle f, a_{(1)} \rangle \langle g, a_{(2)} \rangle, \tag{1.25}$$

$$\langle \Delta(f), a \otimes b \rangle := \langle f, ab \rangle, \tag{1.26}$$

$$\langle 1_{\mathcal{A}^*}, a \rangle := \varepsilon_{\mathcal{A}}(a), \tag{1.27}$$

$$\varepsilon_{\mathcal{A}^*}(f) := \langle f, 1_{\mathcal{A}} \rangle, \tag{1.28}$$

$$\langle S(f), a \rangle := \langle f, S(a) \rangle. \tag{1.29}$$

However, $\Delta(f)$, as defined by (1.26), will be an element of $(\mathcal{A} \otimes \mathcal{A})^*$ and not necessarily of $\mathcal{A}^* \otimes \mathcal{A}^*$. Still, with a suitable adaptation of the definition of Hopf algebra, the Hopf algebra axioms for $\mathcal{A}^*$ with the above operations can be verified in a straightforward way.

If $\mathcal{A}$ is moreover a Hopf $*$-algebra then we can define an involution on $\mathcal{A}^*$ by

$$f^*(a) := \overline{\langle f, (S(a))^* \rangle}. \tag{1.30}$$

Note that (1.24) ensures that $(f^*)^* = f$. The antipode is needed in (1.30) in order to ensure that $(fg)^* = g^* f^*$. (The more simple definition $f^*(a) := \overline{\langle f, a^* \rangle}$ would make the involution on $\mathcal{A}^*$ multiplicative rather than anti-multiplicative.) The reader should verify that $\mathcal{A}^*$ thus indeed becomes a Hopf $*$-algebra, with suitably adapted definition because of the fact that Δ does not necessarily map into $\mathcal{A}^* \otimes \mathcal{A}^*$.

Let $\mathcal{A}$ be a Hopf algebra. We define left and right algebra actions of $\mathcal{A}^*$ on $\mathcal{A}$:

$$f.a := (\mathrm{id} \otimes f)(\Delta(a)) = \sum_{(a)} f(a_{(2)}) a_{(1)}, \tag{1.31}$$

$$a.f := (f \otimes \mathrm{id})(\Delta(a)) = \sum_{(a)} f(a_{(1)}) a_{(2)}. \tag{1.32}$$

It is indeed an easy exercise to verify that

$$(fg).a = f.(g.a), \quad a.(fg) = (a.f).g, \quad 1_{\mathcal{A}^*}.a = a.1_{\mathcal{A}^*} = a.$$

An important property is the following. If $f \in \mathcal{A}^*$ such that $\Delta(f) \in \mathcal{A}^* \otimes \mathcal{A}^*$ and if we write $\Delta(f) = \sum_{(f)} f_{(1)} \otimes f_{(2)}$ then

$$f.(ab) = \sum_{(f)} (f_{(1)}.a)(f_{(2)}.b), \quad a, b \in \mathcal{A}, \tag{1.33}$$

$$(ab).f = \sum_{(f)} (a.f_{(1)})(b.f_{(2)}), \quad a, b \in \mathcal{A}. \tag{1.34}$$

We give the proof of (1.33); the proof of (1.34) is similar.

$$f.(ab) = \sum_{(a),(b)} a_{(1)} b_{(1)} f(a_{(2)}b_{(2)}) = \sum_{(a),(b),(f)} a_{(1)} f_{(1)}(a_{(2)}) b_{(1)} f_{(2)}(b_{(2)})$$

$$= \sum_{(f)} (f_{(1)}.a)(f_{(2)}.b).$$

Definition 1.13. Two Hopf algebras $\mathcal{U}$ and $\mathcal{A}$ are said to be *Hopf algebras in duality* if there is a bilinear mapping $(u, a) \mapsto \langle u, a \rangle : \mathcal{U} \times \mathcal{A} \to \mathbb{C}$ such that (1.25)–(1.29) are satisfied when $\mathcal{U}$ is read instead of $\mathcal{A}^*$.

If $\mathcal{U}$ and $\mathcal{A}$ are moreover Hopf $*$-algebras and if (1.30) also holds then we speak about *Hopf $*$-algebras in duality*.

Two Hopf $(*-)$algebras in duality $\mathcal{A}$ and $\mathcal{U}$ are said to be *Hopf $(*-)$algebras in nondegenerate duality* (in fact in doubly nondegenerate duality) if the two following implications moreover hold: (i) $(\forall a \in \mathcal{A} \ \langle u, a \rangle = 0) \Longrightarrow u = 0$, and (ii) $(\forall u \in \mathcal{U} \ \langle u, a \rangle = 0) \Longrightarrow a = 0$.

If $\mathcal{U}$ and $\mathcal{A}$ are Hopf algebras in duality (not necessarily nondegenerate) then left and right actions of $\mathcal{U}$ on $\mathcal{A}$ as in (1.31) and (1.32) can still be defined.

If $\mathcal{A}$ is a Hopf $(*-)$algebra and if $\mathcal{U}$ is a Hopf $(*-)$subalgebra of $\mathcal{A}^*$, i.e., if $\mathcal{U}$ is a unital subalgebra of $\mathcal{A}^*$ such that $\Delta(\mathcal{U}) \subset \mathcal{U} \otimes \mathcal{U}$, $S(\mathcal{U}) \subset \mathcal{U}$ and, in case of a Hopf $*$-subalgebra, $\mathcal{U}^* \subset \mathcal{U}$, then $\mathcal{U}$ and $\mathcal{A}$ are obviously Hopf $(*-)$algebras in duality, while the duality is already nondegenerate on one side: if $u \in \mathcal{U}$ and if $\langle u, a \rangle = 0$ for all $a \in \mathcal{A}$ then $u = 0$.

If $\mathcal{U}$ and $\mathcal{A}$ are Hopf $(*-)$algebras in nondegenerate duality then $\mathcal{U}$ can be viewed as a Hopf $(*-)$subalgebra of $\mathcal{A}^*$ and $\mathcal{A}$ as a Hopf $(*-)$subalgebra of $\mathcal{U}^*$. So we can write $\langle u, a \rangle = u(a) = a(u)$ for $u \in \mathcal{U}$ and $a \in \mathcal{A}$.

Example 1.14. Let G be a complex Lie group given as a closed connected subgroup of $SL(n, \mathbb{C})$ and let $\mathfrak{g}$ be its (complex) Lie algebra and $\mathcal{U} := \mathcal{U}(\mathfrak{g})$ the universal enveloping algebra of $\mathfrak{g}$. Then $\mathcal{A} := \mathrm{Pol}(G)$ and $\mathcal{U}$ are Hopf algebras (cf. Examples 1.4 and 1.7). These Hopf algebras are naturally in duality with each other. For $X \in \mathfrak{g}$ and $a \in \mathrm{Pol}(G)$ the pairing is given by

$$\langle X, a \rangle = \frac{d}{dt}\Big|_{t=0} a(\exp(tX)). \tag{1.35}$$

Then, for $f := X \in \mathfrak{g}$, equations (1.26)–(1.29) are satisfied in view of (1.35) and (1.23).

For the left and right actions of $X \in \mathfrak{g}$ on $a \in \mathrm{Pol}(G)$ (cf. (1.31) and (1.32)) we then obtain

$$(X.a)(x) = \frac{d}{dt}\Big|_{t=0} a(x \exp(tX)), \quad (a.X)(x) = \frac{d}{dt}\Big|_{t=0} a(\exp(tX)\,x).$$

If $X := X_1 X_2 \ldots X_k \in \mathcal{U}$, with $X_1, \ldots, X_k \in \mathfrak{g}$, then

$$(X.a)(x) = \frac{\partial^k}{\partial t_1 \ldots \partial t_k}\Big|_{t_1,\ldots,t_k=0} a(x \exp(t_1 X_1) \ldots \exp(t_k X_k)),$$

$$(a.X)(x) = \frac{\partial^k}{\partial t_1 \ldots \partial t_k}\Big|_{t_1,\ldots,t_k=0} a(\exp(t_1 X_1) \ldots \exp(t_k X_k)\,x).$$

So we get the familiar left or right action of $\mathcal{U}(\mathfrak{g})$ on smooth functions on G by left or right invariant differential operators. In particular, either of the two last formulas, when evaluated at $x := e$, yields the pairing $\langle X_1 \ldots X_k, a \rangle$. The formula thus obtained is compatible with (1.35) and (1.25). The double non-degeneracy of the pairing can be easily verified.

Example 1.15. Let G, $\mathfrak{g}$, $\mathcal{A}$ and $\mathcal{U}$ be as in the previous example. Let the connected real Lie group G_0 be a real form of G and let $*$ be the corresponding involution on $\mathcal{A}$, by which $\mathcal{A}$ becomes a Hopf $*$-algebra (cf. Example 1.12). Let $\mathfrak{g}_0$ be the (real) Lie algebra of G_0. It is a real form of the complex Lie algebra $\mathfrak{g}$. Let us compute X^* (as defined by (1.30)) for $X \in \mathfrak{g}_0$ and let us see if $X^* \in \mathcal{U}$. For $a \in \mathcal{A}$ we have

$$\langle X^*, a \rangle = \overline{\langle X, (S(a))^* \rangle} = \overline{\frac{d}{dt}\Big|_{t=0} a(\exp(-tX))} = -\langle X, a \rangle.$$

Here we used that $a^*(x) = \overline{a(x)}$ if $x \in G_0$ (cf. Example 1.12). Thus $X^* = -X$ for $X \in \mathfrak{g}_0$. In particular, $*$ maps the Lie algebra $\mathfrak{g}_0$ to itself. It follows that $*$ maps $\mathcal{U}$ to itself, so $\mathcal{U}$ is a Hopf $*$-algebra in duality with $\mathcal{A}$.

Note that in the case $G_0 \subset SU(n)$ we have $\mathfrak{g}_0 \subset su(n)$, the Lie algebra of skew-hermitian matrices. So then the $*$ on $\mathfrak{g}$ induced by the $*$-structure on $\mathcal{A}$ has the same effect as taking the adjoint of the matrix $X \in \mathfrak{g}$. But this is not necessarily true if $G_0 \subset SL(n, \mathbb{R})$.

1.3. Corepresentations.

Definition 1.16. Let $\mathcal{A}$ be a coalgebra. A *corepresentation* of $\mathcal{A}$ on a complex vector space V is defined as a linear mapping $t\colon V \to V \otimes \mathcal{A}$ such that

$$(t \otimes \mathrm{id}) \circ t = (\mathrm{id} \otimes \Delta) \circ t, \quad (\mathrm{id} \otimes \varepsilon) \circ t = \mathrm{id}. \tag{1.36}$$

Suppose that V is a finite dimensional vector space with basis $e_1, \ldots, e_n$. Then t determines elements t_{ij} of $\mathcal{A}$ such that

$$t(e_j) = \sum_i e_i \otimes t_{ij}. \tag{1.37}$$

Then

$$\sum_i e_i \otimes \Delta(t_{ik}) = (\mathrm{id} \otimes \Delta)(t(e_k)) = (t \otimes \mathrm{id})(t(e_k)) = \sum_{j,k} e_i \otimes t_{ij} \otimes t_{jk}$$

and

$$\sum_i \varepsilon(t_{ij}) \, e_i = (\mathrm{id} \otimes \varepsilon)(t(e_j)) = e_j.$$

Hence

$$\Delta(t_{ij}) = \sum_k t_{ik} \otimes t_{kj}, \quad \varepsilon(t_{ij}) = \delta_{ij}. \tag{1.38}$$

Conversely, if elements t_{ij} of $\mathcal{A}$ satisfy (1.38) and if t is defined by (1.37) on the vector space having the e_i as a basis then t is a corepresentation of $\mathcal{A}$ according to Definition 1.16. We call a matrix $(t_{ij})_{i,j=1,\ldots,n}$ satisfying (1.38) a *matrix corepresentation of* $\mathcal{A}$.

Example 1.17. Let G be a group and π a representation of G on some vector space V. Let $\mathrm{Fun}(G)$ be as in Example 1.4. To some extent, we can identify $V \otimes \mathrm{Fun}(G)$ with $\mathrm{Fun}(G; V)$ (the space of V-valued functions on G). Just write $(v \otimes a)(x) := a(x)v$ if $v \in V$, $a \in \mathcal{A}$, $x \in G$. This identification holds in the sense of algebraic tensor products if V is finite dimensional. Similarly, identify $V \otimes \mathrm{Fun}(G) \otimes \mathrm{Fun}(G)$ with $\mathrm{Fun}(G \times G; V)$ by $(v \otimes a \otimes b)(x,y) := a(x)\,b(y)\,v$. Define $t(v) \in V \otimes \mathrm{Fun}(G)$ $(v \in V)$ as the V-valued function given by

$$(t(v))(x) := \pi(x)v, \quad x \in G.$$

Then t is a corepresentation of $\mathrm{Fun}(G)$ on V. For this purpose, check equations (1.36). For instance, for the proof of the first identity of (1.36) we can show that

$$(((t \otimes \mathrm{id}) \circ t)(v))(x,y) = \pi(x)\,\pi(y)\,v \quad \text{and} \quad (((\mathrm{id} \otimes \Delta) \circ t)(v))(x,y) = \pi(xy)\,v.$$

Let us show the first of these two identities. Fix v. Then there are $v_j \in V$ and $a_j \in \mathrm{Fun}(G)$ such that $(t(v))(x) = \pi(x)v = \sum_j a_j(x)\,v_j$, so $t(v) = \sum_j v_j \otimes a_j$. Then

$$((t \otimes \mathrm{id}) \circ t)(v) = \sum_j (t \otimes \mathrm{id})(v_j \otimes a_j) = \sum_j t(v_j) \otimes a_j.$$

Hence $(((t \otimes \mathrm{id}) \circ t)(v))(x,y)$

$$= \sum_j (t(v_j) \otimes a_j)(x,y) = \sum_j a_j(y)\,(t(v_j))(x) = \sum_j a_j(y)\,\pi(x)\,v_j = \pi(x)\,\pi(y)\,v.$$

Example 1.18. If t is a matrix representation of a group G then the matrix elements t_{ij} can be viewed as elements of $\mathrm{Fun}(G)$. They satisfy

$$t_{ij}(xy) = \sum_k t_{ik}(x)\,t_{kj}(y), \qquad t_{ij}(e) = \delta_{ij}.$$

Hence, by (1.38), t is a matrix corepresentation of $\mathrm{Fun}(G)$. If t is a matrix representation of a group $G \subset SL(n,\mathbb{C})$ and if the t_{ij} are in $\mathrm{Pol}(G)$ then t is a matrix corepresentation of $\mathrm{Pol}(G)$. This holds in particular for the natural matrix representation of G defined by $t_{ij}(x) := x_{ij}$ ($x \in G$), cf. the end of Example 1.4.

If t is a matrix corepresentation of a Hopf algebra $\mathcal{A}$ then combination of (1.38) and (1.6) shows that

$$\sum_k S(t_{ik})\,t_{kj} = \delta_{ij}\,1 = \sum_k t_{ik}\,S(t_{kj}). \tag{1.39}$$

For a corepresentation t of $\mathcal{A}$ on V we may use a symbolic notation analogous to (1.14):

$$t(v) = \sum_{(v)} v_{(1)} \otimes v_{(2)}, \qquad v \in V, \tag{1.40}$$

which means that we have a sum of elements $t_{(1)i} \otimes t_{(2)i}$, where $t_{(1)i} \in V$ and $t_{(2)i} \in \mathcal{A}$. Then (1.36) allows us to combine (1.40) with (1.14) and to write

$$((t \otimes \mathrm{id}) \circ t)(v) = \sum_{(v)} v_{(1)} \otimes v_{(2)} \otimes v_{(3)} = ((\mathrm{id} \otimes \Delta) \circ t)(v).$$

Definition 1.19. Let t be a corepresentation of a Hopf $*$-algebra $\mathcal{A}$ on a vector space V with hermitian inner product $(.,.)$. We call the corepresentation t *unitary* if

$$\sum_{(v),(w)} (v_{(1)}, w_{(1)})\,w_{(2)}^*\,v_{(2)} = (v,w)\,1, \qquad v, w \in V. \tag{1.41}$$

Condition (1.41) can be equivalently stated as

$$\sum_{(v)} (v_{(1)}, w)\, S(v_{(2)}) = \sum_{(w)} (v, w_{(1)})\, w^*_{(2)}, \quad v, w \in V. \tag{1.42}$$

We prove the implication (1.41)$\Rightarrow$(1.42), and leave the other direction to the reader. Indeed,

$$\sum_{(v)} (v_{(1)}, w)\, S(v_{(2)}) = \sum_{(v),(w)} (v_{(1)}, w_{(1)})\, w^*_{(2)}\, v_{(2)}\, S(v_{(3)})$$

$$= \sum_{(v),(w)} (v_{(1)}, w_{(1)})\, w^*_{(2)}\, \varepsilon(v_{(2)}) = \sum_{(w)} (v, w_{(1)})\, w^*_{(2)}.$$

If V is finite dimensional with an orthonormal basis $e_1, \ldots, e_n$ then (1.41) implies for the corresponding matrix corepresentation (t_{ij}) that

$$\sum_i t^*_{il}\, t_{ik} = \sum_{i,j} (e_i, e_j)\, t^*_{jl}\, t_{ik} = (e_k, e_l)\, 1 = \delta_{kl}\, 1.$$

In combination with (1.39) we readily obtain:

Proposition 1.20. Let (t_{ij}) be a matrix corepresentation of a Hopf $*$-algebra $\mathcal{A}$ and let t be the corresponding corepresentation of $\mathcal{A}$ on the inner product space which has $e_1, \ldots, e_n$ as an orthonormal basis. Then the following conditions are equivalent.
(a) t is a unitary corepresentation.
(b) $\sum_k t^*_{ki}\, t_{kj} = \delta_{ij}\, 1$.
(c) $S(t_{ij}) = t^*_{ji}$.
(d) $\sum_k t_{ik}\, t^*_{jk} = \delta_{ij}\, 1$.

In Example 1.18 we viewed a matrix representation t of a group G as a matrix corepresentation of $\mathrm{Fun}(G)$. If we have a $*$-operation on $\mathrm{Fun}(G)$ as in Example 1.11, then it follows from Proposition 1.20 that t is unitary as a matrix representation of G iff it is unitary as a matrix corepresentation of $\mathrm{Fun}(G)$.

If π is a unitary representation of G on an inner product space V and if t is the corepresentation of $\mathrm{Fun}(G)$ associated with π acording to Example 1.17 then equations (1.41) and (1.42) are rewritten versions of

$$(\pi(x)\, v, \pi(x)\, w) = (v, w) \quad \text{and} \quad (\pi(x^{-1})\, v, w) = (v, \pi(x)\, w), \quad x \in G_0.$$

Remark 1.21. If t is a corepresentation of $\mathcal{A}$ on a vector space V then t gives rise to an algebra representation π of $\mathcal{A}^*$ on V:

$$\pi(f)\, v := (\mathrm{id} \otimes f)(t(v)) = \sum_{(v)} f(v_{(2)})\, v_{(1)}. \tag{1.43}$$

Suppose V is finite dimensional with basis $e_1, \ldots, e_n$ and let the corresponding matrix elements t_{ij} be given by (1.37). Combination with (1.43) gives that

$$\pi(f) e_j = \sum_i f(t_{ij}) e_i.$$

Hence π can then be written as a matrix representation $\pi \colon f \mapsto (\pi_{ij}(f))$ of $\mathcal{A}$ with $\pi_{ij}(f) := f(t_{ij})$.

If t is a unitary matrix corepresentation of a Hopf $*$-algebra $\mathcal{A}$ and if π is the corresponding matrix representation of $\mathcal{A}^*$ then π is a $*$-representation. Indeed,

$$\pi_{ij}(f^*) = f^*(t_{ij}) = \overline{f((S(t_{ij}))^*)} = \overline{f(t_{ji})} = \overline{\pi_{ji}(f)}.$$

Definition 1.22. Let t be a corepresentation of a coalgebra $\mathcal{A}$ on a vector space V.
(a) A linear subspace W of V is called *invariant* if $t(W) \subset W \otimes \mathcal{A}$.
(b) The corepresentation t is called *irreducible* if V and $\{0\}$ are the only invariant subspaces of V
(c) Let s be another corepresentation of $\mathcal{A}$ on a vector space W. A linear operator $L \colon V \to W$ is called an *intertwining operator* for t and s if

$$s \circ L = (L \otimes \mathrm{id}) \circ t. \tag{1.44}$$

The corepresentations t and s are called *equivalent* if there exists a bijective intertwining operator for t and s.

Observe that the relation of equivalence between corepresentations is an equivalence relation. One easily checks that two matrix corepresentations (t_{ij}) and (s_{ij}) of a coalgebra $\mathcal{A}$ are equivalent iff there is an invertible square complex matrix L such that $Lt = sL$ (where the products are matrix products). Furthermore, the matrix corepresentation t is irreducible iff t is not equivalent to a matrix corepresentation of block form $\begin{pmatrix} * & * \\ 0 & * \end{pmatrix}$.

Proposition 1.23. Let t be a unitary corepresentation of a Hopf $*$-algebra on a finite dimensional Hilbert space V.
(a) Let W be an invariant subspace of V. Then the orthoplement of W in V is also invariant.
(b) V is a direct sum of invariant subspaces on each of which the restriction of t is an irreducible corepresentation of $\mathcal{A}$.

Proof. For the proof of (a) assume that v is orthogonal to W. Write $t(v) = \sum_{(v)} v_{(1)} \otimes v_{(2)}$ such that the $v_{(2)}$ are linearly independent. Apply (1.42) with $w \in W$. Then, by invariance of W, the right hand side of (1.42) will be 0. Hence the left hand side will be 0, so $\sum_{(v)} (v_{(1)}, w) v_{(2)} = 0$. Thus the $(v_{(1)}, w)$ will be zero. So the $v_{(1)}$ will be in the orthoplement of W. Part (b) follows by iteration of (a). $\square$

Lemma 1.24. Let $L: V \to W$ be an intertwining operator for corepresentations t and s of a coalgebra $\mathcal{A}$ on finite dimensional spaces V resp. W.

(a) The image $L(V)$ and the null space $L^{-1}(0)$ are invariant subspaces of W and V, respectively.

(b) (*first Schur lemma*) If t and s are irreducible then $L = 0$ or L is bijective.

(c) (*second Schur lemma*) If $V = W$, $t = s$ and t is irreducible then $L = \lambda I$ for some complex λ.

Proof. In (a) the invariance of $L(V)$ follows immediately from (1.44). In order to prove the invariance of $L^{-1}(0)$, let $Lv = 0$. Then, by (1.44) and (1.40),

$$0 = (L \otimes \mathrm{id})(t(v)) = \sum_{(v)} L(v_{(1)}) \otimes v_{(2)}.$$

Now choose the $v_{(2)}$ linearly independent. Then the $L(v_{(1)})$ will be 0. Hence, the $v_{(1)}$ in (1.40) will belong to $L^{-1}(0)$. Parts (b) and (c) now follow easily, completely analogous to the proof of the classical Schur lemmas. $\square$

Lemma 1.25. Let $\mathcal{A}$ be a coalgebra. Let t be a corepresentation of $\mathcal{A}$ on a finite dimensional vector space V. Suppose that V is a direct sum of subspaces V_i ($i = 1, \ldots, n$) and that each V_i is a direct sum of subspaces W_{ij} ($j = 1, \ldots, m_i$) and that there are mutually inequivalent irreducible corepresentations $t_1, \ldots, t_n$ such that each subspace W_{ij} is invariant and t restricted to W_{ij} is equivalent to t_i. Let U be a nonzero invariant subspace of V such that t restricted to U is an irreducible corepresentation s. Then, for some i, $U \subset V_i$ and s is equivalent to t_i.

Proof. Let $P_{ij}: V \to W_{ij}$ be the projection operator which is identity on W_{ij} and 0 on the other W_{kl}. Let π_{ij} be the restriction of P_{ij} to U. Then $\pi_{ij}: U \to W_{ij}$ is an intertwining operator for s and t_i. Hence, by Lemma 1.24(b), it is either 0 or bijective. Hence, if π_{ij} and π_{kl} are nonzero, then t_i, s and t_k must be equivalent corepresentations, so $i = k$. Hence, there is an i such that

$$U \subset \bigoplus_{k,l} \pi_{kl}(U) = \bigoplus_{\substack{k,l \\ \pi_{kl} \neq 0}} W_{kl} \subset \bigoplus_{j=1}^{m_i} W_{ij} = V_i$$

and s is equivalent to t_i. $\square$

If $\mathcal{A}$ and $\mathcal{B}$ are coalgebras then $\mathcal{A} \otimes \mathcal{B}$ becomes a coalgebra with comultiplication given by $\Delta(a \otimes b) := \sum_{(a),(b)} (a_{(1)} \otimes b_{(1)}) \otimes (a_{(2)} \otimes b_{(2)})$.

Lemma 1.26. Let $\mathcal{A}$ and $\mathcal{B}$ be coalgebras. Let s and t be irreducible corepresentations of $\mathcal{A}$ resp. $\mathcal{B}$ on finite dimensional vector spaces V and W. Define the corepresentation r of the coalgebra $\mathcal{A} \otimes \mathcal{B}$ on $V \otimes W$ as the tensor product of the corepresentations s and t:

$$r(v \otimes w) = \sum_{(v),(w)} (v_{(1)} \otimes w_{(1)}) \otimes (v_{(2)} \otimes w_{(2)}). \tag{1.45}$$

Then r is irreducible.

Proof. Let U be a nonzero invariant irreducible subspace of $V \otimes W$ with respect to r. It will readily follow that $U = V \otimes W$ if we can show that U contains a nonzero element of the form $v \otimes w$. Define the corepresentation $\widetilde{s}$ of $\mathcal{A}$ on $V \otimes W$ by

$$\widetilde{s}(v \otimes w) := \sum_{(v)} (v_{(1)} \otimes w) \otimes v_{(2)}.$$

Now, because $\widetilde{s} = (\mathrm{id} \otimes \mathrm{id} \otimes \mathrm{id} \otimes \varepsilon_{\mathcal{B}}) \circ r$, we see that U is invariant with respect to $\widetilde{s}$, so U contains an invariant subspace U_0 on which the restriction s_0 of $\widetilde{s}$ is an irreducible corepresentation of $\mathcal{A}$. Choose a basis $f_1, \ldots, f_n$ for W. Let $P_j : V \otimes W \to V$ be the operator wich sends $v \otimes f_j$ to v and is 0 on the other $V \otimes f_k$. Let π_j be the restriction of P_j to U_0. Then π_j is an intertwining operator for s_0 and s. By Lemma 1.24(b) π_j is bijective or 0. Now π_j is nonzero for some j; after rearranging the f_j we may assume that $\pi_1 \neq 0$. Then, for each j, $\pi_j \circ \pi_1^{-1} : V \to V$ is an intertwining operator for s and s. Thus, by Lemma 1.24(c), $\pi_j \circ \pi_1^{-1} = \lambda_j I$ for some complex λ. We conclude that $u \in U_0$ can be written as $u = \sum_{j=1}^{n} \pi_j(u) \otimes f_j = \pi_1(u) \otimes \left(\sum_{j=1}^{n} \lambda_j f_j \right)$. $\qquad \square$

Definition 1.27. Let $(t_{ij})_{i,j=1,\ldots,n}$ be a matrix corepresentation of a Hopf algebra $\mathcal{A}$. Then, because of (1.20) and (1.22), $(S(t_{ji}))_{i,j=1,\ldots,n}$ is also a matrix corepresentation of $\mathcal{A}$. We denote it by (t'_{ij}) and we call it the *contragredient corepresentation* of t.

Note that, if the matrix corepresentations t and s are equivalent with intertwining operator given by the matrix A then t' and s' are equivalent with intertwining operator given by the matrix ${}^t A^{-1}$. If S is invertible (for instance if $\mathcal{A}$ is a Hopf $*$-algebra) and if t is a matrix corepresentation then t is irreducible iff t' is irreducible. (For the proof, suppose that one of both corepresentations is not irreducible and bring it, by equivalence, in suitable block matrix form.)

Let $\mathcal{A}$ be a Hopf algebra. Because of (1.4), the mapping $\Delta : \mathcal{A} \to \mathcal{A} \otimes \mathcal{A}$ is a corepresentation of $\mathcal{A}$ on $\mathcal{A}$, the so-called *right regular corepresentation*. Similarly, the mapping $\tau \circ (S \otimes \mathrm{id}) \circ \Delta : \mathcal{A} \to \mathcal{A} \otimes \mathcal{A}$ defines the *left regular corepresentation* on $\mathcal{A}$. This is indeed a corepresentation:

$$(\tau \circ (S \otimes \mathrm{id}) \circ \Delta)(a) = \sum_{(a)} a_{(2)} \otimes S(a_{(1)}), \quad a \in \mathcal{A}.$$

Then both $(\tau \circ (S \otimes \mathrm{id}) \circ \Delta) \otimes \mathrm{id}$ and $\mathrm{id} \otimes \Delta$ applied to the above right hand side yield

$$\sum_{(a)} a_{(3)} \otimes S(a_{(2)}) \otimes S(a_{(1)}).$$

Proposition 1.28. Let $\mathcal{A}$ be a Hopf algebra with invertible antipode. Let $\{t^{\alpha}\}_{\alpha \in I}$ (I some index set) be a collection of mutually inequivalent irreducible matrix corepresentations of $\mathcal{A}$. Then the set of all matrix elements t_{ij}^{α} is a set of linearly independent elements.

Proof. Consider first a single irreducible d-dimensional matrix corepresentation t of $\mathcal{A}$. We have

$$\Delta(t_{ij}) = \sum_k t_{ik} \otimes t_{kj}, \tag{1.46}$$

$$(\tau \circ (S \otimes \mathrm{id}) \circ \Delta)(t_{ij}) = \sum_k t_{kj} \otimes t'_{ki}. \tag{1.47}$$

Let $\tilde{\Delta}$ be the corepresentation of $\mathcal{A} \otimes \mathcal{A}$ on $\mathcal{A}$ which is defined by

$$\tilde{\Delta} := ((\tau \circ (S \otimes \mathrm{id}) \circ \Delta) \otimes \mathrm{id}) \circ \Delta : a \mapsto \sum_{(a)} S(a_{(2)}) \otimes a_{(1)} \otimes a_{(3)} : \mathcal{A} \to \mathcal{A} \otimes \mathcal{A} \otimes \mathcal{A}.$$

Then, by (1.46) and (1.47), we have

$$\tilde{\Delta}(t_{ij}) = \sum_{k,l} t_{kl} \otimes t'_{ki} \otimes t_{lj}.$$

Thus $\mathrm{Span}\{t_{ij}\}$ is an invariant subspace of $\mathcal{A}$ with respect to $\tilde{\Delta}$. Denote the restriction of the corepresentation $\tilde{\Delta}$ to this subspace by $\hat{t}$.

Let $e_1, \ldots, e_d$ be the standard basis of $\mathbb{C}^d$. We can consider both t and t' as corepresentations of $\mathcal{A}$ on $\mathbb{C}^d$, cf. (1.37). Let $\tilde{t}$ be the corepresentation of $\mathcal{A} \otimes \mathcal{A}$ on $\mathbb{C}^d \otimes \mathbb{C}^d$ which is the tensor product of the corepresentations t' and t of $\mathcal{A}$ on $\mathbb{C}^d$. By Lemma 1.26 this corepresentation is irreducible. Then

$$\tilde{t}(e_i \otimes e_j) = \sum_{k,l} (e_k \otimes e_l) \otimes t'_{ki} \otimes t_{lj}.$$

Thus the surjective linear mapping $A : \mathbb{C}^d \otimes \mathbb{C}^d \to \mathrm{Span}\{t_{ij}\}$ defined by $A(e_i \otimes e_j) := t_{ij}$ is an intertwining operator for the corepresentations $\tilde{t}$ and $\hat{t}$. Then $A^{-1}(0)$ is an invariant subspace of $\mathbb{C}^d \otimes \mathbb{C}^d$ with respect to $\tilde{t}$, certainly a proper subspace, since otherwise $t_{11} = 0$, while $\varepsilon(t_{11}) = 1$. Because of the irreducibility of $\tilde{t}$ we then have $A^{-1}(0) = \{0\}$. Thus A is bijective. Hence the t_{ij} are linearly independent.

Next consider m mutually inequivalent irreducible matrix corepresentations t^l ($l = 1, \ldots, m$) of $\mathcal{A}$, where $t^l = (t_{ij}^l)_{i,j=1,\ldots,d_l}$. Denote the linear span of all elements t_{ij}^l by V. Then, with the above notation, V is an invariant subspace of $\mathcal{A}$ with respect to the corepresentation $\tilde{\Delta}$ of $\mathcal{A} \otimes \mathcal{A}$ on $\mathcal{A}$. Let $W := \oplus_{l=1}^m W_l$, where $W_l := \mathbb{C}^{d_l} \otimes \mathbb{C}^{d_l}$. On W we have the corepresentation of $\mathcal{A} \otimes \mathcal{A}$ which is the direct sum of the tensor product corepresentations of $(t^l)'$ with t^l and the W_l are invariant irreducible subspaces of W. The irreducible corepresentations of $\mathcal{A} \otimes \mathcal{A}$ on the W_l are mutually inequivalent.

Similarly as above we have a surjective intertwining operator $A \colon W \to V$ with respect to these two corepresentations of $\mathcal{A} \otimes \mathcal{A}$. Then $A^{-1}(0)$ is an invariant subspace of W. Suppose that $A^{-1}(0) \neq \{0\}$. Then we can take an invariant irreducible subspace U of $A^{-1}(0)$. By Lemma 1.25, $U = W_l$ for some l. This implies that, for this l, $t_{ij}^l = 0$, which contradicts that $\varepsilon(t_{11}^l) = 1$. □

Lemma 1.29. Let $\mathcal{A}$ be a Hopf algebra with invertible antipode. Let s and t be irreducible matrix corepresentations of $\mathcal{A}$. Then s and t are equivalent iff $\mathrm{Span}\{s_{ij}\} = \mathrm{Span}\{t_{ij}\}$.

Proof. Clearly, if s and t are equivalent then the two spans coincide. Conversely, let $V := \mathrm{Span}\{s_{ij}\} = \mathrm{Span}\{t_{ij}\}$. Then V is an invariant subspace of $\mathcal{A}$ with respect to the corepresentation Δ of $\mathcal{A}$ on $\mathcal{A}$. By Proposition 1.28 and formula (1.46) the restriction of Δ to V is a corepresentation of $\mathcal{A}$ which is a direct sum of d copies of s and also a direct sum of d copies of t. By Lemma 1.25 we conclude that s is equivalent to t. □

Definition 1.30. Let s and t be corepresentations of a Hopf algebra $\mathcal{A}$ on vector spaces V and W, respectively. Then the *tensor product* $s \otimes t$ of s and t is defined as the corepresentation of $\mathcal{A}$ on $V \otimes W$ given by the mapping $(\mathrm{id} \otimes \mathrm{id} \otimes m) \circ (\mathrm{id} \otimes \tau \otimes \mathrm{id}) \circ (s \otimes t) \colon V \otimes W \to (V \otimes W) \otimes \mathcal{A}$, i.e.,

$$(s \otimes t)(v \otimes w) = \sum_{(v),(w)} v_{(1)} \otimes w_{(1)} \otimes v_{(2)} w_{(2)}.$$

If s and t are matrix corepresentations then $s \otimes t$ can also be considered as a matrix corepresentation:

$$(s \otimes t)_{ij,kl} := s_{ik}\, t_{jl}.$$

Note that this tensor product is associative, but not commutative. The present tensor product should not be confused with the corepresentation of $\mathcal{A} \otimes \mathcal{A}$ obtained as a tensor product of s and t.

Definition 1.31. We call a corepresentation of a Hopf $*$-algebra *unitarizable* if it is equivalent to a unitary corepresentation. In particular, a matrix corepresentation t of $\mathcal{A}$ is unitarizable iff there is an invertible complex square matrix A such that $s := AtA^{-1}$ satisfies $S(s_{ij}) = s_{ji}^*$.

Note that tensor products of unitary (resp. unitarizable) matrix corepresentations are again unitary (resp. unitarizable).

2. COMPACT QUANTUM GROUPS

This section gives a new approach to the theory of compact matrix quantum groups as developed by Woronowicz [41], [42], [43]. While Woronowicz's definition already involves C^*-algebras and his further development of the theory also heavily uses C^*-algebra theory, the present approach only uses Hopf $*$-algebras, and mentions the connection with C^*-algebras at a much later stage, more as a side remark. This new approach was developed in cooperation with M. S. Dijkhuizen [15], [16]. The relationship with other work on compact quantum groups will be discussed at the end of this section. In particular, we acknowledge that Effros and Ruan [17] introduced CQG algebras (named differently by them) earlier. However, they developed their theory in a very different way.

2.1. CQG algebras and CMQG algebras.

Let G be a compact group. Let $\Pi(G)$ be the linear span of all matrix elements of irreducible unitary (hence finite dimensional) representations of G. Then $\Pi(G)$ is a commutative Hopf $*$-algebra with Hopf $*$-algebra operations as described in Examples 1.4 and 1.11. We speak of a *compact matrix group* if, for some n, G is (isomorphic to) a closed subgroup of the group $U(n)$ of $n \times n$ unitary marices. For a compact group G there are the following equivalent statements: G is a compact matrix group iff G is a compact Lie group iff $\Pi(G)$ is finitely generated. This suggests the following definitions.

Definition 2.1. A Hopf $*$-algebra $\mathcal{A}$ is said to be associated with a *compact quantum group* if it is the linear span of the matrix elements of the unitary (finite - dimensional) matrix corepresentations of $\mathcal{A}$. Then $\mathcal{A}$ is called a *CQG algebra*.

Because of Proposition 1.23(b), a CQG algebra $\mathcal{A}$ is also the linear span of the matrix elements of its irreducible unitary matrix corepresentations. Denote by $\widehat{\mathcal{A}}$ the collection of all equivalence classes of irreducible unitary matrix corepresentations of $\mathcal{A}$. Choose, for each $\alpha \in \widehat{\mathcal{A}}$, a unitary matrix corepresentation $(t_{ij}^\alpha)_{i,j=1,\dots,d_\alpha}$ belonging to class α. Then, by Proposition 1.28, the set of all t_{ij}^α forms a basis of $\mathcal{A}$. Write

$$\mathcal{A}_\alpha := \mathrm{Span}\{t_{ij}^\alpha\}_{i,j=1,\dots,d_\alpha}, \quad \alpha \in \widehat{\mathcal{A}}.$$

Then $\mathcal{A}$ is the direct sum of the $\mathcal{A}_\alpha$. Let 1 be the element of $\widehat{\mathcal{A}}$ for which t^1 is the one-dimensional matrix corepresentation (1). For $\alpha \in \widehat{\mathcal{A}}$ let α' be the element of $\widehat{\mathcal{A}}$ such that $(t^\alpha)'$ is equivalent to $t^{\alpha'}$.

Proposition 2.2. Let $\mathcal{A}$ be a CQG algebra and let the t^α be as above.
(a) Each irreducible matrix corepresentation of $\mathcal{A}$ is equivalent to some t^α ($\alpha \in \widehat{\mathcal{A}}$).
(b) Each irreducible matrix corepresentation of $\mathcal{A}$ is unitarizable.
(c) If s is a unitarizable matrix corepresentation of $\mathcal{A}$ then so is s'.

Proof. (b) immediately follows from (a). For (c) observe that a unitarizable s is a direct sum of irreducible corepresentations. Hence s' is also a direct sum of irreducible corepresentations. Then apply (b). Thus we have to prove (a). Let s be an irreducible matrix corepresentation of $\mathcal{A}$. The spaces $\mathcal{A}_\alpha$ are invariant irreducible subspaces of $\mathcal{A}$ with respect to the corepresentation of $\mathcal{A} \otimes \mathcal{A}$ on $\mathcal{A}$ which is the tensor product of the left regular and the right regular corepresentation (cf. the proof of Proposition 1.28), and so is $V := \mathrm{Span}\{s_{ij}\}$. Moreover, the corepresentations obtained by restriction to the various $\mathcal{A}_\alpha$ are mutually inequivalent. By Lemma 1.25 $V = \mathcal{A}_\alpha$ for some $\alpha \in \hat{\mathcal{A}}$. By Lemma 1.29 we see that s is equivalent to t^α. $\square$

Definition 2.3. A Hopf $*$-algebra $\mathcal{A}$ is said to be associated with a *compact matrix quantum group* if it is a finitely generated CQG algebra. Then $\mathcal{A}$ is called a *CMQG algebra*.

Proposition 2.4. For a Hopf $*$-algebra the following conditions are equivalent.
(i) $\mathcal{A}$ is a CMQG algebra.
(ii) There is a unitary matrix corepresentation v of $\mathcal{A}$ such that $\mathcal{A}$ is generated as an algebra by the matrix coefficients of v.
(iii) There is a matrix corepresentation u of $\mathcal{A}$ such that both u and u' are unitarizable and $\mathcal{A}$ as an algebra is generated by the matrix elements of u and u'.

Proof.
(i)$\Rightarrow$(ii)$\Rightarrow$(iii): Assume (i). Each of the generators is a linear combination of matrix elements of unitary matrix corepresentations of $\mathcal{A}$. Hence, by taking direct sums, there is a unitary matrix corepresentation u of $\mathcal{A}$ of which the matrix elements generate $\mathcal{A}$. Then u' is unitarizable by Proposition 2.2.

(iii)$\Rightarrow$(ii)$\Rightarrow$(i): Assume (iii). Let v be the direct sum of two unitary matrix corepresentations which are equivalent to u respectively u'. Then $\mathcal{A}$ is generated by the matrix elements of v. Each product of matrix elements of v is a matrix element of some multiple tensor product of v. Such tensor products are again unitary corepresentations. Thus $\mathcal{A}$ is a finitely generated CQG algebra. $\square$

2.2. The Haar functional. Define on a CQG algebra $\mathcal{A}$ the *Haar functional* h as the linear mapping $h \colon \mathcal{A} \to \mathbb{C}$ such that

$$h(a) := \begin{cases} 0 & \text{if } a \in \mathcal{A}_\alpha,\ \alpha \neq 1, \\ 1 & \text{if } a = 1, \end{cases} \tag{2.1}$$

The functional h satisfies

$$(h \otimes \mathrm{id})(\Delta(a)) = h(a)\,1 = (\mathrm{id} \otimes h)(\Delta(a)), \quad a \in \mathcal{A}. \tag{2.2}$$

Indeed, if a equals some basis element t^α_{ij} of $\mathcal{A}$ then (2.2) takes the form

$$\sum_k h(t^\alpha_{ik}) t^\alpha_{kj} = h(t^\alpha_{ij}) 1 = \sum_k h(t^\alpha_{kj}) t^\alpha_{ik}, \tag{2.3}$$

and these identities are indeed implied by (2.1).

On the other hand, if h would be any linear functional on $\mathcal{A}$ satisfying the first (or the second) identity in (2.2) then we obtain from (2.3) by linear independence of the t^α_{ij} that $h(t^\alpha_{ij}) = 0$ for $\alpha \neq 1$. Thus, h is determined up to a constant factor by each of the identities in (2.2) and h can next be normalized by

$$h(1) = 1. \tag{2.4}$$

Since $S(\mathcal{A}_\alpha) = \mathcal{A}_{\alpha'}$, (2.1) implies that

$$h(S(a)) = h(a), \quad a \in \mathcal{A}. \tag{2.5}$$

If G is a compact group then the normalized Haar measure, as a linear functional on $\Pi(G)$, can be interpreted as a Haar functional. Indeed,

$$h(a) := \int_G a(x)\,dx, \quad a \in \Pi(G),$$

and

$$((h \otimes \mathrm{id})(\Delta(a)))(y) = \int_G a(xy)\,dx = \int_G a(x)\,dx = h(a)\,1(y),$$

and similarly for the other identity. In this case we also have that $h(aa^*) = \int_G |a(x)|^2\,dx > 0$ for $a \neq 0$. We want to prove this positivity result also for h in the case of a CQG algebra.

The following Lemma will be crucial. For the moment we will assume that $\mathcal{A}$ is a Hopf algebra with invertible antipode (not necessarily a Hopf $*$-algebra) and that h is a linear functional on $\mathcal{A}$ satisfying (2.2), (2.4) and (2.5).

Lemma 2.5. Let r and s be matrix corepresentations of $\mathcal{A}$. Then

$$\sum_l h(s_{ij} S(r_{kl})) r_{lm} = \sum_l s_{il} h(s_{lj} S(r_{km})), \tag{2.6}$$

$$\sum_l h(S(r_{ij}) s_{kl}) s_{lm} = \sum_l r_{il} h(S(r_{lj}) s_{km}). \tag{2.7}$$

With the notation

$$A^{(j,k)}_{il} := h(s_{ij} S(r_{kl})), \quad B^{(j,k)}_{il} := h(S(r_{ij}) s_{kl}), \tag{2.8}$$

the identities (2.6) and (2.7) can be rewritten as

$$A^{(j,k)} r = s A^{(j,k)}, \quad B^{(j,k)} s = r B^{(j,k)}. \tag{2.9}$$

Thus $A^{(j,k)}$ is an intertwining operator for r and s and $B^{(j,k)}$ is an intertwining operator for s and r.

Proof. For the proof of (2.6) write

$$
\begin{aligned}
h(s_{ij} S(r_{kl})) 1 &= (\mathrm{id} \otimes h)(\Delta(s_{ij} S(r_{kl}))) \\
&= \sum_{p,n} (\mathrm{id} \otimes h)((s_{ip} \otimes s_{pj})(S(r_{nl}) \otimes S(r_{kn}))) = \sum_{p,n} h(s_{pj} S(r_{kn})) s_{ip} S(r_{nl}).
\end{aligned}
$$

Substitute this equality in the left hand side of (2.6) and next use that $\sum_l S(r_{nl}) r_{lm} = \delta_{nm} 1$. This settles (2.6). Similarly, (2.7) is obtained from a substitution of

$$h(S(r_{lj}) s_{km}) 1 = (h \otimes \mathrm{id})(\Delta(S(r_{lj}) s_{km}))$$

in the right hand side of (2.7). $\qquad\square$

Let us assume that r and s are irreducible matrix corepresentations of $\mathcal{A}$. We will consider three special cases of Lemma 2.5: (i) r and s are inequivalent, (ii) $s = r$, (iii) $s = r''$ (i.e., $s_{ij} = S^2(r_{ij})$). For each case we apply one of the quantum Schur lemmas (Lemma 1.24(b),(c)).

Case (i) If r and s are inequivalent then (2.9) yields that $A_{il}^{(j,k)} = 0$ and $B_{il}^{(j,k)} = 0$. Hence, by (2.8),

$$h(s_{ij} S(r_{kl})) = 0, \quad h(S(r_{ij}) s_{kl}) = 0.$$

Case (ii) If $s = r$ then (2.9) yields that there are complex constants α_{jk} and β_{jk} such that $A^{(j,k)} = \alpha_{jk} I$ and $B_{il}^{(j,k)} = \beta_{jk} I$. Hence, by (2.8)

$$A_{il}^{(j,k)} = h(r_{ij} S(r_{kl})) = \alpha_{jk} \delta_{il}, \tag{2.10}$$

$$B_{il}^{(j,k)} = h(S(r_{ij}) r_{kl}) = \beta_{jk} \delta_{il}. \tag{2.11}$$

Moreover, if we fix $i = l$ and sum over $j = k$ then we obtain

$$\sum_j \alpha_{jj} = 1 = \sum_j \beta_{jj}. \tag{2.12}$$

Case (iii) If $s = r''$ then (2.6) and (2.7) together with (2.5) yield

$$\sum_l h(r_{kl} S(r_{ij})) r_{lm} = \sum_l S^2(r_{il}) h(r_{km} S(r_{lj})),$$

$$\sum_l h(S(r_{kl}) r_{ij}) S^2(r_{lm}) = \sum_l r_{il} h(S(r_{km}) r_{lj}).$$

Put

$$\widetilde{A}_{il}^{(j,k)} := h(r_{kl}\, S(r_{ij})) = A_{kj}^{(l,i)} = \alpha_{li}\,\delta_{kj}, \tag{2.13}$$

$$\widetilde{B}_{il}^{(j,k)} := h(S(r_{kl})\, r_{ij}) = B_{kj}^{(l,i)} = \beta_{li}\,\delta_{kj}, \tag{2.14}$$

where we refer to (2.10), (2.11). Then $\widetilde{A}^{(j,k)}\, r = r''\, \widetilde{A}^{(j,k)}$ and $\widetilde{B}^{(j,k)}\, r'' = r\, B^{(j,k)}$. Moreover, when we sum the first equality in (2.13) and in (2.14) over $i = l$ then we obtain that

$$\operatorname{tr}\widetilde{A}^{(j,k)} = \delta_{jk}, \quad \operatorname{tr}\widetilde{B}^{(j,k)} = \delta_{jk}.$$

Hence, there exists a nonzero intertwining operator F for r and r''. Thus r and r'' are equivalent irreducible corepresentations (since S is assumed to be invertible, the irreducibility of r implies the irreducibility of r''), and F is an invertible operator, unique up to a constant factor and satisfying $\operatorname{tr} F \neq 0$ and $\operatorname{tr} F^{-1} \neq 0$. Hence there are complex constants $\widetilde{\alpha}_{jk}$ and $\widetilde{\beta}_{jk}$ such that

$$\widetilde{A}_{il}^{(j,k)} = \widetilde{\alpha}_{jk}\, F_{il} = \alpha_{li}\,\delta_{kj},$$

$$\widetilde{B}_{il}^{(j,k)} = \widetilde{\beta}_{jk}\,(F^{-1})_{il} = \beta_{li}\,\delta_{kj},$$

where, in each line, the second equality follows from (2.13). Combination with (2.12) yields that

$$\widetilde{\alpha}_{jk}\,\operatorname{tr} F = \delta_{kj}, \quad \widetilde{\beta}_{jk}\,\operatorname{tr} F^{-1} = \delta_{kj}.$$

Hence

$$\widetilde{A}_{il}^{(j,k)} = \delta_{kj}\,\frac{F_{il}}{\operatorname{tr} F}, \quad \widetilde{B}_{il}^{(j,k)} = \delta_{kj}\,\frac{(F^{-1})_{il}}{\operatorname{tr} F^{-1}}. \tag{2.15}$$

Let us summarize the obtained results in the following Proposition.

Proposition 2.6. Let $\mathcal{A}$ be a Hopf algebra with invertible antipode and let h be a linear functional on $\mathcal{A}$ satisfying (2.2), (2.4) and (2.5). Let r be an irreducible matrix corepresentation of $\mathcal{A}$.

(a) If s is an irreducible matrix corepresentation of $\mathcal{A}$ which is not equivalent to r then

$$h(s_{kl}\, S(r_{ij})) = 0, \quad h(S(r_{kl})\, s_{ij}) = 0. \tag{2.16}$$

(b) r is equivalent to r''. Let F be an invertible intertwining operator for r and r''. Then $\operatorname{tr} F \neq 0$ and $\operatorname{tr} F^{-1} \neq 0$ and

$$h(r_{kl}\, S(r_{ij})) = \delta_{kj}\,\frac{F_{il}}{\operatorname{tr} F}, \tag{2.17}$$

$$h(S(r_{kl})\, r_{ij}) = \delta_{kj}\,\frac{(F^{-1})_{il}}{\operatorname{tr} F^{-1}}. \tag{2.18}$$

Proof. (2.17) and (2.18) were obtained from (2.13), (2.14) and (2.15). $\qquad\square$

Now we assume that $\mathcal{A}$ is a CQG algebra.

Proposition 2.7. Let r be an irreducible unitary matrix corepresentation of the CQG algebra $\mathcal{A}$. Let the matrix F be an invertible intertwining operator for r and r'' as in Proposition 2.6. Then F is a constant multiple of a positive definite matrix. It can be uniquely normalized such that $\operatorname{tr} F = \operatorname{tr} F^{-1} > 0$.

Proof. By Proposition 2.2(c) there is a unitary matrix corepresentation s such that r' is equivalent to s. Thus there is an invertible complex matrix A such that $sA = Ar'$. Let $\overline{A}$ be the matrix for which $(\overline{A})_{ij} := \overline{A_{ij}}$ and let ^tA be the transpose of A. Since $s'_{ij} = (s_{ij})^* = S(s_{ji})$ and $r'_{ij} = (r_{ij})^*$, $r''_{ij} = S(r'_{ji})$, we conclude from $sA = Ar'$ that $s'\overline{A} = \overline{A}r$ and $^tA s' = r'' \, ^tA$. Hence $^tA \overline{A} r = r'' \, ^tA \overline{A}$. Thus $F = \operatorname{const.} {}^tA \overline{A}$ and $^tA \overline{A}$ is positive definite. $\qquad\square$

Theorem 2.8. The Haar functional on a CQG algebra satisfies

$$h(aa^*) > 0 \quad \text{if } a \neq 0,$$

Proof. Propositions 2.6 and 2.7 imply that

$$h(t^\alpha_{kl} \, (t^\beta_{ij})^*) = \delta_{\alpha\beta} \, \delta_{ki} \, G^\alpha_{lj}, \quad \alpha, \beta \in \widehat{\mathcal{A}},$$

for certain positive definite matrices G^α. Let $a := \sum_{\alpha, k, l} c^\alpha_{kl} \, t^\alpha_{kl}$ be an arbitrary element of $\mathcal{A}$. Then

$$h(aa^*) = \sum_{\alpha, k} \sum_{l, j} c^\alpha_{kl} \, \overline{c^\alpha_{kj}} \, G^\alpha_{lj} \geq 0.$$

If $h(aa^*) = 0$ then $\sum_{l,j} c^\alpha_{kl} \, \overline{c^\alpha_{kj}} \, G^\alpha_{lj} = 0$ for all α and k. By positive definiteness of the G^α this implies that all coefficients c^α_{kl} are 0, i.e., $a = 0$. $\qquad\square$

Corollary 2.9. Every finite dimensional corepresentation of a CQG algebra $\mathcal{A}$ is unitarizable, and hence decomposable as a direct sum of irreducible finite dimensional corepresentations.

Proof. Let r be a corepresentation of $\mathcal{A}$ on a vector space V with some inner product $(\, . \, , . \,)$. Define a new sesquilinear form on V by

$$\langle v, w \rangle := \sum_{(v),(w)} (v_{(1)}, w_{(1)}) \, h(w^*_{(2)} \, v_{(2)}).$$

Then r is unitary in the sense of (1.41) with respect to this sesquilinear form. Indeed,

$$\sum_{(v),(w)} \langle v_{(1)}, w_{(1)} \rangle \, w^*_{(2)} \, v_{(2)} = \sum_{(v),(w)} (v_{(1)}, w_{(1)}) \, h(w^*_{(2)} \, v_{(2)}) \, w^*_{(3)} \, v_{(3)}$$

$$= \sum_{(v),(w)} (v_{(1)}, w_{(1)}) \, h(w^*_{(2)} \, v_{(2)}) \, 1 = \langle v, w \rangle \, 1.$$

It is left to show that $\langle . , . \rangle$ is positive definite. Let $e_1, \ldots, e_n$ be an orthonormal basis of V with respect to the inner product $(. , .)$ and write r as a matrix corepresentation with respect to this basis. Then

$$\langle e_i, e_j \rangle = \sum_{k,l} (e_k, e_l)\, h(r_{lj}^* \, r_{ki}) = \sum_k h(r_{kj}^* \, r_{ki}).$$

Hence

$$\langle \textstyle\sum_i c_i e_i, \sum_j c_j e_j \rangle = \sum_k h\big((\sum_j c_j r_{kj})^* (\sum_i c_i r_{ki}) \big) \geq 0,$$

and if the left hand side equals 0 then $\sum_i c_i\, r_{ki} = 0$ for all k, hence $c_k = \varepsilon(\sum_i c_i\, r_{ki})$ $= 0$ for al k. $\qquad\square$

2.3. The C^*-completion of a CQG algebra. Recall that a C^*-algebra is a Banach algebra A which is also a $*$-algebra such that $||aa^*|| = ||a||^2$ for all $a \in A$. In particular, the space $\mathcal{L}(\mathcal{H})$ of all bounded linear operators on a Hilbert space $\mathcal{H}$ forms a C^*-algebra with identity. By a $*$-representation of a $*$-algebra $\mathcal{A}$ on a Hilbert space $\mathcal{H}$ we mean a $*$-algebra homomorphism $\pi \colon \mathcal{A} \to \mathcal{L}(\mathcal{H})$.

On a compact group G the $*$-algebra $\Pi(G)$ can be equipped with the sup norm. The space $C(G)$ of all continuous functions on G is the completion of $\Pi(G)$ with respect to this norm and $C(G)$ is a commutative C^*-algebra with identity. Both for the $*$-algebra $\Pi(G)$ and for the C^*-algebra $C(G)$ the irreducible $*$-representations on a Hilbert space are precisely the one-dimensional $*$-homomorphisms $a \mapsto a(x)$ $(x \in G)$.

This suggests the following strategy for the construction of an analogue of $C(G)$ for a CQG algebra $\mathcal{A}$:

1) Define a seminorm

$$||a|| := \sup_\pi ||\pi(a)|| \tag{2.19}$$

on $\mathcal{A}$, where the supremum runs over all $*$-representations π of the $*$-algebra $\mathcal{A}$.

2) Show that this seminorm is a norm.

3) Let A be the completion of $\mathcal{A}$ with respect to this norm.

First observe that $||a||$ defined by (2.19) is finite for all $a \in \mathcal{A}$. Indeed, if a is expanded in terms of the basis of $\mathcal{A}$ by the finite sum

$$a = \sum_{\alpha, i, j} c_{ij}^\alpha\, t_{ij}^\alpha \tag{2.20}$$

then $||a|| \leq \sum_{\alpha, i, j} |c_{ij}^\alpha|\, ||t_{ij}^\alpha||$. Now use the following lemma.

Lemma 2.10. Let r be a unitary matrix corepresentation of a CQG algebra $\mathcal{A}$. Let V be a complex vector space with hermitian inner product and let $|v| := (v,v)^{\frac{1}{2}}$ define the norm on V. Let π be an algebra homomorphism from $\mathcal{A}$ to the algebra of all linear operators on V such that $(\pi(a)v, w) = (v, \pi(a^*)w)$ for all $a \in \mathcal{A}$ and all $v, w \in V$. Then $\pi(r_{ij})$ is a bounded linear operator on V of norm ≤ 1 for all i, j. In particular, π uniquely extends to a $*$-algebra representation of $\mathcal{A}$ on the Hilbert space completion of V.

Proof. Since r is unitary, we have $\sum_k r_{kj}^* r_{kj} = 1$. Hence, for all $v \in V$,

$$||v||^2 = (v,v) = \sum_k (\pi(r_{kj}^* r_{kj})v, v) = \sum_k (\pi(r_{kj})v, \pi(r_{kj})v) \geq ||\pi(r_{ij})v||^2. \quad \square$$

One consequence of this Lemma is that, for $a \in \mathcal{A}$ with expansion (2.20), all norms $||\pi(a)||$, and hence $||a||$, are bounded by $\sum_{\alpha, i, j} |c_{ij}^\alpha|$.

Another consequence concerns the left regular representation of $\mathcal{A}$ on $\mathcal{A}$, which is defined by $\lambda(a)\,b := ab$. This is an algebra representation, which is *faithful*, i.e., if $\lambda(a) = 0$ then $a = 0$. Now equip $\mathcal{A}$ with the inner product $(a, b) := h(b^*a)$. (By the properties of h this is indeed an inner product.) Clearly, $(\lambda(a)\,b, c) = (b, \lambda(a^*)\,c)$. Since each $a \in \mathcal{A}$ is a linear combination of matrix elements of unitary corepresentations, it follows by Lemma 2.10 that each $\lambda(a)$ is a bounded linear operator on the inner product space $\mathcal{A}$. Hence λ extends to a $*$-representation of $\mathcal{A}$ on the Hilbert space completion of this inner product space. (I thank P. Podleś for this observation.) By faithfulness it follows that the seminorm (2.19) is a norm.

Because $||\pi(ab)|| \leq ||\pi(a)||\,||\pi(b)||$ and $||\pi(a)||^2 = ||\pi(aa^*)||$ for each $*$-representation π of $\mathcal{A}$, it follows that the norm (2.19) is a C^*-norm, i.e., $||ab|| \leq ||a||\,||b||$ and $||aa^*|| = ||a||^2$. Thus the norm completion A of $\mathcal{A}$ is a well-defined C^*-algebra with identity.

Since, for each element a of a C^*-algebra A, there exists an irreducible $*$-representation π of A for which $||\pi(a)|| = ||a||$ (cf. for instance Arveson [6, Corollary to Theorem 1.7.2]), the norm definition (2.19) does not change when we take the supremum there only over the irreducible $*$-representations of $\mathcal{A}$.

The counit and comultiplication on $\mathcal{A}$ have unique continuous extensions to A. For the counit this is obvious, since ε is a one-dimensional $*$-representation of $\mathcal{A}$, so $|\varepsilon(a)| \leq ||a||$.

For the C^*-extension of Δ we need a suitable C^*-norm on the algebraic tensor product $A \otimes A$. We choose the *injective cross norm*, i.e., for $a \in A \otimes A$, the norm $||a||$ is defined as the supremum of the numbers $||(\pi_1 \otimes \pi_2)(a)||$, where π_1 and π_2 run over the $*$-representations of the C^*-algebra A. Then Δ continuously extends to a mapping from A to the completion of $A \otimes A$ with respect to this norm, since $a \mapsto (\pi_1 \otimes \pi_2)(\Delta(a))$ is a $*$-representation of $\mathcal{A}$ for any two $*$-representations π_1 and π_2 of $\mathcal{A}$, so $||(\pi_1 \otimes \pi_2)(\Delta(a))|| \leq ||a||$. (The $*$-representation $a \mapsto (\pi_1 \otimes \pi_2)(\Delta(a))$ of $\mathcal{A}$ is called the *tensor product* $\pi_1 \otimes \pi_2$ of the $*$-representations π_1 and π_2, cf. Remark 1.8.)

2.4. A class of multiplicative linear functionals. Let $\mathcal{A}$ be a CQG algebra. According to Proposition 2.7 we can choose for each $\alpha \in \hat{\mathcal{A}}$ a positive definite complex matrix F_α which is an intertwining operator for the matrix corepresentations t^α and $(t^\alpha)''$ and which is uniquely normalized such that $\operatorname{tr} F_\alpha = \operatorname{tr}(F_\alpha)^{-1}$. We can now rewrite (2.16), (2.17) and (2.18) as

$$h(t^\beta_{kl} (t^\alpha_{ji})^*) = \delta_{\alpha\beta}\, \delta_{kj}\, \frac{(F_\alpha)_{il}}{\operatorname{tr} F_\alpha}, \tag{2.21}$$

$$h((t^\beta_{lk})^*\, t^\alpha_{ij}) = \delta_{\alpha\beta}\, \delta_{kj}\, \frac{(F_\alpha^{-1})_{il}}{\operatorname{tr} F_\alpha^{-1}}. \tag{2.22}$$

These can be considered as the analogues for CQG algebras of Schur's orthogonality relations. Woronowicz, for the case of his compact matrix quantum groups, gave these quantum Schur orthogonality relations in [41, (5.14), (5.15)].

The remainder of this subsection is an account of Theorem 5.6 in Woronowicz [41]. It describes the so-called modular properties of the Haar functional h. Write F_α as

$$F_\alpha = U \begin{pmatrix} \lambda_1 & & 0 \\ & \ddots & \\ 0 & & \lambda_{d_\alpha} \end{pmatrix} U^{-1},$$

where U is a unitary matrix and $\lambda_1, \ldots, \lambda_{d_\alpha} > 0$. Now define complex powers F_α^z $(z \in \mathbb{C})$ of F_α by

$$F_\alpha^z := U \begin{pmatrix} \lambda_1^z & & 0 \\ & \ddots & \\ 0 & & \lambda_{d_\alpha}^z \end{pmatrix} U^{-1}.$$

Then the matrix elements $(F_\alpha^z)_{ij}$ are entire analytic functions of z and there are constants $M > 0$ and $\mu \in \mathbb{R}$ such that

$$|(F_\alpha^z)_{ij}| \leq M\, e^{\mu \operatorname{Re} z} \quad \text{for all } z \in \mathbb{C}.$$

Note also that $F_\alpha^z F_\alpha^{z'} = F_\alpha^{z+z'}$.

Remark 2.11. If f is an entire analytic function satisfying the estimate $|f(z)| \leq M\, e^{\mu \operatorname{Re} z}$ for some $M > 0$ and $\mu \in \mathbb{R}$ then we can find a real constant c such that the function $g(z) := e^{cz}\, f(z)$ satisfies the estimate

$$\exists M > 0 \quad \exists \mu \in (0, \pi) \quad \operatorname{Re} z \geq 0 \Longrightarrow |g(z)| \leq M\, e^{\mu|z|}. \tag{2.23}$$

Carlson's theorem (cf. for instance Titchmarsh [38, §5.81]) states that, if a holomorphic function g on $\{z \mid \operatorname{Re} z \geq 0\}$ satisfies (2.23) and if moreover $g(1) = g(2) = \cdots = 0$, then g will vanish identically. Hence, if the above function f will vanish on the set $\{1, 2, \ldots\}$ then it will also vanish identically.

Now define, for each $z \in \mathbb{C}$, the element $f_z \in \mathcal{A}^*$ by specifying it on our linear basis of $\mathcal{A}$:

$$f_z(t_{ij}^\alpha) := (F_\alpha^z)_{ij}.$$

Note that this definition is independent of the choice of the matrix corepresentation t^α in the class $\alpha \in \overline{\mathcal{A}}$. The following properties of the f_z quickly follow from the definition of f_z.

Proposition 2.12.
(a) $f_1(t_{ij}^\alpha) = (F_\alpha)_{ij}, \quad f_{-1}(t_{ij}^\alpha) = ((F_\alpha)^{-1})_{ij}.$
(b) $f_0 = \varepsilon.$
(c) $f_z(1) = 1.$
(d) $f_z\, f_{z'} = f_{z+z'}.$
(e) For each $a \in \mathcal{A}$ the function $z \mapsto f_z(a)$ is an entire analytic function and there are constants $M > 0$ and $\mu \in \mathbb{R}$ such that $|f_z(a)| \le M\, e^{\mu \operatorname{Re} z}.$

For some further properties we will give the proof.

Proposition 2.13. $S^2(a) = f_{-1}.a.f_1.$

Proof. It follows from Proposition 2.7 that

$$S^2(t_{ij}^\alpha) = \sum_{k,l} (F_\alpha)_{ik}\, t_{kl}^\alpha\, (F_\alpha^{-1})_{lj}$$
$$= (f_1 \otimes \mathrm{id} \otimes f_{-1})((\Delta \otimes \mathrm{id})\Delta(t_{ij}^\alpha))$$
$$= f_{-1}.t_{ij}^\alpha.f_1. \qquad \square$$

Proposition 2.14. If $a, b \in \mathcal{A}$ then

$$h(ab) = h(b(f_1.a.f_1)). \qquad (2.24)$$

Proof. Let $\alpha, \beta \in \overline{\mathcal{A}}$. It is enough to verify (2.24) for $a := t_{kl}^\beta$ and $b := (t_{ij}^\alpha)^*$. We apply (2.21) and (2.22). If $\alpha \ne \beta$ then both sides of (2.24) are 0. If $\alpha = \beta$ then

$$h((t_{ij}^\alpha)^* (f_1.t_{kl}^\alpha.f_1)) = \sum_{m,n} (F_\alpha)_{km}\, (F_\alpha)_{nl}\, h((t_{ij}^\alpha)^*\, t_{mn}^\alpha)$$
$$= \sum_{m,n} (F_\alpha)_{km}\, (F_\alpha)_{nl}\, \frac{\delta_{jn}\, (F_\alpha^{-1})_{mi}}{\operatorname{tr} F_\alpha^{-1}}$$
$$= \frac{\delta_{ik}\, (F_\alpha)_{jl}}{\operatorname{tr} F_\alpha^{-1}} = \frac{\delta_{ik}\, (F_\alpha)_{jl}}{\operatorname{tr} F_\alpha} = h(t_{kl}^\alpha\, (t_{ij}^\alpha)^*),$$

where we also used the chosen normalization for F_α. $\qquad \square$

74

The last result justifies the normalization of the F_α. Note on the other hand that Propositions 2.13 and 2.14, together with the property $f_1 f_{-1} = \varepsilon = f_{-1} f_1$ completely determine f_1 and f_{-1} as elements of $\mathcal{A}^*$. The other f_z are then determined by Proposition 2.12(d),(e), in view of Remark 2.11.

A linear functional f on an algebra $\mathcal{A}$ is called *central* if $f(ab) = f(ba)$ for all $a, b \in \mathcal{A}$. Thus h is not necessarily central, but its non-centrality is well controlled by (2.24).

Lemma 2.15. For $a, b \in \mathcal{A}$ we have

$$f_1.(ab).f_1 = (f_1.a.f_1)(f_1.b.f_1). \tag{2.25}$$

Proof. Apply (2.24):

$$h(c(f_1.(ab).f_1)) = h(abc) = h(bc(f_1.a.f_1)) = h(c(f_1.a.f_1)(f_1.b.f_1)).$$

Now use that $d \in \mathcal{A}$ equals 0 if $h(cd) = 0$ for all $c \in \mathcal{A}$. $\qquad \square$

A linear functional $f \in \mathcal{A}^*$ is called *multiplicative* if $f(ab) = f(a) f(b)$ and $f(1) = 1$. The multiplicative linear functionals form again an algebra. (This is true for $\mathcal{A}$ an arbitrary bialgebra.) Indeed, if $f, g \in \mathcal{A}$ are multiplicative then

$$(fg)(ab) = \sum_{(a),(b)} f(a_{(1)} b_{(1)}) g(a_{(2)} b_{(2)})$$

$$= \sum_{(a),(b)} f(a_{(1)}) g(a_{(2)}) f(b_{(1)}) g(b_{(2)}) = (fg)(a)(fg)(b).$$

Proposition 2.16. For all $z \in \mathbb{C}$ f_z is a multiplicative linear functional:

$$f_z(ab) = f_z(a) f_z(b), \quad a, b \in \mathcal{A}. \tag{2.26}$$

Proof. Apply ε to both sides of (2.25). Then we get $f_1^2(ab) = f_1^2(a) f_1^2(b)$. Hence (2.26) is true for $z = 2$. By the above remark, (2.26) is then true for $z = 2, 4, 6, \ldots$. Now apply Proposition 2.12(e) and Remark 2.11. $\qquad \square$

Proposition 2.16 tells us in particular that f_1 is multiplicative. I wonder if a more algebraic proof, not using Carlson's theorem, is possible for this fact.

Remark 2.17. As a consequence of this Proposition, f_1 and f_{-1} are already determined in a CMQG algebra by their values on the matrix elements v_{ij}, where v is as in Proposition 2.4(ii). This also determines the right hand sides of (2.21) and (2.22), since $(F_\alpha^{\pm 1})_{ij} = f_{\pm 1}(t_{ij}^\alpha)$.

Let us finally consider expressions for $S(f_z)$ and $(f_z)^*$. Note that a nonzero multiplicative linear functional $f \in \mathcal{A}^*$ has two-sided inverse $S(f)$. (This is true for $\mathcal{A}$ an arbitary Hopf algebra.) Indeed,

$$\varepsilon(a) = f(\varepsilon(a)1) = f(m \circ (S \otimes \mathrm{id})\Delta(a))$$
$$= (f \otimes f)((S \otimes \mathrm{id})\Delta(a)) = (S(f) \otimes f)(\Delta(a)) = (S(f) f)(a).$$

This shows that
$$S(f_z) = f_{-z}. \tag{2.27}$$

In order to compute $(f_z)^*$ observe that

$$(f_z)^*(t_{ij}^\alpha) = \overline{f_z((S(t_{ij}^\alpha))^*)} = \overline{f_z(t_{ji}^\alpha)} = \overline{(F_\alpha^z)_{ji}} = (F_\alpha^{\bar z})_{ij} = f_{\bar z}(t_{ij}^\alpha).$$

Hence
$$(f_z)^* = f_{\bar z}. \tag{2.28}$$

2.5. Comparison with other literature.

(a) *Woronowicz [41], [42], [43]*
Woronowicz, in his influential 1987 paper [41], gives the following definition of a *compact matrix quantum group* (originally called *compact matrix pseudogroup*). It is a pair (A, u), where A is a unital C^*-algebra and $u = (u_{ij})_{i,j=1,\ldots,N}$ is an $N \times N$ matrix with entries in A, such that the following properties hold.
1) The unital *-subalgebra $\mathcal{A}$ of A generated by the entries of u is dense in A.
2) There exists a (necessarily unique) C^*-homomorphism $\Delta \colon A \to A \otimes A$ such that $\Delta(u_{ij}) = \sum_{k=1}^{N} u_{ik} \otimes u_{kj}$.
3) There exists a (necessarily unique) linear antimultiplicative mapping $S \colon \mathcal{A} \to \mathcal{A}$ such that $S \circ * \circ S \circ * = \mathrm{id}$ on $\mathcal{A}$ and $\sum_{k=1}^{N} S(u_{ik}) u_{kj} = \delta_{ij} 1 = \sum_{k=1}^{N} u_{ik} S(u_{kj})$.
In his note [43] Woronowicz shows that, instead of 3), we may equivalently require:
3') The matrix u and its transpose are invertible.

Woronowicz now essentially shows (cf. [41, Prop. 1.8]) that there exists a (necessarily unique) *-homomorphism $\varepsilon \colon \mathcal{A} \to \mathbb{C}$ such that $\varepsilon(u_{ij}) = \delta_{ij}$ and that $\mathcal{A}$ becomes a Hopf *-algebra with comultiplication Δ, counit ε and antipode S. In [41] the notation Φ, e, κ is used instead of our Δ, ε, S, respectively. Note that the above *-algebra $\mathcal{A}$ is very close to what we have defined as a CMQG algebra (cf. Proposition 2.2). However, it is not postulated and not yet obvious in the beginning of [41] that the corepresentations u and u' are unitarizable.
    A central result in the paper (see [41, Theorem 4.2]) is the existence of a *state* (normalized positive linear functional) h on the C^*-algebra A such that $(h \otimes \mathrm{id})(\Delta(a)) = h(a)1 = (\mathrm{id} \otimes h)(\Delta(a))$ for all $a \in A$. This state is necessarily unique and it is faithful on $\mathcal{A}$. Then h may be called the Haar functional.
    Woronowicz [41, §2] defines a *representation* of the compact matrix quantum group (A, u) on a finite dimensional vector space V as a linear mapping $t \colon V \to V \otimes A$

such that $(t \otimes \mathrm{id}) \circ t = (\mathrm{id} \otimes \Delta) \circ t$. If $t(v) = 0$ implies $v = 0$ then the representation is called *non-degenerate* and if $t(V) \subset V \otimes \mathcal{A}$ then the representation is called *smooth*. A smooth representation is non-degenerate iff $(\mathrm{id} \otimes \varepsilon) \circ t = \mathrm{id}$. Thus corepresentations of $\mathcal{A}$ on finite dimensional vector spaces, as defined in §1.3 of the present paper, correspond to nondegenerate smooth representations of (A, u) in [41].

As a consequence of the existence of the Haar functional, it is now shown (see [41, Theorem 5.2 and Prop. 3.2]) that nondegenerate smooth representations of (A, u) are unitarizable. Then we can conclude that the dense *-algebra $\mathcal{A}$ of A is a CMQG algebra.

Conversely, if we start with a CMQG algebra $\mathcal{A}$ with fundamental corepresentation u as in Proposition 2.2 then it is possible to show the existence of a Haar functional h (cf. Theorem 2.8) without using C^*-algebras. Next a C^*-completion A can be made which uses the existence of the Haar functional (cf. §2.3). Then it is clear that the pair (A, u) is a compact matrix quantum group in the sense of Woronowicz. However, the C^*-algebra A as constructed from $\mathcal{A}$ in §2.3, is canonically determined by $\mathcal{A}$, but this is not the case with the compact matrix quantum groups (A, u) of Woronowicz. In the present paper we have taken the completion of $\mathcal{A}$ with respect to the largest C^*-seminorm on $\mathcal{A}$, while, in [41], the norm induced by A on $\mathcal{A}$ may not be the largest C^*-seminorm on $\mathcal{A}$. In that case, the counit $\varepsilon \colon \mathcal{A} \to \mathbb{C}$ does not necessarily have a continuous extension to a linear functional on A (cf. [41, second Remark to Prop. 1.8]).

In his paper [42], Woronowicz starts with a compact matrix quantum group (A, u) in the sense of [41], then constructs out of its finite dimensional unitary representations a so-called complete concrete monoidal W^*-category (see [42, Theorem 1.2]) and next constructs from any such category a compact matrix quantum group (B, u). Then B is not necessarily isomorphic to A, but B is obtained canonically from the CMQG algebra $\mathcal{A}$ generated by the entries of u, as in §2.3.

Both in [41] and in the present paper there is a similar key result [41, last statement of Theorem 5.4] resp. Proposition 2.7. We got the idea of the statement and proof of Proposition 2.7 from [41], but in the present paper, different from [41], the positivity and faithfulness of the Haar functional on $\mathcal{A}$ is a corollary rather than a prerequisite. As soon as these matters are settled, the quantum Schur orthogonality relations and the modular property of the Haar functional can be obtained in the same way in both papers, cf. [41, pp. 653–656] resp. §2.4 in the present paper.

(b) *Woronowicz [44] and S. Wang [39]*

Woronowicz [44] defines a *compact quantum group* as a pair (A, Δ), where A is a separable unital C^*-algebra and $\Delta \colon A \to A \otimes A$ is a C^*-homomorphism, such that the following properties hold.

1) $(\Delta \otimes \mathrm{id}) \circ \Delta = (\mathrm{id} \otimes \Delta) \circ \Delta$.
2) $\mathrm{Span}\{(b \otimes 1)\Delta(c) \mid b, c \in A\}$ and $\mathrm{Span}\{(1 \otimes b)\Delta(c) \mid b, c \in A\}$ are dense subspaces of $A \otimes A$.

In particular, if (A, u) is a compact matrix pseudogroup as defined in [41] and if Δ is the corresponding comultiplication then (A, Δ) is a compact quantum group as just defined. Conversely, it is shown in [44] that, if (A, Δ) is a compact quantum group and if $\mathcal{A}$ is the set of all linear combinations of matrix elements of finite dimensional unitary representations of (A, Δ) then $\mathcal{A}$ is a dense $*$-subalgebra of A and $\mathcal{A}$ is a Hopf $*$-algebra. The existence of a Haar functional is also shown. It is observed that the representation theory as developed in [41] can be formulated in a similar way for compact quantum groups.

It is pointed out in Wang [39, Remark 2.2] that the results of [44] remain true if separability of the C^*-algebra A is no longer required, but if it is assumed instead that the C^*-algebra A has a faithful state. This observation would imply that a compact quantum group (A, Δ) in the sense of Wang gives rise to a CQG algebra $\mathcal{A}$ ($\mathcal{A}$ being constructed from A as in the previous paragraph), and that conversely each CQG algebra $\mathcal{A}$ would give rise to a compact quantum group (A, Δ) (A being completion of $\mathcal{A}$ with respect to maximal C^*-seminorm).

(c) *Effros and Ruan [17]*

In different terminology, CQG algebras were earlier introduced by Effros and Ruan [17]. They defined these algebras as cosemisimple Hopf algebras with a so-called standard $*$-operation and they called these structures *discrete quantum groups*. This name was motivated by the fact that special examples of these algebras are provided by the group algebra of a discrete group, while the name CQG algebra comes from the class of examples, where we deal with the algebra of representative functions on a compact group. In the final section of [17] the authors define a *compact quantum group* as a natural generalization of the compact matrix quantum groups defined in [41]. Their definition involves a unital C^*-algebra A with a dense unital $*$-subalgebra $\mathcal{A}$, where $\mathcal{A}$ is a CQG-algebra (in the terminology of the present paper) and the comultiplication on $\mathcal{A}$ extends continuously to A. Conversely, they show that a CQG algebra $\mathcal{A}$ gives rise to a compact quantum group according to their definition. This involves a C^*-completion, for which a Haar functional h on $\mathcal{A}$ is needed. This Haar functional is obtained in a way very different from the method in the present paper. The authors first show the existence of a left Haar functional ϕ on a certain subspace of the linear dual of $\mathcal{A}$. Then h is constructed in terms of ϕ.

3. Basic hypergeometric functions

A standard reference for this section is the recent book by Gasper and Rahman [20]. In particular, the present section is quite parallel to their introductory Chapter 1. See also the useful compendia of formulas in the Appendices to that book. The foreword to [20] by R. Askey gives a succinct historical introduction to the subject. For first reading on the subject I can also recommend Andrews [3].

From now on the notations $\mathbb{N} := \{1, 2, \ldots\}$ and $\mathbb{Z}_+ := \{0, 1, 2, \ldots\}$ will be used.

3.1. Preliminaries. We start with briefly recalling the definition of the general hypergeometric series (see Erdélyi e.a. [18, Ch. 4] or Bailey [13, Ch. 2]). For $a \in \mathbb{C}$ the *shifted factorial* or *Pochhammer symbol* is defined by $(a)_0 := 1$ and

$$(a)_k := a(a+1)\ldots(a+k-1), \quad k = 1, 2, \ldots .$$

The general *hypergeometric series* is defined by

$$_rF_s \left[\begin{matrix} a_1, \ldots, a_r \\ b_1, \ldots, b_s \end{matrix} ; z \right] := \sum_{k=0}^{\infty} \frac{(a_1)_k \ldots (a_r)_k}{(b_1)_k \ldots (b_s)_k \, k!} z^k. \tag{3.1}$$

Here $r, s \in \mathbb{Z}_+$ and the upper parameters $a_1, \ldots, a_r$, the lower parameters $b_1, \ldots, b_s$ and the argument z are in $\mathbb{C}$. However, in order to avoid zeros in the denominator, we require that $b_1, \ldots, b_s \notin \{0, -1, -2, \ldots\}$. If, for some $i = 1, \ldots, r$, a_i is a non-positive integer then the series (3.1) is terminating. Otherwise, we have an infinite power series with radius of convergence equal to 0, 1 or ∞ according to whether $r - s - 1 > 0$, $= 0$ or < 0, respectively. On the right hand side of (3.1) we have that

$$\frac{(k+1)\text{th term}}{k\text{th term}} = \frac{(k+a_1)\ldots(k+a_r)z}{(k+b_1)\ldots(k+b_s)(k+1)} \tag{3.2}$$

is rational in k. Conversely, any rational function in k can be written in the form of the right hand side of (3.2). Hence, any series $\sum_{k=0}^{\infty} c_k$ with $c_0 = 1$ and c_{k+1}/c_k rational in k is of the form of a hypergeometric series (3.1).

The cases $_0F_0$ and $_1F_0$ are elementary: exponential resp. binomial series. The case $_2F_1$ is the familiar Gaussian hypergeometric series, cf. [18, Ch. 2].

We next give the definition of basic hypergeometric series. For $a, q \in \mathbb{C}$ define the *q-shifted factorial* by $(a; q)_0 := 1$ and

$$(a; q)_k := (1-a)(1-aq)\ldots(1-aq^{k-1}), \quad k = 1, 2, \ldots .$$

For $|q| < 1$ put

$$(a; q)_\infty := \prod_{k=0}^{\infty} (1 - aq^k).$$

We also write

$$(a_1, a_2, \ldots, a_r; q)_k := (a_1; q)_k (a_2; q)_k \ldots (a_r; q)_k, \quad k = 0, 1, 2, \ldots \text{ or } \infty.$$

Then a *basic hypergeometric series* or *q-hypergeometric series* is defined by

$$_r\phi_s \left[\begin{matrix} a_1, \ldots, a_r \\ b_1, \ldots, b_s \end{matrix} ; q, z \right] = {}_r\phi_s(a_1, \ldots, a_r; b_1, \ldots, b_s; q, z)$$

$$:= \sum_{k=0}^{\infty} \frac{(a_1, \ldots, a_r; q)_k}{(b_1, \ldots, b_s, q; q)_k} \left((-1)^k q^{k(k-1)/2} \right)^{1+s-r} z^k, \quad r, s \in \mathbb{Z}_+. \tag{3.3}$$

On the right hand side of (3.3) we have that

$$\frac{(k+1)\text{th term}}{k\text{th term}} = \frac{(1 - a_1 q^k) \ldots (1 - a_r q^k)(-q^k)^{1+s-r} z}{(1 - b_1 q^k) \ldots (1 - b_s q^k)(1 - q^{k+1})} \tag{3.4}$$

is rational in q^k. Conversely, any rational function in q^k can be written in the form of the right hand side of (3.4). Hence, any series $\sum_{k=0}^{\infty} c_k$ with $c_0 = 1$ and c_{k+1}/c_k rational in q^k is of the form of a q-hypergeometric series (3.3). This characterization is one explanation why we allow q raised to a power quadratic in k in (3.3).

Because of the easily verified relation

$$(a; q^{-1})_k = (-1)^k a^k q^{-k(k-1)/2} (a^{-1}; q)_k ,$$

any series (3.3) can be transformed into a series with base q^{-1}. Hence, it is sufficient to study series (3.3) with $|q| \leq 1$. The tricky case $|q| = 1$ has not yet been studied much and will be ignored by us. Therefore we may assume that $|q| < 1$. In fact, for convenience we will always assume that $0 < q < 1$, unless it is otherwise stated.

In order to have a well-defined series (3.3), we require that

$$b_1, \ldots, b_s \neq 1, q^{-1}, q^{-2}, \ldots .$$

The series (3.3) will terminate iff, for some $i = 1, \ldots, r$, we have $a_i \in \{1, q^{-1}, q^{-2}, \ldots\}$. If $a_i = q^{-n}$ $(n = 0, 1, 2, \ldots)$ then all terms in the series with $k > n$ will vanish. In the non-vanishing case, the convergence behaviour of (3.3) can be read off from (3.4) by use of the ratio test. We conclude:

$$\text{convergence radius of (3.3)} = \begin{cases} \infty & \text{if } r < s + 1, \\ 1 & \text{if } r = s + 1, \\ 0 & \text{if } r > s + 1. \end{cases}$$

We can view the q-shifted factorial as a q-analogue of the shifted factorial by the limit formula

$$\lim_{q \to 1} \frac{(q^a; q)_k}{(1 - q)^k} = (a)_k := a(a + 1) \ldots (a + k - 1).$$

Hence $_r\phi_s$ is a q-analogue of $_rF_s$ by the formal (termwise) limit

$$\lim_{q \uparrow 1} {}_r\phi_s \left[\begin{matrix} q^{a_1}, \ldots, q^{a_r} \\ q^{b_1}, \ldots, q^{b_s} \end{matrix}; q, (q-1)^{1+s-r} z \right] = {}_rF_s \left[\begin{matrix} a_1, \ldots, a_r \\ b_1, \ldots, b_s \end{matrix}; z \right]. \tag{3.5}$$

However, we get many other q-analogues of $_rF_s$ by adding upper or lower parameters to the left hand side of (3.5) which are equal to 0 or which depend on q in such a way that they tend to a limit $\neq 1$ as $q \uparrow 1$. Note that the notation (3.3) has the drawback that some rescaling is necessary before we can take limits for $q \to 1$. On the other hand, parameters can be put equal to zero without problem in (3.3), which would not be the case if we worked with $a_1, \ldots, a_r$, $b_1, \ldots, b_s$ as in (3.5). In general, q-hypergeometric series can be studied in their own right, without much consideration for the $q = 1$ limit case. This philosophy is often (but not always) reflected in the notation generally in use.

It is well known that the confluent $_1F_1$ hypergeometric function can be obtained from the Gaussian $_2F_1$ hypergeometric function by a limit process called *confluence*. A similar phenomenon occurs for q-series. Formally, by taking termwise limits we have

$$\lim_{a_r \to \infty} {}_r\phi_s \left[\begin{matrix} a_1, \ldots, a_r \\ b_1, \ldots, b_s \end{matrix} ; q, \frac{z}{a_r} \right] = {}_{r-1}\phi_s \left[\begin{matrix} a_1, \ldots, a_{r-1} \\ b_1, \ldots, b_s \end{matrix} ; q, z \right]. \tag{3.6}$$

In particular, this explains the particular choice of the factor $\left((-1)^k q^{k(k-1)/2}\right)^{1+s-r}$ in (3.3). If $r = s + 1$ this factor is lacking and for $r < s + 1$ it is naturally obtained by the confluence process. The proof of (3.6) is by the following lemma.

Lemma 3.1. Let the complex numbers a_k, $k \in \mathbb{Z}_+$, satisfy the estimate $|a_k| \leq R^{-k}$ for some $R > 0$. Let

$$F(b; q, z) := \sum_{k=0}^{\infty} a_k (b; q)_k z^k, \quad |z| < R, \quad b \in \mathbb{C}, \quad 0 < q < 1.$$

Then

$$\lim_{b \to \infty} F(b; q, z/b) = \sum_{k=0}^{\infty} a_k (-1)^k q^{k(k-1)/2} z^k, \tag{3.7}$$

uniformly for z in compact subsets of $\mathbb{C}$.

Proof. For the kth term of the series on the left hand side of (3.7) we have

$$c_k := a_k (b; q)_k (z/b)^k$$
$$= a_k (b^{-1} - 1)(b^{-1} - q) \ldots (b^{-1} - q^{k-1}) z^k.$$

Now let $|b| > 1$ and let $N \in \mathbb{N}$ be such that $q^{N-1} > |b|^{-1} \geq q^N$. If $k \leq N$ then

$$|c_k| \leq |a_k| q^{k(k-1)/2} (2|z|)^k.$$

If $k > N$ then

$$|(b^{-1} - 1)(b^{-1} - q) \ldots (b^{-1} - q^{k-1})| \leq 2^k q^{N(N-1)/2} |b|^{N-k}$$
$$\leq 2^k |b|^{(1-N)/2} |b|^{N-k}$$
$$\leq 2^k |b|^{-k/2},$$

so

$$|c_k| \leq |a_k| (2 |z| |b|^{-1/2})^k.$$

Now fix $M > 0$. Then, for each $\varepsilon > 0$ we can find $K > 0$ and $B > 1$ such that

$$\sum_{k=K}^{\infty} |c_k| < \varepsilon \quad \text{if } |z| < M \text{ and } |b| > B.$$

Combination with the termwise limit result completes the proof. $\square$

It can be observed quite often in literature on q-special functions that no rigorous limit proofs are given, but only formal limit statements like (3.6) and (3.5). Sometimes, in heuristic reasonings, this is acceptable and productive for quickly finding new results. But in general, I would say that rigorous limit proofs have to be added.

Any terminating power series

$$\sum_{k=0}^{n} c_k z^k$$

can also be written as

$$z^n \sum_{k=0}^{n} c_{n-k} (1/z)^k.$$

When we want to do this for a terminating q-hypergeometric series, we have to use that

$$(a;q)_{n-k} = \frac{(a;q)_n}{(q^{n-1}a;q^{-1})_k} = (a;q)_n \frac{(-1)^k q^{k(k-1)/2} (a^{-1}q^{1-n})^k}{(q^{1-n}a^{-1};q)_k}$$

and

$$\frac{(q^{-n};q)_{n-k}}{(q;q)_{n-k}} = \frac{(q^{-n};q)_n}{(q;q)_n} \frac{(q^n;q^{-1})_k}{(q^{-1};q^{-1})_k} = (-1)^n q^{-n(n+1)/2} \frac{(q^{-n};q)_k}{(q;q)_k} q^{(n+1)k}.$$

Thus we obtain, for $n \in \mathbb{Z}_+$,

$$_{s+1}\phi_s \left[\begin{matrix} q^{-n}, a_1, \ldots, a_s \\ b_1, \ldots, b_s \end{matrix} ; q, z \right] = (-1)^n q^{-n(n+1)/2} \frac{(a_1, \ldots, a_s; q)_n}{(b_1, \ldots, b_s; q)_n} z^n$$

$$\times {}_{s+1}\phi_s \left[\begin{matrix} q^{-n}, q^{-n+1}b_1^{-1}, \ldots, q^{-n+1}b_s^{-1} \\ q^{-n+1}a_1^{-1}, \ldots, q^{-n+1}a_s^{-1} \end{matrix} ; q, \frac{q^{n+1}b_1 \ldots b_s}{a_1 \ldots a_s z} \right]. \quad (3.8)$$

Similar identities can be derived for other $_r\phi_s$ and for cases where some of the parameters are 0.

Thus, any explicit evaluation of a terminating q-hypergeometric series immediately implies a second one by inversion of the direction of summation in the series, while any identity between two terminating q-hypergeometric series implies three other identities.

3.2. The q-integral. Standard operations of classical analysis like differentiation and integration do not fit very well with q-hypergeometric series and can be better replaced by q-derivative and q-integral.

The q-derivative $D_q f$ of a function f on an open real interval is given by

$$(D_q f)(x) := \frac{f(x) - f(qx)}{(1-q)x}, \quad x \neq 0, \quad (3.9)$$

and $(D_q f)(0) := f'(0)$ by continuity, provided $f'(0)$ exists. Note that $\lim_{q \uparrow 1}(D_q f)(x) = f'(x)$ if f is differentiable. Note also that, analogous to $d/dx\,(1-x)^n = -n\,(1-x)^{n-1}$, we have

$$f(x) = (x;q)_n \implies (D_q f)(x) = -\frac{1-q^n}{1-q}\,(qx;q)_{n-1}. \tag{3.10}$$

Now recall that $0 < q < 1$. If $D_q F = f$ and f is continuous then, for real a,

$$F(a) - F(0) = a\,(1-q) \sum_{k=0}^{\infty} f(aq^k)\,q^k \tag{3.11}$$

This suggests the definition of the *q-integral*

$$\int_0^a f(x)\,d_q x := a\,(1-q) \sum_{k=0}^{\infty} f(aq^k)\,q^k. \tag{3.12}$$

Note that it can be viewed as an infinite Riemann sum with nonequidistant mesh widths. In the limit, as $q \uparrow 1$, the right hand side of (3.12) will tend to the classical integral $\int_0^a f(x)\,dx$.

From (3.11) we can also obtain $F(a) - F(b)$ expressed in terms of f. This suggests the definition

$$\int_a^b f(x)\,d_q x := \int_0^a f(x)\,d_q x - \int_0^b f(x)\,d_q x. \tag{3.13}$$

Note that (3.12) and (3.13) remain valid if a or b is negative.

There is no unique canonical choice for the q-integral from 0 to ∞. We will put

$$\int_0^{\infty} f(x)\,d_q x := (1-q) \sum_{k=-\infty}^{\infty} f(q^k)\,q^k$$

(provided the sum converges absolutely). The other natural choices are then expressed by

$$a \int_0^{\infty} f(ax)\,d_q x = a\,(1-q) \sum_{k=-\infty}^{\infty} f(aq^k)\,q^k, \quad a > 0.$$

Note that the above expression remains invariant when we replace a by aq^n $(n \in \mathbb{Z})$.

As an example consider

$$\int_0^1 x^{\alpha}\,d_q x = (1-q) \sum_{k=0}^{\infty} q^{k(\alpha+1)} = \frac{1-q}{1-q^{\alpha+1}}, \quad \text{Re}\,\alpha > -1, \tag{3.14}$$

which tends, for $q \uparrow 1$, to

$$\frac{1}{\alpha + 1} = \int_0^1 x^\alpha \, dx. \tag{3.15}$$

From the point of view of explicitly computing definite integrals by Riemann sum approximation this is much more efficient than the approximation with equidistant mesh widths

$$\frac{1}{n} \sum_{k=1}^n \left(\frac{k}{n}\right)^\alpha.$$

The q-approximation (3.14) of the definite integral (3.15) (viewed as an area) goes essentially back to Fermat.

3.3. Elementary examples. The q-binomial series is defined by

$$_1\phi_0(a; -; q, z) := \sum_{k=0}^\infty \frac{(a; q)_k}{(q; q)_k} z^k, \quad |z| < 1. \tag{3.16}$$

Here the '$-$' in a $_r\phi_s$ expression denotes an empty parameter list. The name "q-binomial" is justified since (3.16), with a replaced by q^a, formally tends, as $q \uparrow 1$, to the binomial series

$$_1F_0(a; z) := \sum_{k=0}^\infty \frac{(a)_k}{k!} z^k = (1 - z)^{-a}, \quad |z| < 1. \tag{3.17}$$

The limit transition is made rigorous in [27, Appendix A]. It becomes elementary in the case of a terminating series ($a := q^{-n}$ in (3.16) and $a := -n$ in (3.17), where $n \in \mathbb{Z}_+$).

The q-analogue of the series evaluation in (3.17) is as follows.

Proposition 3.2. We have

$$_1\phi_0(a; -; q, z) = \frac{(az; q)_\infty}{(z; q)_\infty}, \quad |z| < 1. \tag{3.18}$$

In particular,

$$_1\phi_0(q^{-n}; -; q, z) = (q^{-n}z; q)_n. \tag{3.19}$$

Proof. Put $h_a(z) := {}_1\phi_0(a; -; q, z)$. Then

$$(1 - z) h_a(z) = (1 - az) h_a(qz), \quad \text{hence} \quad h_a(z) = \frac{(1 - az)}{(1 - z)} h_a(qz).$$

Iteration gives

$$h_a(z) = \frac{(az; q)_n}{(z; q)_n} h_a(q^n z), \quad n \in \mathbb{Z}_+.$$

Now use that h_a is analytic and therefore continuous at 0 and use that $h_a(0) = 1$. Thus, for $n \to \infty$ we obtain (3.18). $\qquad\square$

84

The two most elementary q-hypergeometric series are the two q-exponential series

$$e_q(z) := {}_1\phi_0(0; -; q, z) = \sum_{k=0}^{\infty} \frac{z^k}{(q; q)_k} = \frac{1}{(z; q)_\infty}, \quad |z| < 1, \tag{3.20}$$

and

$$E_q(z) := {}_0\phi_0(-; -; q, -z) = \sum_{k=0}^{\infty} \frac{q^{k(k-1)/2} z^k}{(q; q)_k} = (-z; q)_\infty, \quad z \in \mathbb{C}. \tag{3.21}$$

The evaluation in (3.20) is a specialization of (3.18). On the other hand, (3.21) is a confluent limit of (3.18):

$$\lim_{a \to \infty} {}_1\phi_0(a; -; q, -z/a) = {}_0\phi_0(-; -; q, -z)$$

(cf. (3.5)). The limit is uniform for z in compact subsets of $\mathbb{C}$, see Lemma 3.1. Thus the evaluation in (3.21) follows also from (3.18).

It follows from (3.20) and (3.21) that

$$e_q(z) E_q(-z) = 1. \tag{3.22}$$

This identity is a q-analogue of $e^z e^{-z} = 1$. Indeed, the two q-exponential series are q-analogues of the exponential series by the limit formulas

$$\lim_{q \uparrow 1} E_q((1 - q)z) = e^z = \lim_{q \uparrow 1} e_q((1 - q)z).$$

The first limit is uniform on compacta of $\mathbb{C}$ by the majorization

$$\left| \frac{q^{k(k-1)/2} (1 - q)^k z^k}{(q; q)_k} \right| \le \frac{|z|^k}{k!}.$$

The second limit then follows by use of (3.22).

Although we assumed the convention $0 < q < 1$, it is often useful to find out for a given q-hypergeometric series what will be obtained by changing q into q^{-1} and then rewriting things again in base q. This will establish a kind of duality for q-hypergeometric series. For instance, we have

$$e_{q^{-1}}(z) = E_q(-qz),$$

which can be seen from the power series definitions.

The following four identities, including (3.20), are obtained from each other by trivial rewriting.

$$\sum_{k=0}^{\infty} \frac{(1-q)^k z^k}{(q;q)_k} = \frac{1}{((1-q)z;q)_\infty},$$

$$\sum_{k=0}^{\infty} \frac{z^k}{(q;q)_k} = \frac{1}{(z;q)_\infty},$$

$$(1-q)^{1-b} \sum_{k=0}^{\infty} q^{kb}(q^{k+1};q)_\infty = \frac{(q;q)_\infty}{(q^b;q)_\infty (1-q)^{b-1}},$$

$$\int_0^{(1-q)^{-1}} t^{b-1} ((1-q)qt;q)_\infty \, d_q t = \frac{(q;q)_\infty}{(q^b;q)_\infty (1-q)^{b-1}}, \quad \operatorname{Re} b > 0. \quad (3.23)$$

As $q \uparrow 1$, the first identity tends to

$$\sum_{k=0}^{\infty} \frac{z^k}{k!} = e^z,$$

while the left hand side of (3.23) tends formally to

$$\int_0^{\infty} t^{b-1} e^{-t} \, dt.$$

Since this last integral can be evaluated as $\Gamma(b)$, it is tempting to consider the right hand side of (3.23) as a the q-gamma function. Thus we put

$$\Gamma_q(z) := \frac{(q;q)_\infty}{(q^z;q)_\infty (1-q)^{z-1}} = \int_0^{(1-q)^{-1}} t^{z-1} E_q(-(1-q)qt) \, d_q t, \quad \operatorname{Re} z > 0.$$

where the last identity follows from (3.23). It was proved in [27, Appendix B] that

$$\lim_{q \uparrow 1} \Gamma_q(z) = \Gamma(z), \quad z \neq 0, -1, -2, \ldots. \quad (3.24)$$

We have just seen an example how an identity for q-hypergeometric series can have two completely different limit cases as $q \uparrow 1$. Of course, this is achieved by different rescaling. In particular, reconsideration of a power series as a q-integral is often helpful and suggestive for obtaining distinct limits.

Regarding Γ_q it can yet be remarked that it satisfies the functional equation

$$\Gamma_q(z+1) = \frac{1-q^z}{1-q} \Gamma_q(z)$$

and that
$$\Gamma_q(n+1) = \frac{(q;q)_n}{(1-q)^n}, \quad n \in \mathbb{Z}_+.$$

Similarly to the chain of equivalent identities including (3.20) we have a chain including (3.18):

$$\sum_{k=0}^{\infty} \frac{(q^a;q)_k\, z^k}{(q;q)_k} = \frac{(q^a z;q)_\infty}{(z;q)_\infty},$$

$$(1-q)\sum_{k=0}^{\infty} q^{kb}\frac{(q^{k+1};q)_\infty}{(q^{k+a};q)_\infty} = \frac{(1-q)\,(q,q^{a+b};q)_\infty}{(q^a,q^b;q)_\infty},$$

$$\int_0^1 t^{b-1}\frac{(qt;q)_\infty}{(q^a t;q)_\infty}\,d_q t = \frac{\Gamma_q(a)\,\Gamma_q(b)}{\Gamma_q(a+b)}, \quad \operatorname{Re} b > 0 \qquad (3.25)$$

In the limit, as $q \uparrow 1$, the first identity tends to

$$\sum_{k=0}^{\infty} \frac{(a)_k\, z^k}{k!} = (1-z)^{-a},$$

while (3.25) formally tends to

$$\int_0^1 t^{b-1}\,(1-t)^{a-1}\,dt = \frac{\Gamma(a)\,\Gamma(b)}{\Gamma(a+b)}. \qquad (3.26)$$

Thus (3.25) can be considered as a q-beta integral and we define the *q-beta function* by

$$B_q(a,b) := \frac{\Gamma_q(a)\,\Gamma_q(b)}{\Gamma_q(a+b)} = \frac{(1-q)\,(q,q^{a+b};q)_\infty}{(q^a,q^b;q)_\infty} = \int_0^1 t^{b-1}\frac{(qt;q)_\infty}{(q^a t;q)_\infty}\,d_q t. \qquad (3.27)$$

3.4. Heine's $_2\phi_1$ series. Euler's integral representation for the $_2F_1$ hypergeometric function

$$_2F_1(a,b;c;z) = \frac{\Gamma(c)}{\Gamma(b)\Gamma(c-b)}\int_0^1 t^{b-1}\,(1-t)^{c-b-1}\,(1-tz)^{-a}\,dt,$$
$$\operatorname{Re} c > \operatorname{Re} b > 0, \ |\arg(1-z)| < \pi, \qquad (3.28)$$

(cf. [18, 2.1(10)]) has the following q-analogue due to Heine.

$$_2\phi_1(q^a,q^b;q^c;q,z) = \frac{\Gamma_q(c)}{\Gamma_q(b)\Gamma_q(c-b)}\int_0^1 t^{b-1}\frac{(tq;q)_\infty}{(tq^{c-b};q)_\infty}\frac{(tzq^a;q)_\infty}{(tz;q)_\infty}\,d_q t,$$
$$\operatorname{Re} b > 0, \ |z| < 1. \qquad (3.29)$$

Note that the left hand side and right hand side of (3.29) tend formally to the corresponding sides of (3.28). The proof of (3.29) is also analogous to the proof of (3.28). Expand $(tzq^a; q)_\infty/(tz; q)_\infty$ as a power series in tz by (3.16), interchange summation and q-integration, and evaluate the resulting q-integrals by (3.25).

If we rewrite the q-integral in (3.29) as a series according to the definition (3.12), and if we replace q^a, q^b, q^c by a, b, c then we obtain the following transformation formula:

$$2\phi_1(a, b; c; q, z) = \frac{(az; q)_\infty}{(z; q)_\infty} \frac{(b; q)_\infty}{(c; q)_\infty} 2\phi_1(c/b, z; az; q, b). \tag{3.30}$$

Although (3.29) and (3.30) are equivalent, they look quite different. In fact, in its form (3.30) the identity has no classical analogue. We see a new phenomenon, not occurring for $_2F_1$, namely that the independent variable z of the left hand side mixes on the right hand side with the parameters. So, rather than having a function of z with parameters a, b, c, we deal with a function of a, b, c, z satisfying certain symmetries.

Just as in the classical case (cf. [18, 2.1(14)]), substitution of some special value of z in the q-integral representation (3.29) reduces it to a q-beta integral which can be explicitly evaluated. We obtain

$$2\phi_1(a, b; c; q, c/(ab)) = \frac{(c/a, c/b; q)_\infty}{(c, c/(ab); q)_\infty}, \quad |c/(ab)| < 1, \tag{3.31}$$

where the more relaxed bounds on the parameters are obtained by analytic continuation. The terminating case of (3.31) is

$$2\phi_1(q^{-n}, b; c; q, cq^n/b) = \frac{(c/b; q)_n}{(c; q)_n}, \quad n \in \mathbb{Z}_+. \tag{3.32}$$

The two fundamental transformation formulas

$$2F_1(a, b; c; z) = (1 - z)^{-a} {}_2F_1(a, c - b; c; z/(z - 1)) \tag{3.33}$$
$$= (1 - z)^{c-a-b} {}_2F_1(c - a, c - b; c; z)$$

(cf. [18, 2.1(22) and (23)]) have the following q-analogues.

$$2\phi_1(a, b; c; q, z) = \frac{(az; q)_\infty}{(z; q)_\infty} 2\phi_2(a, c/b; c, az; q, bz) \tag{3.34}$$

$$= \frac{(abz/c; q)_\infty}{(z; q)_\infty} 2\phi_1(c/a, c/b; c; q, abz/c). \tag{3.35}$$

Formula (3.35) can be proved either by threefold iteration of (3.30) or by twofold iteration of (3.34). The proof of (3.34) is more involved. Write both sides as power series in z. Then make both sides into double series by substituting for

$(b;q)_k/(c;q)_k$ on the left hand side a terminating $_2\phi_1$ (cf. (3.32)) and by substituting for $(aq^kz;q)_\infty/(z;q)_\infty$ on the right hand side a q-binomial series (cf. (3.16)). The result follows by some rearrangement of series. See [20, §1.5] for the details.

Observe the difference between (3.33) and its q-analogue (3.34). The argument $z/(z-1)$ in (3.33) no longer occurs in (3.34) as a rational function of z, but the z-variable is distributed over the argument and one of the lower parameters of the $_2\phi_2$. Also we do not stay within the realm of $_2\phi_1$ functions.

Equation (3.8) for $s := 1$ becomes

$$_2\phi_1(q^{-n}, b; c; q, z) = q^{-n(n+1)/2} \frac{(b;q)_n}{(c;q)_n} (-z)^n \, _2\phi_1\left(q^{-n}, \frac{q^{-n+1}}{c}; \frac{q^{-n+1}}{b}; q, \frac{q^{n+1}c}{bz}\right),$$

$$n \in \mathbb{Z}_+.$$
$$(3.36)$$

We may apply (3.36) to the preceding evaluation and transformation formulas for $_2\phi_1$ in order to obtain new ones in the terminating case. From (3.32) we obtain

$$_2\phi_1(q^{-n}, b; c; q, q) = \frac{(c/b;q)_n \, b^n}{(c;q)_n}, \quad n \in \mathbb{Z}_+. \tag{3.37}$$

By inversion of direction of summation on both sides of (3.34) (with $a := q^{-n}$) we obtain

$$_2\phi_1(q^{-n}, b; c; q, z) = \frac{(c/b;q)_n}{(c;q)_n} \, _3\phi_2\left[\begin{matrix} q^{-n}, b, bzq^{-n}/c \\ bq^{1-n}/c, 0 \end{matrix}; q, q\right], \quad n \in \mathbb{Z}_+. \tag{3.38}$$

A terminating $_2\phi_1$ can also be transformed into a terminating $_3\phi_2$ with one of the upper parameters zero (result of Jackson):

$$_2\phi_1(q^{-n}, b; c; q, z) = (q^{-n}bz/c; q)_n \, _3\phi_2\left[\begin{matrix} q^{-n}, c/b, 0 \\ c, cqb^{-1}z^{-1} \end{matrix}; q, q\right] \tag{3.39}$$

This formula can be proved by applying (3.19) to the factor $(cq^{k+1}b^{-1}z^{-1}; q)_{n-k}$ occuring in the kth term of the right hand side. Then interchange summation in the resulting double sum and substitute (3.37) for the inner sum. See [26, p.101] for the details.

3.5. A three-term transformation formula.

Formula (3.36) has the following generalization for non-terminating $_2\phi_1$:

$$_2\phi_1(a, b; c; q, z) + \frac{(a, q/c, c/b, bz/q, q^2/bz; q)_\infty}{(c/q, aq/c, q/b, bz/c, cq/(bz); q)_\infty} \, _2\phi_1\left(\frac{aq}{c}, \frac{bq}{c}; \frac{q^2}{c}; q, z\right)$$

$$= \frac{(abz/c, q/c, aq/b, cq/(abz); q)_\infty}{(bz/c, q/b, aq/c, cq/(bz); q)_\infty} \, _2\phi_1\left(a, \frac{aq}{c}; \frac{aq}{b}; q, \frac{cq}{abz}\right), \quad \left|\frac{cq}{ab}\right| < |z| < 1. \tag{3.40}$$

This identity is a q-analogue of [18, 2.1(17)]. In the following proposition we rewrite (3.40) in an equivalent form and next sketch the elegant proof due to [33].

Proposition 3.3. Suppose $\{q^n \mid n \in \mathbb{Z}_+\}$ is disjoint from $\{a^{-1}q^{-n}, b^{-1}q^{-n} \mid n \in \mathbb{Z}_+\}$. Then

$$(z, qz^{-1}; q)_\infty \, {}_2\phi_1(a, b; c; q, z) = (az, qa^{-1}z^{-1}; q)_\infty \frac{(c/a, b; q)_\infty}{(c, b/a; q)_\infty}$$

$$\times {}_2\phi_1(a, qa/c; qa/b; q, qc/(abz)) + (a \longleftrightarrow b). \quad (3.41)$$

Proof. Consider the function

$$F(w) := \frac{(a, b, cw, q, qwz^{-1}, w^{-1}z; q)_\infty}{(aw, bw, c, w^{-1}; q)_\infty} \frac{1}{w}.$$

Its residue at q^n $(n \in \mathbb{Z}_+)$ equals the nth term of the series on the left hand side of (3.41). The negative of its residue at $a^{-1}q^{-n}$ $(n \in \mathbb{Z}_+)$ equals the nth term of the first series on the right hand side of (3.41), and the residue at $b^{-1}q^{-n}$ is similarly related to the second series on the right hand side. Let $\mathcal{C}$ be a positively oriented closed curve around 0 in $\mathbb{C}$ which separates the two sets mentioned in the Proposition. Then $(2\pi i)^{-1} \int_{\mathcal{C}} F(w)\,dw$ can be expressed in two ways as an infinite sum of residues: either by letting the contour shrink to $\{0\}$ or by blowing it up to $\{\infty\}$. □

If we put $z := q$ in (3.40) and substitute (3.31) then we obtain a generalization of (3.37) for non-terminating ${}_2\phi_1$:

$${}_2\phi_1(a, b; c; q, q) + \frac{(a, b, q/c; q)_\infty}{(aq/c, bq/c, c/q; q)_\infty} \, {}_2\phi_1\left(\frac{aq}{c}, \frac{bq}{c}; \frac{q^2}{c}; q, q\right) = \frac{(abq/c, q/c; q)_\infty}{(aq/c, bq/c; q)_\infty}.$$

$$(3.42)$$

By (3.13) this identity (3.42) can be equivalently written in q-integral form:

$$\int_{qc-1}^{1} \frac{(ct, qt; q)_\infty}{(at, bt; q)_\infty} \, d_q t = \frac{(1-q)(abq/c, q/c, c, q; q)_\infty}{(aq/c, bq/c, a, b; q)_\infty}. \quad (3.43)$$

Replace in (3.43) c, a, b by $-q^c$, $-q^a$, q^b, respectively, and let $q \uparrow 1$. Then we obtain formally

$$\int_{-1}^{1} (1+t)^{a-c}(1-t)^{b-1}\, dt = \frac{2^{b+a-c}\,\Gamma(a-c+1)\,\Gamma(b)}{\Gamma(a+b-c+1)}. \quad (3.44)$$

Note that, although it is trivial to obtain (3.44) from (3.26) by an affine transformation of the integration variable, their q-analogues (3.25) and (3.43) are by no means trivially equivalent.

90

3.6. Bilateral series. Put $a := q$ in (3.40) and substitute (3.16). Then we can combine the two series from 0 to ∞ into a series from $-\infty$ to ∞ without "discontinuity" in the summand:

$$\sum_{k=-\infty}^{\infty} \frac{(b;q)_k}{(c;q)_k} z^k = \frac{(q, c/b, bz, q/(bz); q)_\infty}{(c, q/b, z, c/(bz); q)_\infty}, \qquad |c/b| < |z| < 1. \tag{3.45}$$

Here we have extended the definition of $(b;q)_k$ by

$$(b;q)_k := \frac{(b;q)_\infty}{(bq^k;q)_\infty}, \qquad k \in \mathbb{Z}.$$

Formula (3.45) was first obtained by Ramanujan (*Ramanujan's ${}_1\psi_1$-summation formula*). It reduces to the q-binomial formula (3.18) for $c := q^n$ ($n \in \mathbb{Z}_+$). In fact, this observation can be used for a proof of (3.45), cf. [23], [3, Appendix C]. Formula (3.45) is a q-analogue of the explicit Fourier series evaluation

$$\sum_{k=-\infty}^{\infty} \frac{(b)_k}{(c)_k} e^{ik\theta} = \frac{\Gamma(c)\,\Gamma(1-b)}{\Gamma(c-b)} e^{i(1-c)(\theta-\pi)} (1 - e^{i\theta})^{c-b-1},$$

$$0 < \theta < 2\pi, \ \operatorname{Re}(c-b-1) > 0,$$

where $(b)_k := \Gamma(b+k)/\Gamma(b)$, also for $k = -1, -2, \ldots$.
Define *bilateral q-hypergeometric series* by

$${}_r\psi_s \begin{bmatrix} a_1, \ldots, a_r \\ b_1, \ldots, b_s \end{bmatrix} ; q, z \end{bmatrix}$$

$$:= \sum_{k=-\infty}^{\infty} \frac{(a_1, \ldots, a_r; q)_k}{(b_1, \ldots, b_s; q)_k} \left((-1)^k q^{k(k-1)/2} \right)^{s-r} z^k \tag{3.46}$$

$$= {}_{r+1}\phi_s \begin{bmatrix} a_1, \ldots, a_r, q \\ b_1, \ldots, b_s \end{bmatrix} ; q, z \end{bmatrix}$$

$$+ \frac{(b_1 - q) \ldots (b_s - q)}{(a_1 - q) \ldots (a_r - q) z} {}_{s+1}\phi_s \begin{bmatrix} q^2/b_1, \ldots, q^2/b_s, q \\ q^2/a_1, \ldots, q^2/a_r, 0, \ldots, 0 \end{bmatrix} ; q, \frac{b_1 \ldots b_s}{a_1 \ldots a_r z} \end{bmatrix}, \tag{3.47}$$

where $a_1, \ldots a_r, b_1, \ldots, b_s \neq 0$ and $s \geq r$. The Laurent series in (3.46) is convergent for

$$\left| \frac{b_1 \ldots b_s}{a_1 \ldots a_r} \right| < |z| \quad \text{if} \quad s > r$$

and for

$$\left| \frac{b_1 \ldots b_s}{a_1 \ldots a_r} \right| < |z| < 1 \quad \text{if} \quad s = r.$$

91

If some of the lower parameters in the $_r\psi_s$ are 0 then a suitable confluent limit has to be taken in the $_{s+1}\phi_s$ of (3.47).

Thus we can write (3.45) as

$$_1\psi_1(b;c;q,z) = \frac{(q,c/b,bz,q/(bz);q)_\infty}{(c,q/b,z,c/(bz);q)_\infty}, \qquad |c/b| < |z| < 1. \tag{3.48}$$

Replace in (3.48) z by z/b, substitute (3.47), let $b \to \infty$, and apply (3.6) Then we obtain

$$_0\psi_1(-;c;q,z) := \sum_{k=-\infty}^{\infty} \frac{(-1)^k q^{k(k-1)/2} z^k}{(c;q)_k} = \frac{(q,z,q/z;q)_\infty}{(c,c/z;q)_\infty}, \qquad |z| > |c|. \tag{3.49}$$

In particular, for $c = 0$, we get the *Jacobi triple product identity*

$$_0\psi_1(-;0;q,z) := \sum_{k=-\infty}^{\infty} (-1)^k q^{k(k-1)/2} z^k = (q,z,q/z;q)_\infty, \qquad z \neq 0. \tag{3.50}$$

The series in (3.49) is essentially a *theta function*. With the notation [19, 13.19(9)] we get

$$\theta_4(x;q) := \sum_{k=-\infty}^{\infty} (-1)^k q^{k^2} e^{2\pi i k x}$$
$$=_0\psi_1(-;0;q^2,q\,e^{2\pi i x})$$
$$=(q^2, q\,e^{2\pi i x}, q\,e^{-2\pi i x}; q^2)_\infty$$
$$=\prod_{k=1}^{\infty}(1 - q^{2k})(1 - 2q^{2k-1}\cos(2\pi x) + q^{4k-2}),$$

and similarly for the other theta functions $\theta_i(x;q)$ $(i = 1, 2, 3)$.

3.7. The q-hypergeometric q-difference equation. Just as the *hypergeometric differential equation*

$$z(1 - z)\,u''(z) + (c - (a + b + 1)z)\,u'(z) - a\,b\,u(z) = 0$$

(cf. [18, 2.1(1)]) has particular solutions

$$u_1(z) := {}_2F_1(a, b; c; z), \qquad u_2(z) := z^{1-c}\,{}_2F_1(a - c + 1, b - c + 1; 2 - c; z),$$

the *q-hypergeometric q-difference equation*

$$z(q^c - q^{a+b+1}z)(D_q^2 u)(z) + \left[\frac{1 - q^c}{1 - q} - \left(q^b\frac{1 - q^a}{1 - q} + q^a\frac{1 - q^{b+1}}{1 - q}\right)z\right](D_q u)(z)$$
$$- \frac{1 - q^a}{1 - q}\frac{1 - q^b}{1 - q}u(z) = 0 \tag{3.51}$$

92

has particular solutions

$$u_1(z) := {}_2\phi_1(q^a, q^b; q^c; q, z), \tag{3.52}$$

$$u_2(z) := z^{1-c} {}_2\phi_1(q^{1+a-c}, q^{1+b-c}; q^{2-c}; q, z). \tag{3.53}$$

There is an underlying theory of q-difference equations with regular singularities, similarly to the theory of differential equations with regular singularities discussed for instance in Olver [35, Ch. 5]. It is not difficult to prove the following proposition.

Proposition 3.4. Let $A(z) := \sum_{k=0}^{\infty} a_k z^k$ and $B(z) := \sum_{k=0}^{\infty} b_k z^k$ be convergent power series. Let $\lambda \in \mathbb{C}$ be such that

$$\frac{(1 - q^{\lambda+k})(1 - q^{\lambda+k-1})}{(1-q)^2} + a_0 \frac{1 - q^{\lambda+k}}{1-q} + b_0 \quad \begin{cases} = 0, & k = 0, \\ \neq 0, & k = 1, 2, \dots. \end{cases} \tag{3.54}$$

Then the q-difference equation

$$z^2 (D_q^2 u)(z) + z A(z)(D_q u)(z) + B(z) u(z) = 0 \tag{3.55}$$

has an (up to a constant factor) unique solution of the form

$$u(z) = \sum_{k=0}^{\infty} c_k z^{\lambda+k}. \tag{3.56}$$

Note that (3.51) can be rewritten in the form (3.55) with $a_0 = (q^{-c} - 1)/(1-q)$, $b_0 = 0$, so (3.54) has solutions $\lambda = 0$ and $-c+1 \pmod{(2\pi i \log q^{-1})\mathbb{Z}}$ provided $c \notin \mathbb{Z}$ $\pmod{(2\pi i \log q^{-1})\mathbb{Z}}$. For the coefficients c_k in (3.56) we find the recursion

$$\frac{c_{k+1}}{c_k} = \frac{(1 - q^{a+\lambda+k})(1 - q^{b+\lambda+k})}{(1 - q^{c+\lambda+k})(1 - q^{\lambda+k+1})}.$$

Thus we obtain solutions u_1, u_2 as given in (3.52), (3.53).

Proposition 3.4 can also be applied to the case $z = \infty$ of (3.51). Just make the transformation $z \mapsto z^{-1}$ in (3.51). One solution then obtained is

$$u_3(z) := z^{-a} {}_2\phi_1\left[\begin{matrix} q^a, q^{a-c+1} \\ q^{a-b+1} \end{matrix}; q, q^{-a-b+c+1} z^{-1}\right].$$

Now (3.40) can be rewritten as

$$u_1(z) + \frac{(q^a, q^{1-c}, q^{c-b}; q)_\infty}{(q^{c-1}, q^{a-c+1}, q^{1-b}; q)_\infty} \frac{(q^{b-1}z, q^{2-b}z^{-1}; q)_\infty z^{c-1}}{(q^{b-c}z, q^{c-b+1}z^{-1}; q)_\infty} u_2(z)$$

$$= \frac{(q^{1-c}, q^{a-b+1}; q)_\infty}{(q^{1-b}, q^{a-c+1}; q)_\infty} \frac{(q^{a+b-c}z, q^{c-a-b+1}z^{-1}; q)_\infty z^a}{(q^{b-c}z, q^{c-b+1}z^{-1}; q)_\infty} u_3(z). \tag{3.57}$$

Note that u_3 is not a linear combination of u_1 and u_2, as the coefficients of u_2 and u_3 in (3.57) depend on z. However, since an expression of the form

$$z \mapsto \frac{(q^\alpha z, q^{1-\alpha} z^{-1}; q)_\infty \, z^{\alpha-\beta}}{(q^\beta z, q^{1-\beta} z^{-1}; q)_\infty}$$

is invariant under transformations $z \mapsto qz$, each term in (3.57) is a solution of (3.51). Thus everything works fine when we restrict ourselves to a subset of the form $\{z_0 \, q^k \mid k \in \mathbb{Z}\}$.

The q-analogue of the regular singularity at $z = 1$ for the ordinary hypergeometric differential equation has to be treated in a different way. We can rewrite (3.51) as

$$(q^c - q^{a+b} z) \, u(qz) + (-q^c - q + (q^a + q^b) z) \, u(z) + (q - z) \, u(q^{-1} z) = 0.$$

It can be expected that the points $z = q^{c-a-b}$ and $z = q$, where the coefficient of $u(qz)$ respectively $u(q^{-1} z)$ vanishes, will replace the classical regular singularity $z = 1$, see also (3.31), (3.37) and (3.42). One can find solutions related to these singularities, for instance

$$u_4(z) := {}_3\phi_2 \left[\begin{matrix} q^a, q^b, q^{a+b-c} z \\ q^{a+b-c+1}, 0 \end{matrix} ; q, q \right],$$

$$u_5(z) := z^{-b} \, {}_3\phi_2 \left[\begin{matrix} q^b, q^{b-c+1}, q z^{-1} \\ q^{a+b-c+1}, 0 \end{matrix} ; q, q \right],$$

and many others. A systematic theory for such singularities has not yet been developed.

3.8. q-Bessel functions. The classical *Bessel function* is defined by

$$J_\nu(z) := (z/2)^\nu \sum_{k=0}^{\infty} \frac{(-1)^k}{\Gamma(\nu + k + 1) \, k!} (z/2)^{2k}$$

$$= \frac{(z/2)^\nu}{\Gamma(\nu + 1)} \, {}_0F_1(-; \nu + 1; -z^2/4),$$

cf. [19, Ch. 7]). Jackson (1905) introduced two q-analogues of the Bessel function:

$$J_\nu^{(1)}(z; q) := \frac{(q^{\nu+1}; q)_\infty}{(q; q)_\infty} (z/2)^\nu \, {}_2\phi_1(0, 0; q^{\nu+1}; q, -z^2/4)$$

$$= (z/2)^\nu \sum_{k=0}^{\infty} \frac{(q^{\nu+k+1}; q)_\infty}{(q; q)_\infty} \frac{(-1)^k (z/2)^{2k}}{(q; q)_k}, \tag{3.58}$$

$$J_\nu^{(2)}(z; q) := \frac{(q^{\nu+1}; q)_\infty}{(q; q)_\infty} (z/2)^\nu \, {}_0\phi_1(-; q^{\nu+1}; q, -q^{\nu+1} z^2/4)$$

$$= (z/2)^\nu \sum_{k=0}^{\infty} \frac{(q^{\nu+k+1}; q)_\infty}{(q; q)_\infty} \frac{q^{k(k+\nu)} (-1)^k (z/2)^{2k}}{(q; q)_k}. \tag{3.59}$$

Formally we have

$$\lim_{q \uparrow 1} J_\nu^{(i)}((1-q)z; q) = J_\nu(z), \quad i = 1,2$$

(cf. (3.24)). The two q-Bessel functions $J_\nu^{(i)}(z; q)$ can be simply expressed in terms of each other. From (3.35) and (3.6) we obtain

$$_2\phi_1(0, b; c; q, z) = \frac{1}{(z; q)_\infty} \, _1\phi_1(c/b; c; q, bz).$$

A further confluence with $b \to 0$ yields

$$_2\phi_1(0, 0; c; q, z) = \frac{1}{(z; q)_\infty} \, _0\phi_1(-; c; q, cz).$$

This can be rewritten as

$$J_\nu^{(2)}(z; q) = (-z^2/4; q)_\infty \, J_\nu^{(1)}(z; q).$$

Yet another q-analogue of the Bessel function is as follows.

$$\begin{aligned} J_\nu(z; q) :=& \frac{(q^{\nu+1}; q)_\infty}{(q; q)_\infty} z^\nu \, _1\phi_1(0; q^{\nu+1}; q, qz^2) \\ =& z^\nu \sum_{k=0}^{\infty} \frac{(q^{\nu+k+1}; q)_\infty}{(q; q)_\infty} \frac{(-1)^k q^{k(k+1)/2} z^{2k}}{(q; q)_k} . \end{aligned} \tag{3.60}$$

Formally we have

$$\lim_{q \uparrow 1} J_\nu((1-q^{1/2})z; q) = J_\nu(z).$$

This q-Bessel function is not simply related to the q-Bessel functions (3.58), (3.59). As far as we know, it was first introduced by Hahn (in a special case) and by Exton (in general). In recent work by Koornwinder and Swarttouw [29] a satisfactory q-analogue of the Hankel transform could be given in terms of the q-Bessel function (3.60).

3.9. Various results. Goursat's list of *quadratic transformations* for Gaussian hypergeometric functions can be found in [18, §2.11]. In a recent paper Rahman and Verma [36] have given a full list of q-analogues of Goursat's table. However, all their formulas involve on at least one of both sides an $_8\phi_7$ series. Moreover, a good foundation from the theory of q-difference equations is not yet available. For terminating series many of their $_8\phi_7$'s will simplify to $_4\phi_3$'s. In section 4 we will meet some natural examples of these transformations coming from orthogonal polynomials.

An important part of the book by Gasper and Rahman [20] deals with the derivation of summation and transformation formulas of $_{s+1}\phi_s$ functions with $s > 1$. A simple example is the q-Saalschütz formula

$$_3\phi_2(a, b, q^{-n}; c, abc^{-1}q^{1-n}; q, q) = \frac{(c/a, c/b; q)_n}{(c, c/(ab); q)_n}, \qquad n \in \mathbb{Z}_+, \qquad (3.61)$$

which follows easily from (3.35) by expanding the quotient on the right hand side of (3.35) with the aid of (3.18), and next comparing equal powers of z at both sides of (3.35). Formula (3.61) is the q-analogue of the *Pfaff-Saalschütz formula*

$$_3F_2(a, b, -n; c, 1 + a + b - c - n; 1) = \frac{(c - a)_n (c - b)_n}{(c)_n (c - a - b)_n}, \qquad n \in \mathbb{Z}_+,$$

which can be proved in an analogous way as (3.61).

An example of a much more involved transformation formula, which has many important special cases, is *Watson's transformation formula*

$$_8\phi_7\left[\begin{matrix} a, qa^{1/2}, -qa^{1/2}, b, c, d, e, q^{-n} \\ a^{1/2}, -a^{1/2}, aq/b, aq/c, aq/d, aq/e, aq^{n+1} \end{matrix}; q, \frac{a^2 q^{2+n}}{bcde}\right]$$
$$= \frac{(aq, aq/(de); q)_n}{(aq/d, aq/e; q)_n} \, _4\phi_3\left[\begin{matrix} q^{-n}, d, e, aq/(bc) \\ aq/b, aq/c, deq^{-n}/a \end{matrix}; q, q\right], \qquad n \in \mathbb{Z}_+, \qquad (3.62)$$

cf. Gasper and Rahman [20, §2.5]. Then, for $a^2 q^{n+1} = bcde$, the right hand side can be evaluated by use of (3.61). The resulting evaluation

$$\frac{(aq, aq/(bc), aq/(bd), aq/(cd); q)_n}{(aq/b, aq/c, aq/d, aq/(bcd); q)_n}$$

of the left hand side of (3.62) subject to the given relation between a, b, c, d, e, q is called *Jackson's summation formula*, cf. [20, §2.6].

The famous *Rogers-Ramanujan identities*

$$_0\phi_1(-; 0; q, q) := \sum_{k=0}^{\infty} \frac{q^{k^2}}{(q; q)_k} = \frac{(q^2, q^3, q^5; q^5)_\infty}{(q; q)_\infty},$$

$$_0\phi_1(-; 0; q, q^2) := \sum_{k=0}^{\infty} \frac{q^{k(k+1)}}{(q; q)_k} = \frac{(q, q^4, q^5; q^5)_\infty}{(q; q)_\infty}$$

have been proved in many different ways (cf. Andrews [3]), not only analytically but also by an interpretation in combinatorics or in the framework of Kac-Moody algebras. A quick analytic proof starting from (3.62) is described in [20, §2.7].

Exercises to §3.

3.1 Prove that
$$(a;q)_n = (-a)^n q^{n(n-1)/2} (a^{-1}q^{1-n};q)_n.$$

3.2 Prove that
$$(a;q)_{2n} = (a;q^2)_n (aq;q^2)_n,$$
$$(a^2;q^2)_n = (a;q)_n (-a;q)_n,$$
$$(a;q)_\infty = (a^{1/2}, -a^{1/2}, (aq)^{1/2}, -(aq)^{1/2};q)_\infty.$$

3.3 Prove the following identity of Euler:
$$(-q;q)_\infty (q;q^2)_\infty = 1.$$

3.4 Prove that
$$\sum_{l=-\infty}^{\infty} \frac{(-1)^{l-m} q^{(l-m)(l-m-1)/2}}{\Gamma_q(n-l+1)\Gamma_q(l-m+1)} = \delta_{n,m}, \quad 0 < q \le 1.$$

Do it first for $q = 1$. Start for instance with $e^z e^{-z} = 1$ and, in the q-case, with $e_q(z) E_q(-z) = 1$.

3.5 Let
$$\frac{1}{(q;q)_\infty} = \sum_{n=0}^{\infty} a_n q^n$$
be the power series expansion of the left hand side in terms of q. Show that a_n equals the number of partitions of n.

Show also that the coefficient $b_{k,n}$ in
$$\frac{q^k}{(q;q)_k} = \sum_{n=k}^{\infty} b_{k,n} q^n$$
is the number of partitions of n with highest part k. Give now a partition theoretic proof of the identity
$$\frac{1}{(q;q)_\infty} = \sum_{k=0}^{\infty} \frac{q^k}{(q;q)_k} .$$

3.6 In the same way, give a partition theoretic proof of the identity
$$\frac{1}{(q^m;q)_\infty} = \sum_{k=0}^{\infty} \frac{q^{mk}}{(q;q)_k} , \quad m \in \mathbb{N}.$$

3.7 Give also a partition theoretic proof of
$$\sum_{k=0}^{\infty} \frac{q^{k(k-1)/2} q^{mk}}{(q;q)_k} = (-q^m;q)_\infty, \quad m \in \mathbb{N}.$$

(Consider the problem first for $m = 1$.)

3.8 Let $GF(p)$ be the finite field with p elements. Let A be a $n \times n$ matrix with entries chosen independently and at random from $GF(p)$, with equal probability for the field elements to be chosen. Let $q := p^{-1}$. Prove that the probability that A is invertible is $(q; q)_n$. (See SIAM News 23 (1990) no.6, p.8.)

3.9 Let $GF(p)$ be as in the previous exercise. Let V be an n-dimensional vector space over $GF(p)$. Prove that the number of k-dimensional linear subspaces of V equals the q-binomial coefficient

$$\begin{bmatrix} n \\ k \end{bmatrix}_p := \frac{(p; p)_n}{(p; p)_k \, (p; p)_{n-k}}.$$

3.10 Show that

$$(ab; q)_n = \sum_{k=0}^{n} \begin{bmatrix} n \\ k \end{bmatrix}_q b^k \, (a; q)_k \, (b; q)_{n-k}.$$

Show that both Newton's binomial formula and the formula

$$(a + b)_n = \sum_{k=0}^{n} \binom{n}{k} (a)_k \, b_{n-k}$$

are limit cases of the above formula.

4. q-ANALOGUES OF THE CLASSICAL ORTHOGONAL POLYNOMIALS

Originally, by classical orthogonal polynomials were only meant the three families of Jacobi, Laguerre and Hermite polynomials, but recent insights consider a much bigger class of polynomials as "classical". On the one hand, there is an extension to hypergeometric orthogonal polynomials up to the $_4F_3$ level and including certain discrete orthogonal polynomials. These are brought together in the Askey tableau, cf. Table 1. On the other hand there are q-analogues of all the families in the Askey tableau, often several q-analogues for one classical family (cf. Table 2 for some of them). The master class of all these q-analogues is formed by the celebrated Askey-Wilson polynomials. They contain all other families described in this chapter as special cases or limit cases. Good references for this chapter are Andrews and Askey [5] and Askey and Wilson [10]. See also Atakishiyev, Rahman and Suslov [11] for a somewhat different approach.

Some parts of this section contain surveys without many proofs. However, the subsections 4.3 and 4.4 on big and little q-Jacobi polynomials and 4.5 and 4.6 on the Askey-Wilson integral and related polynomials are rather self-contained.

4.1. Very classical orthogonal polynomials. An introduction to the traditional classical orthogonal polynomials can be found, for instance, in [19, Ch. 10]. One possible characterization is as systems of orthogonal polynomials $\{p_n\}_{n=0,1,2,\dots}$ which are eigenfunctions of a second order differential operator not involving n with eigenvalues λ_n depending on n:

$$a(x)\,p_n''(x) + b(x)\,p_n'(x) + c(x)\,p_n(x) = \lambda_n\,p_n(x). \tag{4.1}$$

Because we will extend the definition of classical orthogonal polynomials in this section, we will call the orthogonal polynomials satisfying (4.1) *very classical orthogonal polynomials*. The classification shows that, up to an affine transformation of the independent variable, the only cases are as follows.

Jacobi polynomials

$$P_n^{(\alpha,\beta)}(x) := \frac{(\alpha+1)_n}{n!}\,{}_2F_1(-n, n+\alpha+\beta+1; \alpha+1; (1-x)/2), \quad \alpha,\beta > -1, \tag{4.2}$$

orthogonal on $[-1,1]$ with respect to the measure $(1-x)^\alpha\,(1+x)^\beta\,dx$;

Laguerre polynomials

$$L_n^\alpha(x) := \frac{(\alpha+1)_n}{n!}\,{}_1F_1(-n; \alpha+1; x), \quad \alpha > -1, \tag{4.3}$$

orthogonal on $[0,\infty)$ with respect to the measure $x^\alpha\,e^{-x}\,dx$;

Hermite polynomials

$$H_n(x) := (2x)^n\,{}_2F_0(-n/2, (1-n)/2; -; -x^{-2}), \tag{4.4}$$

orthogonal on $(-\infty,\infty)$ with respect to the measure $e^{-x^2}\,dx$.

In fact, Jacobi polynomials are the generic case here, while the other two classes are limit cases of Jacobi polynomials:

$$L_n^\alpha(x) = \lim_{\beta\to\infty} P_n^{(\alpha,\beta)}(1 - 2x/\beta), \tag{4.5}$$

$$H_n(x) = 2^n\,n!\,\lim_{\alpha\to\infty} \alpha^{-n/2}\,P_n^{(\alpha,\alpha)}(\alpha^{-1/2}x). \tag{4.6}$$

Hermite polynomials are also limit cases of Laguerre polynomials:

$$H_n(x) = (-1)^n\,2^{n/2}\,n!\,\lim_{\alpha\to\infty} \alpha^{-n/2}\,L_n^\alpha((2\alpha)^{1/2}x + \alpha). \tag{4.7}$$

The limit (4.5) is immediate from (4.2) and (4.3). The limit (4.6) follows from (4.4) and a similar series representation for *Gegenbauer* or *ultraspherical polynomials* (special Jacobi polynomials with $\alpha = \beta$):

$$P_n^{(\alpha,\alpha)}(x) = \frac{(\alpha+1)_n\,(\alpha+1/2)_n}{(2\alpha+1)_n\,n!}\,(2x)^n\,{}_2F_1(-n/2, (1-n)/2; -\alpha-n+1/2; x^{-2}),$$

cf. [19, 10.9(4) and (18)]. The limit (4.7) cannot be easily derived by comparison of series representations. One method of proof is to rewrite the three term recurrence relation for Laguerre polynomials

$$(n+1)\, L_{n+1}^\alpha(x) - (2n + \alpha + 1 - x)\, L_n^\alpha(x) + (n + \alpha)\, L_{n-1}^\alpha(x) = 0$$

(cf. [19, 10.12(8)]) in terms of the polynomials in x given by the right hand side of (4.7) and, next, to compare it with the three term recurrence relation for Hermite polynomials (cf. [19, 10.13(10)])

$$H_{n+1}(x) - 2x\, H_n(x) + 2n\, H_{n-1}(x) = 0.$$

Very classical orthogonal polynomials have other characterizations, for instance by the existence of a Rodrigues type formula or by the fact that the first derivatives again form a system of orthogonal polynomials (cf. [19, §10.6]). Here we want to point out that, associated with the last two characterizations, there is a pair of differential recurrence relations from which many of the basic properties of the polynomials can be easily derived. For instance, for Jacobi polynomials we have the pair

$$\frac{d}{dx}\, P_n^{(\alpha,\beta)}(x) = \frac{n + \alpha + \beta + 1}{2}\, P_{n-1}^{(\alpha+1,\beta+1)}(x), \tag{4.8}$$

$$(1-x)^{-\alpha}\,(1+x)^{-\beta}\,\frac{d}{dx}\left((1-x)^{\alpha+1}\,(1+x)^{\beta+1}\,P_{n-1}^{(\alpha+1,\beta+1)}(x)\right) = -2n\, P_n^{(\alpha,\beta)}(x). \tag{4.9}$$

The differential operators in (4.8), (4.9) are called *shift operators* because of their parameter shifting property. The differential operator in (4.8) followed by the one in (4.9) yields the second order differential operator of which the Jacobi polynomials are eigenfunctions. If we would have defined Jacobi polynomials only by their orthogonality property, not by their explicit expression, then we would have already been able to derive (4.8), (4.9) up to constant factors just by the remarks that the operators D_- in (4.8) and $D_+^{(\alpha,\beta)}$ in (4.9) satisfy

$$\int_{-1}^1 (D_- f)(x)\, g(x)\, (1-x)^{\alpha+1}\,(1+x)^{\beta+1}\, dx$$

$$= -\int_{-1}^1 f(x)\,(D_+^{(\alpha,\beta)} g)(x)\,(1-x)^\alpha\,(1+x)^\beta\, dx \tag{4.10}$$

and that D_- sends polynomials of degree n to polynomials of degree $n-1$, while $D_+^{(\alpha,\beta)}$ sends polynomials of degree $n-1$ to polynomials of degree n.

The same idea can be applied again and again for the more general classical orthogonal polynomials we will discuss in this section.

It follows from (4.8), (4.9) and (4.10) that

$$\int_{-1}^1 \left(P_n^{(\alpha,\beta)}(x)\right)^2 (1-x)^\alpha\,(1+x)^\beta\, dx = \text{const.}\ \int_{-1}^1 (1-x)^{\alpha+n}\,(1+x)^{\beta+n}\, dx,$$

where the constant on the right hand side can be easily computed. What is left for computation is a beta integral, which of course is elementary. However, we emphasize this reduction of computation of quadratic norms of classical orthogonal polynomials to computation of the integral of the weight function with shifted parameter, because this phenomenon will also return in the generalizations. Usually, the computation of the integral of the weight function is the only nontrivial part. On the other hand, if we have some deep evaluation of a definite integral with positive integrand, then it is worth to explore the polynomials being orthogonal with respect to the weight function given by the integrand.

4.2. The Askey tableau. Similar to the classification discussed in §2.1, one can classify all systems of orthogonal polynomials $\{p_n\}$ which are eigenfunctions of a second order difference operator:

$$a(x)\,p_n(x+1) + b(x)\,p_n(x) + c(x)\,p_n(x-1) = \lambda_n\,p_n(x). \tag{4.11}$$

Here the definition of orthogonal polynomials is relaxed somewhat. We include the possibility that the degree n of the polynomials p_n only takes the values $n = 0, 1, \ldots, N$ and that the orthogonality is with respect to a positive measure having support on a set of $N+1$ points. Hahn [21] studied the q-analogue of this classification (cf. §2.5) and he pointed out how the polynomials satisfying (4.11) come out as limit cases for $q \uparrow 1$ of his classification.

The generic case for this classification is given by the *Hahn polynomials*

$$
Q_n(x;\alpha,\beta,N) := {}_3F_2\left[\begin{matrix} -n, n+\alpha+\beta+1, -x \\ \alpha+1, -N \end{matrix}; 1\right]
$$
$$
= \sum_{k=0}^{n} \frac{(-n)_k\,(n+\alpha+\beta+1)_k\,(-x)_k}{(\alpha+1)_k\,(-N)_k\,k!}. \tag{4.12}
$$

Here we assume $n = 0, 1, \ldots, N$ and we assume the notational convention that

$$
{}_rF_s\left[\begin{matrix} -n, a_2, \ldots, a_r \\ b_1, \ldots, b_s \end{matrix}; z\right] := \sum_{k=0}^{n} \frac{(-n)_k\,(a_2)_k \ldots (a_r)_k}{(b_1)_k \ldots (b_s)_k\,k!} z^k, \quad n \in \mathbb{Z}_+, \tag{4.13}
$$

remains well-defined when some of the b_i are possibly non-positive integer but $\leq -n$. For the notation of Hahn polynomials and other families to be discussed in this subsection we keep to the notation of Askey and Wilson [10, Appendix] and Labelle's poster [30]. There one can also find further references.

Hahn polynomials satisfy orthogonality relations

$$
\sum_{x=0}^{N} Q_n(x)\,Q_m(x)\,\rho(x) = \delta_{n,m}\,\frac{1}{\pi_n}, \quad n, m = 0, 1 \ldots, N, \tag{4.14}
$$

101

where

$$\rho(x) := \binom{N}{x} \frac{(\alpha+1)_x \, (\beta+1)_{N-x}}{(\alpha+\beta+2)_N} \qquad (4.15)$$

and

$$\pi_n := \binom{N}{n} \cdot \frac{2n+\alpha+\beta+1}{\alpha+\beta+1} \, \frac{(\alpha+1)_n \, (\alpha+\beta+1)_n}{(\beta+1)_n \, (N+\alpha+\beta+2)_n} . \qquad (4.16)$$

We get positive weights $\rho(x)$ (or weights of fixed sign) if $\alpha, \beta \in (-1, \infty) \cup (-\infty, -N)$.

The other orthogonal polynomials coming out of this classification are limit cases of Hahn polynomials. We get the following families.

Krawtchouk polynomials

$$K_n(x; p, N) := {}_2F_1(-n, -x; -N; p^{-1}), \quad n = 0, 1, \ldots, N, \quad 0 < p < 1, \qquad (4.17)$$

with orthogonality measure having weights $x \mapsto \binom{N}{x} p^x (1-p)^{N-x}$ on $\{0, 1, \ldots, N\}$;

Meixner polynomials

$$M_n(x; \beta, c) := {}_2F_1(-n, -x; \beta; 1 - c^{-1}), \quad 0 < c < 1, \ \beta > 0, \qquad (4.18)$$

with orthogonality measure having weights $x \mapsto (\beta)_x \, c^x / x!$ on $\mathbb{Z}_+$;

Charlier polynomials

$$C_n(x; a) := {}_2F_0(-n, -x; -; -a^{-1}), \quad a > 0,$$

with orthogonality measure having weights $x \mapsto a^x / x!$ on $\mathbb{Z}_+$.

Note that Krawtchouk polynomials are Meixner polynomials (4.18) with $\beta := q^{-N}$. Krawtchouk and Meixner polynomials are limits of Hahn polynomials, while Charlier polynomials are limits of both Krawtchouk and Meixner polynomials. In a certain sense, the very classical orthogonal polynomials are also contained in the class discussed here, since Jacobi, Laguerre and Hermite polynomials are limits of Hahn, Meixner and Charlier polynomials, respectively. For instance,

$$P_n^{(\alpha,\beta)}(1 - 2x) = \frac{(\alpha+1)_n}{n!} \lim_{N \to \infty} Q_n(Nx; \alpha, \beta, N).$$

See Table 1 (or rather a part of it) for a pictorial representation of these families and their limit transitions.

The Krawtchouk, Meixner and Charlier polynomials are *self-dual*, i.e., they satisfy

$$p_n(x) = p_x(n), \quad x, n \in \mathbb{Z}_+ \text{ or } x, n \in \{0, 1, \ldots, N\}.$$

Thus the orthogonality relations and dual orthogonality relations of these polynomials essentially coincide and the second order difference equation (4.11) becomes the

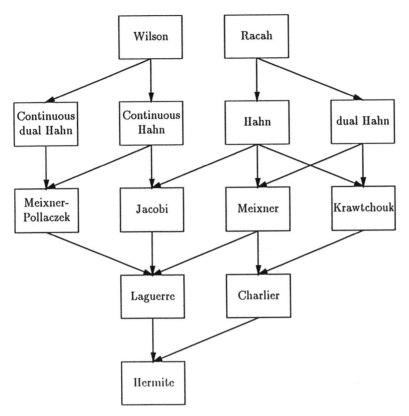

Table 1. The Askey tableau

three term recurrence relation after interchange of x and n. Then it can be arranged that the eigenvalue λ_n in (4.11) becomes n. By the self-duality, the L^2 completeness of the systems in case $N = \infty$ is also clear.

For Hahn polynomials we have no self-duality, so the dual orthogonal system will be different. Observe that, in (4.12), we can write

$$(-n)_k \, (n + \alpha + \beta + 1)_k = \prod_{j=0}^{k-1} (-n(n + \alpha + \beta + 1) + j(j + \alpha + \beta + 1)),$$

which is a polynomial of degree k in $n(n + \alpha + \beta + 1)$. Now define

$$R_n(x(x + \alpha + \beta + 1); \alpha, \beta, N) := Q_x(n; \alpha, \beta, N), \quad n, x = 0, 1, \ldots, N.$$

Then R_n $(n = 0, 1, \ldots, N)$ extends to a polynomial of degree n:

$$R_n(x(x + \alpha + \beta + 1); \alpha, \beta, N) = {}_3F_2 \left[\begin{matrix} -n, -x, x + \alpha + \beta + 1 \\ \alpha + 1, -N \end{matrix} ; 1 \right]$$

$$= \sum_{k=0}^{n} \frac{(-n)_k \, (-x)_k \, (x + \alpha + \beta + 1)_k}{(\alpha + 1)_k \, (-N)_k \, k!} \, .$$

These are called the *dual Hahn polynomials*. The dual orthogonality relations implied by (4.14) are the orthogonality relations for the dual Hahn polynomials:

$$\sum_{x=0}^{N} R_m(x(x + \alpha + \beta + 1)) \, R_n(x(x + \alpha + \beta + 1)) \, \pi_x = \delta_{m,n} \, \frac{1}{\rho(n)} \, .$$

Here π_x and $\rho(n)$ are as in (4.15) and (4.16). Thus the dual Hahn polynomials are also orthogonal polynomials. The three term recurrence relation for Hahn polynomials translates as a second order difference equation which is a slight generalization of (4.11). It has the form

$$a(x) \, p_n(\lambda(x + 1)) + b(x) \, p_n(\lambda(x)) + c(x) \, p_n(\lambda(x - 1)) = \lambda_n \, p_n(\lambda(x)), \qquad (4.19)$$

where $\lambda(x) := x(x + \alpha + \beta + 1)$ is a quadratic function of x. The Krawtchouk and Meixner polynomials are also limit cases of the dual Hahn polynomials.

It is a natural question to ask for other orthogonal polynomials being eigenfunctions of a second order difference equation of the form (4.19). A four-parameter family with this property are the Racah polynomials, essentially known for a long time to the physicists as Racah coefficients, which occur in connection with threefold tensor products of irreducible representations of the group $SU(2)$. However, it was not recognized before the late seventies (cf. Wilson [40]) that orthogonal polynomials are hidden in the Racah coefficients. *Racah polynomials* are defined by

$$R_n(x(x + \gamma + \delta + 1); \alpha, \beta, \gamma, \delta) := {}_4F_3 \left[\begin{array}{c} -n, n + \alpha + \beta + 1, -x, x + \gamma + \delta + 1 \\ \alpha + 1, \beta + \delta + 1, \gamma + 1 \end{array} ; 1 \right],$$
$$(4.20)$$

where $\alpha + 1$ or $\beta + \delta + 1$ or $\gamma + 1 = -N$ for some $N \in \mathbb{Z}_+$, and where $n = 0, 1, \ldots, N$. Similarly as for the dual Hahn polynomials it can be seen that the Racah polynomial R_n is indeed a polynomial of degree n in $\lambda(x) := x(x + \gamma + \delta + 1)$. The Racah polynomials are orthogonal with respect to weights $w(x)$ on the points $\lambda(x)$, $x = 0, 1 \ldots, N$, given by

$$w(x) := \frac{(\gamma + \delta + 1)_x \, ((\gamma + \delta + 3)/2)_x \, (\alpha + 1)_x \, (\beta + \delta + 1)_x \, (\gamma + 1)_x}{x! \, ((\gamma + \delta + 1)/2)_x \, (\gamma + \delta - \alpha + 1)_x \, (\gamma - \beta + 1)_x \, (\delta + 1)_x} \, .$$

It is evident from (4.20) that dual Racah polynomials are again Racah polynomials with α, β interchanged with γ, δ. Hahn and dual Hahn polynomials can be obtained as limit cases of Racah polynomials.

Each orthogonality relation for the Racah polynomials is an explicit evaluation of a finite sum. It is possible to interpret this finite sum as a sum of residues coming from a contour integral in the complex plane. This contour integral can also be considered and evaluated for values of N not necessarily in $\mathbb{Z}_+$. For suitable values of the parameters the contour integral can then be deformed to an integral over the imaginary axis and it gives rise to the orthogonality relations for the *Wilson polynomials* (cf. Wilson [40]) defined by

$$W_n(x^2; a, b, c, d)$$
$$:= (a+b)_n \, (a+c)_n \, (a+d)_n \, {}_4F_3 \left[\begin{array}{c} -n, n+a+b+c+d-1, a+ix, a-ix \\ a+b, a+c, a+d \end{array} ; 1 \right]. \quad (4.21)$$

Apparently, the right hand side defines a polynomial of degree n in x^2. If a, b, c, d have positive real parts and complex parameters appear in conjugate pairs then the functions $x \mapsto W_n(x^2)$ ($n \in \mathbb{Z}_+$) of (4.21) are orthogonal with respect to the measure $w(x)dx$ on $[0, \infty)$, where

$$w(x) := \left| \frac{\Gamma(a+ix)\,\Gamma(b+ix)\,\Gamma(c+ix)\,\Gamma(d+ix)}{\Gamma(2ix)} \right|^2 .$$

The normalization in (4.21) is such that the Wilson polynomials are symmetric in their four parameters a, b, c, d.

The Wilson polynomials satisfy an eigenfunction equation of the form

$$a(x)\,W_n((x+i)^2) + b(x)\,W_n(x^2) + c(x)\,W_n((x-i)^2) = \lambda_n\,W_n(x^2).$$

So we have the new phenomenon that the difference operator at the left hand side shifts into the complex plane, out of the real interval on which the Wilson polynomials are orthogonal. This difference operator can be factorized as a product of two shift operators of similar type. They have properties and applications analogous to the shift operators for Jacobi polynomials discussed at the end of §4.1. They also reduce the evaluation of the quadratic norms to the evaluation of the integral of the weight function, but this last problem is now much less trivial than in the Jacobi case.

Now, we can descend from the Wilson polynomials by limit transitions, just as we did from the Racah polynomials. On the ${}_3F_2$ level we thus get continuous analogues of the Hahn and dual Hahn polynomials as follows.

Continuous dual Hahn polynomials:

$$S_n(x^2; a, b, c) := (a+b)_n \, (a+c)_n \, {}_3F_2 \left[\begin{array}{c} -n, a+ix, a-ix \\ a+b, a+c \end{array} ; 1 \right],$$

where a, b, c have positive real parts; if one of these parameters is not real then one of the other parameters is its complex conjugate. The functions $x \mapsto S_n(x^2)$ are orthogonal with respect to the measure $w(x)dx$ on $[0, \infty)$, where

$$w(x) := \left| \frac{\Gamma(a+ix)\,\Gamma(b+ix)\,\Gamma(c+ix)}{\Gamma(2ix)} \right|^2 .$$

Continuous Hahn polynomials:

$$p_n(x; a, b, \bar{a}, \bar{b}) := i^n \frac{(a + \bar{a})_n (a + \bar{b})_n}{n!} \, {}_3F_2 \left[\begin{array}{c} -n, n + a + \bar{a} + b + \bar{b} - 1, a + ix \\ a + \bar{a}, a + \bar{b} \end{array} ; 1 \right],$$

where a, b have positive real part. The polynomials p_n are orthogonal on $\mathbb{R}$ with respect to the measure $|\Gamma(a + ix)\Gamma(b + ix)|^2 \, dx$. In Askey and Wilson [10, Appendix] only the symmetric case ($a, b > 0$ or $a = \bar{b}$) of these polynomials occurs. The general case was discovered by Atakishiyev and Suslov [12].

Jacobi polynomials are limit cases of continuous Hahn polynomials and also directly of Wilson polynomials (with one pair of complex conjugate parameters). There is one further class of orthogonal polynomials on the ${}_2F_1$ level occurring as limit cases of orthogonal polynomials on the ${}_3F_2$ level:

Meixner-Pollaczek polynomials:

$$P_n^{(a)}(x; \phi) := \frac{(2a)_n}{n!} \, e^{in\phi} \, {}_2F_1(-n, a + ix; 2a; 1 - e^{-2i\phi}), \quad a > 0, \ 0 < \phi < \pi.$$

(Here we have chosen the normalization as in Labelle [30], which is the same as in Pollaczek's original 1950 paper.) They are orthogonal on $\mathbb{R}$ with respect to the weight function $x \mapsto e^{(2\phi - \pi)x} |\Gamma(a + ix)|^2$. They can be considered as continuous analogues of the Meixner and Krawtchouk polynomials. They are limits of both continuous Hahn and continuous dual Hahn polynomials. Laguerre polynomials are limit cases of Meixner-Pollaczek polynomials. Note that the last three families are analytic continuations, both in x and in the parameters, of dual Hahn polynomials, Hahn polynomials and Meixner polynomials, respectively.

All families of orthogonal polynomials discussed until now, together with the limit transitions between them, form the *Askey tableau* (or *scheme* or *chart*) of *hypergeometric orthogonal polynomials*. See Askey and Wilson [10, Appendix], Labelle [30] or Table 1. See also [25] for group theoretic interpretations.

4.3. Big q-Jacobi polynomials.
These polynomials were hinted at by Hahn [21] and explicitly introduced by Andrews and Askey [5]. Here we will show how their basic properties can be derived from a suitable pair of shift operators. We keep the convention of §3 that $0 < q < 1$.

First we introduce q-integration by parts. This will involve *backward* and *forward* q-derivatives:

$$(D_q^- f)(x) := \frac{f(x) - f(qx)}{(1 - q)x}, \quad (D_q^+ f)(x) := \frac{f(q^{-1}x) - f(x)}{(1 - q)x}.$$

Here D_q^- coincides with D_q introduced in (3.9).

Proposition 4.1. If f and g are continuous on $[-d, c]$ $(c, d \geq 0)$ then

$$\int_{-d}^{c} (D_q^- f)(x) \, g(x) \, d_q x = f(c) \, g(q^{-1}c) - f(-d) \, g(-q^{-1}d) - \int_{-d}^{c} f(x) \, (D_q^+ g)(x) \, d_q x.$$

Proof.

$$\int_{0}^{c} (D_q^- f)(x) \, g(x) \, d_q(x) = \sum_{k=0}^{\infty} (f(cq^k) - f(cq^{k+1})) \, g(cq^k)$$

$$= \lim_{N \to \infty} \left\{ f(c) \, g(q^{-1}c) - f(cq^{N+1}) \, g(cq^N) + \sum_{k=0}^{N} f(cq^k) \, (g(cq^k) - g(cq^{k-1})) \right\}$$

$$= f(c) \, g(q^{-1}c) - f(0) \, g(0) + \sum_{k=0}^{\infty} f(cq^k) \, (g(cq^k) - g(cq^{k-1}))$$

$$= f(c) \, g(q^{-1}c) - f(0) \, g(0) - \int_{0}^{c} f(x) \, (D_q^+ g)(x) \, d_q x.$$

Now apply (3.13). $\qquad\qquad\qquad\qquad\qquad\qquad\qquad\qquad\qquad\qquad\qquad\qquad\qquad\quad$ □

Let

$$w(x; a, b, c, d; q) := \frac{(qx/c, -qx/d; q)_\infty}{(qax/c, -qbx/d; q)_\infty} . \qquad (4.22)$$

Note that $w(x; a, b, c, d; q) > 0$ on $[-d, c]$ if $c, d > 0$ and

$$\left[-\frac{c}{dq} < a < \frac{1}{q} \text{ and } -\frac{d}{cq} < b < \frac{1}{q} \right] \quad \text{or} \quad [a = c\alpha \And b = d\bar{\alpha} \text{ for some } \alpha \in \mathbb{C}\backslash\mathbb{R}.] \qquad (4.23)$$

From now on we assume that these inequalities hold. If f is continuous on $[-d, c]$ then

$$\lim_{q \uparrow 1} \int_{-d}^{c} f(x) \, w(x; q^\alpha, q^\beta, c, d; q) \, d_q x = \int_{-d}^{c} f(x) \, (1 - c^{-1}x)^\alpha \, (1 + d^{-1}x)^\beta \, dx, \qquad (4.24)$$

Thus the measure $w(x; a, b, c, d; q) \, d_q x$ can be considered as a q-analogue of the Jacobi polynomial orthogonality measure shifted to an arbitrary finite interval containing 0.

Since

$$w(q^{-1}c; a, b, c, d; q) = 0 = w(-q^{-1}d; a, b, c, d; q),$$

we get from Proposition 4.1 that for any two polynomials f, g the following holds.

$$\int_{-d}^{c} (D_q^- f)(x) \, g(x) \, w(x; qa, qb, c, d; q) \, d_q x$$

$$= -\int_{-d}^{c} f(x) \, [D_q^+ \, (g(.) \, w(.; qa, qb, c, d; q))](x) \, d_q x. \qquad (4.25)$$

Define

$$(D_q^{+,a,b} f)(x) := \frac{[D_q^+ (w(.;qa, qb, c, d; q) f(.))](x)}{w(x; a, b, c, d; q)}$$

$$= (1-q)^{-1} x^{-1}(1 - x/c)(1 + x/d) f(q^{-1}x)$$
$$- (1-q)^{-1} x^{-1} (1 - qax/c)(1 + qbx/d) f(x). \qquad (4.26)$$

In the following $D_q^- f(x)$ or $D_q^-(f(x))$ will mean $(D_q^- f)(x)$. Also $D_q^{+,a,b} f(x)$ or $D_q^{+,a,b}(f(x))$ will mean $(D_q^{+,a,b} f)(x)$. A simple computation yields the following.

$$D_q^- x^n = \frac{1-q^n}{1-q} x^{n-1}, \qquad (4.27)$$

$$D_q^{+,a,b}((q^2 ax/c; q)_{n-1})$$
$$= \frac{q^{-n} a^{-1} - qb}{(1-q)d} (qax/c; q)_n + \text{terms of degree} \leq n - 1 \text{ in } x. \qquad (4.28)$$

Define the *big q-Jacobi polynomial* $\widetilde{P}_n(x; a, b, c, d; q)$ as the monic orthogonal polynomial of degree n in x with respect to the measure $w(x; a, b, c, d; q) \, d_q x$ on $[-d, c]$. Later we will introduce another normalization for these polynomials and we will then write them as P_n instead of $\widetilde{P}_n$. It follows from (4.24) that

$$\lim_{q\uparrow 1} \widetilde{P}_n(x; q^\alpha, q^\beta, c, d; q) = \text{const.} \ P_n^{(\alpha,\beta)} \left(\frac{d - c + 2x}{d + c} \right),$$

a Jacobi polynomial shifted to the interval $[-d, c]$. Big q-Jacobi polynomials with $d = 0$, $c = 1$ are called *little q-Jacobi polynomials*.

Two simple consequences of the definition are:

$$\widetilde{P}_n(-x; a, b, c, d; q) = (-1)^n \ \widetilde{P}_n(x; b, a, d, c; q) \qquad (4.29)$$

and

$$\widetilde{P}_n(\lambda x; a, b, c, d; q) = \lambda^n \ \widetilde{P}_n(x; a, b, \lambda^{-1}c, \lambda^{-1}d; q), \qquad \lambda > 0.$$

It follows from (4.25), (4.27) and (4.28) that D_q^- and $D_q^{+,a,b}$ act as shift operators on the big q-Jacobi polynomials:

$$D_q^- \ \widetilde{P}_n(x; a, b, c, d; q) = \frac{1-q^n}{1-q} \ \widetilde{P}_{n-1}(x; qa, qb, c, d; q), \qquad (4.30)$$

$$D_q^{+,a,b} \ \widetilde{P}_{n-1}(x; qa, qb, c, d; q) = \frac{q^2 ab - q^{-n+1}}{(1-q)cd} \ \widetilde{P}_n(x; a, b, c, d; q). \qquad (4.31)$$

Composition of (4.30) and (4.31) yields a second order q-difference equation for the big q-Jacobi polynomials:

$$D_q^{+,a,b} \ D_q^- \ \widetilde{P}_n(x; a, b, c, d; q) = \frac{q(1 - q^{-n})(1 - q^{n+1}ab)}{(1-q)^2 cd} \ \widetilde{P}_n(x; a, b, c, d; q). \qquad (4.32)$$

The left hand side of (4.32) can be rewritten as

$$A(x)\left((D_q^-)^2 \tilde{P}_n\right)(q^{-1}x) + B(x)(D_q^- \tilde{P}_n)(q^{-1}x)$$
$$= a(x)\tilde{P}_n(q^{-1}x) + b(x)\tilde{P}_n(x) + c(x)\tilde{P}_n(qx) \qquad (4.33)$$

for certain polynomials A, B and a, b, c. Here

$$A(x) = q^{-1}\left(1 - qax/c\right)\left(1 + qbx/d\right),$$
$$B(x) = \frac{(1 - qb)c - (1 - qa)d - (1 - q^2ab)x}{(1 - q)cd}. \qquad (4.34)$$

Compare (4.32), (4.33) with (4.1), (4.11). Apparently we have here another extension of the concept of classical orthogonal polynomials. Two other properties point into the same direction. First, by (4.30) the polynomials $D_q^- \tilde{P}_{n+1}$ ($n = 0, 1, 2, \ldots$) form again a system of orthogonal polynomials. Second, iteration of (4.31) yields a Rodrigues type formula.

It follows from (4.26) that

$$(D_q^{+,a,b} f)\left(\frac{c}{qa}\right) = -\frac{(1 - qa)(1 + qad/c)}{qad(1 - q)} f\left(\frac{c}{q^2 a}\right).$$

Combination with (4.31) yields the recurrence

$$\frac{q^2 ab - q^{-n+1}}{(1 - q)cd} \tilde{P}_n\left(\frac{c}{qa}; a, b, c, d; q\right)$$
$$= -\frac{(1 - qa)(1 + qad/c)}{qad(1 - q)} \tilde{P}_{n-1}\left(\frac{c}{q^2 a}; qa, qb, c, d; q\right).$$

By iteration we get an evaluation of the big q-Jacobi polynomial at a special point:

$$\tilde{P}_n\left(\frac{c}{qa}; a, b, c, d; q\right) = \left(\frac{c}{qa}\right)^n \frac{(qa; q)_n \, (-qad/c; q)_n}{(q^{n+1}ab; q)_n}. \qquad (4.35)$$

Now we will normalize the big q-Jacobi polynomials such that they take the value 1 at $c/(qa)$:

$$P_n(x; a, b, c, d; q) := \frac{\tilde{P}_n(x; a, b, c, d; q)}{\tilde{P}_n(c/(qa); a, b, c, d; q)}. \qquad (4.36)$$

Note that the value at $-d/(qb)$ now follows from (4.29) and (4.35):

$$P_n\left(\frac{-d}{qb}; a, b, c, d; q\right) = \left(-\frac{ad}{bc}\right)^n \frac{(qb; q)_n \, (-qbc/d; q)_n}{(qa; q)_n \, (-qad/c; q)_n}. \qquad (4.37)$$

It follows from (4.29) and (4.37) that

$$P_n(-x; a, b, c, d; q) = \left(-\frac{ad}{bc}\right)^n \frac{(qb; q)_n \, (-qbc/d; q)_n}{(qa; q)_n \, (-qad/c; q)_n} P_n(x; b, a, d, c; q).$$

The following lemma, proved by use of (3.10), gives the q-analogue of a Taylor series expansion.

Lemma 4.2. If $f(x) := \sum_{k=0}^{n} c_k (qax/c; q)_k$ then

$$((D_q^-)^k f)\left(\frac{c}{q^{k+1}a}\right) = c_k (-1)^k \left(\frac{qa}{c}\right)^k q^{k(k-1)/2} \frac{(q; q)_k}{(1-q)^k}.$$

Now put $f(x) := P_n(x; a, b, c, d; q)$ in Lemma 4.2 and substitute (4.30) (iterated) and (4.36). Then the c_k can be found explicitly and we obtain a representation by a q-hypergeometric series:

$$P_n(x; a, b, c, d; q) = \sum_{k=0}^{n} \frac{(q^{-n}; q)_k (q^{n+1}ab; q)_k (qax/c; q)_k q^k}{(qa; q)_k (-qad/c; q)_k (q; q)_k}$$

$$= {}_3\phi_2 \left[\begin{matrix} q^{-n}, q^{n+1}ab, qax/c \\ qa, -qad/c \end{matrix} ; q, q \right]. \qquad (4.38)$$

Andrews & Askey [5, (3,28)] use the notation $P_n^{(\alpha,\beta)}(x; c, d : q)$, which coincides, up to a constant factor, with our $p_n(x; q^\alpha, q^\beta, c, d; q)$.

It follows by combination of (4.29) and (4.38) that

$$\frac{P_n(x; a, b, c, d; q)}{P_n(-d/(qb); a, b, c, d; q)} = {}_3\phi_2 \left[\begin{matrix} q^{-n}, q^{n+1}ab, -qbx/d \\ qb, -qbc/d \end{matrix} ; q, q \right], \qquad (4.39)$$

where the denominator of the left hand side is explicitly given by (4.37).

Next we will compute the quadratic norms. By (4.25), (4.30) and (4.31) we get the recurrence

$$\int_{-d}^{c} (\tilde{P}_n^2 w)(x; a, b, c, d; q) \, d_q x = \frac{q^{n-1}(1 - q^n) cd}{1 - q^{n+1}ab} \int_{-d}^{c} (\tilde{P}_{n-1}^2 w)(x; qa, qb, c, d; q) \, d_q x$$

$$= \frac{q^{n(n-1)/2} (q; q)_n (cd)^n}{(q^{n+1}ab; q)_n} \int_{-d}^{c} w(x; q^n a, q^n b, c, d; q) \, d_q x, \qquad (4.40)$$

where the second equality follows by iteration. Now the q-integral of w from $-d$ to c can be rewritten as a sum of two ${}_2\phi_1$'s of argument q of the form of the left hand side of (3.42). Evaluation by (3.42) yields

$$\int_{-d}^{c} w(x; a, b, c, d; q) \, d_q x = (1 - q)c \frac{(q, -d/c, -qc/d, q^2 ab; q)_\infty}{(qa, qb, -qbc/d, -qad/c; q)_\infty}. \qquad (4.41)$$

Together with (4.40) this yields:

$$\frac{\int_{-d}^{c} (\tilde{P}_n^2 w)(x; a, b, c, d; q) \, d_q x}{\int_{-d}^{c} w(x; a, b, c, d; q) \, d_q x} = q^{n(n-1)/2} (cd)^n \frac{(q, qa, qb, -qbc/d, -qad/c; q)_n}{(q^2 ab; q)_{2n} (q^{n+1}ab; q)_n}.$$

$$(4.42)$$

Now we can compute the coefficients in the three term recurrence relation for the big q-Jacobi polynomials $\tilde{P}_n(x) := \tilde{P}_n(x; a, b, c, d; q)$:

$$x\, \tilde{P}_n(x) = \tilde{P}_{n+1}(x) + B_n\, \tilde{P}_n(x) + C_n\, \tilde{P}_{n-1}(x). \qquad (4.43)$$

Then C_n is the quotient of the right hand sides of (4.42) for degree n and for degree $n - 1$, respectively, so

$$C_n = \frac{q^{n-1}\,(1 - q^n)(1 - q^n a)(1 - q^n b)(1 - q^n ab)(d + q^n bc)(c + q^n ad)}{(1 - q^{2n-1}ab)(1 - q^{2n}ab)^2\,(1 - q^{2n+1}ab)}. \qquad (4.44)$$

Then we obtain B_n by substution of $x := c/(qa)$ in (4.43), in view of (4.35).

4.4. Little q-Jacobi polynomials. These polynomials are the most straightforward q-analogues of the Jacobi polynomials. They were first observed by Hahn [21] and studied in more detail by Andrews and Askey [4]. Here we will study them as a special case of the big q-Jacobi polynomials.

When we specialize the big q-Jacobi polynomials (4.36) to $c = 1$, $d = 0$ and normalize them such that they take the value 1 at 0 then we obtain the *little q-Jacobi polynomials*

$$p_n(x; a, b; q) := \frac{P_n(x; b, a, 1, 0; q)}{P_n(0; b, a, 1, 0; q)}$$

$$= {}_2\phi_1(q^{-n}, q^{n+1}ab; qa; q, qx) \qquad (4.45)$$

$$= (-qb)^{-n}\, q^{-n(n-1)/2}\, \frac{(qb; q)_n}{(qa; q)_n}\, {}_3\phi_2\left[\begin{matrix} q^{-n}, q^{n+1}ab, qbx \\ qb, 0 \end{matrix}; q, q\right]. \qquad (4.46)$$

Here (4.45) follows by letting $d \to 0$ in (4.39), while (4.46) follows from (4.38) and (4.37). With the equality of (4.45) and (4.46) we have reobtained the transformation formula (3.38) for terminating ${}_2\phi_1$ series. From (4.46) we obtain

$$p_n(q^{-1}b^{-1}; a, b; q) = (-qb)^{-n}\, q^{-n(n-1)/2}\, \frac{(qb; q)_n}{(qa; q)_n}.$$

From (4.45) and (3.37) we obtain

$$p_n(1; a, b; q) := (-a)^n\, q^{n(n+1)/2}\, \frac{(qb; q)_n}{(qa; q)_n}.$$

Little q-Jacobi polynomials satisfy the orthogonality relations

$$\frac{1}{B_q(\alpha + 1, \beta + 1)} \int_0^1 p_n(t; q^\alpha, q^\beta; q)\, p_m(t; q^\alpha, q^\beta; q)\, t^\alpha\, \frac{(qt; q)_\infty}{(q^{\beta+1}t; q)_\infty}\, d_q t$$

$$= \delta_{n,m}\, \frac{q^{n(\alpha+1)}\,(1 - q^{\alpha+\beta+1})(q^{\beta+1}; q)_n\,(q; q)_n}{(1 - q^{2n+\alpha+\beta+1})(q^{\alpha+1}; q)_n\,(q^{\alpha+\beta+1}; q)_n}. \qquad (4.47)$$

111

The orthogonality measure is the measure of the q-beta integral (3.27). For positivity and convergence we require that $0 < a < 1$, $b < 1$ (after we have replaced q^α by a and q^β by b in (4.47)). It is maybe not immediately seen that (4.47) is a limit case of (4.42). However, we can establish this by observing the weak convergence as $d \downarrow 0$ of the normalized measure const.$\times w(x; q^\beta, q^\alpha, 1, d; q) d_q x$ on $[-d, 1]$ to the normalized measure in (4.47) on $[0,1]$:

$$\lim_{d\downarrow 0} \frac{\int_{-d}^1 (q^{\beta+1} t; q)_n \, w(t; q^\beta, q^\alpha, 1, d; q) \, d_q t}{\int_{-d}^1 w(t; q^\beta, q^\alpha, 1, d; q) \, d_q t} = \lim_{d\downarrow 0} \frac{(q^{\beta+1}; q)_n \, (-q^{\beta+1} d; q)_n}{(q^{\alpha+\beta+2}; q)_n}$$

$$= \frac{(q^{\beta+1}; q)_n}{(q^{\alpha+\beta+2}; q)_n} = \frac{1}{B_q(\alpha+1, \beta+1)} \int_0^1 (q^{\beta+1} t; q)_n \frac{(qt; q)_\infty}{(q^{\beta+1} t; q)_\infty} t^\alpha \, d_q t.$$

Here the first equality follows from (4.41) and (4.22), while the last equality follows from (3.27).

Little q-Jacobi polynomials are the q-analogues of the Jacobi polynomials shifted to the interval $[0, 1]$:

$$\lim_{q\uparrow 1} p_n(x; q^\alpha, q^\beta; q) = \frac{P_n^{(\alpha,\beta)}(1 - 2x)}{P_n^{(\alpha,\beta)}(1)}.$$

If we fix $b < 1$ (possibly 0 or negative) and put $a = q^\alpha$ then little q-Jacobi polynomials tend for $q \uparrow 1$ to Laguerre polynomials:

$$\lim_{q\uparrow 1} p_n\left(\frac{(1-q)x}{1-b}; q^\alpha, b; q\right) = \frac{L_n^\alpha(x)}{L_n^\alpha(0)}.$$

In particular, little q-Jacobi polynomials $p_n(x; a, 0; q)$, called *Wall polynomials*, are q-analogues of Laguerre polynomials.

Analogous to the quadratic tansformations for Jacobi polynomials (cf. [19, 10.9(4), (21) and (22)]) we can find *quadratic transformations* between little and big q-Jacobi polynomials:

$$P_{2n}(x; a, a, 1, 1; q) = \frac{p_n(x^2; q^{-1}, a^2; q^2)}{p_n((qa)^{-2}; q^{-1}, a^2; q^2)}, \tag{4.48}$$

$$P_{2n+1}(x; a, a, 1, 1; q) = \frac{x \, p_n(x^2; q, a^2; q^2)}{(qa)^{-1} \, p_n((qa)^{-2}; q, a^2; q^2)}. \tag{4.49}$$

The proof is also analogous: by use of the orthogonality properties and normalization of the polynomials.

Just as Jacobi polynomials tend to Bessel functions by

$$\frac{P_{n_N}^{(\alpha,\beta)}(1 - x^2/(2N^2))}{P_{n_N}^{(\alpha,\beta)}(1)} = {}_2F_1\left(-n_N, n_N + \alpha + \beta + 1; \alpha + 1; \frac{x^2}{4N^2}\right)$$

$$\xrightarrow{N\to\infty} {}_0F_1(-; \alpha+1; -(\lambda x/2)^2) = (\lambda x/2)^{-\alpha} \Gamma(\alpha+1) J_\alpha(\lambda x), \quad n_N/N \to \lambda \text{ as } N \to \infty,$$

little q-Jacobi polynomials tend to the Hahn-Exton q-Bessel functions (3.60):

$$\lim_{N\to\infty} p_{N-n}(q^N x; a, b; q) = \lim_{N\to\infty} {}_2\phi_1(q^{-N+n}, abq^{N-n+1}; aq; q, q^{N+1}x)$$
$$= {}_1\phi_1(0; aq; q, q^{n+1}x),$$

cf. Koornwinder and Swarttouw [29, Prop. A.1].

4.5. Hahn's classification. Hahn [21] classified all families of orthogonal polynomials p_n ($n = 0, 1, \ldots, N$ or $n = 0, 1, \ldots$) which are eigenfunctions of a second order q-difference equation, i.e.,

$$A(x)(D_q^2 p_n)(q^{-1}x) + B(x)(D_q p_n)(q^{-1}x) = \lambda_n p_n(x) \qquad (4.50)$$

or equivalently

$$a(x) p_n(q^{-1}x) + b(x) p_n(x) + c(x) p_n(qx) = \lambda_n p_n(x),$$

where A, B and a, b, c are fixed polynomials. Necessarily, A is of degree ≤ 2 and B is of degree ≤ 1. The eigenvalues λ_n will be completely determined by A and B. One distinguishes cases depending on the degrees of A and B and the situation of the zeros of A. For each case one finds a family of q-hypergeometric polynomials satisfying (4.50) which, moreover, satisfies an explicit three term recurrence relation

$$x p_n(x) = p_{n+1}(x) + B_n p_n(x) + C_n p_{n-1}(x).$$

(Here we assumed the p_n to be monic.) Then we will have orthogonal polynomials with respect to a positive orthogonality measure iff $C_n > 0$ for all n. A next problem is to find the explicit orthogonality measure. For a given family of q-hypergeometric polynomials depending on parameters the type of this measure may vary with the parameters. Finally the limit transitions between the various families of orthogonal polynomials can be examined.

In essence this program has been worked out by Hahn [21], but his paper is somewhat sketchy in details. Unfortunately, there is no later publication, where the details have all been filled in. In Table 2 we give a *q-Hahn tableau*: a q-analogue of that part of the Askey tableau (Table 1) which is dominated by the Hahn polynomials. In the ${}_r\phi_s$ formulas in the Table we have omitted the last but one parameter denoting the base except when this is different from q. The arrows denote limit transitions. Below we will give a brief discussion of each case. We do not claim completeness.

Several new phenomena occur here:
1. Within one class of q-hypergeometric polynomials we may obtain, depending on the values of the parameters, either a q-analogue of the Jacobi-Laguerre-Hermite class or of the (discrete) Hahn-Krawtchouk-Meixner-Charlier class.
2. q-Analogues of Jacobi, Laguerre and Hermite polynomials occur in a "little" version (corresponding to Jacobi polynomials on $[0, c]$, Laguerre polynomials on $[0, \infty)$

and Hermite polynomials which are even or odd functions) and a "big" version (q-analogues of Jacobi, Laguerre and Hermite polynomials of arbitrarily shifted argument).

3. There is a duality under the transformation $q \mapsto q^{-1}$, denoted in Table 2 by dashed lines. We insist on the convention $0 < q < 1$, but we can rewrite q^{-1}-hypergeometric orthogonal polynomials as q-hypergeometric orthogonal polynomials and thus sometimes obtain another family.

4. Cases may occur where the orthogonality measure is not uniquely determined.

We now briefly discuss each case occuring in Table 2.

1) *Big q-Jacobi polynomials* $P_n(x; a, b, c, d; q)$, defined by (4.36), form the generic case in this classification. $A(x)$ and $B(x)$ in (4.50) are given by (4.34) and λ_n is as in the right hand side of (4.32). From the explicit expression for C_n in (4.44) we get the values of a, b, c, d for which there is a positive orthogonality measure. For $c, d > 0$ these are given by (4.23).

The *q-Hahn polynomials* can be obtained as special big q-Jacobi polynomials with $-qad/c = q^{-N}$, $n = 0, 1, \ldots, N$ ($N \in \mathbb{Z}_+$). For convenience we may take $c = qa$. The q-Hahn polynomials are usually (cf. [20, (7.2.21)]) notated as

$$Q_n(x; a, b. N; q) := {}_3\phi_2 \left[\begin{matrix} q^{-n}, abq^{n+1}, x \\ aq, q^{-N} \end{matrix} ; q, q \right] = \sum_{k=0}^{n} \frac{(q^{-n}; q)_k \, (abq^{n+1}; q)_k \, (x; q)_k}{(aq; q)_k \, (q^{-N}; q)_k \, (q; q)_k} q^k.$$

Here the convention regarding lower parameters q^{-N} ($N \in \mathbb{Z}_+$) is similar to the convention for (4.13).

q-Hahn polynomials satisfy orthogonality relations

$$\sum_{x=0}^{N} (Q_n Q_m)(q^{-x}; a, b, N; q) \frac{(aq; q)_x \, (bq; q)_{N-x}}{(q; q)_x \, (q; q)_{N-x}} (aq)^{-x} = 0, \quad n \neq m.$$

The weights are positive in one of the three following cases: (i) $b < q^{-1}$ and $0 < a < q^{-1}$; (ii) $a, b > q^{-N}$; (iii) $a < 0$ and $b > q^{-N}$. For $q \uparrow 1$ the polynomials tend to ordinary Hahn polynomials (4.12):

$$\lim_{q \uparrow 1} Q_n(q^{-x}; q^{\alpha}, q^{\beta}, N; q) = Q_n(x; \alpha, \beta, N).$$

2a) *Little q-Jacobi polynomials* $p_n(x; a, b; q)$, given by (4.45), (4.46), were discussed in §2.4. When we put $b := q^{-N-1}$ in (4.46), we obtain q-analogues of Krawtchouk polynomials (4.17): the *q-Krawtchouk polynomials*

$$K_n(x; b, N; q) := {}_3\phi_2 \left[\begin{matrix} q^{-n}, -b^{-1}q^n, x \\ 0, q^{-N} \end{matrix} ; q, q \right] = \lim_{a \to 0} Q_n(x; a, -(qba)^{-1}, N; q)$$

114

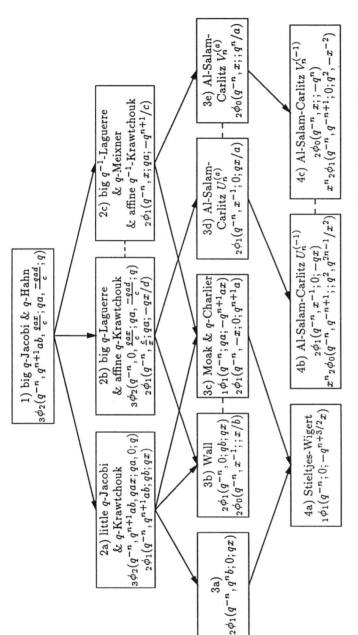

Table 2. The q-Hahn tableau.

with orthogonality relations

$$\sum_{x=0}^{N}(K_n K_m)(q^{-x}; b, N; q)\frac{(q^{-N};q)_x}{(q;q)_x}(-b)^x = 0, \quad n \neq m, \quad n, m = 0, 1, \ldots, N,$$

and limit transition

$$\lim_{q \uparrow 1} K_n(q^{-x}; b, N; q) = K_n(x; b/(b+1), N).$$

See [20, Exercise 7.8(i)] and the reference given there. Note that the q-Krawtchouk polynomials $K_n(q^{-x}; b, N; q)$ are not self-dual under the interchange of x and n.

2b) *Big q-Laguerre polynomials*, these are big q-Jacobi polynomials with $b = 0$:

$$P_n(x; a, 0, c, d; q) = {}_3\phi_2\left[\begin{matrix} q^{-n}, 0, qax/c \\ qa, -qad/c \end{matrix}; q, q\right] \tag{4.51}$$

$$= \frac{1}{(-q^{-n}ca^{-1}d^{-1}; q)_n}\, {}_2\phi_1\left[\begin{matrix} q^{-n}, c/x \\ qa \end{matrix}; q, -qx/d\right].$$

Here the second equality follows by (3.39). Note that

$$\lim_{q \uparrow 1} P_n(x; q^\alpha, 0, c, (1-q)^{-1}; q) = {}_1F_1(-n; \alpha+1; c-x) = \text{const. } L_n^\alpha(c-x).$$

Another family of q-analogues of Krawtchouk polynomials, called *affine q-Krawtchouk polynomials*, can be obtained from (4.51) by putting $-qad/c = q^{-N}$:

$$K_n^{Aff}(x; a, N; q) := {}_3\phi_2\left[\begin{matrix} q^{-n}, 0, x \\ aq, q^{-N} \end{matrix}; q, q\right] = Q_n(x; a, 0, N; q)$$

with orthogonality relations

$$\sum_{x=0}^{N}(K_n^{Aff} K_m^{Aff})(q^{-x}; a, N; q)\frac{(aq; q)_x\,(aq)^{-x}}{(q;q)_x\,(q;q)_{N-x}} = 0, \quad n \neq m, \quad n, m = 0, 1, \ldots, N,$$

and limit transition

$$\lim_{q \uparrow 1} K_n^{Aff}(q^{-x}; a, N; q) = K_n(x; 1-a, N).$$

See [20, Exercise 7.11] and the references given there.

2c) *q-Meixner polynomials*

$$M_n(x; a, c; q) := {}_2\phi_1 \left[\begin{matrix} q^{-n}, x \\ qa \end{matrix} ; q, \frac{-q^{n+1}}{c} \right]. \qquad (4.52)$$

For certain values of the parameters these can be considered as q-analogues of the Meixner polynomials (4.18), cf. [20, Exercise 7.12]. However, when we write these polynomials as

$$M_n(qax; a, \bar{a}^{-1}; q) = {}_2\phi_1 \left[\begin{matrix} q^{-n}, qax \\ qa \end{matrix} ; q, -q^{n+1}\bar{a} \right] = \lim_{d \to \infty} P_n(x; a, \bar{a}d, 1, d; q)$$

and a is complex but not real, then these polynomials also become orthogonal (not documented in the literature) and they can be considered as q-analogues of Laguerre polynomials of shifted argument. Moreover, these polynomials can be obtained from the big q-Laguerre polynomials (4.51) by the transformation $q \mapsto q^{-1}$. Therefore, we call these polynomials also *big q^{-1}-Laguerre polynomials*.

For $a := q^{-N-1}$ the polynomials (4.52) become yet another family of q-analogues of the Krawtchouk polynomials, which we will call *affine q^{-1}-Krawtchouk polynomials*, because they are obtained from affine q-Krawtchouk polynomials by changing q into q^{-1}. These polynomials, written as

$$M_n(x; q^{-N-1}, -b^{-1}; q) := {}_2\phi_1(q^{-n}, x; q^{-N}; q, bq^{n+1}) = \lim_{a \to \infty} Q_n(x; a, b, N; q),$$

where $n = 0, 1, \ldots, N$, have orthogonality relations

$$\sum_{x=0}^{N} (M_n M_m)(q^{-x}; q^{-N-1}, -b^{-1}; q) \frac{(bq; q)_{N-x} (-1)^{N-x} q^{x(x-1)/2}}{(q; q)_x (q; q)_{N-x}} = 0, \quad n \neq m,$$

and limit transition

$$\lim_{q \uparrow 1} M_n(q^{-x}; q^{-N-1}, -b^{-1}; q) = K_n(x; b^{-1}, N).$$

See Koornwinder [26].

3a) These orthogonal polynomials are not documented in the literature.

3b) *Wall polynomials* (cf. Chihara [14, §VI.11]) are special little q-Jacobi polynomials

$$p_n(x; a, 0; q) = {}_2\phi_1(q^{-n}, 0; qa; q, qx)$$
$$= \frac{1}{(q^n a^{-1}; q)_n} \, {}_2\phi_0(q^{-n}, x^{-1}; -; q, x/a).$$

The second equality is a limit case of (3.39). Wall polynomials are q-analogues of Laguerre polynomials on $[0, \infty)$, so they might be called *little q-Laguerre polynomials*.

3c) Moak's [34] *q-Laguerre polynomials* (notation as in [20, Exercise 7.43]) are given by

$$L_n^\alpha(x; q) := \frac{(q^{\alpha+1}; q)_n}{(q; q)_n} \, {}_1\phi_1(q^{-n}; q^{\alpha+1}; q, -xq^{n+\alpha+1})$$

$$= \frac{1}{(q; q)_n} \, {}_2\phi_1(q^{-n}, -x; 0, q, q^{n+\alpha+1}).$$

The second equality is a limit case of (3.38). They can be obtained from the Wall polynomials by replacing q by q^{-1}. Their orthogonality measure is not unique. For instance, there is a continuous orthogonality measure

$$\int_0^\infty L_m^\alpha(x; q) L_n^\alpha(x; q) \, \frac{x^\alpha \, dx}{(-(1-q)x; q)_\infty} = \delta_{m,n}, \quad m \neq n,$$

but also discrete orthogonality measures

$$\int_0^\infty L_m^\alpha(cx; q) L_n^\alpha(cx; q) \, \frac{x^\alpha \, d_q x}{(-c(1-q)x; q)_\infty} = \delta_{m,n}, \quad m \neq n, \quad c > 0.$$

Chihara [14]Ch. VI, §2 calls these polynomials generalized Stieltjes-Wigert polynomials and he uses another notation. For certain parameter values these polynomials may be considered as q-analogues of the Charlier polynomials, see [20, Exercise 7.13].

3d) *Al-Salam-Carlitz polynomials*

$$U_n^{(a)}(x) = U_n^{(a)}(x; q) := (-1)^n \, q^{n(n-1)/2} \, a^n \, {}_2\phi_1(q^{-n}, x^{-1}; 0; q, qx/a)$$

(cf. Al-Salam and Carlitz [2]) satisfy the three term recurrence relation

$$x \, U_n^{(a)}(x) = U_{n+1}^{(a)}(x) + (1+a) \, q^n \, U_n^{(a)}(x) - a \, q^{n-1} \, (1-q^n) U_{n-1}^{(a)}(x).$$

Thus they are orthogonal polynomials for $a < 0$. They can be expressed in terms of big q-Jacobi polynomials by

$$U_n^{(a)}(x) = \tilde{P}_n(x; 0, 0, 1, -a; q),$$

so they are orthogonal with respect to the measure $(qx, qx/a; q)_\infty d_q x$ on $[a, 1]$, cf. (4.22). By the above recurrence relation these polynomials are q-analogues of Hermite polynomials of shifted argument, so they may be considered as "big" q-Hermite polynomials.

3e) *Al-Salam-Carlitz polynomials*

$$V_n^{(a)}(x) = V_n^{(a)}(x; q) := U_n^{(a)}(x; q^{-1}) = (-1)^n \, q^{-n(n-1)/2} \, a^n \, {}_2\phi_0(q^{-n}, x; -; q^n a^{-1})$$

(cf. [2]). For $a > 0$ these form another family of q-analogues of the Charlier polynomials. On the other hand, the polynomials $x \mapsto V_n^{(\alpha/\tilde{\alpha})}(-q\alpha x)$ can be considered as q-analogues of Hermite polynomials of shifted argument.

118

4a) *Stieltjes-Wigert polynomials*

$$S_n(x;q) := (-1)^n \, q^{-n(2n+1)/2} \, {}_1\phi_1(q^{-n};0;q,-q^{n+3/2}x)$$

(cf. Chihara [14, §VI.2]) do not have a unique orthogonality measure. It was already noted by Stieltjes that the corresponding moments are an example of a nondeterminate (Stieltjes) moment problem. After suitable scaling these polynomials tend to Hermite polynomials as $q \uparrow 1$.

4b) *Al-Salam-Carlitz polynomials* $U_n^{(a)}$ with $a := -1$:

$$\begin{aligned}
U_n^{(-1)}(x) &= q^{n(n-1)/2} \, {}_2\phi_1(q^{-n},x^{-1};0;q,-qx) \\
&= \widetilde{P}_n(x;0,0,1,1;q) \\
&= x^n \, {}_2\phi_0(q^{-n},q^{-n+1};-;q^2,q^{2n-1}x^{-2})
\end{aligned}$$

(cf. [2]), q-analogues of Hermite polynomials. By the last equality there is a quadratic transformation between these polynomials and certain Wall polynomials. This is a limit case of the quadratic transformations (4.48) and (4.49).

4c) *Al-Salam-Carlitz polynomials* $V_n^{(a)}$ with $a := -1$:

$$\begin{aligned}
i^{-n} V_n^{(-1)}(ix) &= i^{-n} \, q^{-n(n-1)/2} \, {}_2\phi_0(q^{-n},ix;-;-q^n) \\
&= \lim_{c\to\infty} \widetilde{P}_n(x;ic,ic,qc,qc;q) \\
&= x^n \, {}_2\phi_1(q^{-n},q^{-n+1};0;q^2,-x^{-2})
\end{aligned}$$

(cf. [2]), also q-analogues of Hermite polynomials. By the last equality there is a quadratic transformation between these polynomials and certain of Moak's q-Laguerre polynomials. This is a limit case of (4.48) and (4.49).

4.6. The Askey-Wilson integral. We remarked earlier that, whenever some nontrivial evaluation of an integral can be given, the orthogonal polynomials having the integrand as weight function may be worthwhile to study. In this subsection we will give an evaluation of the integral which corresponds to the Askey-Wilson polynomials. After the original evaluation in Askey and Wilson [10] several easier approaches were given, cf. Gasper and Rahman [20, Ch. 6] and the references given there. Our proof below borrowed ideas from Kalnins and Miller [24] and Miller [32] but is still different from theirs. Fix $0 < q < 1$. Let

$$w_{a,b,c,d}(z) := \frac{(z^2,z^{-2};q)_\infty}{(az,az^{-1},bz,bz^{-1},cz,cz^{-1},dz,dz^{-1};q)_\infty}. \tag{4.53}$$

We want to evaluate the integral

$$I_{a,b,c,d} := \frac{1}{2\pi i} \oint_{|z|=1} w_{a,b,c,d}(z) \, \frac{dz}{z}, \qquad |a|,|b|,|c|,|d| < 1. \tag{4.54}$$

Note that $I_{a,b,c,d}$ is analytic in the four complex variables a,b,c,d when these are bounded in absolute value by 1. It is also symmetric in a,b,c,d.

Lemma 4.3.

$$I_{a,b,c,d} = \frac{1 - abcd}{(1 - ab)(1 - ac)(1 - ad)} I_{qa,b,c,d}. \tag{4.55}$$

Proof. The integral

$$\oint \frac{w_{q^{1/2}a,q^{1/2}b,q^{1/2}c,q^{1/2}d}(z)}{z - z^{-1}} \frac{dz}{z}$$

equals on the one hand

$$\oint \frac{w_{q^{1/2}a,q^{1/2}b,q^{1/2}c,q^{1/2}d}(q^{1/2}z)}{q^{1/2}z - q^{-1/2}z^{-1}} \frac{dz}{z}$$

$$= \oint \frac{(1 - az)(1 - bz)(1 - cz)(1 - dz)\,w_{a,b,c,d}(z)}{q^{1/2}\,z\,(1 - z^2)} \frac{dz}{z}$$

and on the other hand

$$\oint \frac{w_{q^{1/2}a,q^{1/2}b,q^{1/2}c,q^{1/2}d}(q^{-1/2}z)}{q^{-1/2}z - q^{1/2}z^{-1}} \frac{dz}{z}$$

$$= -\oint \frac{(1 - a/z)(1 - b/z)(1 - c/z)(1 - d/z)w_{a,b,c,d}(z)}{q^{1/2}\,z^{-1}\,(1 - z^{-2})} \frac{dz}{z}.$$

Subtraction yields

$$0 = q^{-1/2}a^{-1} \oint \{-(1 - abcd)(1 - az)(1 - az^{-1}) + (1 - ab)(1 - ac)(1 - ad)\}$$

$$\times w_{a,b,c,d}(z) \frac{dz}{z} = 2\pi i q^{-1/2}a^{-1}\{-(1 - abcd)I_{qa,b,c,d} + (1 - ab)(1 - ac)(1 - ad)I_{a,b,c,d}\}.\ \square$$

By iteration of (4.55) and use of the analyticy and symmetry of $I_{a,b,c,d}$ we obtain

$$I_{a,b,c,d} = \frac{(abcd; q)_\infty}{(ab, ac, ad, bc, bd, cd; q)_\infty} I_{0,0,0,0}. \tag{4.56}$$

We might evaluate $I_{0,0,0,0}$ by the Jacobi triple product identity (3.50), but it is easier to observe that $I_{1,q^{1/2},-1,-q^{1/2}} = 1$. (Note that this case can be continuously reached from the domain of definition of $I_{a,b,c,d}$ in (4.54).) Hence (4.56) yields that $I_{0,0,0,0} = 2/(q; q)_\infty$. Thus

$$I_{a,b,c,d} = \frac{2(abcd; q)_\infty}{(ab, ac, ad, bc, bd, cd, q; q)_\infty}. \tag{4.57}$$

By the symmetry of $w_{a,b,c,d}(z)$ under $z \mapsto z^{-1}$ we can rewrite (4.54), (4.57) as

$$\frac{1}{2\pi} \int_0^\pi w_{a,b,c,d}(e^{i\theta})\,d\theta = \frac{(abcd; q)_\infty}{(ab, ac, ad, bc, bd, cd, q; q)_\infty}, \quad |a|, |b|, |c|, |d| < 1. \tag{4.58}$$

Here w is still defined by (4.53). The integral (4.58) is known as the *Askey-Wilson integral*.

4.7. Askey-Wilson polynomials. We now look for an orthogonal system corresponding to the weight function in the Askey-Wilson integral. First observe that

$$\frac{1}{2\pi i} \oint \frac{(az, az^{-1}; q)_k}{(ac, ad; q)_k} \frac{(bz, bz^{-1}; q)_l}{(bc, bd; q)_l} w_{a,b,c,d}(z) \frac{dz}{z} = \frac{I_{q^k a, q^l b, c, d}}{(ac, ad; q)_k (bc, bd; q)_l}$$

$$= \frac{(ab; q)_{k+l}}{(abcd; q)_{k+l}} I_{a,b,c,d}.$$

Thus

$$\frac{1}{I_{a,b,c,d}} \frac{1}{2\pi i} \oint \left\{ \sum_{k=0}^{n} \frac{(q^{-n}, q^{n-1} abcd; q)_k q^k}{(ab, q; q)_k} \frac{(az, az^{-1}; q)_k}{(ac, ad; q)_k} \right\} \frac{(bz, bz^{-1}; q)_l}{(bc, bd; q)_l} w_{a,b,c,d}(z) \frac{dz}{z}$$

$$= \sum_{k=0}^{n} \frac{(q^{-n}, q^{n-1} abcd; q)_k q^k}{(ab, q; q)_k} \frac{(ab; q)_{k+l}}{(abcd; q)_{k+l}}$$

$$= \frac{(ab; q)_l}{(abcd; q)_l} {}_3\phi_2 \left[\begin{matrix} q^{-n}, q^{n-1} abcd, q^l ab \\ ab, q^l abcd \end{matrix}; q, q \right]$$

$$= \frac{(ab; q)_l}{(abcd; q)_l} \frac{(q^{-n+1} c^{-1} d^{-1}, q^{-l}; q)_n}{(ab, q^{-n-l+1}/(abcd); q)_n} = 0, \quad l = 0, 1, \ldots, n-1. \qquad (4.59)$$

Here we used the q-Saalschütz formula (3.61).

The above orthogonality suggests to define *Askey-Wilson polynomials* (Askey & Wilson [10])

$$\frac{p_n(\cos\theta; a, b, c, d \mid q)}{a^{-n} (ab, ac, ad; q)_n} = r_n(\cos\theta; a, b, c, d \mid q) \qquad (4.60)$$

$$:= {}_4\phi_3 \left[\begin{matrix} q^{-n}, q^{n-1} abcd, ae^{i\theta}, ae^{-i\theta} \\ ab, ac, ad \end{matrix}; q, q \right] \qquad (4.61)$$

$$= \sum_{k=0}^{n} \frac{(q^{-n}, q^{n-1} abcd; q)_k q^k}{(ab, q; q)_k} \frac{(ae^{i\theta}, ae^{-i\theta}; q)_k}{(ac, ad; q)_k}.$$

Since

$$(ae^{i\theta}, ae^{-i\theta}; q)_k = \prod_{j=0}^{k-1} (1 - 2aq^j \cos\theta + a^2 q^{2j}),$$

formula (4.61) defines a polynomial of degree n in $\cos\theta$. It follows from (4.59) that the functions $\theta \mapsto p_n(\cos\theta; a, b, c, d \mid q)$ are orthogonal with respect to the measure $w_{a,b,c,d}(e^{i\theta}) \, d\theta$ on $[0, \pi]$, so we are really dealing with orthogonal polynomials. Now

$$p_n(x; a, b, c, d \mid q) = k_n x^n + \text{terms of lower degree}, \quad \text{with } k_n := 2^n (q^{n-1} abcd; q)_n, \qquad (4.62)$$

so the coefficient of x^n is symmetric in a, b, c, d. Since the weight function is also symmetric in a, b, c, d, the Askey-Wilson polynomial will be itself symmetric in a, b, c, d.

Proposition 4.4. Let $|a|, |b|, |c|, |d| < 1$. Then

$$\frac{1}{2\pi} \int_0^\pi p_n(\cos\theta)\, p_m(\cos\theta)\, w(\cos\theta)\, d\theta = \delta_{m,n}\, h_n,$$

where

$$p_n(\cos\theta) = p_n(\cos\theta; a, b, c, d \mid q),$$

$$w(\cos\theta) = \frac{(e^{2i\theta}, e^{-2i\theta}; q)_\infty}{(ae^{i\theta}, ae^{-i\theta}, be^{i\theta}, be^{-i\theta}, ce^{i\theta}, ce^{-i\theta}, de^{i\theta}, de^{-i\theta}; q)_\infty},$$

$$\frac{h_n}{h_0} = \frac{(1 - q^{n-1}abcd)\,(q, ab, ac, ad, bc, bd, cd; q)_n}{(1 - q^{2n-1}abcd)\,(abcd; q)_n},$$

and

$$h_0 = \frac{(abcd; q)_\infty}{(q, ab, ac, ad, bc, bd, cd; q)_\infty}.$$

The orthogonality measure is positive if a, b, c, d are real, or if complex, appear in conjugate pairs,

Proof. Apply (4.59) to

$$\frac{1}{I_{a,b,c,d}} \frac{1}{2\pi i} \oint (p_n p_m)\left((z + z^{-1})/2; a, b, c, d \mid q\right) w_{a,b,c,d}(z) \frac{dz}{z},$$

where p_n is expanded according to (4.61) and p_m similarly, but with a and b interchanged. $\qquad\square$

Note that we can evaluate the Askey-Wilson polynomial $p_n((z + z^{-1})/2)$ for $z = a$ (and by symmetry also for $z = b, c, d$):

$$p_n((a + a^{-1})/2; a, b, c, d \mid q) = a^{-n}\,(ab, ac, ad; q)_n. \tag{4.63}$$

When we write the three term recurrence relation for the Askey-Wilson polynomial as

$$2x\, p_n(x) = A_n\, p_{n+1}(x) + B_n\, p_n(x) + C_n\, p_{n-1}(x), \tag{4.64}$$

then the coefficients A_n and C_n can be computed from

$$A_n = \frac{2k_n}{k_{n+1}}, \qquad C_n = \frac{2k_{n-1}}{k_n}\frac{h_n}{h_{n-1}},$$

where k_n and h_n are given by (4.62) and Proposition 4.4, respectively. Then B_n can next be computed from (4.63) by substituting $x := (a + a^{-1})/2$ in (4.64).

4.8. Various results. Here we collect without proof some further results about Askey-Wilson polynomials and their special cases and limit cases.

The *q-ultraspherical polynomials*

$$C_n(\cos\theta;\beta\mid q):=\frac{(\beta^2;q)_n}{\beta^{n/2}(q;q)_n}\;{}_4\phi_3\left[\begin{matrix}q^{-n},q^n\beta^2,\beta^{1/2}e^{i\theta},\beta^{1/2}e^{-i\theta}\\\beta q^{1/2},-\beta q^{1/2},-\beta\end{matrix};q,q\right]$$

$$=\text{const. } p_n(\cos\theta;\beta^{1/2},\beta^{1/2}q^{1/2},-\beta^{1/2},-\beta^{1/2}q^{1/2}\mid q)$$

$$=\frac{(\beta;q)_n}{(q;q)_n}\sum_{k=0}^{n}\frac{(q^{-n},\beta;q)_k}{(q^{1-n}\beta^{-1},q;q)_k}\,(q/\beta)^k\,e^{i(n-2k)\theta}$$

are special Askey-Wilson polynomials which were already known to Rogers (1894), however not as orthogonal polynomials. See Askey & Ismail [7].

By easy arguments using analytic continuation, contour deformation and taking of residues it can be seen that Askey-Wilson polynomials for more general values of the parameters than in Proposition 4.4 become orthogonal with respect to a measure which contains both a continuous and discrete part (Askey & Wilson [10])

Proposition 4.5. Assume a,b,c,d are real, or if complex, appear in conjugate pairs, and that the pairwise products of a,b,c,d are not ≥ 1, then

$$\frac{1}{2\pi}\int_0^\pi p_n(\cos\theta)\,p_m(\cos\theta)\,w(\cos\theta)\,d\theta + \sum_k p_n(x_k)\,p_m(x_k)\,w_k = \delta_{m,n}\,h_n\,,$$

where $p_n(\cos\theta)$, $w(\cos\theta)$ and h_n are as in Proposition 4.4, while the x_k are the points $(eq^k + e^{-1}q^{-k})/2$ with e any of the parameters a,b,c or d whose absolute value is larger than one, the sum is over the $k \in \mathbb{Z}_+$ with $|eq^k| > 1$ and w_k is $w_k(a;b,c,d)$ as defined by [10, (2.10)] when $x_k = (aq^k + a^{-1}q^{-k})/2$. (Be aware that $(1-aq^{2k})/(1-a)$ should be replaced by $(1-a^2q^{2k})/(1-a^2)$ in [10, (2.10)].)

Both the big and the little q-Jacobi polynomials can be obtained as limit cases of Askey-Wilson polynomials. In order to formulate these limits, let r_n be as in (4.60). Then

$$\lim_{\lambda\to 0} r_n\left(\frac{q^{1/2}x}{2\lambda(cd)^{1/2}};\lambda a(qd/c)^{1/2},\lambda^{-1}(qc/d)^{1/2},-\lambda^{-1}(qd/c)^{1/2},-\lambda b(qc/d)^{1/2}\mid q\right)$$

$$= P_n(x;a,b,c,d;q)$$

and

$$\lim_{\lambda\to 0} r_n\left(\frac{q^{1/2}x}{2\lambda^2};q^{1/2}\lambda^2 a,q^{1/2}\lambda^{-2},-q^{1/2},-q^{1/2}b\mid q\right) = \frac{(qb;q)_n}{(q^{-n}a^{-1};q)_n}\,p_n(x;b,a;q).$$

See Koornwinder [28, §6]. As λ becomes smaller in these two limits, the number of mass points in the orthogonality measure grows, while the support of the continuous part shrinks. In the limit we have only infinitely many mass points and no continuous mass left

An important special class of Askey-Wilson polynomials are the *Al-Salam-Chihara polynomials* $p_n(x; a, b, 0, 0 \mid q)$, cf. [8, Ch. 3]. Both these polynomials and the continuous q-ultraspherical polynomials have the *continuous q-Hermite polynomials* $p_n(x; 0, 0, 0, 0 \mid q)$ (cf. Askey & Ismail [7]) as a limit case.

The Askey-Wilson polynomials are eigenfunctions of a kind of q-difference operator. Write $P_n(e^{i\theta})$ for the expression in (4.61) and put

$$A(z) := \frac{(1 - az)(1 - bz)(1 - cz)(1 - dz)}{(1 - z^2)(1 - qz^2)}.$$

Then

$$A(z)P_n(qz) - (A(z) + A(z^{-1}))P_n(z) + A(z^{-1})P_n(q^{-1}z)$$
$$= -(1 - q^{-n})(1 - q^{n-1}abcd)P_n(z).$$

The operator on the left hand side can be factorized as a product of two shift operators. See Askey & Wilson [10, §5].

An analytic continuation of the Askey-Wilson polynomials gives q-Racah polynomials

$$R_n(q^{-x} + q^{x+1}\gamma\delta) := {}_4\phi_3 \left[\begin{matrix} q^{-n}q^{n+1}\alpha\beta, q^{-x}, q^{x+1}\gamma\delta \\ \alpha q, \beta\delta q, \gamma q \end{matrix} ; q, q \right],$$

where one of αq, $\beta\delta q$ or γq is q^{-N} for some $N \in \mathbb{Z}_+$ and $n = 0, 1, \ldots, N$ (see Askey & Wilson [9]). These satisfy an orthogonality of the form

$$\sum_{x=0}^{N} R_n(\mu(x)) R_m(\mu(x)) w(x) = \delta_{m,n} h_n, \quad m, n = 0, 1, \ldots, N,$$

where $\mu(x) := q^{-x} + q^{x+1}\gamma\delta$. They are eigenfunctions of a second order difference operator in x.

There is nice characterization theorem of Leonard [31] for the q-Racah polynomials. Let $N \in \mathbb{Z}_+$ or $N = \infty$. Let the polynomials p_n ($n \in \mathbb{Z}_+$, $n < N + 1$) be orthogonal with respect to weights on distinct points μ_k ($k \in \mathbb{Z}_+$, $k < N + 1$). Let the polynomials p_n^* be similarly orthogonal with respect to weights on distinct points μ_k^*. Suppose that the two systems are dual in the sense that

$$p_n(\mu_k) = p_k^*(\mu_n^*).$$

Then the p_n are q-Racah polynomials or one of their limit cases.

124

Exercises to §4.

4.1 Show that (at least formally) the generating function

$$e^{2xt-t^2} = \sum_{n=0}^{\infty} \frac{H_n(x)}{n!} t^n$$

for Hermite polynomials (cf. [19, 10.13(19)]) follows from the generating function

$$(1-t)^{-\alpha-1} e^{-tx/(1-t)} = \sum_{n=0}^{\infty} L_n^{\alpha}(x) t^n, \quad |t| < 1,$$

for Laguerre polynomials (cf. [19, 10.12(17)]) by the limit transition

$$H_n(x) = (-1)^n \, 2^{n/2} \, n! \lim_{\alpha \to \infty} \alpha^{-n/2} \, L_n^{\alpha}((2\alpha)^{1/2}x + \alpha)$$

given in (4.7).

4.2 Show that (at least formally) the above generating function for Laguerre polynomials follows from the generating function for Jacobi polynomials

$$\sum_{n=0}^{\infty} P_n^{(\alpha,\beta)}(x) t^n = 2^{\alpha+\beta} R^{-1} (1-t+R)^{-\alpha} (1+t+R)^{-\beta}, \quad |t| < 1,$$

where

$$R := (1 - 2xt + t^2)^{1/2},$$

(cf. [19, 10.8 (29)]) by the limit transition

$$L_n^{\alpha}(x) = \lim_{\beta \to \infty} P_n^{(\alpha,\beta)}(1 - 2x/\beta)$$

given in (4.5).

4.3 Show that (at least formally) the above generating function for Hermite polynomials follows from the specialization $\alpha = \beta$ of the above generating function for Jacobi polynomials by the limit transition

$$H_n(x) = 2^n \, n! \lim_{\alpha \to \infty} \alpha^{-n/2} \, P_n^{(\alpha,\alpha)}(\alpha^{-1/2}x)$$

given in (4.6).

4.4 Prove that the Charlier polynomials

$$C_n(x; a) := {}_2F_0(-n, -x; -; -a^{-1}), \quad a > 0,$$

satisfy the recurrence relation

$$x\, C_n(x; a) = -a\, C_{n+1}(x; a) + (n + a)\, C_n(x; a) - n\, C_{n-1}(x; a).$$

4.5 Prove that

$$\lim_{a \to \infty} (-(2a)^{1/2})^n\, C_n((2a)^{1/2} x + a) = H_n(x)$$

by using the above recurrence relation for Charlier polynomials and the recurrence relation

$$H_{n+1}(x) - 2x\, H_n(x) + 2n\, H_{n-1}(x) = 0$$

for Hermite polynomials.

4.6 Prove the following generating function for Al-Salam-Chihara polynomials (notation as for Askey-Wilson polynomials in (4.60), (4.61)):

$$\frac{(zc, zd; q)_\infty}{(ze^{i\theta}, ze^{-i\theta}; q)_\infty} = \sum_{m=0}^{\infty} z^m \frac{p_m(\cos\theta; 0, 0, c, d \mid q)}{(q; q)_m}, \quad |z| < 1.$$

4.7 Use the above generating function in order to derive the following transformation formula from little q-Jacobi polynomials (notation and definition by (4.45)) to Askey-Wilson polynomials by means of a summation kernel involving Al-Salam-Chihara polynomials:

$$\frac{a^n\, p_n(\cos\theta; a, b, c, d \mid q)}{(ab, ac, ad; q)_n} \frac{(ac, ad; q)_\infty}{(ae^{i\theta}, ae^{-i\theta}; q)_\infty}$$

$$= \sum_{m=0}^{\infty} p_n(q^m; q^{-1}ab, q^{-1}cd; q)\, a^m \frac{p_m(\cos\theta; 0, 0, c, d \mid q)}{(q; q)_m}.$$

REFERENCES

[1] E. Abe, *Hopf algebras*, Cambridge University Press, 1980.

[2] W. A. Al-Salam and L. Carlitz, *Some orthogonal q-polynomials*, Math. Nachr. 30 (1965), 47–61.

[3] G. E. Andrews, *q-Series: their development and application in analysis, number theory, combinatorics, physics, and computer algebra*, Regional Conference Series in Math. 66, Amer. Math. Soc., 1986.

[4] G. E. Andrews, R. Askey, *Enumeration of partitions: The role of Eulerian series and q-orthogonal polynomials*, in *Higher combinatorics*, M. Aigner (ed.), Reidel, 1977, pp. 3–26.

[5] G. E. Andrews, R. Askey, *Classical orthogonal polynomials*, in *Polynômes orthogonaux et applications*, C. Brezinski, A. Draux, A. P. Magnus, P. Maroni, A. Ronveaux (eds.), Lecture Notes in Math. 1171, Springer, 1985, pp. 36–62.

[6] W. Arveson, *An invitation to C^*-algebra*, Springer, 1976.

[7] R. Askey, M. E. H. Ismail, *A generalization of ultraspherical polynomials*, in *Studies in Pure Mathematics*, P. Erdös (ed.), Birkhäuser, 1983, 55–78.

[8] R. Askey, M. E. H. Ismail, *Recurrence relations, continued fractions and orthogonal polynomials*, Mem. Amer. Math. Soc. 49 (1984), no. 300.

[9] R. Askey, J. Wilson, *A set of orthogonal polynomials that generalize the Racah coefficients or 6-j symbols*, SIAM J. Math. Anal. 10 (1979), 1008–1016.

[10] R. Askey, J. Wilson, *Some basic hypergeometric orthogonal polynomials that generalize Jacobi polynomials*, Mem. Amer. Math. Soc. 54 (1985), no. 319.

[11] N. M. Atakishiyev, M. Rahman, S. K. Suslov, *On the classical orthogonal polynomials*, preprint, 1993, to appear in Constr. Approx.

[12] N. M. Atakishiyev, S. K. Suslov, *The Hahn and Meixner polynomials of an imaginary argument and some of their applications*, J. Phys. A 18 (1985), 1583–1596.

[13] W. N. Bailey, *Generalized hypergeometric series*, Cambridge University Press, 1935; reprinted by Hafner Publishing Company, 1972.

[14] T. S. Chihara, *An introduction to orthogonal polynomials*, Gordon and Breach, 1978.

[15] M. S. Dijkhuizen, *On compact quantum groups and quantum homogeneous spaces*, Dissertation, University of Amsterdam, 1994.

[16] M. S. Dijkhuizen, T. H. Koornwinder, *CQG algebras: a direct algebraic approach to compact quantum groups*, Report AM-R9401, CWI, Amsterdam, 1994. Submitted to Lett. Math. Phys.

[17] E. G. Effros, Z.-J. Ruan, *Discrete quantum groups, I. The Haar measure*, preprint, 1993.

[18] A. Erdélyi, W. Magnus, F. Oberhettinger, F. G. Tricomi, *Higher transcendental functions*, Vol. 1, McGraw-Hill, 1953.

[19] A. Erdélyi, W. Magnus, F. Oberhettinger, F. G. Tricomi, *Higher transcendental functions*, Vol. 2, McGraw-Hill, 1953.

[20] G. Gasper, M. Rahman, *Basic hypergeometric series*, Encyclopedia of Mathematics and its Applications 35, Cambridge University Press, 1990.

[21] W. Hahn, *Über Orthogonalpolynome, die q-Differenzengleichungen genügen*, Math. Nachr. 2 (1949), 4–34, 379.

[22] M. Hazewinkel, *Formal groups and applications*, Academic Press, 1978.

[23] M. E. H. Ismail, *A simple proof of Ramanujan's $_1\psi_1$ sum*, Proc. Amer. Math. Soc. 63 (1977), 185–186.

[24] E. G. Kalnins, W. Miller, Jr., *Symmetry techniques for q-series: Askey-Wilson polynomials*, Rocky Mountain J. Math. 19 (1989), 223–230.

[25] T. H. Koornwinder, *Group theoretic interpretations of Askey's scheme of hypergeometric orthogonal polynomials*, in *Orthogonal polynomials and their applications*, M. Alfaro, J. S. Dehesa, F. J. Marcellan, J. L. Rubio de Francia, J. Vinuesa (eds.), Lecture Notes in Math. 1329, Springer, 1988, pp. 46–72.

[26] T. H. Koornwinder, *Representations of the twisted SU(2) quantum group and some q-hypergeometric orthogonal polynomials*, Nederl. Akad. Wetensch. Proc. Ser. A 92 (1989), 97–117.

[27] T. H. Koornwinder, *Jacobi functions as limit cases of q-ultraspherical polynomials*, J. Math. Anal. Appl. 148 (1990), 44–54.

[28] T. H. Koornwinder, *Askey-Wilson polynomials as zonal spherical functions on the SU(2) quantum group*, SIAM J. Math. Anal. 24 (1993), 795–813.

[29] T. H. Koornwinder, R. F. Swarttouw, *On q-Analogues of the Fourier and Hankel transforms*, Trans. Amer. Math. Soc. 333 (1992), 445–461.

[30] J. Labelle, *Askey's scheme of hypergeometric orthogonal polynomials*, poster, Université de Quebec à Montréal, 1990.

[31] D. A. Leonard, *Orthogonal polynomials, duality and association schemes*, SIAM J. Math. Anal. 13 (1982), 656–663.

[32] W. Miller, Jr., *Symmetry techniques and orthogonality for q-Series*, in *q-Series and partitions*, D. Stanton (ed.), IMA Volumes in Math. and its Appl. 18, Springer, 1989, pp. 191–212.

[33] K. Mimachi, *Connection problem in holonomic q-difference system associated with a Jackson integral of Jordan-Pochhammer type*, Nagoya Math. J. 116 (1989), 149–161.

[34] D. S. Moak, *The q-analogue of the Laguerre polynomials*, J. Math. Anal. Appl. 81 (1981), 20–47.

[35] F. W. J. Olver, *Asymptotics and special functions*, Academic Press, 1974.

[36] M. Rahman, A. Verma, *Quadratic transformation formulas for basic hypergeometric series*, Trans. Amer. Math. Soc. 335 (1993), 277–302.

[37] M. E. Sweedler, *Hopf algebras*, Benjamin, 1969.

[38] E. C. Titchmarsh, *The theory of functions*, Oxford University Press, second ed., 1939.

[39] S. Z. Wang, *General constructions of compact quantum groups*, preprint, 1993.

[40] J. A. Wilson, *Some hypergeometric orthogonal polynomials*, SIAM J. Math. Anal. 11 (1980), 690–701.

[41] S. L. Woronowicz, *Compact matrix pseudogroups*, Comm. Math. Phys. 111 (1987), 613–685.

[42] S. L. Woronowicz, *Tannaka-Krein duality for compact matrix pseudogroups. Twisted SU(N) groups*, Invent. Math. 93 (1988), 35–76.

[43] S. L. Woronowicz, *A remark on compact matrix quantum groups*, Lett. Math. Phys. 21 (1991), 35–39.

[44] S. L. Woronowicz, *Compact quantum groups*, preprint, version August 1, 1993.

University of Amsterdam, Faculty of Mathematics and Computer Science, Plantage Muidergracht 24, 1018 TV Amsterdam. The Netherlands

e–mail: thk@fwi.uva.nl

REPRESENTATIONS OF CLASSICAL p-ADIC GROUPS

Marko Tadić

Contents

Preface.

This paper is based on the notes of the course "Representations of classical p-adic groups" given on the European School of Group Theory in C.I.R.M. Luminy-Marseille, in 1991. To the original notes of the course we have added, at some places in this revised version, more detailed explanations. Also, we have corrected a number of misprints which we noted in the original notes. Most of them were in the ninth section. At this point we want to thank to M. Duflo, J. Faraut and J.-L. Waldspurger for an excellent organization of the school, and to the participants of the school for the interest that they have shown.

There exist excellent papers which can be used for an introduction to the representation theory of reductive p-adic groups. Particularly interesting for this paper are the paper [Cs1] of W. Casselman and the series of papers [BeZe1], [BeZe2], [Ze1] of J. Bernstein and A. V. Zelevinsky. In this paper our interest is to introduce only to some of the ideas of the representation theory of reductive p-adic groups. To keep the size of this paper moderate, we have tried to avoid technical details as much as possible. The size of this paper is only a fraction of the size of the papers that we wanted to cover. In general, detailed proofs are omitted in this paper. At the most places we have tried either to explain the ideas of the proofs, or to give a proof in same simple case where one should be able to catch the idea of the general proof. Unfortunately, because of the size of the paper, and of the attempt to keep the paper on an introductory level, we were forced to omit a lot of topics which would be treated naturally with the subject of the representation theory of reductive p-adic groups. Let us mention just a few of them: the structure theory of reductive groups, Tits' systems, Bruhat-Tits' structure theory of reductive p-adic groups, the representation theory of general locally compact groups, the theory of automorphic forms and the Langlands' program, e.t.c. Never the less, these topics are implicitly present in the paper. This omission is a reason for including of a wider list of the references where these other topics are considered. Some of the references are included also because of the historical reasons.

We shall describe now the content of this paper in more details. In the first section we introduce the classical groups whose representation theory we shall consider. We have also described the fields over which the classical groups will be considered. The second section introduces the notion of the parabolic induction and discusses the place of this notion in the representation theory of reductive groups over local fields.

The theory of admissible representations of reductive p-adic groups is essentially the language of the theory of general locally compact groups adapted to the reductive p-adic groups. We introduce this theory in the third section. The Jacquet modules, which are crucial in the study of the parabolically induced representations, are studied in the fourth section.

The fifth section is devoted to the computation of the composition series of the Jacquet modules of the parabolically induced representations of $SL(2)$. This illustrates Casselman's, and also Bernstein's and Zelevinsky's calculation in the general case ([Cs1], [BeZe2]). We gave an example of a representation which can not be reached by the parabolic induction in the sixth section, namely an example of a cuspidal representation. We have also shown how one can use the calculation of the fifth section to get a preliminary classification of irreducible representations of $GL(2)$ over a finite field. This is a simple introduction to the analysis of the parabolically induced representations of $SL(2)$ and $GL(2)$ which has been done in the seventh section. This analysis implies a preliminary classification of the irreducible representations of these groups. In this section we have also explained the classification of the irreducible unitary representations of these groups. In the eighth section we wrote down the general consequences which may be obtained from the computation of composition series of the Jacquet modules of the parabolically induced representations of general reductive p-adic groups. We have proved that results for the case of $SL(2)$ in the previous section.

The ninth section considers $GL(n)$. We follow here Bernstein and Zelevinsky. They have used the Hopf algebra structure to get a control of the Jacquet modules of the parabolically induced representations of $GL(n)$. The algebra structure is defined using the parabolic induction, while the definition of the coalgebra structure uses the Jacquet modules. We tried to explain here how one can use this Hopf algebra structure in the representation theory of p-adic $GL(n)$. At the end of this section we describe the Langlands' classification for $GL(n)$. The unitary dual is also described there ([Td6]). This section is a good introduction to the tenth section where the symplectic groups are studied. Representations of these groups are studied as modules over the representations of the general linear groups, via the parabolic induction. The Jacquet modules define a comodule structure. The connection of these two structures is explained here. These two structures are used here to construct some new square integrable representations. The last section explains how one can use these structures in the study of the irreducibility of the parabolically induced representations of the symplectic groups.

The first nine sections can be used as an introduction to the already mentioned introductory papers of Casselman and of Bernstein and Zelevinsky. A complete proofs of the facts discussed in these sections could be found mainly in the papers [Cs], [BeZe1], [BeZe2] and [Ze1]. Most of these facts this author has learned either from that papers, or from several people, primarily from D. Miličić. For the unitary dual of $GL(n)$ one should consult [Td6]. The last two sections, together with the ninth section, introduces to the ideas of the papers [Td13]-[Td15], [SaTd] and forthcoming papers on the representations of classical groups. We hope that these ideas will play a role in a number of unsolved problems of the representation theory of classical p-adic groups. Most of the problems of the classifications for the classical p-adic groups other than $GL(n)$ are still unsolved.

1. CLASSICAL GROUPS.

Let G be a locally compact group. Suppose that V is a complex vector space. Let π be a homomorphism of G into the group of all invertible operators on V. Then (π, V), or simply π, is called a *representation* of G. Suppose that (π, H) is a representation of G where H is a Hilbert space. Suppose that the mapping

$$(g, v) \mapsto \pi(g)v$$

from $G \times H$ to H is continuous. Then (π, H) is called a *continuous representation* of G. A continuous representation (π, H) is called *irreducible* if there does not exist non-trivial closed subspaces of H which are invariant for all $\pi(g)$ when $g \in G$. Sometimes such representations are called also *topologically irreducible*. A continuous representation on a one-dimensional space is called a *character* of G. Clearly, a character is always an irreducible representation. A *unitary representation* of G is a continuous representation of G such that all operators $\pi(g), g \in G$, are unitary operators. Let (π_1, H_1) and (π_2, H_2) be unitary representations of G. They are called *unitarily equivalent* if there exists a Hilbert space isomorphism $\varphi : H_1 \to H_2$ such that $\pi_2(g)\varphi = \varphi\pi_1(g)$ for all $g \in G$. Denote by $\hat{G}$ the set of all unitarily equivalence classes of irreducible unitary representations of G on non-trivial Hilbert spaces. Then $\hat{G}$ is called the *dual space* of G, or the *unitary dual* of G.

The first step in building of harmonic analysis on G is classification of irreducible unitary representations of G. The next step is description of interesting and important unitary representations of G in terms of irreducible unitary representations, i.e. in terms of $\hat{G}$.

We shall be interested in the problem of classification of irreducible unitary representations of G mostly when G is a classical simple group over a locally compact non-discrete field F. Very soon we shall restrict ourselves to the case when F is not isomorphic to $\mathbb{R}$ or $\mathbb{C}$.

A locally compact non-discrete field will be called a *local field*. First, we are going to describe local fields. Each connected local field is isomorphic either to $\mathbb{R}$ or to $\mathbb{C}$. These fields are called *archimedean*. A local field which is not isomorphic to $\mathbb{R}$ or $\mathbb{C}$ will be called *non-archimedean*.

Let p be a prime rational integer. Write $q \in \mathbb{Q}^\times$ as $q = p^\alpha \frac{u}{v}$, where $u, v, \alpha \in \mathbb{Z}$ and where p does not divide uv. Set

$$|q|_p = p^{-\alpha},$$

and $|0|_p = 0$. Denote by $\mathbb{Q}_p$ the completion of $\mathbb{Q}$ with respect to $|\ |_p$. Then $\mathbb{Q}_p$ is called the *field of p-adic numbers*. Let E be a finite field extension of $\mathbb{Q}_p$. Then E has a natural topology of vector space over $\mathbb{Q}_p$. With this topology, E becomes a local field. The topology on E can be introduced with the following absolute value

$$|x|_E = |N_{E/\mathbb{Q}_p}(x)|_p,$$

where $N_{E/\mathbb{Q}_p}$ denotes the norm map of the extension $E \supset \mathbb{Q}_p$. Note that this absolute value does not extend in general the absolute value $|\ |_p$ from $\mathbb{Q}_p$. The absolute value which extends $|\ |_p$ is

$$|x|'_E = |N_E/\mathbb{Q}_p(x)|^{\frac{1}{[E:\mathbb{Q}_p]}}.$$

For our purposes it is more convenient to deal with $|\ |_E$ instead of $|\ |'_E$. We shall see soon what is the reason. These absolute values are *ultrametric*, i.e. the following inequality holds

$$|x + y|_E \leqslant \max\{|x|_E, |y|_E\}, \quad x, y \in E.$$

This inequality is stronger then the usual triangle inequality. Obviously, for $t \geqslant 0$, the ball

$$\{x \in E; |x|_p \leqslant t\}$$

is an additive subgroup. If $t \leqslant 1$, then it is moreover closed for multiplication. Therefore, E is a totally disconnected topological space. Each non-archimedean local field of characteristic 0 is isomorphic to some finite extension E of some $\mathbb{Q}_p$.

We could come to these fields also in the following more arithmetic way. Let K be a number field (i.e., a finite extension of $\mathbb{Q}$). Let $\mathcal{O}_K$ be the ring of integers of K

(i.e. the set of elements of K integral over $\mathbb{Z} \subseteq K$). Let $\mathfrak{p}$ be a prime ideal in $\mathcal{O}_K$. Consider the topology on K having

$$\mathfrak{p}^i, \quad i = 1, 2, 3, \ldots$$

for a basis of neighborhoods of 0. Then the completion of K with respect to corresponding uniform structure is a local non-archimedean field.

Let $\mathbb{F} = \mathbb{F}_q$ be a finite field with q elements. Let $\mathbb{F}((X))$ be the field of formal power series over $\mathbb{F}$. For $f \in \mathbb{F}((X))^{\times}$ set

$$|f|_{\mathbb{F}((X))} = q^{-n},$$

where

$$f = \sum_{k=n}^{\infty} a_k X^k$$

and $a_n \neq 0$. One sees directly that $|\ |_{\mathbb{F}((X))}$ is an ultrametric absolute value. Now $\mathbb{F}((X))$ with respect to the above absolute value is a local non-archimedean field. Each non-archimedean field of positive characteristic is isomorphic to some $\mathbb{F}((X))$.

For the above classification of local fields one can consult the first chapter of the fundamental Weil's book [We2].

On a local field F there exists a positive measure

$$f \mapsto \int_F f(x)dx \tag{1.1}$$

which is invariant under translations by elements of F. Such measure is unique up to a multiplication with a positive constant. If $a \in F$ and f is a compactly supported continuous function on F, then

$$\int f(ax)dx = |a|_F^{-1} \int_F f(x)dx.$$

In the case of $F = \mathbb{R}$ one takes for $|\ |_{\mathbb{R}}$ the usual absolute value. If $F = \mathbb{C}$, then one takes $|x + iy|_{\mathbb{C}} = x^2 + y^2, x, y \in \mathbb{R}$.

We shall now describe the groups.

Let V be a finite dimensional vector space over a local field F. The group of all regular linear operators on V is denoted by $GL(V)$. This group is called the *general linear group* of V. The *special linear group* of V consists of operators in $GL(V)$ which have determinant equal to one. It is denoted by $SL(V)$. If V possesses a non-degenerate symplectic form, then the subgroup of $GL(V)$ of operators which preserves this form is called the *symplectic group* of V. It is denoted by $Sp(V)$. Suppose that V is supplied with a non-degenerate orthogonal form which possesses isotropic subspaces of maximal possible dimension. Then the subgroup of $SL(V)$ of all operators which preserve the orthogonal form is denoted by $SO(V)$. It is called the *orthogonal group* of V.

We have the following matrix realizations of these groups. First, $GL(n, F)$ denotes the group of all $n \times n$ regular matrices with entries in F. Further, $SL(n, F)$ denotes the subgroup of those matrices which have determinant equal to one. Consider the following $n \times n$ matrix

$$J_n = \begin{bmatrix} 00 & \cdots & 01 \\ 00 & \cdots & 10 \\ \vdots & & \\ 10 & \cdots & 0 \end{bmatrix}.$$

Let

$$Sp(n, F) = \left\{ S \in GL(2n, F); \ {}^t S \begin{bmatrix} 0 & J_n \\ -J_n & 0 \end{bmatrix} S = \begin{bmatrix} 0 & J_n \\ -J_n & 0 \end{bmatrix} \right\}.$$

Here ${}^t S$ denotes the transposed matrix of S. We shall denote by ${}^\tau S$ the transposed matrix of S with respect to the second diagonal. Denote by I_n the identity matrix in $GL(n, F)$. Let

$$SO(n, F) = \{ S \in SL(n, F); \ {}^\tau S S = I_n \}.$$

We shall always work with matrix forms of classical groups.

These groups are topological groups in the natural way. If F is a non-archimedean field, then these groups are totally disconnected. One can write a basis of neighborhoods of identity consisting of open (and then also closed) compact subgroups.

We shall say now a few words about the structure of these groups. Let G be either $GL(n, F)$ or $Sp(n, F)$ or $SO(n, F)$. We shall now introduce subgroups of G which play a very important role in the representation theory of G. Roughly, these subgroups enable a reduction of some of the problems of the representation theory of G to a groups of similar type of lower dimension (and complexity).

The subgroup of all upper triangular matrices in G will be denoted by $P_{\min}$. It will be called the *standard minimal parabolic subgroup*. Note that $P_{\min}$ is a solvable group. The standard minimal parabolic subgroup is maximal with respect to this property. A subgroup of G containing $P_{\min}$ is called a *standard parabolic subgroup*.

One can describe standard parabolic subgroups of $GL(n, F)$ in the following way. Let $\alpha = (n_1, ..., n_k)$ be an ordered partition of n into positive integers. Look at elements of $GL(n, F)$ as block matrix with blocks of sizes $n_i \times n_j$. Let P_α (resp. M_α), be the upper block-triangular matrices (resp. block-diagonal matrices) in $GL(n, F)$. Let N_α be the matrices in P_α which have identity matrices on the block-diagonal. Then

$$\alpha \mapsto P_\alpha$$

is an one-to-one mapping of the set of all partitions of n onto the set of all standard parabolic subgroups of $GL(n, F)$. One has also $P_\alpha = M_\alpha N_\alpha$. More precisely,

$$P_\alpha = M_\alpha \ltimes N_\alpha.$$

This will be called the *standard Levi decomposition* of P_α. The group M_α is called the *standard Levi factor* of P_α. The group N_α is called the *unipotent radical* of P_α. Note that M_α is a direct product of some general linear groups.

Let

$$\alpha = (n_1, ..., n_k)$$

be a partition of some $0 \leqslant m \leqslant n$. In the case of the group $Sp(n, F)$ we denote

$$\alpha' = (n_1, ..., n_k, 2n - 2m, n_k, ..., n_1).$$

In the case of the group $SO(2n + 1, F)$ we denote

$$\alpha' = (n_1, ..., n_k, 2n + 1 - 2m, n_k, ..., n_1).$$

Let G denotes either the group $Sp(n, F)$ or $SO(2n + 1, F)$. Set

$$P_\alpha^S = P_{\alpha'} \cap G,$$

$$M_\alpha^S = M_{\alpha'} \cap G,$$

$$N_\alpha^S = N_{\alpha'} \cap G.$$

Then $\alpha \mapsto P_\alpha^S$ is a parametrization of all standard parabolic subgroups of G. One has also

$$P_\alpha^S = M_\alpha^S \ltimes N_\alpha^S.$$

This is called the *standard Levi decomposition* of P_α^S, M_α^S is called the *standard Levi factor* of P_α^S, N_α^S is called the *unipotent radical* of P_α^S. If $G = Sp(n, F)$ (resp. $G = SO(2n + 1, F)$), then M_α^S is a direct product of a group of Sp-type (resp. $SO(2k + 1, F)$-type) and of general linear groups.

For $SO(2n, F)$ there is a slightly different situation in parametrization of the standard parabolic subgroups. We shall omit this case here.

Let G be either $GL(n, F)$, or $Sp(n, F)$, or $SO(2n + 1, F)$. *Parabolic subgroups* of G are conjugates of the standard parabolic subgroups of G. *Levi decompositions* are conjugates of the corresponding standard Levi decompositions of the standard parabolic subgroups.

A Levi factor of a parabolic subgroup of the group G is either a direct product of general linear groups if G is a general linear group, or in the other case, a direct product of a classical group of the same type with general linear groups. We define *parabolic subgroups* (resp. *standard parabolic subgroups*) of a Levi factor (resp. standard Levi factor) to be direct products of parabolic subgroups (resp. standard parabolic subgroups). Analogously, we define *Levi decompositions* (*standard Levi decompositions*) of such parabolic subgroups (resp. standard parabolic subgroups), etc.

For a general locally compact group G there exists a positive measure on G which is invariant under right translations. Such a measure will be called a *right Haar measure* on G. Two right Haar measures on G are proportional. The integral of a continuous compactly supported function f with respect to a fixed right Haar measure will be denoted by

$$\int_{G.} f(g)dg.$$

There exists a continuous positive-valued character Δ_G of G such that

$$\int_G f(xg)dg = \Delta_G(x)^{-1} \int_G f(g)dg$$

for each continuous compactly supported function f on G and each $x \in G$. The character Δ_G is called the *modular character* of G. If $\Delta_G \equiv 1_G$, then one says that G is a *unimodular group* (by 1_X we shall denote in this paper the constant function on X which is equal to 1 everywhere).

For a proof of these facts one may consult [Bb3]. Let me mention that the definition of the modular character in [Bb3] is diferent from the one that we use here. The modular character in [Bb3] corresponds to Δ_G^{-1} in our notation. The proof of existence of Haar measures is very simple for totally disconnected groups (see [BeZe1]).

1.1. Examples.
(i) The (right) Haar measure on $(F, +)$ is the invariant measure (1.1).
(ii) We have the following Haar measure on F^m

$$\int_{F^m} f(x)dx = \int_F \cdots \int_F f(x_1, \ldots, x_m)dx_1, \ldots dx_m.$$

(iii) The Haar measure on $GL(n, F)$ is

$$\int_{GL(n,F)} f(g)dg = \int_{F^{(n^2)}} f(x)dx/|\det(g)|_F^n.$$

As one can see easily from (iii), the group $GL(n, F)$ is unimodular. Moreover, all classical groups are unimodular. In general, proper parabolic subgroups are not unimodular. Never the less, Levi factors and unipotent radicals are unimodular groups.

1.2. Remark.
A good framework when one works with representations of classical groups are *reductive groups*. The classical groups are the most important examples of reductive groups. We have already used some of the general notions from the theory of reductive groups (a linear algebraic group is reductive if it does not contain a normal unipotent subgroup of positive dimension). Usually in this paper we shall not give general definitions, but we shall rather present objects explicitly.

136

The theory that we present here is directed to the representations of classical groups over local fields (after the following section we shall assume that the local field is non-archimedean). For the technical reasons it is useful to develop the theory for products of classical groups of type $Sp(n)$ or $SO(n)$ with general linear groups. This is useful in order to include in the theory the Levi factors of parabolic subgroups. In the sequel, we shall denote by G one of such groups. Most of the general results that we present in these notes apply to general reductive groups. For a theory of reductive groups one may consult [BlTi1] and [BlTi2].

2. PARABOLIC INDUCTION.

If one wants to classify $\hat{G}$, one should have some method of construction of irreducible unitary representations of G, or at least for the first step, of irreducible continuous representations of G.

Let us consider for a moment a finite group G and a subgroup $P \leqslant G$. One of the simplest functors which one can consider on representations of G is the restriction functor to the subgroup P. An interesting question is usually does some functor have an adjoint functor. The right adjoint functor to the restriction functor is the induction functor from P to G. Clearly, the induction functor assigns to representations of the smaller group P, representations of the whole group G.

The notion of induction can be generalized to the case of locally compact groups. For reductive groups over local fields a particular type of induction is pretty simple, but a very powerful tool of producing of irreducible continuous representations. This tool is parabolic induction. We shall now define this notion.

Let G be one of the classical groups over local fields introduced in the first section, or one of the Levi factors of parabolic subgroups of such classical groups. Let P be a (standard) parabolic subgroup with (the standard) Levi decomposition $P = MN$. Then there exists a maximal compact subgroup K_o of G such that

$$ G = P_{\min} K_o. $$

This is called the *Iwasawa decomposition* of G.

2.1. Examples.

(i) If $G = GL(n, F)$ and if F is a non-archimedean field, then one may take $K_o = GL(n, \mathcal{O}_F)$ where $\mathcal{O}_F = \{x \in F; |x|_F \leqslant 1\}$ is the *ring of integers of F*. If $F = \mathbb{R}$ (resp. $\mathbb{C}$) one may take for K_o the group of orthogonal (resp. unitary) matrices in $GL(n, F)$.

(ii) Let F be a local non-archimedean field. Since $GL(n, \mathcal{O}_F)$ is an open (and then closed) subgroup of $GL(n, F)$, the restriction of the Haar measure of $GL(n, F)$ (see Examples 1.1., (i)) to $GL(n, \mathcal{O}_F)$, is a Haar measure on $GL(n, \mathcal{O}_F)$. Note that this Haar measure on $GL(n, \mathcal{O}_F)$ is just a restriction of the standard invariant measure on $F^{(n^2)}$ to $GL(n, \mathcal{O}_F)$.

Let (σ, H) be a continuous representation of M. Denote by $\mathrm{Ind}_P^G(\sigma)$ the Hilbert space of all (classes of) measurable functions

$$f : G \longrightarrow H$$

which satisfy

(i) $f(mng) = \Delta_P(m)^{1/2}\sigma(m)f(g)$, for any $m \in M, n \in N$ and $g \in G$;
(ii) $\int_{K_o} \|f(k)\|^2 dk < \infty$.
For $g \in G$ and $f \in \mathrm{Ind}_P^G(\sigma)$ denote by $R_g f$ the function

$$(R_g f)(x) = f(xg). \tag{2.1}$$

We shall say that G acts on $\mathrm{Ind}_P^G(\sigma)$ by the *right translations*. Then $R_g f \in \mathrm{Ind}_P^G(\sigma)$ and $(R, \mathrm{Ind}_P^G(\sigma))$ is a continuous representation of G. If σ was a unitary representation, then $\mathrm{Ind}_P^G(\sigma)$ is also a unitary representation. The representation $\mathrm{Ind}_P^G(\sigma)$ is called a *parabolically induced representation* of G by σ from P.

The integration over K_o which was a condition in the definition of the parabolically induced representations may appear a little bit mysterious for somebody. For an explanation of this condition one may consult Remarks 2.2, (i). There is also an explanation of the factor $\Delta_P^{1/2}$ which appears in the definition of the parabolically induced representations.

Let $(\pi, H) \in \hat{G}$. Then π, as a representation of K_o, decomposes into a direct sum of irreducible representation

$$\pi|K_o \cong \bigoplus_{\delta \in \hat{K}_o} n_\delta \delta,$$

where $n_\delta \in \mathbb{Z}_+ \cup \{\infty\}$. A fundamental property of the representation theory of G is that $n_\delta \in \mathbb{Z}_+$, i.e. K_o-multiplicities are finite. This fact has a very important technical consequences, as well as qualitative consequences for the harmonic analysis on G (G is a type I group, see [Dx]). Roughly, this finiteness condition is a consequence of the fact that certain convolution algebras of functions on G are not too far from being commutative. This result was proved by Harish-Chandra for the archimedean fields (see 4.5. in [Wr]), and by J. Bernstein in [Be1] for the non-archimedean fields. In the following two sections there is an outline of the proof of that Bernstein's result. Because of this Harish-Chandra's and Bernstein's result, we shall assume at the rest of this section that the irreducible continuous representations of G that we consider, have all K_o-multiplicities finite.

Suppose for a moment that F is a non-archimedean field. Since each $GL(n, \mathbb{C})$ has a neighborhood of identity which does not contain a non-trivial subgroup, each irreducible unitary representation of K_o factors through a representation of K_o/K where K is an open normal subgroup of K_o. It is easy to see from this that the above fact about K_o-multiplicities is equivalent to the following fact: for any $(\pi, H) \in \hat{G}$ and for any open compact subgroup K of G the space of K-invariants

$$H^K = \{v \in H; \pi(k)v = v \text{ for any } k \in K\}$$

is finite dimensional.

The problem of the classification of the irreducible unitary representations of G appeared to be much harder than it was expected. An easier problem appeared to be the problem of the classification of the irreducible continuous representations of G, which have finite K_o-multiplicities. To get the unitary dual $\hat{G}$ one needs then to extract from irreducible representations those representations which are actually unitary.

From the point of view of the parabolic induction, there are two classes of irreducible continuous representations of G. The first class is formed of the representations which are equivalent to the irreducible subrepresentations of the parabolically induced representations from the proper parabolic subgroups by the irreducible continuous representations. Certainly, it is very important to make precise what means equivalent. Since we shall pass very soon to an algebraic treatment of the theory, we shall omit the precise definition here (see Remark 2.2 (ii)). The second class consists of remaining irreducible representations. Let us call for a moment the representations from the second class primitive.

One approach to the irreducible continuous representations of G may be first to classify primitive representations of Levi factors of parabolic subgroups, and then to classify representations in the first class. In this way the representations in the first class reduce to the primitive representations of smaller reductive groups, modulo an understanding of parabolic induction. Our principal aim in this lectures will be to describe some methods of the study of the parabolic induction in the non-archimedean case. One further step in the strategy that we have described here is introduction of the Langlands classification (see the fourth section).

Suppose that F is an archimedean field. Then the Casselman's subrepresentation theorem (Theorem 8.21. in [CsMi]), which generalizes the famous Haris-Chandra's subquotient theorem, tells that each irreducible continuous representation is equivalent to a subrepresentation of $\mathrm{Ind}_{P_{\min}}^{G}(\sigma)$, where σ is an irreducible continuous representation of (the standard) Levi factor of $P_{\min}$. In the case of the classical groups, σ is a character.

Let us take a look at one of the simplest non-trivial cases, the case of $SL(2, \mathbb{R})$. Here one may take

$$K_o = \left\{ \begin{bmatrix} \cos(\varphi) & -\sin(\varphi) \\ \sin(\varphi) & \cos(\varphi) \end{bmatrix} ; \varphi \in \mathbb{R}/2\pi\mathbb{Z} \right\}.$$

Note that K_o is commutative. Because of that, each irreducible unitary representation of K_o is one dimensional. Now $\hat{K}_o = \{\delta_n; n \in \mathbb{Z}\}$, where

$$\delta_n : \begin{bmatrix} \cos(\varphi) & -\sin(\varphi) \\ \sin(\varphi) & \cos(\varphi) \end{bmatrix} \mapsto e^{in\varphi}.$$

Take

$$P_{\min} = \left\{ \begin{bmatrix} a & b \\ 0 & a^{-1} \end{bmatrix} ; a \in \mathbb{R}^{\times}, b \in \mathbb{R} \right\},$$

$$M_{\min} = \left\{ \begin{bmatrix} a & 0 \\ 0 & a^{-1} \end{bmatrix} ; a \in \mathbb{R}^\times \right\}.$$

Let χ be a character of $M_{\min}$. Restriction to K_o gives an isomorphism between $\operatorname{Ind}_{P_{\min}}^{SL(2,\mathbb{R})}(\sigma)$ and

$$\bigoplus_{n \in 2\mathbb{Z}} \delta_n$$

$$(\text{resp.} \bigoplus_{n \in (2\mathbb{Z}+1)} \delta_n)$$

if

$$\sigma\left(\begin{bmatrix} -1 & 0 \\ 0 & -1 \end{bmatrix} \right) = 1$$

$$(\text{resp.} \ \sigma\left(\begin{bmatrix} -1 & 0 \\ 0 & -1 \end{bmatrix} \right) = -1)$$

as representations of K_o. Suppose that H is a (closed) irreducible subrepresentation of $\operatorname{Ind}_{P_{\min}}^{SL(2,\mathbb{R})}(\sigma)$. Then it is completely determined by $X \subseteq 2\mathbb{Z}$ (resp. $X \subseteq (2\mathbb{Z}+1)$). Denote by $\operatorname{Ind}_{P_{\min}}^{SL(2,\mathbb{R})}(\sigma)$ the algebraic span of irreducible K_o-subrepresentations of $\operatorname{Ind}_{P_{\min}}^{SL(2,\mathbb{R})}(\sigma)$. Then

$$\pi(X)f = \frac{d}{dt}(\pi(\exp tX)f)|_{t=0}$$

defines an action of the Lie algebra

$$\mathfrak{sl}(2,\mathbb{R}) = \left\{ \begin{bmatrix} a & b \\ c & d \end{bmatrix} ; a,b,c,d \in \mathbb{R}, a+d = 0 \right\}$$

of the Lie group $SL(2,\mathbb{R})$ on $\operatorname{Ind}_{P_{\min}}^{SL(2,\mathbb{R})}(\sigma)$. One can find basis of $\mathfrak{sl}(2,\mathbb{R})$ in such a way that the formulas for the action of the elements of the basis are particularly simple (see §5. of ch. VI of [Lang]). Certainly, $H \cap \operatorname{Ind}_{P_{\min}}^{SL(2,\mathbb{R})}(\sigma)$ is invariant for the action of the Lie algebra and it is dense in H. In this way, the problem of the classification of the irreducible continuous representations reduces to a combinatorial problem which is now not very hard to solve. For general reductive groups over archimedean fields, such type of approach leads to the theory of $(\mathfrak{g}, K)$-modules (see [Vo2]).

In the case of the non-archimedean fields, such type of approach to parabolically induced representations fails. First of all, we do not know what is $\hat{K}_o$ here. Then, we do not have the action of the Lie algebra. In the non-archimedean case the most powerful tool in the study of the parabolically induced representations are Jacquet modules. Before we introduce them, we shall introduce an algebraic version of the representation theory of reductive groups over non-archimedean fields which is very convenient in the study of the questions related to the problems of the classifications of the unitary duals.

2.2. Remarks.

(i) In the definition of the parabolically induced representations in the theory of admissible representations, which will be given in the following section, we shall not need integration over K_o. Nevertheless, this type of integration will appear at some other places related to the parabolic induction. Therefore, we shall explain the background of this integration. The parabolic induction extends the notion of the unitary induction. In general, the unitary induction assigns to a unitary representation of a subgroup a unitary representation of the whole group. The first problem that one faces with the unitary induction is that there does not need to exist a non-trivial measure on $P\backslash G$ which is invariant for the right translations by elements of G. But there is another useful form which we shall describe now. Let $C_c(G)$ be the vector space of all compactly supported continuous functions on G. Denote by X the space of all continuous functions on G which satisfy $f(pg) = \Delta_P(p)f(g)$ for any $p \in P$ and $g \in G$. The mapping

$$q : C_c(G) \to X$$

defined by

$$(qf)(g) = \int_P f(pg)\Delta_P^{-1}(p)dp$$

is surjective (here the measure that we consider on P is a right Haar measure). If $qf_1 = qf_2$, then

$$\int_G f_1(g)dg = \int_G f_2(g)dg.$$

Therefore, we can define integration of elements of X by

$$\int_{P\backslash G} (qf)(x)dx = \int_G f(g)dg. \tag{2.2}$$

In this way we get a positive form on X which is obviously invariant for right translations since q commutes with right translations and the measure on G is invariant for right translations. A positive linear form on X invariant for right translations is unique up to a positive multiple. Sometimes, it is called the Haar measure on $P\backslash G$.

It is possible to build the Haar measure on G from the Haar measures on K_o and P by the formula

$$\int_G f(g)dg = \int_{K_o} \left(\int_P f(pk)\Delta_P(p)^{-1}dp \right) dk$$

i.e.

$$\int_G f(g)dg = \int_{K_o} (qf)(k)dk. \tag{2.3}$$

Because of the definition of the Haar measure on $P\backslash G$, from (2.2) and (2.3) we have

$$\int_{P\backslash G} \varphi(x)dx = \int_{K_o} \varphi(k)dk$$

for $\varphi \in X$. Thus, integration over K_o is simply the Haar measure on $P\backslash G$. For more details and proofs concerning these facts one can consult [Bb3].

Let σ be a unitary representation of M. Take a continuous functions $f_1, f_2 \in \text{Ind}_P^G(\sigma)$. Then the function

$$g \mapsto (f_1(g), f_2(g))$$

belongs to X. Therefore,

$$(f_1, f_2) \mapsto \int_{K_o} (f_1(k), f_2(g))dk$$

defines a G-invariant inner product on $\text{Ind}_P^G(\sigma)$ and because of that, $\text{Ind}_P^G(\sigma)$ is unitary.

(ii) The equivalence that we have mentioned when we were talking about the classification of continuous irreducible representations, is the Naimark equivalence. For a precise definition of this notion one should consult [Wr]. Roughly, two irreducible continuous representations are Naimark equivalent if there is densely defined closed intertwining between them. The important fact is that two irreducible unitary representations which are Naimark equivalent are unitarily equivalent.

3. ADMISSIBLE REPRESENTATIONS.

Suppose that G is a totally disconnected locally compact group. Then G has a basis of neighborhoods of identity consisting of open compact subgroups.

Let (π, V) be a representation of G. If U is a subspace of V invariant for all $\pi(g), g \in G$, then U is called a *subrepresentation* of V. If V does not contain subrepresentations different from $\{0\}$ and V, then one says that V is an *irreducible representations*.

For two representations (π_1, V_1) and (π_2, V_2) a linear map $\varphi : V_1 \to V_2$ is called *G-intertwining* or *morphism of G-modules* if

$$\varphi \pi_1(g) = \pi_2(g)\varphi$$

for all $g \in G$. Representations π_1 and π_2 are called *isomorphic* or *equivalent* if there exists a G-intertwining φ which is a one-to-one mapping onto.

We may talk of *representations of finite length* in a usual way (at least, representations of G are just modules over suitably defined group algebras).

Let (π, V) be a representation of G. A vector $v \in V$ is called *smooth* if there exists an open subgroup K of G such that $\pi(k)v = v$ for all $k \in K$. The vector subspace of all smooth vectors in V will be denoted by V^∞. Then V^∞ is invariant for the action of G. The representation of G on V^∞ is denoted by π^∞. Then (π^∞, V^∞) is called the *smooth part* of (π, V). A representation (π, V) is called a *smooth representation* if $V = V^\infty$. If K is an open compact subgroup of G, then we denote K-*invariants* by

$$V^K = \{v \in V; \pi(k)v = v \text{ for any } k \in K\}.$$

It is easy to see that for a compact subgroup K of G

$$(\pi, V) \mapsto V^K$$

is an exact functor from the category of all smooth representations and G-intertwinings into the category of complex vector spaces.

A smooth representation (π, V) of G is called *admissible* if

$$\dim_{\mathbb{C}} V^K < \infty$$

for any open compact subgroup K of G.

An admissible representation (π, V) of G is called *unitarizable* if there exists an inner product $(\,,\,)$ on V such that

$$(\pi(g)v, \pi(g)w) = (v, w)$$

for all $v, w \in V$ and $g \in G$. Each unitarizable representation is completely reducible. We shall see soon that for an irreducible admissible representation (π, V) of G the space of a G-invariant Hermitian forms on V is at most one dimensional.

We shall assume in further that G is one of the groups introduced in the first section and that it is defined over a non-archimedean local field.

Denote by $\tilde{G}$ the set of all equivalence classes of non-zero irreducible admissible representations of G. The set $\tilde{G}$ is called the *non-unitary dual* or the *admissible dual* of G. From the Bernstein's result [Be1] it follows that the mapping

$$(\pi, H) \mapsto (\pi^\infty, H^\infty)$$

is a one-to-one mapping of $\hat{G}$ into $\tilde{G}$ (see Remarks 4.2 (iv)). Moreover, it maps $\hat{G}$ onto the set of all unitarizable classes in $\tilde{G}$. Because of that, we shall in further identify the unitary dual $\hat{G}$ with the subset of all unitarizable classes in $\tilde{G}$. It means that we assume in further

$$\hat{G} \subseteq \tilde{G}.$$

In this way the problem of the classification of the unitary dual of G splits into two parts. The first part is the problem of the classification of the non-unitary dual $\tilde{G}$. The second part is the *unitarizability problem*: determination of the subset $\hat{G}$ of $\tilde{G}$.

We shall say now a few words about representations of certain algebras which are very useful in the study of the smooth representations. Denote by $C_c^\infty(G)$ the space of all compactly supported locally constant functions on G. It is an associative algebra for the convolution which is defined by

$$(f_1 * f_2)(x) = \int_G f_1(xg^{-1})f_2(g)dg.$$

This algebra does not have identity. For an open compact subgroup K of G denote by $C_c(G//K)$ the vector space of all compactly supported functions f which satisfy

$$f(k_1 g k_2) = f(g)$$

for all $k_1, k_2 \in K$ and $g \in G$. Then $C_c(G//K)$ is a subalgebra of $C_c^\infty(G)$. Denote by Ξ_K the characteristic function of the set K, divided by the (Haar) measure of K. Then $\Xi_K \in C_c(G//K)$ and it is the identity of the algebra $C_c(G//K)$. Moreover

$$C_c(G//K) = \Xi_K * C_c^\infty(G) * \Xi_K. \tag{3.1}$$

Note that each $f \in C_c^\infty(G)$ is in some $C_c(G//K)$.

Algebra $C_c(G//K)$ is called the *Hecke algebra* of G with respect to K. There are also a more general Hecke algebras (see [HoMo]).

Let (π, V) be a smooth representation of G. Take $f \in C_c^\infty(G)$ and $v \in V$. Since the function

$$g \mapsto f(g)\pi(g)v$$

is compactly supported locally constant, we can find open compact subsets $K_1, \cdots, K_n$ such that the above function is constant on each K_i, and that it vanishes outside the union of all K_i's. Set

$$\pi(f)v = \sum_{i=1}^{n} f(g_i) \left(\int_{K_i} dg \right) \pi(g_i)v$$

where g_i is some element of K_i. Then $\pi(f)$ is a linear operator on V and we write

$$\pi(f) = \int_G f(g)\pi(g)dg.$$

The operator $\pi(f)$ is characterized by the condition that for any $v \in V$ and any linear form v^* on V we have

$$v^*(\pi(f)v) = \int_K f(g)v^*(\pi(g)v)dg.$$

In this way one gets a representation of the algebra $C_c^\infty(G)$ on V.

It is easy to get that a subspace $W \subseteq V$ is a G-subrepresentation if and only if it is a $C_c^\infty(G)$-submodule. Also, a linear map φ between two representations of

144

G is a G-intertwining if and only if it is a homomorphism of a $C_c^\infty(G)$-modules. The $C_c^\infty(G)$-modules which are coming from the smooth representations of G are characterized among all $C_c^\infty(G)$-modules by the condition

$$\mathrm{span}_{\mathbb{C}} \, \{fv; f \in C_c^\infty(G), v \in V\} = V. \tag{3.2}$$

The $C_c^\infty(G)$-modules satisfying the above condition are called *non-degenerate* $C_c^\infty(G)$-*modules*. Suppose that K is an open compact subgroup of G. For a smooth representation (π, V) of G we have

$$V^K = \pi(\Xi_K)V.$$

Moreover,
$$V^K = \pi(C_c(G//K))V.$$

Thus, V^K is a $C_c(G//K)$-module. Since $\pi(\Xi_K)$ is an idempotent, we have

$$V = \mathrm{Im}\,\pi(\Xi_K) \oplus \mathrm{Ker}\,\pi(\Xi_K).$$

Also
$$\mathrm{Ker}\,\pi(\Xi_K) = \mathrm{span}_{\mathbb{C}} \, \{\pi(k)v - v; k \in K, v \in V\}.$$

We shall prove now that if π is irreducible and $V^K \neq \{0\}$, then V^K is an irreducible $C_c(G//K)$-module. Let $\{0\} \neq W \subseteq V^K$ be a $C_c(G//K)$-submodule. Take $v \in V^K$. Let $w \in W$, $w \neq 0$. Since V is irreducible $\mathbb{C}_c^\infty(G)$-module, there exists $f \in C_c(G)$ such that $v = \pi(f)w$. Now

$$v = \pi(\Xi_K)v = \pi(\Xi_K)\pi(f)w = \pi(\Xi_K)\pi(f)\pi(\Xi_K)w = \pi(\Xi_K * f * \Xi_K)w.$$

Since $\Xi_K * f * \Xi_K \in C_c(G//K)$ and W is a $C_c(G//K)$-submodule, we have $v \in W$. Thus $W = V^K$, what proves the irreducibility.

Suppose that (π, V) is an irreducible admissible representation of G. Let φ be in the commutator of the representation π Then $\varphi(V^K) \subseteq V^K$. Since V^K is irreducible $C_c(G//K)$-module, φ is a scalar operator on V^K. From this one gets that φ is a scalar operator on the whole V.

Algebras $C_c(G//K)$ play a very important role in the representation theory of G. It can be easily shown that each irreducible $C_c(G//K)$-module is isomorphic to a $C_c(G//K)$-module V^K for some irreducible smooth representation (π, V) of G (see the Proposition 2.10. of [BeZe1]).

Suppose that $f \in C_c^\infty(G//K)$ acts trivially in every irreducible $C_c(G//K)$-module. Then also $f^* * f$ acts trivially, where $f^*(g) = \overline{f(g^{-1})}$. The theory of non-commutative rings implies that $f * f^*$ is nilpotent. Since $\varphi \in C_c^\infty(G)$ and $\varphi \neq 0$ implies $(\varphi * \varphi^*)(1) \neq 0$, we have that $f = 0$. Thus, every $f \in C_c(G//K)$, $f \neq 0$, acts non-trivially in same irreducible $C_c(G//K)$-module.

If (π, V_1) and (π_2, V_2) are two irreducible smooth representations such that their $C_c(G//K)$-modules are non-trivial and isomorphic, then $\pi_1 \cong \pi_2$.

For an admissible representation (π, V) of G, $\pi(f)$ is an operator of finite rank for $f \in C_c^\infty(G)$. Thus

$$f \mapsto \text{Trace } \pi(f)$$

defines a linear form on $C_c^\infty(G)$. This linear form is called the *character of the representation* π. Characters of representations in $\tilde{G}$ are linearly independent. Therefore, if two admissible representations of finite lengths have equal characters, then they have the same Jordan-Hölder series.

The problem of computation of the characters of $\tilde{G}$, in particular of $\hat{G}$, is a very important problem. In general, not too much is known about this problem. We shall not discuss more about this problem in these notes.

Let (π, V) be a smooth representation of G. By $(\overline{\pi}, \overline{V})$ we denote the *complex conjugate representation* of the representation (π, V). It is the representation by the same operators on the same space, except that the vector space structure is conjugate to the previous one. The new multiplication with scalars is given by

$$z \cdot v = \overline{z}v.$$

Clearly, π is irreducible (resp. admissible) if and only if $\overline{\pi}$ is irreducible (resp. admissible).

Denote by V^* the dual space of V. Define a representation π^* on V^* by

$$[\pi^*(g)(v^*)](v) = v^*(\pi(g^{-1})v).$$

Let $(\tilde{\pi}, \tilde{V})$ be the smooth part of (π^*, V^*). Then $(\tilde{\pi}, \tilde{V})$ is called the *contragredient representation* of (π, V). Now

$$(\pi, V) \mapsto (\tilde{\pi}, \tilde{V})$$

becomes a contravariant functor in a natural way on the category of all smooth representations of G and G-intertwinings.

Let K be an open compact subgroup of G. One can see directly that the linear map

$$f \mapsto f \circ \pi(\Xi_K), \tag{3.3}$$

goes from $(V^K)^*$ to $(V^*)^K = (\tilde{V})^K$ and that it is an isomorphism of vector spaces. Since the natural linear map

$$\varphi : V \mapsto \tilde{\tilde{V}}$$

defined by $[\varphi(v)](\tilde{v}) = \tilde{v}(v), \tilde{v} \in \tilde{V}$, is G-intertwining, we have

$$(\pi, V) \cong (\tilde{\tilde{\pi}}, \tilde{\tilde{V}})$$

if (π, V) is an admissible representation of G.

The functor $\pi \mapsto \tilde{\pi}$ is an exact functor on the category of all smooth representations of G.

In general, for a smooth representation (π, V) of G the form

$$(\tilde{v}, v) \mapsto \tilde{v}(v)$$

on $\tilde{V} \times V$ is a non-degenerate bilinear G-invariant form.

Suppose that $W \neq \{0\}$ is a proper subrepresentation of a smooth representation (π, V). Then

$$W^{\perp} = \left\{ \tilde{v} \in \tilde{V}; \tilde{v}(w) = 0 \ \text{for any} \ w \in W \right\}$$

is a proper non-zero subrepresentation of $\tilde{\pi}$. Therefore, for an admissible representation π we have that π is irreducible if and only if $\tilde{\pi}$ is irreducible. Moreover, π is an admissible representation of length n if and only if $\tilde{\pi}$ is an admissible representation of length n.

Suppose that (π, V) is an irreducible unitarizable admissible representation of G. Then

$$v \mapsto \overline{(\cdot, v)}$$

is a non-trivial G-intertwining from V to $\overline{\tilde{V}}$. Thus

$$\pi \cong \overline{\tilde{\pi}}.$$

An irreducible admissible representation (π, V) is called *Hermitian* if $\overline{\tilde{\pi}} \cong \pi$. Thus, every $\pi \in \hat{G}$ is Hermitian.

For each irreducible admissible representation (π, V) of G with a non-trivial G-invariant Hermitian form Ψ, the mapping

$$v \mapsto \overline{\Psi(\cdot, v)}$$

is an isomorphism of V onto $\overline{\tilde{V}}$. Since the commutator of an irreducible representation consists of scalars only, we see that each two G-invariant Hermitian forms on V are proportional. This explains also why a G-invariant inner product on $(\pi, V) \in \hat{G}$ is unique up to a positive multiple, if it exists.

Let $P = MN$ be a parabolic subgroup of G. Take a smooth representation (σ, U) of M. Let $\underline{\mathrm{Ind}}_P^G(\sigma)$ be the space of all functions

$$f \colon G \to U$$

such that

$$f(nmg) = \Delta_P(m)^{1/2} \sigma(m) f(g)$$

147

for all $m \in M$, $n \in N$, $g \in G$. Define the action R of G on $\underline{\mathrm{Ind}}_P^G(\sigma)$ by the right translations

$$(R_g f)(x) = f(xg),$$

where $x, g \in G$. The smooth part of this representation is denoted by

$$(R, \mathrm{Ind}_P^G(\sigma)).$$

The representation $\mathrm{Ind}_P^G(\sigma)$ is called a *parabolically induced representation* of G from P by σ. It is easy to see that for a continuous representation (σ, H) of M we have

$$\left(\mathrm{Ind}_P^G(\sigma)\right)^\infty \cong \mathrm{Ind}_P^G(\sigma^\infty).$$

From this we see that there is a natural relationship between the two parabolic inductions that we have introduced. Namely, the parabolic induction that we have just introduced is just an algebraic version of the parabolic induction that we have introduced in the second section.

Suppose that K is an open compact subgroup of G. Then $f \in (\mathrm{Ind}_P^G(\sigma))^K$ is completely determined by values on any set of representatives for $P \backslash G / K$. A consequence of the Iwasawa decomposition is that $P \backslash G$ is compact. Since K is open, $P \backslash G / K$ is a finite set. If σ is admissible, then the values of $f \in (\mathrm{Ind}_P^G(\sigma))^K$ are contained in a certain finite dimensional subspaces of invariants in U. This implies that $\mathrm{Ind}_P^G(\sigma)$ is admissible if σ is admissible. Moreover, if σ is an admissible representation of finite length, then $\mathrm{Ind}_P^G(\sigma)$ has finite length (for additional comments about this fact see the section eight).

Suppose that

$$\varphi : U_1 \to U_2$$

is an M-intertwining between smooth M-representations σ_1 and σ_2. Define

$$\mathrm{Ind}_P^G(\varphi) : \mathrm{Ind}_P^G(\sigma_1) \to \mathrm{Ind}_P^G(\sigma_2),$$

by the formula

$$f \mapsto \varphi \circ f.$$

It is easy to see that $\mathrm{Ind}_P^G(\varphi)$ is G-intertwining. In this way Ind_P^G becomes a functor from the category of all smooth representations of M to the category of all smooth representations of G. Considering a description of K-invariants in induced representations, it is easy to see that Ind_P^G is an exact functor. Further, if $\varphi \neq 0$ then $\mathrm{Ind}_P^G(\varphi) \neq 0$. Moreover, if $\mathrm{Ind}_P^G(\varphi)$ is a one-to-one mapping (resp. mapping onto), then φ is also one-to-one mapping (resp. mapping onto).

Let (σ, U) be a smooth representation of M. Let $f \in \mathrm{Ind}_P^G(\sigma)$ and $\tilde{f} \in \mathrm{Ind}_P^G(\tilde{\sigma})$. One directly checks that the function

$$g \mapsto [\tilde{f}(g)](f(g))$$

belongs to the space X introduced in Remarks 2.2, (i). Therefore the formula

$$\Psi(\tilde{f}, f) = \int_{K_o} [\tilde{f}(k)](f(k)) dk$$

defines a G-invariant bilinear form on $\mathrm{Ind}_P^G(\tilde{\sigma}) \times \mathrm{Ind}_P^G(\sigma)$. A direct consequence is that the mapping

$$\tilde{f} \mapsto \Psi(\tilde{f}, \cdot)$$

defines a G-intertwining

$$\mathrm{Ind}_P^G(\tilde{\sigma}) \to (\mathrm{Ind}_P^G(\sigma))\tilde{}.$$

It is not hard to show that the above intertwining is an isomorphism. Thus

$$(\mathrm{Ind}_P^G(\sigma))\tilde{} \cong \mathrm{Ind}_P^G(\tilde{\sigma}).$$

In the same way one gets that $\mathrm{Ind}_P^G(\sigma)$ is unitarizable if σ is unitarizable (see Remark 2.2 (i)). These two facts are the reason why $\Delta_P^{1/2}$ appears in the definition of induced representations. Clearly

$$(\mathrm{Ind}_P^G(\sigma))^- \cong \mathrm{Ind}_P^G(\overline{\sigma}).$$

Suppose that one has a parabolic subgroup $P = MN$ in G and a smooth representation (σ, U) of M. Let $g \in G$. Let σ' be an admissible representation of gMg^{-1} given by

$$\sigma'(gmg^{-1}) = \sigma(m), \quad m \in M.$$

Then it is easy to see that representations $\mathrm{Ind}_P^G(\sigma)$ and $\mathrm{Ind}_{gPg^{-1}}^G(\sigma')$ are equivalent. We shall say that pairs (P, σ) and (gPg^{-1}, σ') are *conjugate*.

Let P be a standard parabolic subgroup of G with the standard Levi decomposition $P = MN$. Suppose that P' is a standard Levi subgroup of M with the standard Levi decomposition $P' = M'N'$. Let σ be an admissible representation of M'. There exists a standard parabolic subgroup P'' in G whose standard Levi factor is M'. Then it is not hard to prove that

$$\mathrm{Ind}_{P''}^G(\sigma) \cong \mathrm{Ind}_P^G(\mathrm{Ind}_{P'}^M(\sigma)).$$

We may say that the parabolic induction does not depend on the stages of induction. This property will be illustrated on examples in the ninth and tenth sections.

Suppose that $P_1 = M_1 N_1$ and $P_2 = M_2 N_2$ are parabolic subgroups in G. Suppose that $M_1 = M_2$ and that σ is an admissible representation of M_1 of finite length. Then representations $\mathrm{Ind}_{P_1}^G(\sigma)$ and $\mathrm{Ind}_{P_2}^G(\sigma)$ have the same Jordan-Hölder sequences ([BeDeKz]). This fact is much harder to prove than the previous one. It follows from the equality of the characters of two induced representations. We shall prove this fact for $SL(2, F)$ in the seventh section. If additionally (P', σ') (resp. P') is conjugated to (P_2, σ) (resp. P_2), then we say that (P_1, σ) and (P', σ') (resp. P_1 and P') are *associate*. Therefore, the parabolic induction from associate pairs gives the same Jordan-Hölder series.

3.1. Remarks.

(i) An interesting question may be what are the irreducible smooth representations of G (without the assumption of the admissibility). The answer is very simple: each irreducible smooth representation is admissible. This fact, which was first proved by H. Jacquet as far as this author knows, is also important in the proof of the Bernstein's result that each irreducible unitary representation has finite K_o-multiplicities. For more explanations regarding these topics one should consult Remarks 4.2 (ii), in the following section.

(ii) One could prove also directly (without use of the Hecke algebras $C_c(G//K)$) that the commutator of an irreducible smooth representation of G consists only of the scalars operators. It follows directly from the fact from the linear algebra that the commutator of an irreducible family of linear operators on a countable dimensional vector space V over $\mathbb{C}$ consists of scalar operators. Let us outline the proof. Suppose that C is the commutator of such a family. Since kernels and images are invariant subspaces, C is a division ring. Suppose, that $L \in C$ is not scalar. Then $L - \lambda \cdot \mathrm{id}_V \in C$ and it is different from 0. Therefore

$$P(L)v \neq 0$$

for any polynomial $P \neq 0$ and any $v \neq 0$. Fix $v \neq 0$. Since

$$(L - \lambda \,\mathrm{id}_V)^{-1}v, \quad \lambda \in \mathbb{C},$$

is a linearly dependent set, there exist $\lambda_1, \cdots, \lambda_k \in \mathbb{C}$, mutually different, and $\mu_1, \ldots, \mu_k \in \mathbb{C}$ which are not all equal to 0, such that

$$\sum_{i=1}^{k} \mu_i (L - \lambda_i \,\mathrm{id}_V)^{-1}v = 0.$$

Acting on the last relation by

$$\prod_{i=1}^{k} (L - \lambda_i \,\mathrm{id}_V),$$

one gets that $P(L)v = 0$ for a polynomial $P \neq 0$, what is a contradiction. This proves that only the scalar operators can be in the commutator of an irreducible smooth representation of G.

4. JACQUET MODULES AND CUSPIDAL REPRESENTATIONS.

We shall now define a left adjoint functor to the functor Ind_P^G. Let (π, V) be a smooth representation of G and let $P = MN$ be a parabolic subgroup of G. Set

$$V(N) = \mathrm{span}_{\mathbb{C}} \{\pi(n)v - v; n \in N, v \in V\}.$$

The group N has the property that it is the union of its open compact subgroups. In the case of $SL(2, F)$ or $GL(2, F)$, and a proper parabolic subgroup $P = MN$, we have $N \cong F$. Therefore the above property is evident in these cases. The above property of N has for a consequence that

$$V(N) = \bigcup \operatorname{Ker} \pi \left(\Xi_{N_o} \right) \tag{4.1}$$

when N_o runs over all open compact subgroups of N. In the above formula we consider π also as a representation of N only. Therefore, $\pi \left(\Xi_{N_o} \right)$ is well defined. Note that we have also

$$V(gNg^{-1}) = \pi(g)V(N)$$

for $g \in G$. Since M normalizes N, $V(N)$ is invariant for the action of M. Set

$$V_N = V/V(N).$$

We consider the natural quotient action of M on V_N

$$\pi_N(m)(v + V(N)) = \pi(m)v + V(N).$$

The M-representation (π_N, V_N) is called the *Jacquet module* of (π, V) with respect to $P = MN$. It is easy to see that

$$(\pi, V) \mapsto (\pi_N, V_N)$$

is a functor from the category of smooth G-representations to the category of smooth M-representations. It is not hard to show that this functor is exact.

One could consider the Jacquet functor as a functor from the category of smooth P-representations to the category of smooth M-representations. In this setting, it is also an exact functor.

If (π, V) is a finitely generated smooth G-representation, then one can prove directly that $(\pi|P, V)$ is a finitely generated P-representation . This is a consequence of the compactness of $P \backslash G$ and the smoothness of the action of G. By the above observation, the Jacquet functor carries finite generated G-representations to a finite generated M-representations. A little bit more work is required to prove that the Jacquet functor carries the admissible representations to the admissible ones. Moreover, one has that the Jacquet functor carries the admissible representations of finite length again to the representation of finite length (see (8.4)).

Let π be a smooth representation of G and let σ be a smooth representation of M. Then we have a canonical isomorphism

$$\operatorname{Hom}_G \left(\pi, \operatorname{Ind}_P^G(\sigma) \right) \cong \operatorname{Hom}_M \left(\pi_N, \Delta_P^{1/2} \sigma \right).$$

This isomorphism is called the *Frobenius reciprocity.* Let us explain how one gets it. Denote by

$$\Lambda : \operatorname{Ind}_P^G(\sigma) \to \Delta_P^{1/2}\sigma$$

the mapping

$$f \to f(1).$$

Then composition with Λ gives

$$\operatorname{Hom}_G\left(\pi, \operatorname{Ind}_P^G(\sigma)\right) \cong \operatorname{Hom}_P\left(\pi, \Delta_P^{1/2}\sigma\right).$$

Since N acts trivially on σ, one gets $\operatorname{Hom}_P\left(\pi, \Delta_P^{1/2}\sigma\right) \cong \operatorname{Hom}_M\left(\pi_N, \Delta_P^{1/2}\sigma\right)$.

Suppose that we have a standard Levi subgroup P in G, the standard Levi decomposition $P = MN$ of P, a standard Levi subgroup P' of M with the standard Levi decomposition $P' = M'N'$. Let P'' be the standard parabolic subgroup of G which has M' for the standard Levi factor. Let $P'' = M'N''$ be the standard Levi decomposition of P''. Suppose that π is a smooth representation of G. Then we have the following transitivity of Jacquet modules

$$\pi_{N''} \cong (\pi_N)_{N'}.$$

In a number of applications it is more convenient to work with normalized Jacquet modules. Denote by

$$\left(r_M^G(\pi), r_M^G(V)\right)$$

the representation $\Delta_P^{-1/2}\pi_N$ on V_N. The representation $\left(r_M^G(\pi), r_M^G(V)\right)$ is called the *normalized Jacquet module* of (π, V) with respect to $P = MN$. Now the Frobenius reciprocity becomes

$$\operatorname{Hom}_G\left(\pi, \operatorname{Ind}_P^G(\sigma)\right) \cong \operatorname{Hom}_M\left(r_M^G(\pi), \sigma\right).$$

Normalized Jacquet modules have again the above transitivity property.

The Frobenius reciprocity indicates how interesting is to understand the Jacquet modules. But this is only one of the very important information that are contained in the Jacquet module. Very soon we shall see some of the others.

An admissible representation (π, V) of G is called *cuspidal* (or *supercuspidal,* or *absolutely cuspidal* by some authors) if for any proper parabolic subgroup $P = MN$ of G and for any smooth representation σ of M we have

$$\operatorname{Hom}_G(\pi, \operatorname{Ind}_P^G(\sigma)) = 0.$$

By the Frobenius reciprocity (π, V) is cuspidal if and only if

$$V_N = 0$$

for any proper parabolic subgroup $P = MN$ of G. Because of the transitivity of the Jacquet modules, it is enough to prove for cuspidality that $V_N = 0$ for all maximal proper parabolic subgroups. If π is irreducible, then the cuspidality of π is equivalent to the following fact: π is not equivalent to a subrepresentation of $\text{Ind}_P^G(\sigma)$ for any proper parabolic subgroup $P = MN$ and any smooth representation σ of M.

Suppose that $(\pi, V) \in \tilde{G}$. Then there exists a parabolic subgroup $P = MN$ of G such that $r_M^G(\pi) \neq 0$. The case of $P = G$ is not excluded. Choose a minimal P with the property that $r_M^G(\pi) \neq 0$. Then the transitivity of the Jacquet modules implies that $r_{M'}^M(r_M^G(\pi)) = 0$ for any proper parabolic subgroup $P' = M'N'$ of M. Since $r_M^G(\pi)$ is finitely generated, it has an irreducible quotient, say σ. Since $r_M^G(\pi)$ is admissible, σ is admissible. The exactness of the Jacquet functor implies that σ is a cuspidal representation of M. Now from the Frobenius reciprocity one obtains that for irreducible admissible representation π of G there exist a parabolic subgroup $P = MN$ of G and an irreducible cuspidal representation σ of M such that π is equivalent to a subrepresentation of $\text{Ind}_P^G(\sigma)$.

Let (π, V) be a smooth representation of G. A character ω of the center $Z(G)$ of G is called a central character of V if

$$\pi(z) = \omega(z)\,\text{id}_V$$

for all $z \in Z(G)$. The central character of π, if it exists, is denoted ω_π. We have seen that each irreducible admissible representation of G has a central character.

Suppose that π is a cuspidal representation of G which has a central character, say ω. Then π is a projective object in the category of all smooth representations which have the central character equal to ω (see Remarks 4.2 (i) in the end of this section). Using the contragredient functor one gets that π is also an injective object in the same category. These facts imply that $\text{Ind}_P^G(\sigma)$ does not have cuspidal subquotients if P is a proper parabolic subgroup.

These facts about projectivety and injectivety of cuspidal representations imply directly the following fact. Let (π, V) be a smooth representation of G. Assume that the center of G is compact (then it must be finite in our case). Then there exists a decomposition of $V = V_c \oplus V_n$ as a representation of G such that each irreducible subquotient of V_c is cuspidal while no one irreducible subquotient of V_n is cuspidal. Such decomposition is unique. A little additional analysis gives that the above result holds without the assumption of the compactness of the center of G. This decomposition is just one of many decompositions which may be obtained using [BeDe] (see also [Td10]).

Let (π, V) be a smooth representation of G. Take $v \in V$ and $\tilde{v} \in \tilde{V}$. The function

$$c_{v,\tilde{v}} : g \mapsto \tilde{v}(\pi(g)v)$$

is called a matrix coefficient of G. There is a nice description of the cuspidal representations of G in terms of the matrix coefficients of G ([Jc1],[Cs]):

4.1.Theorem. *An admissible representation of G is cuspidal if and only if all matrix coefficients are compactly supported functions on G modulo the center (i.e. for each matrix coefficient c there exists a compact subset X of G such that the support of c is contained in $XZ(G)$.*

We may say that a vanishing of the Jacquet modules forces a vanishing of the matrix coefficients. This holds without assumption of admissibility. Thus, each smooth representation whose Jacquet modules for all proper parabolic subgroups are trivial, has matrix coefficients compactly supported modulo the center.

To give an idea of the relationship between the Jacquet modules and the matrix coefficients we shall prove in the case of $SL(2, F)$ that a vanishing of the Jacquet modules implies a compactness of the supports of the matrix coefficients (the proof for a general reductive group G is the same, modulo the structure of the group). We shall first fix some notation for $SL(2, F)$.

Let

$$P = P_{\min} \tag{4.2}$$

be the parabolic subgroup of all upper triangular matrices in $SL(2, F)$. Let

$$M = M_{\min} = \left\{ \begin{bmatrix} a & 0 \\ 0 & a^{-1} \end{bmatrix} ; a \in F^\times \right\}. \tag{4.3}$$

We shall often identify M with $F^\times$ using the isomorphism

$$a \mapsto \begin{bmatrix} a & 0 \\ 0 & a^{-1} \end{bmatrix}.$$

Denote

$$N_{\min} = N = \left\{ \begin{bmatrix} 1 & x \\ 0 & 1 \end{bmatrix} ; x \in F \right\}.$$

Set $K_o = SL(2, \mathcal{O}_F)$. Then we have the *Cartan decomposition* for $SL(2, F)$

$$SL(2, F) = K_o A^- K_o,$$

where

$$A^- = \left\{ \begin{bmatrix} a & 0 \\ 0 & a^{-1} \end{bmatrix} ; a \in F \quad \text{and} \quad |a|_F \leqslant 1 \right\}.$$

We shall also use the following notation in the further calculations:

$$\mathbf{d}(a) = \begin{bmatrix} a & 0 \\ 0 & a^{-1} \end{bmatrix},$$

for $a \in F^\times$.

We can now present the proof of the above implication. Let (π, V) be a smooth representation of $SL(2, F)$ such that $V_N = 0$. Take $v \in V$ and $\tilde{v} \in \tilde{V}$. Since $\pi(K_o)v$ is a finite subset of $V(N) = V$, (4.1) implies that there exists an open compact subgroup N_1 of N such that $\pi(K_o)v \subseteq \mathrm{Ker}\pi(\Xi_{N_1})$. Take an open compact subgroup N_2 of N such that $\tilde{\pi}(K_o)\tilde{v} \subseteq \tilde{V}^{N_2}$. One can prove directly that there exists $t > 0$ such that if $a \in F$ and $|a| < t$, then $\mathbf{d}(a)N_1\mathbf{d}(a)^{-1} \subseteq N_2$. Take now $k_1, k_2 \in K_o$ and $a \in F$ such that $|a|_F < t$. In the following calculations we shall write $x \backsim y$ for $x, y \in \mathbb{C}$, if there exists $z \in \mathbb{C}^\times$ such that $x = zy$. A simple calculation gives

$$c_{v,\tilde{v}}\left(k_1\mathbf{d}(a)k_2\right) = \tilde{v}\left(\pi\left(k_1\mathbf{d}(a)k_2\right)v\right) =$$

$$\left(\tilde{\pi}(k_1^{-1})\tilde{v}\right)\left(\pi(\mathbf{d}(a)k_2)v\right) =$$

$$\left(\tilde{\pi}\left(\Xi_{N_2}\right)\tilde{\pi}(k_1^{-1})\tilde{v}\right)\left(\pi(\mathbf{d}(a)k_2)v\right) \backsim$$

$$\left(\left(\int_{N_2}\tilde{\pi}(n)dn\right)\tilde{\pi}(k_1^{-1})\tilde{v}\right)\left(\pi(\mathbf{d}(a)k_2)v\right) =$$

$$\int_{N_2}\left(\tilde{\pi}(n)\tilde{\pi}(k_1^{-1})\tilde{v}\right)\left(\pi(\mathbf{d}(a)k_2)v\right)dn =$$

$$\int_{N_2}\left(\tilde{\pi}(k_1^{-1})\tilde{v}\right)\left(\pi(n^{-1})\pi(\mathbf{d}(a)k_2)v\right)dn =$$

$$\int_{N_2}\left(\tilde{\pi}(k_1^{-1})\tilde{v}\right)\left(\pi(n)\pi(\mathbf{d}(a)k_2)v\right)dn =$$

$$\left(\tilde{\pi}(k_1^{-1})\tilde{v}\right)\left(\left(\int_{N_2}\pi(n)dn\right)\pi(\mathbf{d}(a)k_2)v\right) =$$

$$\left(\tilde{\pi}(k_1^{-1})\tilde{v}\right)\left(\pi(\mathbf{d}(a))\left(\int_{N_2}\pi\left(\mathbf{d}(a)^{-1}n\mathbf{d}(a)\right)dn\right)\pi(k_2)v\right) \backsim$$

$$\left(\tilde{\pi}(k_1^{-1})\tilde{v}\right)\left(\pi(\mathbf{d}(a))\left(\int_{\mathbf{d}(a^{-1})N_2\mathbf{d}(a)}\pi(n)dn\right)\pi(k_2)v\right) \backsim$$

$$\left(\tilde{\pi}(k_1^{-1})\tilde{v}\right)\left(\pi(\mathbf{d}(a))\pi\left(\Xi_{\mathbf{d}(a^{-1})N_2\mathbf{d}(a)}\right)\pi(k_2)v\right).$$

A direct calculation shows that $N_1 \subseteq \mathbf{d}(a)^{-1}N_2\mathbf{d}(a)$ implies

$$\Xi_{\mathbf{d}(a^{-1})N_2\mathbf{d}(a)} * \Xi_{N_1} = \Xi_{\mathbf{d}(a^{-1})N_2\mathbf{d}(a)}.$$

Since $\pi(k_2)v \in \mathrm{Ker}\,\pi(\Xi_{N_1})$ we get $c_{v,\tilde{v}}(k_1\mathbf{d}(a)k_2) = 0$. By the Cartan decomposition the support of $c_{v,\tilde{v}}$ is contained in

$$K_o\{\mathbf{d}(a); a \in F \text{ and } t \leqslant |a|_F \leqslant 1\}K_o,$$

155

which is a compact set. This finishes the proof of the implication.

The proof of the other implication in the theorem is more technical.

There exists a strong connection between the asymptotic properties of the matrix coefficients and the Jacquet modules. A nice elaboration of that connection can be found in the fourth section of [Cs1]. An application of this connection to the square integrable representations will be given now. Let us first define the square integrable representations.

Suppose that (π, V) is an admissible representation of G which has a unitary central character. Then the absolute value of each matrix coefficient

$$|c_{v,\tilde{v}}| \; : g \mapsto |c_{v,\tilde{v}}(g)|$$

is a function on $G/Z(G)$. The representation π is called *square integrable* if all functions $|c_{v,\tilde{v}}|$ are square integrable functions on $G/Z(G)$. If for an admissible representation τ of G there exists a character χ of G such that the representation

$$\chi\tau : g \mapsto \chi(g)\tau(g)$$

is square integrable, then τ will be called an *essentially square integrable representation*. It is easy to see that each irreducible cuspidal representation is an essentially square integrable representation.

Suppose that (π, V) is an irreducible square integrable representation of G. Take $\tilde{v}_o \in \tilde{V}$, $\tilde{v}_o \neq 0$. For $u, v \in V$ define

$$(u, v) = \int_{G/Z(G)} \tilde{v}_o(\pi(g)u)\overline{\tilde{v}_o(\pi(g)v)}dg.$$

This defines a G-invariant inner product on V. Thus, each irreducible square integrable representation is unitarizable.

There is a very useful criterion for the square integrability ([Cs1]). Let us explain it in the case of $SL(2, F)$. Take $\pi \in SL(2, F)\tilde{}$. Then $r_M^{SL(2,F)}(\pi)$ is a finite dimensional representation. Irreducible subquotients of $r_M^{SL(2,F)}(\pi)$ are characters of $M = M_{\min}$. Let p be a generator of the unique maximal ideal $\mathfrak{p}_F = \{x \in F : |x|_F < 1\}$ in the ring of integers $\mathcal{O}_F = \{x \in F; |x|_F \leqslant 1\}$ in F. Then π is square integrable if and only if for each irreducible subquotient χ of $r_M^{SL(2,F)}(\pi)$ we have

$$\left| \chi \left(\begin{bmatrix} p & 0 \\ 0 & p^{-1} \end{bmatrix} \right) \right|_F < 1. \tag{SI}$$

An irreducible admissible representation π of G will be called an irreducible tempered representation of G, if there exist a parabolic subgroup $P = MN$ and a square integrable representation δ of M such that π is equivalent to a subrepresentation of

$\operatorname{Ind}_P^G(\delta)$. The Langlands classification ([BlWh], [Si1]) reduces parametrization of $\tilde{G}$ to the classification of the tempered representations of the standard Levi factors of the standard parabolic subgroups. More precisely, for each standard parabolic subgroup P with the standard Levi decomposition $P = MN$, each irreducible tempered representation τ of M, and each positive valued character χ of M satisfying certain "positiveness condition", the representation $\operatorname{Ind}_P^G(\chi\sigma)$ has a unique irreducible quotient, say $L(\chi\sigma)$. If $L(\chi\sigma) = L(\chi'\sigma')$, then $P = P', \sigma \cong \sigma'$ and $\chi = \chi'$. Also, each irreducible admissible representation is equivalent to some $L(\chi\sigma)$, for some P, σ and χ as above. The "positiveness condition" will be described explicitly for the general linear groups in the ninth section.

4.2. Remarks.

(i) Suppose that G has a compact center (then the center is finite in our case). Introduction of this condition is done only in order to avoid dealing with the central characters. As it is well known, an irreducible unitary representation π of G is square integrable if and only if π is unitarily equivalent to a subrepresentation of $L^2(G)$, where G acts by right translations. Suppose that (π, V) is an irreducible cuspidal representation of G. Choose any $\tilde{v} \in \tilde{V}, \tilde{v} \neq 0$. Then

$$v \to c_{v,\tilde{v}}$$

defines a non trivial G-intertwining of V into $C_c^\infty(G)$, where G acts an $C_c^\infty(G)$ by right translations. Also, each irreducible subrepresentation of $C_c^\infty(G)$ is cuspidal (this follows from the following remark and the Theorem 4.1.). Therefore, irreducible cuspidal representations of G are exactly the representations which are equivalent to irreducible subrepresentations of $C_c^\infty(G)$. Therefore, we may say that irreducible cuspidal representations are exactly the representations which appear discretely in $C_c^\infty(G)$. Roughly, this fact explains why irreducible cuspidal representations are projective objects.

(ii) We have noted that each irreducible admissible representation π of G is equivalent to a subrepresentation of $\operatorname{Ind}_P^G(\sigma)$, where σ is an irreducible cuspidal representation of M. In the same way it follows that each irreducible smooth representation π of G is equivalent to a subrepresentation of $\operatorname{Ind}_P^G(\sigma)$ where (σ, U) is an irreducible smooth representations of M such that $\sigma_{N'} = 0$ for any proper parabolic subgroup $P' = M'N'$ of M. To prove that π is admissible, it is enough to prove that σ is admissible. Note that by a previous remark σ has matrix coefficients compactly supported modulo the center $Z(M)$. Suppose that σ is not admissible. Chose an open compact subgroup K of M such that U^K is not finite dimensional (certainly, it is of countable dimension). Take $u \in U^K$, $u \neq 0$. Then $\sigma(m)u, m \in M$ generates U. Thus $\sigma(\Xi_K)\sigma(m)u, m \in M$ generates U^K. Choose a sequence (m_k) in M such that the sequence $\sigma(\Xi_K)\sigma(m_k)u$ form a basis of U^K when k runs over positive integers. Choose $\tilde{u}' \in (U^K)^*$ such that $\tilde{u}'(\sigma(\Xi_K)\sigma(m_k)u) \neq 0$ for all $k \geqslant 1$. Then

$\tilde{u} = \tilde{u}' \circ \sigma(\Xi_K) \in \tilde{U}^K$ by (3.3) and further

$$0 \neq \tilde{u}'(\sigma(\Xi_K)\sigma(m_k)u) = \tilde{u}(\sigma(\Xi_K)\sigma(m_k)u) =$$

$$(\tilde{\sigma}(\Xi_K)\tilde{u})(\sigma(m_k)u) = \tilde{u}(\sigma(m_k)u) = c_{u,\tilde{u}}(m_k).$$

Now using Remarks 3.1., (ii), one obtains

$$\bigcup_{k=1}^{\infty} Z(G)Km_kK \subseteq \text{supp } c_{u,\tilde{u}}.$$

Note that $Z(G)Km_{k_1}K$ and $Z(G)Km_{k_2}K$ are disjoint by the same remark for $k_1 \neq k_2$ because the elements $\sigma(\Xi_K)\sigma(m_k)u$, $k \geq 1$, form a basis of U^K. Therefore, the support of $c_{u,\tilde{u}}$ is not compact modulo the center. This contradiction proves the admissibility of σ, and further, the admissibility of π. So, each irreducible smooth representation of G is admissible.

(iii) Let us now sketch the proof of the Bernstein's result that for any open compact subgroup K of G, the spaces of the K-invariants are finite dimensional in the topologically irreducible unitary representations of G (in Hilbert spaces). From the second section and the previous remark it follows that irreducible $C_c(G//K)$-modules are finite dimensional. The following crucial step in the proof of the Bernstein's result is to show that dimensions of irreducible $C_c(G//K)$-modules are bounded. Note that for a proof of this, it is enough to prove such statement for some open subgroup K' of K.

For a positive integer k set

$$K'_k = \{g \in GL(n,F); g \equiv I_n(\text{mod } \mathfrak{p}_F^k)\}.$$

It is possible to embed G in some $GL(n,F)$ in a such way that the following property holds, for any n. Set $K_k = K'_k \cap G$. Then the algebra $C_c(G//K_k)$ satisfies the following condition. There exist $f_1, \dots, f_q \in C_c(G//K_k)$ and a commutative finitely generated subalgebra A of $C_c(G//K_k)$ such that

$$C_c(G//K_k) = \sum_{i,j=1}^{q} f_i * A * f_j. \tag{4.4}$$

To simplify the notation, we shall denote K_k by K in further. One gets now that the dimensions of the irreducible $C_c(G//K)$-modules are bounded from the following interesting lemma from the linear algebra: there exists a function $g \mapsto p(g)$ from the set of positive integers into the strictly positive real numbers such that if $\mathcal{A}$ is a commutative subalgebra of the algebra of all endomorphisms of an m-dimensional complex vector space generated by g generators, then

$$\dim_{\mathbb{C}} \mathcal{A} \leqslant m^{2-p(g)}$$

(Lemma 4.10. of [BeZe1]). Note that for $g = 1$ one can take $p(1) = 1$. This follows directly from the Hamilton-Cayley Theorem. Thus, in general $0 < p(g) \leqslant 1$. Let g be the cardinality of some generating set of the algebra A in (4.4) which is finite. Suppose that (τ, W) is an irreducible $(C_c(G//K)$-module. Then

$$\tau(C_c(G//K)) = \operatorname{End}_{\mathbb{C}} W.$$

Together with (4.4), this implies

$$(\dim_{\mathbb{C}} W)^2 \leqslant q^2 (\dim_{\mathbb{C}} W)^{2 - p(g)}.$$

Thus

$$\dim_{\mathbb{C}} W \leqslant q^{2/p(g)}.$$

An algebra R is called n-commutative if

$$\sum (-1)^{p(\sigma)} x_{\sigma(1)} x_{\sigma(2)} \cdots x_{\sigma(n)} = 0,$$

for any $x_1, \cdots, x_n \in R$, where the sum runs over all the permutations σ of the set $\{1, 2, \cdots, n\}$ ($p(\sigma)$ denotes the parity of a permutation σ). It is easy to see that the algebra of $n \times n$ matrices is $(n^2 + 1)$-commutative, but it is not $(n-1)$-commutative. Since each $f \in C_c(G//K)$ acts non-trivially in some irreducible $C_c(G//K)$-module, $C_c(G//K)$ is $(c_K^2 + 1)$-commutative.

Suppose that (π, H) is an irreducible unitary representation of G. Then

$$f \mapsto \pi(f) = \int_G f(g) \pi(g) dg$$

defines a topologically irreducible *-representation of $C_c(G//K)$ on H^K, where $C_c(G//K)$ is a *-algebra for the involution $f^*(g) = \overline{f(g^{-1})}$. Now there is a standard strategy to see that the dimension of H^K is less than or equal to any uniform bound of dimensions of irreducible representations of $C_c(G//K)$. Roughly, $\pi(C_c(G//K))$ must be dense in the space of all bounded linear operators on H^K with respect to the strong operator topology. This follows from the von Neumann density theorem (which may be viewed as a topological version of the Jacobsen density theorem). This implies that the algebra of all bounded linear operators on H^K is n-commutative for some n. This implies that H^K is finite dimensional.

(iv) Suppose that (π, H) is an irreducible unitary representation of G. For an open compact subgroup K of G, the space H^K is a finite dimensional topologically irreducible $C_c(G//K)$-module. This implies that H^K is (algebraically) irreducible. Since the smooth part H^∞ is the union of all such spaces of invariants, H^∞ is irreducible $C_c^\infty(G)$-module. Thus $\pi^\infty \in \tilde{G}$. It is easy to see that H^∞ is dense in H.

(v) Let us try to explain why parabolic induction is so interesting in the constructions of elements of $\hat{G}$, or more generally, of $\tilde{G}$. The previous observations about the finiteness of the dimensions of the spaces o f the invariants of the irreducible unitary representations tell us that we are interested in inductions that produce admissible representations (smooth representation which are not admissible are never of finite length). If one induces from a "too small" algebraic subgroup of G, one gets very big and highly reducible representations. One way to provide that algebraic subgroup Q is big in G, is to ask that G/Q is a projective variety. Actually, this is the general definition of parabolic subgroups.

If one induces with a smooth representation σ of a parabolic subgroup $P = MN$ which is not admissible, then the induced representation is never admissible (σ does not need to be trivial on N in this considerations). One can easily show that an admissible representation σ of P must be trivial on N, i.e. σ is essentially the representation of M. This is particularly simple to prove for $G = SL(2, F)$. This explain why it is natural to consider parabolic induction.

Parabolic subgroups are cocompact in the group G. There exist also other "big" subgroups. For example, open compact subgroups are in a certain way also big. Namely, their interior is non-empty (this was not the case for the proper parabolic subgroups). They are not algebraic subgroups. Induction from such subgroups may also produce admissible representations. In the case of non-compact center, one induces from compact modulo center subgroups. Let us suppose for the simplicity that the center is compact. Then an induced admissible representation is unitarizable with a natural inner product and matrix coefficients are compactly supported. Thus, in this way one gets cuspidal representations of G. One such example is outlined in the sixth section.

The proof that multiplicities of irreducible representations of the maximal compact subgroups in the irreducible unitary representations of G are finite, is a nice example of application of non-unitary representations in the study of the unitary ones.

5. COMPOSITION SERIES OF INDUCED REPRESENTATIONS OF $SL(2, F)$ AND $GL(2, F)$.

The considerations of the previous section imply that in the classification of $\tilde{G}$ (and further, of $\hat{G}$), one should classify the irreducible cuspidal representations of the standard Levi factors and then one should classify irreducible subrepresentations of parabolically induced representations by cuspidal ones. Having in mind Langlands classification, it is crucial to have methods for analysis of induced representations $\mathrm{Ind}_P^G(\sigma)$ not only when σ is cuspidal.

In rest of the paper we shall present some methods of the analysis of the parabolically induced representations. Crucial tool in this analysis are Jacquet modules. In general, it is hard to give explicitly Jacquet modules of parabolically induced representations. But there is a result of W. Casselman ([Cs1]), and also of J. Bernstein and A.V. Zelevinsky ([BeZe1]) which enables one to compute subquotients of some filtrations of Jacquet modules of parabolically induced representations. This result was also obtained by Harish-Chandra ([Si2]). We shall explain this result on two simple examples. In the calculations in this section we shall follow mainly [Cs1]. Note that Jacquet modules were already helpful in the proof of the Bernstein's result about finiteness of K_o-multiplicities in irreducible unitary representations of G.

Let X be a totally disconnected locally compact topological space. The space of all locally constant compactly supported functions on X is denoted by $C_c^\infty(X)$. This space has very often the role played by the space of all compactly supported C^∞-functions on a real manifold. But there are also some essential differences. The following example illustrates it.

Let Y be a closed subset of X. It is easy to see that the sequence

$$0 \to C_c^\infty(X \backslash Y) \hookrightarrow C_c^\infty(X) \xrightarrow{\text{restrict.}} C_c^\infty(Y) \tag{5.1}$$

is exact. Clearly, this does not hold in general for a submanifold of a real manifold.

Such type of exactness arises also in the setting of representations of reductive groups over local non-archimedean fields. We shall explain now the exactness at this setting. This exactness enables computation of subquotients of some filtrations of the Jacquet modules of parabolically induced representations.

We shall consider one of the lowest dimensional non-trivial cases, the case of $SL(2, F)$. We have already fixed subgroups $P = P_{\min}$, $M = M_{\min}$ and $N = N_{\min}$ in $SL(2, F)$ (see the preceding section). We shall denote $G = SL(2, F)$ in further. Let

$$w = \begin{bmatrix} 0 & 1 \\ -1 & 0 \end{bmatrix}.$$

Then

$$G = P \cup PwP.$$

This is called the *Bruhat decomposition* for $SL(2, F)$.

Let χ be a character of M. Let X be a subset of G such that $PX = X$. Denote by $I(X)$ the space of all locally constant functions f on X which satisfy $f(mnx) = \Delta_P^{1/2}(m)\chi(m)f(x)$, for all $m \in M$, $n \in N$, $x \in X$ and for which there exists a compact subset C of X such that $\text{supp}(f) \subseteq PC$. Now the following sequence is well-defined

$$0 \to I(PwP) \hookrightarrow \text{Ind}_P^G(\chi) \xrightarrow{\text{restrict.}} I(P) \to 0. \tag{5.2}$$

Similar reasons that imply the exactness of the sequence (5.1), imply also that the above sequence is exact ([Cs1], for example). The group P acts on $I(PwP)$ and $I(P)$ by right translations. In this way the above exact sequence is an exact sequence of P-representations. Because of the exactness of the Jacquet functor, to compute the Jordan-Hölder series of $(\text{Ind}_P^G(\chi))_N$ it is enough to compute the Jordan-Hölder series of $I(PwP)_N$ and $I(P)_N$. Clearly, $I(P)$ is one dimensional. Also, N acts trivially on $I(P)$. Thus $I(P) \cong I(P)_N$ as M-representations. One checks directly that $f \mapsto f(1)$ gives

$$I(P)_N \cong \Delta_P^{1/2}\chi.$$

A more delicate problem is to examine $I(PwP)_N$. Note first that $f \in I(PwP)$ is completely determined by $f|wP$. Define for $f \in I(PwP)$ a function Φ_f on P by the formula

$$\Phi_f(p) = f(wp) \ , \quad p \in P.$$

Then one gets directly

$$\Phi_f(mp) = f(wmp) = f(wmw^{-1}wp) = f(m^{-1}wp) =$$

$$\Delta_P^{-1/2}(m)\chi(m^{-1})f(wp) = \Delta_P^{-1/2}(m)\chi^{-1}(m)\Phi_f(p)$$

for $m \in M$ and $p \in P$. Denote by J the space of all locally constant functions φ on P which satisfy $\varphi(mp) = \Delta_P^{-1/2}(m)\chi^{-1}(m)\varphi(p)$ for all $m \in M$, $p \in P$, and for which there exists a compact subset $C \subseteq P$ such that $\operatorname{supp}\varphi \subseteq MC$. The group P acts on J by right translations. If $f \in I(PwP)$, then we have seen that $\Phi_f \in J$. Moreover, one can see easily that

$$I(PwP) \cong J$$

as representations of P. Consider the following mapping from J to functions on P

$$\Psi_f(p) = \int_N f(np)dn = \int_F f\left(\begin{bmatrix} 1 & x \\ 0 & 1 \end{bmatrix} p\right) dx.$$

Obviously, Ψ_f is a function on $N\backslash P$. Further, for $m = \begin{bmatrix} a & 0 \\ 0 & a^{-1} \end{bmatrix} \in M$

$$\Psi_f(m) = \int_N f(nm)\,dn = \int_F f\left(\begin{bmatrix} 1 & x \\ 0 & 1 \end{bmatrix}\begin{bmatrix} a & 0 \\ 0 & a^{-1} \end{bmatrix}\right) dx =$$

$$\int_F f\left(\begin{bmatrix} a & 0 \\ 0 & a^{-1} \end{bmatrix}\begin{bmatrix} 1 & a^{-2}x \\ 0 & 1 \end{bmatrix}\right) dx =$$

$$\Delta_P^{-1/2}(m)\chi^{-1}(m) \int_F f\left(\begin{bmatrix} 1 & a^{-2} & x \\ 0 & 1 & \end{bmatrix}\right) dx =$$

$$\Delta_P^{-1/2}(m)\chi^{-1}(m)|a|_F^2 \int_F f\left(\begin{bmatrix} 1 & x \\ 0 & 1 \end{bmatrix}\right) dx =$$

$$\Delta_P^{-1/2}(m)\chi^{-1}(m)|a|_F^2 \Psi_f\left(\begin{bmatrix} 1 & 0 \\ 0 & 1 \end{bmatrix}\right).$$

We shall see in the following section that $\Delta_P\left(\begin{bmatrix} a & 0 \\ 0 & a^{-1} \end{bmatrix}\right) = |a|_F^2$. Thus

$$\Psi_f(m) = \Delta_P^{1/2}(m)\chi^{-1}(m)\Psi_f\left(\begin{bmatrix} 1 & 0 \\ 0 & 1 \end{bmatrix}\right). \tag{5.3}$$

Therefore, Ψ_f is completely determined by $\Psi_f \left(\begin{bmatrix} 1 & 0 \\ 0 & 1 \end{bmatrix} \right)$. Define

$$\hat{\Psi} : J \to \mathbb{C}$$

by

$$\hat{\Psi}(f) = \Psi_f \left(\begin{bmatrix} 1 & 0 \\ 0 & 1 \end{bmatrix} \right).$$

Since

$$\hat{\Psi}(R_m f) = \Psi_{R_m f} \left(\begin{bmatrix} 1 & 0 \\ 0 & 1 \end{bmatrix} \right) =$$

$$\int_N (R_m f)(n) dn = \int_N f(nm) dn = \Psi_f(m),$$

we have

$$\hat{\Psi}(R_m f) = \Delta_P^{1/2}(m) \chi^{-1}(m) \hat{\Psi}(f). \tag{5.4}$$

Thus the above formula implies that we have a P-intertwining

$$\hat{\Psi} : J \to \Delta_P^{1/2} \chi^{-1}, \tag{5.5}$$

where N acts trivially on the right hand side. It is easy to see that $\hat{\Psi}$ is surjective (one constructs explicitly a function $f \in J$ such that $\hat{\Psi}(f) \neq 0$). Exactness of the Jacquet functor implies that

$$\hat{\Psi}_N : J_N \to \Delta_P^{1/2} \chi^{-1}$$

is a mapping onto. The last step is proving that $\hat{\Psi}_N$ is an isomorphism. For that one needs to prove

$$\mathrm{Ker} \hat{\Psi} = J(N).$$

Since N acts trivially on the right hand side of (5.5), we have $J(N) \subseteq \mathrm{Ker} \hat{\Psi}$. Let us take a look at $\hat{\Psi}$. By the definition of $\hat{\Psi}$ we have

$$\hat{\Psi}(f) = \int_N f(n) \, dn.$$

It means that $\hat{\Psi}$ is the Haar measure on N (recall that $f \in J$ is completely determined by $f|N$). The kernel of the Haar measure, considered on the functions from $C_c^\infty(G)$, consists of the span of all $R_n f - f$, $f \in C_c^\infty(G)$, $n \in N$. Thus, $\mathrm{Ker} \hat{\Psi} \subseteq J(N)$ because a form on $C_c^\infty(G)$ which is invariant for the (right) translations by the elements of N, must be proportional to the Haar measure on N.

So, we have shown at the end that there exists the following exact sequence

$$0 \to \Delta_P^{1/2}\chi^{-1} \to \left(\mathrm{Ind}_P^{SL(2,F)}(\chi)\right)_N \to \Delta_P^{1/2}\chi \to 0.$$

In terms of the normalized Jacquet modules we have the following exact sequence

$$0 \to \chi^{-1} \to r_M^{SL(2,F)}\left(\mathrm{Ind}_P^{SL(2,F)}(\chi)\right) \to \chi \to 0.$$

Consider the case of $GL(2,F)$. Set

$$M = \left\{ \begin{bmatrix} a & 0 \\ 0 & b \end{bmatrix} ; \ a,b \in F^\times \right\}, \tag{5.6}$$

$$N = \left\{ \begin{bmatrix} 1 & n \\ 0 & 1 \end{bmatrix} ; n \in F \right\} \tag{5.7}$$

and

$$P = MN.$$

For characters χ_1 and χ_2 of $F^\times$ we denote by $\chi_1 \otimes \chi_2$ the character

$$(\chi_1 \otimes \chi_2)\left(\begin{bmatrix} a & 0 \\ 0 & b \end{bmatrix} \right) = \chi_1(a)\chi_2(b)$$

of M. Then the same type of calculation as it was described for $SL(2,F)$, gives the following exact sequence

$$0 \to \chi_2 \otimes \chi_1 \to r_M^{GL(2,F)}\left(\mathrm{Ind}_P^{GL(2,F)}(\chi_1 \otimes \chi_2)\right) \to \chi_1 \otimes \chi_2 \to 0.$$

In general, let $P = MN$ and $P' = M'N'$ be parabolic subgroups in a reductive group G. Suppose that σ is an admissible representation of M. Then the same type of considerations as we did for $SL(2,F)$ give a description of $(\mathrm{Ind}_P^G(\sigma))_{N'}$ in the following way. There exist M'-subrepresentations

$$\{0\} = V_o \subseteq V_1 \subseteq \cdots \subseteq V_k = (\mathrm{Ind}_P^G(\sigma))_{N'}$$

such that it is possible to describe subquotients V_{i+1}/V_i as certain induced representations from suitable Jacquet modules of σ. These M'-representations are indexed by the double cosets

$$P\backslash G/P'.$$

One proceeds similarly as it was done in the case of $SL(2,F)$. There exists an open double class Pw_1P' in G. Then one defines $I(Pw_1P')$ and $I(G\backslash Pw_1P')$ in a similar way as it was done for $SL(2,F)$. One has the exact sequence

$$0 \to I(Pw_1P') \hookrightarrow \mathrm{Ind}_P^G(\sigma) \xrightarrow{\text{restrict.}} I(G\backslash Pw_1P') \to 0.$$

With a similar analysis as before, one can describe $I(PwP')_{N'}$ as a certain parabolically induced representation from suitable Jacquet module of σ. Then one can pick another double coset Pw_2P' which is open in $G\backslash Pw_1P'$. One proceeds in a similar way. One finishes when one comes to the double coset PP'.

6. SOME EXAMPLES.

Modular characters of $SL(2,F)$ and $GL(2,F)$, reducibility points: Because of the definition of $\mathrm{Ind}_P^G(\sigma)$, the trivial representation is always a subrepresentations of $\mathrm{Ind}_P^G(\Delta_P^{-1/2})$. One of the topics that interests us is the reducibility of $\mathrm{Ind}_P^G(\sigma)$. If $P \neq G$, then $\mathrm{Ind}_P^G(\Delta_P^{-1/2})$ is reducible. We shall now calculate explicitly this reducibility point for $SL(2,F)$.

We shall use now the notation for $SL(2,F)$ which was introduced in the last section. Let us write some Haar measures. The Haar measures on N and M are

$$\int_N f(n)dn = \int_F f\left(\begin{bmatrix} 1 & x \\ 0 & 1 \end{bmatrix}\right) dx$$

and

$$\int_M f(m)dm = \int_{F^\times} f\left(\begin{bmatrix} a & 0 \\ 0 & a^{-1} \end{bmatrix}\right) d^\times a$$

respectively. Here $d^\times a$ denotes a Haar measure on the multiplicative group $F^\times$. The definition of $|\ |_F$ implies that

$$\int_M f(m)dm = \int_F f\left(\begin{bmatrix} a & 0 \\ 0 & a^{-1} \end{bmatrix}\right) \frac{da}{|a|_F}.$$

Consider the measure

$$f \rightarrow \int_N \int_M f(nm)\, dn\, dm$$

on $P = MN$. It is obvious that the above measure is invariant for right translations by elements of M. Also, for $n' \in N$

$$\int_N \int_M f(nmn')\, dn\, dm =$$

$$\int_N \int_M f\left(n(mn'm^{-1})m\right)\, dn\, dm =$$

$$\int_N \int_M f(nm)\, dn\, dm$$

since M normalizes N. We have used the Fubini's theorem in the above manipulations. Thus

$$f \mapsto \int_N \int_M f(nm)\, dn\, dm$$

is a right Haar measure on P. Since for $n' \in N$

$$\int_M \int_N f(n'nm) \, dn \, dm = \int_M \int_N f(nm) \, dn \, dm$$

we have $\Delta_P(n) = 1$ for $n \in N$. Let $m' = \begin{bmatrix} a & 0 \\ 0 & a^{-1} \end{bmatrix} \in M$. Then

$$\int_M \int_N f(m'nm) \, dn \, dm =$$

$$\int_M \int_F f\left(\begin{bmatrix} a & 0 \\ 0 & a^{-1} \end{bmatrix} \begin{bmatrix} 1 & x \\ 0 & 1 \end{bmatrix} m \right) dx \, dm =$$

$$\int_M \int_F f\left(\begin{bmatrix} a & 0 \\ 0 & a^{-1} \end{bmatrix} \begin{bmatrix} 1 & x \\ 0 & 1 \end{bmatrix} \begin{bmatrix} a^{-1} & 0 \\ 0 & a \end{bmatrix} m \right) dx \, dm =$$

$$\int_M \int_F f\left(\begin{bmatrix} 1 & a^2 & x \\ 0 & 1 \end{bmatrix} m \right) dx \, dm =$$

$$|a|_F^{-2} \int_M \int_F f\left(\begin{bmatrix} 1 & x \\ 0 & 1 \end{bmatrix} m \right) dx \, dm =$$

$$|a|_F^{-2} \int_M \int_N f(nm) \, dn \, dm.$$

Since

$$\int_M \int_N f(m'nm) \, dn \, dm = \Delta_P^{-1}(m') \int_M \int_N f(nm) \, dn \, dm,$$

we have

$$\Delta_P \left(\begin{bmatrix} a & 0 \\ 0 & a^{-1} \end{bmatrix} \right) = |a|_F^2.$$

A similar calculation for $GL(2, F)$ gives

$$\Delta_{P_{\min}} \left(\begin{bmatrix} a & 0 \\ 0 & b \end{bmatrix} \right) = |a|_F |b|_F^{-1}.$$

$GL(2)$ **over finite field:** We shall see how the calculations done in the previous section can be used in a relatively simple case, in the study of the representation theory of $GL(2)$ over a finite field $\mathbb{F} = \mathbb{F}_q$. We keep the notation which was introduced in the last section also for $GL(2)$ over the finite field.

In the case of the finite fields, one defines parabolically induced representations, Jacquet modules and cuspidal representations in the same way as it was done in the case of local non-archimedean fields. The same form of Frobenius reciprocity holds for finite fields. Then the same calculations as in the last section give the Jacquet modules of the parabolically induced representation. Note that here representations are completely reducible and all modular functions are trivial.

Let χ_1 and χ_2 be characters of $\mathbb{F}^\times$. Then

$$\left(\mathrm{Ind}_P^{GL(2,\mathbb{F})}(\chi_1 \otimes \chi_2)\right)_N = (\chi_1 \otimes \chi_2) \oplus (\chi_2 \otimes \chi_1).$$

First of all,

$$\dim_{\mathbb{C}} \mathrm{Ind}_P^{GL(2,\mathbb{F})}(\chi_1 \otimes \chi_2) = \frac{\mathrm{card}\ GL(2,\mathbb{F})}{\mathrm{card}\ P} = \frac{(q^2-1)(q^2-q)}{(q-1)^2 q} = q+1.$$

The Frobenius reciprocity implies that $\mathrm{Ind}_P^{GL(2,\mathbb{F})}(\chi_1 \otimes \chi_2)$ is irreducible if and only if $\chi_1 \neq \chi_2$. If $\chi_1 = \chi_2 = \chi$, then $\chi \circ \det$ is a subrepresentation of $\mathrm{Ind}_P^{GL(2,\mathbb{F})}(\chi \otimes \chi)$. Frobenius reciprocity gives that $\mathrm{Ind}_P^{GL(2,\mathbb{F})}(\chi \otimes \chi)$ is a sum of two irreducible representations which are clearly not isomorphic. One of them is one dimensional and the other one is not one dimensional. If $\mathrm{Ind}_P^{GL(2,\mathbb{F})}(\chi_1 \otimes \chi_2)$ and $\mathrm{Ind}_P^{GL(2,\mathbb{F})}(\chi_1' \otimes \chi_2')$ have irreducible subrepresentations which are equivalent, then the Frobenius reciprocity implies

$$\chi_1' \otimes \chi_2' = \chi_1 \otimes \chi_2 \quad \text{or}\quad \chi_1' \otimes \chi_2' = \chi_2 \otimes \chi_1.$$

·Therefore, the irreducible subrepresentations of $\mathrm{Ind}_P^{GL(2,\mathbb{F})}(\chi_1 \otimes \chi_2)$ that we have obtained are the following

(i) $[(q-1)^2 - (q-1)]/2 = (q-1)(q-2)/2$ $(q+1)$-dimensional representations,
(ii) $(q-1)$ q-dimensional representations, (iii) $(q-1)$ one dimensional representations. The above representations are not equivalent.

We shall say a few words about cuspidal representations following [PS]. For more details one should consult that nice introductory book. We shall use two well known facts from the representations th eory of finite groups. The first fact is that the number of equivalence classes of irreducible representations of a finite group is equal to the number of conjugacy classes of the group. The second fact is that the sum of squares of the dimensions of the equivalence classes of the irreducible representations of a finite group is equal to the cardinality of the group.

Set

$$T = \left\{ \begin{bmatrix} a & b \\ 0 & 1 \end{bmatrix} ; a \in \mathbb{F}^\times, b \in \mathbb{F} \right\}.$$

Then one checks directly that T has q conjugacy classes. Note that we have $(q-1)$ characters

$$\begin{bmatrix} a & b \\ 0 & 1 \end{bmatrix} \mapsto \chi(a), \quad \chi \in (\mathbb{F}^\times)\hat{\ }.$$

167

Thus, we have only one additional irreducible representation. Denote this representation by τ_o. Since the sum of squares of all (classes of) irreducible representations is equal to the order of the group, we get

$$\dim_{\mathbb{C}} \tau_o = \sqrt{(q-1)q - (q-1) \cdot 1} = q - 1.$$

One can realize τ_o as an induced representation of T from N by any non-trivial character of N.

Let ρ be an irreducible cuspidal representation of $GL(2, \mathbb{F})$. Since N acts trivially in the representations $\begin{bmatrix} a & b \\ 0 & 1 \end{bmatrix} \mapsto \chi(a)$, $\chi \in F^{\times}$, we have $\rho|T \cong n\tau_o$ for some positive integer n. A simple calculation gives that $GL(2, \mathbb{F})$ has $q^2 - 1$ conjugacy classes. Thus $GL(2, \mathbb{F})$ has

$$(q^2 - 1) - (q-1)(q-2)/2 - 2(q-1) = (q^2 - q)/2$$

classes of irreducible cuspidal representations. So, irreducible cuspidal representations for finite fields exist. The sum of squares of their dimensions is

$$(q^2 - 1)(q^2 - q) - \frac{(q-1)(q-2)}{2}(q+1)^2 - (q-1)q^2 - (q-1)1 = (q-1)^2 \frac{q^2 - q}{2}.$$

This immediately implies that all irreducible cuspidal representations are $(q-1)$-dimen-sional. An explicit construction of cuspidal representations of $GL(2, \mathbb{F})$ one can found in [PS]. The cuspidal representations are parametrized with the primitive characters of the multiplicative group of the quadratic extension of $\mathbb{F}$, modulo the action of the Galois group. We can give one example easily. The group $GL(2, \mathbb{F}_2) = SL(2, \mathbb{F}_2)$ is not commutative and has 6 elements. This implies that dimensions of irreducible representations are 2,1,1. Now the non-trivial character (which is of order two) is a cuspidal representation. This is the only cuspidal representation of $GL(2, \mathbb{F}_2)$.

For a nice introduction to representations of general $GL(n, \mathbb{F})$ one can consult [HoMo] (see the appendices in that book).

A cuspidal representation: We shall see now that non-trivial cuspidal representations do exist also in the case of a local non-archimedean field F. We shall give an example of a cuspidal representation of $SL(2, F)$. This example was done by F. Mautner.

We have denoted by $\mathfrak{p}_F = \{x \in F; |x|_F < 1\}$ the only non-zero prime ideal in $\mathcal{O}_F$. Denote by $\mathbb{F} = \mathcal{O}_F/\mathfrak{p}_F$ the *residual field* of F. Clearly, it is a finite field. Set $K_o = SL(2, \mathcal{O}_F)$. Then the projection $\mathcal{O}_F \to \mathbb{F}$ induces a group-homomorphism

$$SL(2, \mathcal{O}_F) \to SL(2, \mathbb{F}).$$

Let (σ, U) be an irreducible cuspidal representation of $SL(2, \mathbb{F})$. We shall consider σ as a representation of K_o. Consider the space of all compactly supported functions

$$f : SL(2, F) \to U$$

which satisfy

$$f(kg) = \sigma(k)f(g), \quad \text{for all } k \in K_o \text{ and } g \in SL(2, F).$$

The group $SL(2, F)$ acts on this space by right translations. Let $\operatorname{Ind}_{K_o}^{SL(2,F)}(\sigma)$ be the smooth part of that representation. Then $\operatorname{Ind}_{K_o}^{SL(2,F)}(\sigma)$ is an irreducible cuspidal representation of $SL(2, F)$. Let us explain briefly the argument that gives that.
Fix a $SL(2, \mathbb{F})$-invariant inner product $(\ ,\ )$ on U. Then

$$< f_1, f_2 >= \int_{SL(2,F)} (f_1(g), f_2(g)) \, dg \tag{6.1}$$

is an $SL(2, F)$-invariant inner product on $\operatorname{Ind}_{K_o}^{SL(2,F)}(\sigma)$.
For a positive integer n set

$$K_n = \left\{ \begin{bmatrix} a & b \\ c & d \end{bmatrix} \in K_o; \begin{bmatrix} a & b \\ c & d \end{bmatrix} \equiv \begin{bmatrix} 1 & 0 \\ 0 & 1 \end{bmatrix} \pmod{\mathfrak{p}_F^n} \right\}.$$

Then K_n is a normal subgroup of K_o. Write

$$K_o = \bigcup_{i=1}^{m} k_i K_n.$$

Suppose that $f \in (\operatorname{Ind}_{K_o}^{SL(2,F)}(\sigma))^{K_n}$. Then for $x \in \mathcal{O}_F$ we have

$$\sigma\left(\begin{bmatrix} 1 & x \\ 0 & 1 \end{bmatrix}\right) f\left(\begin{bmatrix} a & 0 \\ 0 & a^{-1} \end{bmatrix} k_i\right) =$$

$$f\left(\begin{bmatrix} 1 & x \\ 0 & 1 \end{bmatrix}\begin{bmatrix} a & 0 \\ 0 & a^{-1} \end{bmatrix} k_i\right) =$$

$$f\left(\begin{bmatrix} a & 0 \\ 0 & a^{-1} \end{bmatrix}\begin{bmatrix} 1 & xa^{-2} \\ 0 & 1 \end{bmatrix} k_i\right) =$$

$$f\left(\begin{bmatrix} a & 0 \\ 0 & a^{-1} \end{bmatrix} k_i k_i^{-1}\begin{bmatrix} 1 & a^{-2}x \\ 0 & 1 \end{bmatrix} k_i\right).$$

One can find $t > 1$ such that $a^{-2}\mathcal{O}_F \subseteq \mathfrak{p}_F^n$ if $|a|_F \geqslant t$. Thus

$$f\left(\begin{bmatrix} a & 0 \\ 0 & a^{-1} \end{bmatrix} k_i\right) = 0$$

since σ is cuspidal. The Cartan decomposition of $SL(2, F)$ implies that the support of f is contained in

$$K_o \left\{ \begin{bmatrix} a & 0 \\ 0 & a^{-1} \end{bmatrix} ; a \in F \text{ and } 1 \leqslant |a|_F \leqslant t \right\} K_o.$$

Thus, the supports of the functions in $\left(\operatorname{Ind}_{K_o}^{SL(2,F)}(\sigma)\right)^{K_n}$ are contained in the fixed compact subset. Since each function f from that space of the K_n-invariants must take values in a certain finite dimensional space, and it is determined on representatives of $K_o\backslash SL(2,F)/K_n$, we get that the spaces of the K_n-invariants are finite dimensional. This implies the admissibility of the representation. In particular, $\operatorname{Ind}_{K_o}^{SL(2,F)}(\sigma)$ is unitarizable. Because all the matrix coefficients of a unitarizable representation (π, V) are of the form

$$g \mapsto (\pi(g)v_1, v_2), v_1, v_2 \in V,$$

we get that

$$x \mapsto \int_{SL(2,F)} (f_1(gx), f_2(g)) \, dg, \quad f_1, f_2 \in \operatorname{Ind}_{K_o}^{SL(2,F)}(\sigma),$$

are all matrix coefficients of $\operatorname{Ind}_{K_o}^{SL(2,F)}(\sigma)$. Clearly, they are compactly supported. Thus, $\operatorname{Ind}_{K_o}^{SL(2,F)}(\sigma)$ is cuspidal.

Note that for $f \in \left(\operatorname{Ind}_{K_o}^{SL(2,F)}(\sigma)\right)^{K_1}$ we have that $\operatorname{sup} f \subseteq K_o$. Thus

$$\left(\operatorname{Ind}_{K_o}^{SL(2,F)}(\sigma)\right)^{K_1} \cong U.$$

This implies that the multiplicity of σ in $\operatorname{Ind}_{K_o}^{SL(2,F)}(\sigma)$ as K_o-representation is less than or equal to one. Actually, it is one because $f \mapsto f(1)$ is a K_o-intertwining of $\operatorname{Ind}_{K_o}^{SL(2,F)}(\sigma)$ onto U.

To get the irreducibility observe that we have a linear map

$$\Lambda \mapsto \Lambda',$$

$$\operatorname{Hom}_G\left(\operatorname{Ind}_{K_o}^{SL(2,F)}(\sigma), \operatorname{Ind}_{K_o}^{SL(2,F)}(\sigma)\right) \to \operatorname{Hom}_{K_o}\left(\operatorname{Ind}_{K_o}^{SL(2,F)}(\sigma), \sigma\right),$$

given by $\Lambda' f = (\Lambda f)(1)$. One sees directly that $\Lambda \neq 0$ implies $\Lambda' \neq 0$. Since the multiplicity of σ in $\operatorname{Ind}_{K_o}^{SL(2,F)}(\sigma)$ is one, we have that the commutator of the representation $\operatorname{Ind}_{K_o}^{SL(2,F)}(\sigma)$ consists only of the scalar operators. Since the representation is completely reducible, we get the irreducibility.

There is a conjecture that each cuspidal representation of a reductive p-adic group can be induced from a compact modulo center subgroup. For such constructions of cuspidal representations one may consult [Ho], [Cy]. More informations about expectations in that direction one can find in [Ku2].

7. PARABOLICALLY INDUCED REPRESENTATIONS OF $SL(2,F)$ AND $GL(2,F)$.

We shall make an elementary analysis of the case of the parabolically induced representations of the group $SL(2,F)$.

Let χ be a character of $F^{\times}$. Then we shall consider χ also as a character of M since we have identified $F^{\times}$ with M. Recall of the exact sequence

$$0 \to \chi^{-1} \to r_M^{SL(2,F)}\left(\operatorname{Ind}_P^{SL(2,F)}(\chi)\right) \to \chi \to 0. \tag{7.1}$$

The above exact sequence implies that $r_M^{SL(2,F)}\left(\operatorname{Ind}_P^{SL(2,F)}(\chi)\right)$ is of length two. Note that there is only one conjugacy class of proper parabolic subgroups in $SL(2,F)$. Since $\operatorname{Ind}_P^{SL(2,F)}(\chi)$ does not contain cuspidal subquotients, $\operatorname{Ind}_P^{SL(2,F)}(\chi)$ is at most of length two. The Jacquet modules imply that if $\operatorname{Ind}_P^{SL(2,F)}(\chi)$ and $\operatorname{Ind}_P^{SL(2,F)}(\chi')$ have irreducible subquotients which are isomorphic, then

$$\chi = \chi' \quad \text{or} \quad \chi^{-1} = \chi'. \tag{7.2}$$

Recall of the Frobenius reciprocity in this situation. For characters χ and χ' of $F^{\times}$ we have

$$\operatorname{Hom}_{SL(2,F)}\left(\operatorname{Ind}_P^{SL(2,F)}(\chi), \operatorname{Ind}_P^{SL(2,F)}(\chi')\right) \cong$$
$$\cong \operatorname{Hom}_M\left(r_M^{SL(2,F)}\left(\operatorname{Ind}_P^{SL(2,F)}(\chi)\right), \chi'\right). \tag{7.3}$$

The Frobenius reciprocity and the exact sequence (7.1) imply that

$$\dim_{\mathbb{C}} \operatorname{End}_{SL(2,F)}\left(\operatorname{Ind}_P^{SL(2,F)}(\chi)\right) \leqslant 2. \tag{7.4}$$

Suppose that $\operatorname{Ind}_P^{SL(2,F)}(\chi)$ is reducible and that it is not a multiplicity one representation. Then $\chi = \chi^{-1}$. This implies $\chi^2 = 1$. Thus, χ is a unitary character. Therefore, $\operatorname{Ind}_P^{SL(2,F)}(\chi)$ is a unitarizable representation. Since such representations are completely reducible, we get

$$\dim_{\mathbb{C}} \operatorname{End}_{SL(2,F)}\left(\operatorname{Ind}_P^{SL(2,F)}(\chi)\right) = 4.$$

This is impossible by (7.4). The last contradiction implies that $\mathrm{Ind}_P^{SL(2,F)}(\chi)$ is always a multiplicity one representation.

A character χ of M is called *regular* if $\chi \neq \chi^{-1}$. For a regular character χ we have

$$r_M^{SL(2,F)}\left(\mathrm{Ind}_P^{SL(2,F)}(\chi)\right) = \chi \oplus \chi^{-1}.$$

This implies that

$$\dim_{\mathbb{C}} \mathrm{End}_{SL(2,F)}\left(\mathrm{Ind}_P^{SL(2,F)}(\chi)\right) = 1 \qquad (7.5)$$

if χ is a regular character.

Suppose that χ is a unitary regular character. Since $\mathrm{Ind}_P^{SL(2,F)}(\chi)$ is a unitarizable representation, it is completely reducible. Thus, $\mathrm{Ind}_P^{SL(2,F)}(\chi)$ is an irreducible unitarizable representation. This is a special case of a general Bruhat result.

Up to now we have seen what happens with $\mathrm{Ind}_P^{SL(2,F)}(\chi)$ when χ is a unitary character which satisfies $\chi^2 \neq 1_{F^\times}$.

Suppose now that χ is not unitary and that $\mathrm{Ind}_P^{SL(2,F)}(\chi)$ splits. Note that χ is a regular character because it is not unitary. Let π_1 and π_2 be different irreducible subquotients. Then, say

$$r_M^{SL(2,F)}(\pi_1) = \chi, \quad r_M^{SL(2,F)}(\pi_2) = \chi^{-1}.$$

Now the square integrability criterion (SI) implies that either π_1 or π_2 is square integrable. Without a lost of generality, we can suppose that π_1 is square integrable. Therefore, π_1 is unitarizable and $\tilde{\pi}_1 \cong \pi_1$ is a subquotient of $\mathrm{Ind}_P^{SL(2,F)}(\tilde{\chi})$. Thus, $\mathrm{Ind}_P^{SL(2,F)}((\tilde{\chi})^{-1})$ and $\mathrm{Ind}_P^{SL(2,F)}(\chi)$ have non-disjoint Jordan-Hölder series. Then we know by (7.2) that $\chi = (\tilde{\chi})^{-1}$ or $\chi^{-1} = (\tilde{\chi})^{-1}$. The first relation is equivalent to $\chi\tilde{\chi} = 1$ what means that χ is unitary. Thus $\chi = \tilde{\chi}$. In other words, χ must be a real-valued character.

Clearly, constant functions are contained in $\mathrm{Ind}_P^{SL(2,F)}(\Delta_P^{-1/2})$. It follows that $\mathrm{Ind}_P^{SL(2,F)}(\Delta_P^{-1/2})$ is reducible. One irreducible subquotient is the trivial representation while the other irreducible subquotient is a square integrable representation. This square integrable representation is called the *Steinberg representation* of $SL(2,F)$. Since

$$(\mathrm{Ind}_P^{SL(2,F)}(\Delta_P^{1/2}))^\sim \cong \mathrm{Ind}_P^{SL(2,F)}(\Delta_P^{-1/2}),$$

$\mathrm{Ind}_P^{SL(2,F)}(\Delta_P^{-1/2})$ is also reducible.

If $\chi = \chi'$ or $\chi^{-1} = \chi'$, then $\mathrm{Ind}_P^{SL(2,F)}(\chi)$ and $\mathrm{Ind}_P^{SL(2,F)}(\chi')$ have the same Jordan-Hölder sequences by the general result about the parabolic induction from the associate pairs which was mentioned in the third section. We shall outline the proof of this fact for $SL(2,F)$.

Let χ be any character of $F^{\times}$. Note that $(\text{Ind}_P^{SL(2,F)}(\chi))^{\sim} \cong \text{Ind}_P^{SL(2,F)}(\chi^{-1})$. Thus $\text{Ind}_P^{SL(2,F)}(\chi)$ is irreducible if and only if $\text{Ind}_P^{SL(2,F)}(\chi^{-1})$ is irreducible. From the Frobenius reciprocity (7.3) and the exact sequence (7.1) one obtains that

$$\text{Hom}_{SL(2,F)}\left(\text{Ind}_P^{SL(2,F)}(\chi), \text{Ind}_P^{SL(2,F)}(\chi^{-1})\right) \neq 0. \tag{7.6}$$

Therefore we have a non-zero intertwining

$$\Lambda_{\chi} : \text{Ind}_P^{SL(2,F)}(\chi) \to \text{Ind}_P^{SL(2,F)}(\chi^{-1}).$$

Thus

$$\text{Ind}_P^{SL(2,F)}(\chi) \cong \text{Ind}_P^{SL(2,F)}(\chi^{-1}), \tag{7.7}$$

if $\text{Ind}_P^{SL(2,F)}(\chi)$ is irreducible. If χ is unitary and if $\text{Ind}_P^{SL(2,F)}(\chi)$ is reducible, then we know that $\chi^2 = 1$, i.e. that $\chi = \chi^{-1}$. Thus (7.7) holds for any unitary character χ. Suppose now that $\text{Ind}_P^{SL(2,F)}(\chi)$ reduces and that $\chi \neq \chi^{-1}$. Then (7.5) implies that $\text{Ind}_P^{SL(2,F)}(\chi)$ has a unique irreducible subrepresentation, say V_1, and a unique irreducible quotient. They are not isomorphic. Since $V_1 \hookrightarrow \text{Ind}_P^{SL(2,F)}(\chi)$, the Frobenius reciprocity implies that the Jacquet module of the irreducible subrepresentation is χ, while the irreducible quotient has χ^{-1} for the Jacquet module.

Let $\varphi \neq 0$ be an intertwining mapping from the space (7.6). If $\text{Ker} \varphi = \{0\}$, then $\text{Ind}_P^{SL(2,F)}(\chi^{-1})$ has an irreducible subrepresentation whose Jacquet module is χ. This is impossible by the previous remarks if we apply them to $\text{Ind}_P^{SL(2,F)}(\chi^{-1})$. Therefore the irreducible quotient of $\text{Ind}_P^{SL(2,F)}(\chi)$ is isomorphic to the irreducible subrepresentation of $\text{Ind}_P^{SL(2,F)}(\chi^{-1})$. One has that the representations $\text{Ind}_P^{SL(2,F)}(\chi)$ and $\text{Ind}_P^{SL(2,F)}(\chi^{-1})$ have the same Jordan-Hölder series by applying the same observation to $\text{Ind}_P^{SL(2,F)}(\chi^{-1})$. So, we have proved this result for arbitrary character χ.

Let us now describe Casselman's method for the study of the irreducibility of the parabolically induced representations. Suppose that χ is a character of $M \cong F^{\times}$ such that $\chi^2 \neq 1_{F^{\times}}$. Consider the intertwinings Λ_{χ} from the spaces (7.6). If $\text{Ind}_P^{SL(2,F)}(\chi)$ is irreducible, then by the Schur's lemma there exists $c(\chi) \in \mathbb{C}^{\times}$ such that

$$\Lambda_{\chi^{-1}} \Lambda_{\chi} = c(\chi) \, \text{id}_{\text{Ind}_P^{SL(2,F)}(\chi)}.$$

Note that $c(\chi) \in \mathbb{C}$ depends on the choice of Λ_{χ} and $\Lambda_{\chi^{-1}}$. Suppose that the representation $\text{Ind}_P^{SL(2,F)}(\chi)$ reduces. We have seen that Λ_{χ} and $\Lambda_{\chi^{-1}}$ have non-trivial kernels. Also representations $\text{Ind}_P^{SL(2,F)}(\chi)$ and $\text{Ind}_P^{SL(2,F)}(\chi^{-1})$ have unique irreducible subrepresentations. Thus $\Lambda_{\chi^{-1}} \Lambda_{\chi} = 0$. One can say that in this situation we have $c(\chi) = 0$. Therefore, for a regular character χ, $\text{Ind}_P^{SL(2,F)}(\chi)$ is irreducible if and only if $c(\chi) \neq 0$. The delicate part in this method is an explicit computation of $c(\chi)$.

Casselman has computed $c(|\ |_F^\alpha)$ in [Cs1]. That computation gives that $c(|\ |_F^\alpha) \neq 0$ for $\alpha \in \mathbb{R} \backslash \{-1, 0, 1\}$. Thus, $\text{Ind}_P^{SL(2,F)}(|\ |_F^\alpha)$ is irreducible for $\alpha \in \mathbb{R} \backslash \{-1, 0, 1\}$. Note that the space

$$Q = \left\{ f | SL(2, \mathcal{O}_F); f \in \text{Ind}_P^{SL(2,F)}(|\ |^\alpha) \right\}$$

does not depend on α. Therefore, one can realize all these representations on Q. Denote the action of $SL(2, F)$ on Q which corresponds to $\text{Ind}_P^{SL(2,F)}(|\ |^\alpha)$ by π_α. One can now normalize the intertwinings $\Lambda_{|\ |^\alpha}$ in such a way that $\Lambda_{|\ |^\alpha}$, and thus also $c(|\ |^\alpha)$, depends analytically on α. Then, it is enough to compute $(\Lambda_{\chi^{-1}} \Lambda_\chi f)(k_o)$ for some function f from Q on a non-empty open subset of $\mathbb{C}$. One needs to assume only that $f(k_o) \neq 0$. This is computed in [Cs1].

We shall see now what one can obtain by the similar analysis of the Jacquet modules in the case of $GL(2, F)$. In the same way as for $SL(2, F)$ one obtains the following results. If

$$\text{Ind}_P^{GL(2,F)}(\chi_1 \otimes \chi_2) \quad \text{and} \quad \text{Ind}_P^{GL(2,F)}(\chi_1' \otimes \chi_2')$$

have non-disjoint Jordan-Hölder series, then

$$\chi_1 \otimes \chi_2 = \chi_1' \otimes \chi_2' \quad \text{or} \quad \chi_1 \otimes \chi_2 = \chi_2' \otimes \chi_1'. \tag{7.8}$$

Also, $\text{Ind}_P^{GL(2,F)}(\chi_1 \otimes \chi_2)$ and $\text{Ind}_P^{GL(2,F)}(\chi_2 \otimes \chi_1)$ have the same Jordan-Hölder series. The Frobenius reciprocity implies that $\text{Ind}_P^{GL(2,F)}(\chi_1 \otimes \chi_2)$ is irreducible if $\chi_1 \neq \chi_2$ and if χ_1/χ_2 is unitary. If χ_1/χ_2 is not unitary and if $\text{Ind}_P^{GL(2,F)}(\chi_1 \otimes \chi_2)$ reduces, then the square integrability criterion implies that $\chi_2 = (\overline{\chi}_1)^{-1}$. This implies that $\chi_1 = \chi_2 |\ |_F^\alpha$ for some $\alpha \in \mathbb{R}^\times$. Thus $\chi_1 \otimes \chi_2 = \chi_2 |\ |_F^\alpha \otimes \chi_2$. The restriction of the functions on $GL(2, F)$ from $\text{Ind}_P^{GL(2,F)}(\chi_2 |\ |_F^\alpha \otimes \chi_2)$ to $SL(2, F)$ gives an isomorphism of $\text{Ind}_P^{GL(2,F)}(\chi_2 |\ |_F^\alpha \otimes \chi_2)$ onto $\text{Ind}_P^{SL(2,F)}(|\ |_F^\alpha)$, as representations of $SL(2, F)$. Since $\text{Ind}_P^{SL(2,F)}(|\ |_F^\alpha)$ is irreducible for $\alpha \in \mathbb{R} \backslash \{1, 0, -1\}$, we have that $\text{Ind}_P^{GL(2,F)}(\chi_2 |\ |_F^\alpha \otimes \chi_2)$ is also irreducible for such α.

To have a complete analysis of the parabolically induced representations of $GL(2, F)$, one should see what happens with $\text{Ind}_P^{GL(2,F)}(\chi \otimes \chi)$. Since

$$(\chi \circ \det) \text{Ind}_P^{GL(2,F)}(\chi_1 \otimes \chi_2) \cong \text{Ind}_P^{GL(2,F)}(\chi\chi_1 \otimes \chi\chi_2), \tag{7.9}$$

one should check what happens with $\text{Ind}_P^{GL(2,F)}(1_{F^\times} \otimes 1_{F^\times})$. Set $N^- = {}^t N$. Then PN^- has a full measure in $GL(2, F)$ (the complement has the Haar measure equal to 0). In the same way as it was explained in Remark 2.2 (i), we get that

$$\int_{N^-} f(n^-) dn^-$$

is a $GL(2, F)$-invariant measure on the space X from Remarks 2.2, (i). Thus

$$f \mapsto f|N^-$$

defines an isomorphism of $\mathrm{Ind}_P^{GL(2,F)}(1_{F^\times} \otimes 1_{F^\times})$ onto $L^2(N^-)$. We shall identify these two spaces.

We identify N^- with F by the identification

$$\begin{bmatrix} 1 & 0 \\ x & 1 \end{bmatrix} \leftrightarrows x.$$

The action of $GL(2, F)$ on $L^2(F)$ will be denoted by π. Now a simple computation gives

$$\left(\pi \left(\begin{bmatrix} a & 0 \\ 0 & 1 \end{bmatrix} \right) f \right)(x) = \left(|a|_F^{1/2} f \right)(ax) \tag{7.10}$$

and

$$\left(\pi \left(\begin{bmatrix} 1 & 0 \\ y & 1 \end{bmatrix} \right) f \right)(x) = f(x + y). \tag{7.11}$$

Let ψ be a non-trivial character of F. Denote by $\psi_a, a \in F$, the character defined by $\psi_a(x) = \psi(ax)$. Then $a \mapsto \psi_a$ is an isomorphism of F onto $\hat{F}$. The Fourier transform $\mathcal{F}$ is defined by

$$\hat{f}(a) = \int_F f(x)\psi_a(x)dx.$$

We define a representation $\hat{\pi}$ of $GL(2, F)$ on $L^2(F)$ by the formula

$$\hat{\pi}(g) = \mathcal{F}\pi(g)\mathcal{F}^{-1}.$$

Such representation $\hat{\pi}$ is called the *Gelfand-Naimark model* of the representation $\mathrm{Ind}_P^{GL(2,F)}(1_{F^\times} \otimes 1_{F^\times})$. Now the formulas (7.10.) and (7.11) imply

$$\left(\hat{\pi} \left(\begin{bmatrix} a & 0 \\ 0 & 1 \end{bmatrix} \right) f \right)(x) = f(a^{-1}x) \tag{7.12}$$

and

$$\left(\hat{\pi} \left(\begin{bmatrix} 1 & 0 \\ y & 1 \end{bmatrix} \right) f \right)(x) = \overline{\psi(yx)}f(x). \tag{7.13}$$

We shall need now a little bit of Fourier analysis on F. Let T be a continuous operator on $L^2(F)$ which is in the commutator of the representation $\hat{\pi}$. By (7.13) T commutes with all multiplications with characters of F. Therefore, T commutes with all multiplications with functions on F. This implies that T itself is a multiplication with a function, say φ. Now (7.12) implies that φ must be a constant function.

Thus T is a scalar operator. Since the representation $\hat{\pi}$ is unitary, it is completely reducible. Thus, $\hat{\pi}$ is irreducible. Since

$$\left(\operatorname{Ind}_P^{GL(2,F)}(1_{F^\times} \otimes 1_{F^\times})\right)^\infty \cong \operatorname{Ind}_P^{GL(2,F)}(1_{F^\times} \otimes 1_{F^\times}),$$

we have the irreducibility of $\operatorname{Ind}_P^{GL(2,F)}(1_{F^\times} \otimes 1_{F^\times})$. This completes the analysis of the induced representations of $GL(2,F)$. We have seen that $\operatorname{Ind}_P^{GL(2,F)}(\chi_1 \otimes \chi_2)$ is reducible if and only if

$$\chi_1 = |\ |_F \chi_2 \text{ or } \chi_1 = |\ |_F^{-1} \chi_2.$$

From the $GL(2,F)$-case one can settle now the case of $SL(2,F)$. Similarly to the Clifford theory for finite groups, there exists the Clifford theory for p-adic groups. It is developed by S. Gelbart and A.W. Knapp ([GbKn]). Let π be an irreducible representation of $GL(2,F)$. Then, as in the case of the finite groups, we have

$$\dim_{\mathbb{C}} \operatorname{End}_{SL(2,F)}(\pi) = \operatorname{card} \{\chi \in (F^\times)\hat{\ } ; \chi\pi \cong \pi\}.$$

Note that

$$\operatorname{Ind}_P^{SL(2,F)}(\chi) \cong \operatorname{Ind}_P^{GL(2,F)}(\chi \otimes 1_{F^\times})$$

as representations of $SL(2,F)$. Now (7.9) and (7.8) give that $\operatorname{Ind}_P^{SL(2,F)}(\chi)$ is reducible if and only if $\chi = \Delta_P^{\pm 1/2}$, or if χ is a character of order two (i.e. $\chi^2 = 1_{F^\times}$ and $\chi \neq 1_{F^\times}$).

At the end of this section we shall say a few words about the unitary duals of $SL(2,F)$ and $GL(2,F)$. We start with the case of $SL(2,F)$. The irreducible cuspidal representations are obviously in $SL(2,F)\hat{\ }$. Clearly, all the irreducible subrepresentations of the parabolically induced representations by the unitary characters are also in $SL(2,F)\hat{\ }$. The Steinberg representation and the trivial representation are in $SL(2,F)\hat{\ }$. Suppose that some induced representation $\operatorname{Ind}_P^{SL(2,F)}(\chi)$ by a non-unitary character χ has a unitarizable subquotient π. Since $\bar{\bar{\pi}} \cong \pi$, $\operatorname{Ind}_P^{SL(2,F)}(\chi)$ and $\operatorname{Ind}_P^{SL(\mathbb{Q},F)}((\bar{\chi})^{-1})$ have non-disjoint Jordan-Hölder series. Now (7.2) implies $(\bar{\chi})^{-1} = \chi^{-1}$, i.e. χ must be a real-valued character. We have seen that if $\operatorname{Ind}^{SL(2,F)}(\chi)$ reduces, then all irreducible subquotients are unitarizable. Suppose therefore that χ is a real valued non-unitary character of $F^\times$ such that $\operatorname{Ind}_P^{SL(2,F)}(\chi)$ is irreducible. We have seen that

$$\operatorname{Ind}_P^{SL(2,F)}(\chi) \cong \operatorname{Ind}_P^{SL(2,F)}(\chi^{-1}).$$

Further $\bar{\chi} = \chi$ implies

$$\left(\operatorname{Ind}_P^{SL(2,F)}(\chi)\right)^{\bar{z}} \cong \operatorname{Ind}_P^{SL(2,F)}(\chi^{-1}).$$

This implies that there exists a non-degenerate $SL(2, F)$-invariant Hermitian form on $\text{Ind}_P^{SL(2,F)}(\chi)$. Actually, our previous observations imply that the form is given by

$$(f_1, f_2) = \int_{SL(2,O_F)} f_1(k)\overline{(\Lambda_\chi f_2)(k)}dk. \tag{7.14}$$

The question of the unitarizability of $\text{Ind}_P^{SL(2,F)}(\chi)$ is the question if the above form is positive definite.

Consider the case of $\chi = |\ |_F^\alpha$ where $\alpha \in \mathbb{R}$. We have the irreducibility for $\alpha \neq \pm 1$. It can be shown that operators Λ_χ do not have a "singularity" at $\alpha = 0$. Therefore, one obtains a continuous family of Hermitian forms on $\text{Ind}_P^{SL(2,F)}(|\ |^\alpha)$, $-1 < \alpha < 1$. Since the set of parameters α is connected, all representations $\text{Ind}_P^{SL(2,F)}(|\ |^\alpha)$, $-1 < \alpha < 1$, are unitarizable. This is a consequence of the following simple fact from the linear algebra. If we have a continuous family of non-degenerate Hermitian forms on a fixed finite dimensional complex vector space, which is parametrized by a connected set, and if one of that forms is positive definite, then all of them are positive definite. In this case we have a positive definiteness which is coming from $\text{Ind}_P^{SL(2,F)}(1_M)$. The above unitarizable representations are called *complementary series*.

Suppose that $|\alpha| > 1$. Then the connection between the asymptotics of the matrix coefficients and the Jacquet modules, and the explicit computation of the Jacquet modules, imply that the matrix coefficients of $\text{Ind}_P^{SL(2,F)}(|\ |^\alpha)$ are not bounded functions. Thus, $\text{Ind}_P^{SL(2,F)}(|\ |^\alpha)$ is not unitarizable for $|\alpha|_F > 1$, since obviously the matrix coefficients of the unitarizable representations are bounded functions.

Suppose that χ is a real valued character. We can write $\chi = \chi_o|\ |_F^\alpha$ where χ_o is a unitary character and $\alpha \in \mathbb{R}$. Since $\chi_o = \overline{\chi}_o$, we have that $\chi_o^2 = 1_{F^\times}$. We shall assume that $\chi_o \neq 1_{F^\times}$. Since the matrix coefficients of unitarizable representations are bounded, $\text{Ind}_P^{SL(2,F)}(\chi_o|\ |_F^\alpha)$ is not unitarizable if $|\alpha| > 1$. Because on the representations

$$\text{Ind}_P^{SL(2,F)}(\chi_o|\ |_F^\alpha), \quad \alpha > 0$$

we have a continuous family of Hermitian forms, they are not positive definite. This ends the description of the unitary dual of $SL(2, F)$.

One gets the unitary dual of $GL(2, F)$ in the same manner. The unitary dual of $GL(2, F)$ consists of the square integrable representations, the irreducible subrepresentations of the parabolically induced representations by the unitary characters, the unitary characters of the group and the complementary series

$$\text{Ind}_P^{GL(2,F)}(\nu^\alpha\chi \otimes \nu^{-\alpha}\chi), \ 0 < \alpha < 1/2, \ \chi \in (F^\times)\widehat{\ }.$$

8. SOME GENERAL CONSEQUENCES.

We shall return now to the general case. One can compute Jordan-Hölder series of the Jacquet modules of the parabolically induced representations using similar ideas to the ideas that were explained in the case of $SL(2, F)$. This computions gives a similar consequences for a general reductive group G to the consequences that we have obtained for $SL(2, F)$ and $GL(2, F)$.

The quotient of the normalizer in G of the standard Levi factor $M_{\min}$ by itself, will be denoted by W_G. Then W_G is called the *Weyl group* of G.

Let $P = MN$ be a parabolic subgroup in G and let σ be an irreducible cuspidal representation of M. For $g \in G$ denote by $g\sigma$ a representation of gMg^{-1} given by the formula

$$(g\sigma)(gmg^{-1}) = \sigma(m),$$

for $m \in M$. Then we have the following result of Bernstein and Zelevinsky, and of Casselman.

8.1. Theorem.

   (i) *The Jordan-Hölder series of $r_M^G(\mathrm{Ind}_P^G(\sigma))$ consists of all $\omega\sigma$ when ω runs over all representatives of $W_M \backslash W_G / W_M$ which normalize M.*

   (ii) *If the Jacquet module of $\mathrm{Ind}_P^G(\sigma)$ for a parabolic subgroup P' has a cuspidal subquotient, then P and P' are associate.*

This theorem has a number of interesting direct consequences. Let us explain some of them.

(**8.2**) Let $P_1 = M_1 N_1$ be a parabolic subgroup of G associate to P. Using the fact that the parabolic induction from the associate pairs gives the same Jordan-Hölder sequences, and the exactness of the Jacquet functor, one gets that the Jordan-Hölder series of $r_{M_1}^G(\mathrm{Ind}_P^G(\sigma))$ is obtained from the Jordan-Hölder series of $r_M^G(\mathrm{Ind}_P^G(\sigma))$ by the conjugation with a suitable element of the group. Thus, the theorem gives also the Jordan-Hölder series of the Jacquet modules for the parabolic subgroups which are associate to P.

(**8.3**) The transitivity of the Jacquet modules implies that each irreducible subquotient of $\mathrm{Ind}_P^G(\sigma)$ has some Jacquet module which is cuspidal. Therefore, the length of $\mathrm{Ind}_P^G(\sigma)$ is finite. Recall that each irreducible admissible representation of G is equivalent to a subrepresentation of some representation $\mathrm{Ind}_P^G(\sigma)$ where σ is an irreducible cuspidal representation (see the fourth section). Now the property of the parabolic induction that it does not depend on the stages of induction, and the exactness of the induction functor imply that the parabolic induction carries the representations of the finite length to the representations of the finite length again.

(**8.4**) Similarly, one gets that the Jacquet functor carries the representations of finite length again to the representations of finite length.

(**8.5**) At this point it is easy to prove that each finitely generated admissible representation of G has a finite length (see Theorem 6.3.10 of [Cs1]).

(**8.6**) Suppose that $P' = M'N'$ and $P'' = M''N''$ are two parabolic subgroups of G. Let σ' and σ'' be an irreducible cuspidal representations of M' and M'' respectively.

If $\mathrm{Ind}_P^G(\sigma')$ and $\mathrm{Ind}_{P''}^G(\sigma'')$ have non-disjoint Jordan-Hölder sequences, then (P', σ'') and (P'', σ'') must be associate.

(**8.7**) Suppose that σ is an irreducible cuspidal representation of a Levi factor M of a parabolic subgroup P. Suppose that the Jacquet module

$$r_{M'}^G \left(\mathrm{Ind}_P^G(\sigma) \right),$$

for a parabolic subgroup $P' = M'N'$, has an irreducible cuspidal subquotient. Then the Theorem 8.1. and (8.2) imply that (P, σ) and (P', σ') are associate pairs.

(**8.8**) Suppose that τ is an irreducible admissible representation of a Levi factor M of a parabolic subgroup $P = MN$. Let ρ' (resp. ρ'') be an irreducible cuspidal subquotient of

$$r_{M'}^G \left(\mathrm{Ind}_P^G(\tau) \right) \quad \left(\text{resp. } r_{M''}^G \left(\mathrm{Ind}_P^G(\tau) \right) \right)$$

for a parabolic subgroup $P' = M'N'$ (resp. $P'' = M''N''$). Then (P', ρ'') and (P'', ρ'') are associate.

The following theorem can be very useful. For a proof one may consult [Cs1].

8.9. Theorem. *Let σ be an irreducible cuspidal representation of a Levi factor M of a parabolic subgroup P of G. Let π be an irreducible subquotient of $\mathrm{Ind}_P^G(\sigma)$. Then there exists $w \in G$ which normalizes M such that π is isomorphic to a subrepresentation of $\mathrm{Ind}_P^G(w\sigma)$.*

Note that we have seen that the above theorem holds for $SL(2, F)$. We shall list now some useful consequences of the above theorem:

(**8.10**) With the same notation as in the above theorem, the Frobenius reciprocity implies that $r_M^G(\pi)$ is a non-zero cuspidal representation. We can conclude further. Suppose that $P' = M'N'$ is a parabolic subgroup such that $r_{M'}^G \left(\mathrm{Ind}_P^G(\sigma) \right)$ has an irreducible cuspidal quotient. Then $r_{M'}^G(\pi) \neq 0$ and $r_{M'}^G \left(\mathrm{Ind}_P^G(\sigma) \right)$ is a cuspidal representation.

(**8.11**) Suppose now that τ is an irreducible admissible representation of a Levi factor M of a parabolic subgroup P. Let π be an irreducible subquotient of $\mathrm{Ind}_P^G(\tau)$. Suppose that $r_{M'}^G \left(\mathrm{Ind}_P^G(\tau) \right) \neq 0$ for some parabolic subgroup $P' = M'N'$ of G. The transitivity of the Jacquet modules implies that

$$r_{M'}^G(\pi) \neq 0.$$

All the time we are using the exactness of the Jacquet functor.

Now we have directly

8.12. Lemma. *Let τ be an irreducible admissible representation of a Levi factor M of a parabolic subgroup P of G. If there exists a parabolic subgroup $P' = M'N'$ of G such that $r^G_{M'}\left(\operatorname{Ind}^G_P(\tau)\right)$ is a non-zero irreducible representation, then $\operatorname{Ind}^G_P(\tau)$ is an irreducible representation.*

Note that in this section we have used only a small part of the information contained in the Jacquet modules. Namely, we have used only the facts coming from the calculation of the Jacquet modules which correspond to the parabolic subgroups which are minimal among all the parabolic subgroups for which the Jacquet modules are non-trivial. A very important information are contained also in the Jacquet modules for the other parabolic subgroups. This can be seen from the last lemma and also from the following one. We shall see in the sequel how one can have a useful control of these other Jacquet modules for some series of groups.

The last lemma is a special case of the following more general lemma

8.13. Lemma. *Let τ, M and P be as in the above lemma. Suppose that there exists a standard Levi subgroup P' of G with the standard Levi decomposition $P' = M'N'$ such that $r^G_{M'}\left(\operatorname{Ind}^G_P(\tau)\right)$ is a multiplicity-one representation. Suppose that for each two different irreducible subquotients π_1 and π_2 of $r^G_{M'}\left(\operatorname{Ind}^G_P(\sigma)\right)$ there exists a standard parabolic subgroup P'' of G with the standard Levi decomposition $P'' = M''N''$ such that $M' \subseteq M''$ and that the following condition holds: there exists an irreducible subquotient ρ of $r^G_{M''}\left(\operatorname{Ind}^G_P(\tau)\right)$ such that π_1 and π_2 are subquotients of $r^{M''}_{M'}(\rho)$. Then $\operatorname{Ind}^G_P(\tau)$ is an irreducible representation.*

We shall use in this paper only the Lemma 8.12., not the above one. There is a modification of the Lemma 8.13. to the non-multiplicity-one case, which was used in the proofs of the irreducibilities announced in [Td13]. We shall see in the sequel how Jacquet modules can be used to get also reducibilities.

One can get easily from the computation of the Jacquet modules in Theorem 8.1 and the square integrability criterion, the following result ([Cs1]).

8.14. Proposition. *Suppose that $P = MN$ is a maximal proper parabolic subgroup of G and suppose that G has compact center. If σ is a non-unitarizable cuspidal representation of M such that $\operatorname{Ind}^G_P(\sigma)$ reduces, then the length of $\operatorname{Ind}^G_P(\sigma)$ is two and one irreducible subquotient is a square integrable representation.*

The assumption on the center of G is not essential. In the non-compact case it is slightly more complicated to describe the corresponding condition on σ.

9. $GL(n, F)$.

For admissible representations σ_1 of $GL(n_1, F)$ and σ_2 of $GL(n_2, F)$ set

$$\sigma_1 \times \sigma_2 = \operatorname{Ind}^{GL(n_1+n_2,F)}_{P_{(n_1,n_2)}}(\sigma_1 \otimes \sigma_2).$$

Since parabolic induction does not depend on the stages of induction, we have

$$(\sigma_1 \times \sigma_2) \times \sigma_3 \cong \sigma_1 \times (\sigma_2 \times \sigma_3).$$

Having in mind the above fact about induction in stages, each induced representation of $GL(n)$ from a standard parabolic subgroup by an irreducible admissible representation, can be expressed in terms of $\times$. Note that parabolic induction from other parabolic subgroups does not provide new irreducible subquotients.

Concerning the Jacquet modules, one would like to have a reasonably simple way to compute

$$(\sigma_1 \times \cdots \times \sigma_n)_N \, ,$$

or at least, to have some other information about these Jacquet modules. Having in mind the transitivity of the Jacquet modules and induction in stages, this reduces to the question about

$$(\sigma_1 \times \sigma_2)_{N_{\max}},$$

where $N_{\max}$ is a maximal proper standard parabolic subgroup.

In [Ze1] this question was solved in the following way. Denote by $R[G]$ the Grothendieck group of the category of all admissible representations of some reductive group G, which are of the finite length. It is simply a free $\mathbb{Z}$-module over the basis $\tilde{G}$. A natural mapping which assigns to an admissible representation of finite length its Jordan-Hölder sequence (together with multiplicities), which we consider as an element of $R[G]$, is denoted by s.s.

Set

$$R_n = R[GL(n, F)].$$

Consider that $GL(0, F)$ is the trivial group. First of all, we lift $\times$ to a biadditive mapping

$$\times : R_n \times R_m \to R_{n+m},$$

$$\left(\sum_{i=1}^{k} p_i \sigma_i \right) \times \left(\sum_{j=1}^{\ell} q_j \tau_j \right) = \sum_{i=1}^{k} \sum_{j=1}^{\ell} p_i q_j \text{ s.s.}(\sigma_i \times \tau_j),$$

Set

$$R = \bigoplus_{n \in \mathbb{Z}_+} R_n \, .$$

Then we can lift $\times$ to an operation on R

$$\times : R \times R \to R \, .$$

Clearly, $(R, +, \times)$ is an associative ring. Moreover, since the parabolic induction from associate pairs gives the same Jordan-Hölder sequences, the ring R is commutative. We can factor in a natural way $\times$ through $R \otimes R$. Denote the induced map by

$$m : R \otimes R \to R \, .$$

For $\pi \in GL(n,F)\tilde{}$ set

$$m^*(\pi) = \sum_{k=0}^{n} \text{s.s.} \left(r_{M_{(k,n-k)}}^{GL(n,F)}(\pi) \right) .$$

Note that we can consider

$$\text{s.s.} \left(r_{M_{(k,n-k)}}^{GL(n,F)}(\pi) \right) \in R_k \otimes R_{n-k}$$

since for each $\tau \in (GL(k,F) \times GL(n-k,F))\tilde{}$ there exist unique $\tau_k \in GL(k,F)\tilde{}$ and $\tau_{n-k} \in GL(n-k,F)\tilde{}$ such that $\tau \cong \tau_k \otimes \tau_{n-k}$. Thus, we may consider

$$m^*(\pi) \in R \otimes R .$$

Lift m^* to an additive mapping

$$m^*: R \to R \otimes R .$$

The mapping m^* is a dual notion to the multiplication m. It defines the structure of a coalgebra on R. This coalgebra is coassociative i.e.

$$(1 \otimes m^*) \circ m^* = (m^* \otimes 1) \circ m^* .$$

Define a multiplication on $R \otimes R$ in a natural way by

$$\left(\sum \pi_i \otimes \rho_i \right) \times \left(\sum \pi'_j \otimes \rho'_j \right) = \sum_i \sum_j (\pi_i \times \pi'_j) \otimes (\rho_i \times \rho'_j) .$$

A simple computation of the composition series of the Jacquet modules of the parabolically induced representations is now enabled by the following nice formula

$$m^*(\pi_1 \times \pi_2) = m^*(\pi_1) \times m^*(\pi_2).$$

So, we have a description of the composition $m^* \circ m$. The proof of the above formula is done in [Ze1]. The above formula implies that R is a Hopf algebra.

We shall show now two applications of this structure.

Denote by $\nu_n = \nu$ the character $|\det|_F$ of $GL(n,F)$. We have from the sixth section that for $GL(2,F)$, $\nu_1 \otimes \nu_1^{-1}$ is the restriction of the modular character of P to M. Let χ be a character of $F^\times = GL(1,F)$. Then

$$\chi \circ \det \in \text{Ind}_P^{GL(2,F)} \left(\nu_1^{-1/2} \chi \otimes \nu_1^{1/2} \chi \right) .$$

Therefore, as we have observed already, the induced representation is reducible. Further

$$\mathrm{Ind}_P^{GL(2,F)}\left(\nu_1^{-1/2}\chi \otimes \nu_1^{1/2}\chi\right)/\left(\mathbb{C}(\chi \circ \det)\right)$$

is essentially square integrable by the already mentioned criterion for the square integrability. Denote it by $\delta\left([\nu_1^{-1/2}\chi, \nu_1^{1/2}\chi]\right)$. Since on $\mathbb{C}(\chi \circ \det)$ the representation is $\chi \circ \det$, we have

$$(\chi \circ \det)_N = (\chi \circ \det)\mid M = \chi \otimes \chi.$$

Therefore

$$\delta\left([\nu_1^{-1/2}\chi, \nu_1^{1/2}\chi]\right)_N = \nu_1\chi \otimes \nu_1^{-1}\chi.$$

In terms of the normalized Jacquet modules, it means that

$$r_M^{GL(2,F)}\left(\delta\left([\nu_1^{-1/2}\chi, \nu_1^{1/2}\chi]\right)\right) = \nu_1^{1/2}\chi \otimes \nu_1^{-1/2}\chi.$$

Thus

$$m^*\left(\delta([\chi, \nu\chi])\right) = 1 \otimes \delta\left([\chi, \nu\chi]\right) + \nu\chi \otimes \chi + \delta\left([\chi, \nu\chi]\right) \otimes 1 \ .$$

For $n \in \mathbb{Z}_+$ denote

$$[\chi, \nu^n\chi] = \{\nu^k\chi; \ k \in \mathbb{Z}_+, \ k \leqslant n\}\ .$$

Consider now the representation

$$\chi \times \nu\chi \times \nu^2\chi.$$

We have $m^*(\chi) = 1 \otimes \chi + \chi \otimes 1$ since χ is a cuspidal representation. Now we have

$$m^*(\chi \times \nu\chi) = 1 \otimes \chi \times \nu\chi + \chi \otimes \nu\chi + \nu\chi \otimes \chi + \chi \times \nu\chi \otimes 1\ .$$

Further

$$m^*(\chi \times \nu\chi \times \nu^2\chi) = 1 \otimes \chi \times \nu\chi \times \nu^2\chi+$$

$$\chi \otimes \nu\chi \times \nu^2\chi + \nu\chi \otimes \chi \times \nu^2\chi + \nu^2\chi \otimes \chi \times \nu\chi+$$

$$\nu\chi^2 \times \chi \otimes \nu\chi + \nu^2\chi \times \nu\chi \otimes \chi + \chi \times \nu\chi \otimes \nu^2\chi+$$

$$\chi \times \nu\chi \times \nu^2\chi \otimes 1\ .$$

From the above formula we see also that each irreducible subquotient has a nontrivial Jacquet module for the minimal parabolic subgroup. Further,

$$r_{P_{(1,1,1)}}^{GL(3,F)}(\chi \times \nu\chi \times \nu^2\chi) = \nu\chi^2 \otimes \chi \otimes \nu\chi + \chi \otimes \nu\chi^2 \otimes \nu\chi+$$

$$\nu^2\chi \otimes \nu\chi \otimes \chi + \nu\chi \otimes \nu^2\chi \otimes \chi + \chi \otimes \nu\chi \otimes \nu^2\chi + \nu\chi \otimes \chi \otimes \nu^2\chi.$$

In the Grothendieck groups that we have introduced, we have a natural partial order. One writes $x \leqslant y$ if and only if there exist irreducible representations $\pi_1, \ldots, \pi_k$ and $n_1, \ldots, n_k \in \mathbb{Z}_+$ such that

$$y - x = \sum_{i=1}^{k} n_i \pi_i .$$

This partial order can be very useful in constructions of new interesting representations.

Consider $\delta([\chi, \nu\chi]) \times \nu^2\chi$. Then

$$m^* \left(\delta([\chi, \nu\chi]) \times \nu^2\chi \right) = 1 \otimes \delta([\chi, \nu\chi]) \times \nu^2\chi +$$

$$\nu^2\chi \otimes \delta([\chi, \nu\chi]) + \nu\chi \otimes \chi \times \nu^2\chi +$$

$$\nu^2\chi \times \nu\chi \otimes \chi + \delta([\chi, \nu\chi]) \otimes \nu^2\chi +$$

$$\nu^2\chi \times \delta([\chi, \nu\chi]) \otimes 1.$$

Thus

$$r_{P_{(1,1,1)}}^{GL(3,F)} \left(\delta([\chi, \nu\chi]) \times \nu^2\chi \right) = \nu^2\chi \otimes \nu\chi \otimes \chi + \nu\chi \otimes \chi \otimes \nu^2\chi + \nu\chi \otimes \nu^2\chi \otimes \chi.$$

Analogously

$$r_{P_{(1,1,1)}}^{GL(3,F)} \left(\chi \times \delta([\nu\chi, \nu^2\chi]) \right) = \nu^2\chi \otimes \nu\chi \otimes \chi + \chi \otimes \nu^2\chi \otimes \nu\chi + \nu^2\chi \otimes \chi \otimes \nu\chi.$$

Since $r_{P_{(1,1,1)}}^{GL(3,F)} \left(\chi \times \nu\chi \times \nu^2\chi \right)$ is a multiplicity one representation, we see that representations

$$\delta([\chi, \nu\chi]) \times \nu^2\chi \quad \text{and} \quad \chi \times \delta([\nu\chi, \nu^2\chi])$$

have exactly one irreducible subquotient in common. Denote it by $\delta([\chi, \nu^2\chi])$. It is easy to read from the above formulas that

$$m^* \left(\delta([\chi, \nu^2\chi]) \right) = 1 \otimes \delta([\chi, \nu^2\chi]) +$$

$$\nu^2\chi \otimes \delta([\chi, \nu\chi]) + \delta([\nu\chi, \nu^2\chi]) \otimes \chi + \delta([\chi, \nu^2\chi]) \otimes 1.$$

From the criteria for the square integrability one can obtain directly that $\delta([\chi, \nu^2\chi])$ is an essentially square integrable representation. Therefore, we have proved that for $n = 1$, and for $n = 2$, and for any character χ of $F^\times$ there exists a unique subquotient $\delta([\chi, \nu^n\chi])$ of $\chi \times \nu\chi \times \nu^2\chi \times \cdots \times \nu^n\chi$ such that

$$m^* \left(\delta([\chi, \nu^n\chi]) \right) = 1 \otimes \delta([\chi, \nu^n\chi]) + \nu^n\chi \otimes \delta([\chi, \nu^{n-1}\chi]) + \tag{9.1}$$

$$\delta([\nu^{n-1}\chi, \nu^n \chi]) \otimes \delta([\chi, \nu^{n-2}\chi]) + \cdots + \delta([\chi, \nu^n \chi]) \otimes 1.$$

In a similar way, by induction, one can prove this statement for general integer $n \in \mathbb{Z}_+$. Representations $\delta([\chi, \nu^n \chi])$ are essentially square integrable representations.

For an irreducible cuspidal representation ρ of $GL(k, F)$, $\rho \times \nu\rho$ reduces. This very non-trivial fact is proved in [BeZe1]. This fact was also proved by F. Shahidi using L-functions. One can construct now representations $\delta([\rho, \nu^n \rho])$ in the same way as were constructed representations $\delta([\chi, \nu^n \chi])$ before. The formula (9.1) holds for them if one writes ρ instead of χ there. They are essentially square integrable representations. In this way one gets all essentially square integrable representations of general linear groups. If

$$\delta([\rho, \nu^n \rho]) \cong \delta([\rho', \nu^{n'} \rho']),$$

then $n = n'$ and $\rho \cong \rho'$. For more details one should consult the original paper [Ze1] where these representations were constructed.

The above essentially square integrable representations are generalizations, for $GL(n)$, of the Steinberg representation, which was constructed by W. Casselman for any reductive group group G ([Cs1]). In $\mathrm{Ind}_{P_{\min}}^{G}\left(\Delta_{P_{\min}}^{-1/2}\right)$ one generates a subrepresentation V generated by all

$$\mathrm{Ind}_P^G \left(\Delta_P^{-1/2}\right)$$

where $P_{\min} \subsetneqq P \subseteq G$. Then

$$\mathrm{Ind}_{P_{\min}}^{G}\left(\Delta_{P_{\min}}^{-1/2}\right)/V$$

is the *Steinberg representation* of G. It is a square integrable representation.

9.1. Example.

Suppose that ρ_1 and ρ_2 are cuspidal representations of $GL(n_1, F)$ and $GL(n_2, F)$ respectively. Suppose that $\rho_1 \times \rho_2$ splits. Then the square integrability criterion and the Frobenius reciprocity give that

$$\rho_2 = \nu^\alpha \rho_1$$

for some $\alpha \in \mathbb{R}$. Thus if $n_1 \neq n_2$, then obviously $\rho_1 \times \rho_2$ is irreducible. We shall see now how one can get a stronger result very easy from the Hopf algebra structure.

For the simplicity we shall assume that χ is a character of $GL(1, F) = F^\times$. Let ρ be an irreducible cuspidal representation of $GL(m, F)$ with $m > 1$. Take $n \geqslant 0$. Now

$$m^* \left(\chi \times \delta([\rho, \nu^n \rho])\right) = 1 \otimes \chi \times \delta([\rho, \nu^n \rho])$$

$$+\chi \otimes \delta([\rho, \nu^n \rho]) + \nu^n \rho \otimes \chi \times \rho([\rho, \nu^{n-1}\rho])+$$

$$\cdots + \delta([\nu\rho, \nu^n \rho]) \otimes \chi \times \rho + \chi \times \delta([\nu\rho, \nu^n \rho]) \otimes \rho +$$

$$\boxed{\delta([\rho, \nu^n \rho]) \otimes \chi} + \chi \times \delta([\rho, \nu^n \rho]) \otimes 1.$$

Lemma 8.12. implies irreducibility (see the member in the frame).

A complete description of the reducibilities of the representations

$$\delta([\rho_1, \nu^{n_1} \rho]) \times \delta([\rho_2, \nu^{n_2} \rho])$$

is obtained in [Ze1].

For a much less trivial application of this Hopf algebra structure one should consult the paper [Td12].

One very interesting application of this Hopf algebra structure was done by A.V. Zelevinsky in [Ze2] for $GL(n)$ over a finite field $\mathbb{F}$. The structure theory of this Hopf algebra gives a reduction of the classification of the irreducible representations of $GL(n, \mathbb{F})$ to irreducible cuspidal representations of $GL(m, \mathbb{F})$'s where $m \leq n$.

At this point we shall present the Langlands classification for $GL(n)$ ([BlWh], [Si1], [Ze1]).

Denote by T^u the union of all the equivalence classes of the irreducible tempered representations of all $GL(n, F)$ with $n \geq 1$. If δ_i is a square integrable representation of $GL(n_i, F), n_i > 0$, for $i = 1, \ldots, k$, then

$$\delta_1 \times \cdots \times \delta_k$$

is an irreducible tempered representation. If

$$\delta_1 \times \cdots \times \delta_k = \delta'_1 \times \cdots \times \delta'_{k'},$$

then $k = k'$ and sequences $\delta_1, \ldots, \delta_k$ and $\delta'_1, \ldots, \delta'_k$ differ up to a permutation. Each element of T^u can be obtained in the above way.

Let $\tau_1, \ldots, \tau_k \in T^u$. Take $\alpha_1, \ldots, \alpha_k \in \mathbb{R}$ such that

$$\alpha_1 > \alpha_2 > \cdots > \alpha_k$$

Then the representation

$$\nu^{\alpha_1} \tau_1 \times \cdots \times \nu^{\alpha_k} \tau_k$$

has a unique irreducible quotient. Denote it by $L(\nu^{\alpha_1} \tau_1 \otimes \cdots \otimes \nu^{\alpha_k} \tau_k)$. Each irreducible representation of a general linear group is isomorphic to some $L(\nu^{\alpha_1} \tau_1 \otimes \cdots \otimes \nu^{\alpha_k} \tau_k)$. In this way one gets parametrization of $GL(n, F)^{\sim}$ by irreducible cuspidal representations of $GL(m, F)$'s with $m \leq n$. If

$$L(\nu^{\alpha_1} \tau_1 \otimes \cdots \otimes \nu^{\alpha_k} \tau_k) \cong L(\nu^{\alpha'_1} \tau'_1 \otimes \cdots \otimes \nu^{\alpha'_{k'}} \tau'_{k'})$$

with τ_i, $\tau_j' \in T^u$, $\alpha_1 > \cdots > \alpha_k$ and $\alpha_1' > \cdots > \alpha_{k'}'$, then $k = k'$, $\alpha_1 = \alpha_1', \ldots,\ \alpha_k = \alpha_k'$ and $\tau_1 \cong \tau_1', \ldots,\ \tau_k \cong \tau_k'$.

We shall finish this section with the unitary dual of $GL(n, F)$. For the proofs one should consult [Td6]. In general, the unitary duals of the reductive groups over the local fields are still pretty mysterious objects, very often even in the cases when they were determined explicitly. For more explanations concerning the unitary duals one may consult [Td16].

Denote by D^u the set of all equivalence classes of irreducible square integrable representations of all $GL(n, F), n \geq 1$. For $\delta \in D^u$ and $m \geq 1$ set

$$u(\delta, m) = L\left(\nu^{\frac{m-1}{2}}\delta \otimes \nu^{\frac{m-3}{2}}\delta \otimes \cdots \otimes \nu^{-\frac{m-1}{2}}\delta\right).$$

For $0 < \alpha < 1/2$ denote

$$\pi(u(\delta, m), \alpha) = \nu^\alpha u(\delta, m) \times \nu^{-\alpha} u(\delta, m).$$

Representations $u(\delta, m)$ and $\pi(u(\delta, m), \alpha)$ are unitarizable. Denote by B the set of all such representations. If $\pi_1, \ldots, \pi_k \in B$, then $\pi_1 \times \cdots \times \pi_k$ is an irreducible unitarizable representation. If

$$\pi_1 \times \cdots \times \pi_k = \pi_1' \times \cdots \times \pi_{k'}'$$

then $k = k'$ and sequences of representations

$$\pi_1, \ldots, \pi_k \quad \text{and} \quad \pi_1', \ldots, \pi_k'$$

differ up to a permutation. Each irreducible unitarizable representation of a general linear group can be obtained as $\pi_1 \times \cdots \times \pi_k$ for a suitable choice of $\pi_i \in B$.

10. $GSp(n, F)$.

It is convenient to work first with a slightly bigger group than the group $Sp(n, F)$, even if one is interested just in $Sp(n, F)$. We shall describe now that group.

Denote by $GSp(n, F)$ the group of all $S \in GL(2n, F)$ for which there exists $\psi(S) \in F^\times$ such that

$${}^t S \begin{bmatrix} 0 & J_n \\ -J_n & 0 \end{bmatrix} S = \psi(S) \begin{bmatrix} 0 & J_n \\ -J_n & 0 \end{bmatrix}.$$

Then ψ is a homomorphism and $\operatorname{Ker}\psi = Sp(n, F)$. We can identify the characters of $GSp(n, F)$ with the characters of $F^\times$ using ψ. Take $GSp(0, F)$ to be $F^\times$. Note that

$$GSp(1, F) = GL(2, F).$$

187

To a partition α of $m \leqslant n$ we attach a parabolic subgroup and its Levi decomposition in a similar way as it was done for $Sp(n, F)$. That parabolic subgroups will be denoted by $P_\alpha^G = M_\alpha^G N_\alpha$.

Let $m \leqslant n$. For $g \in GL(k, F)$ we have denoted by ${}^\tau g$ the transposed matrix of g with respect to the second diagonal. Then

$$M_{(m)}^S = \left\{ \begin{bmatrix} g & 0 & 0 \\ 0 & h & 0 \\ 0 & 0 & {}^\tau g^{-1} \end{bmatrix} ; g \in GL(m, F), h \in Sp(n - m, F) \right\},$$

$$M_{(m)}^G = \left\{ \begin{bmatrix} g & 0 & 0 \\ 0 & h & 0 \\ 0 & 0 & \psi(h)\,{}^\tau g^{-1} \end{bmatrix} ; g \in GL(m, F), h \in GSp(n - m, F) \right\}.$$

Note that

$$M_{(m)}^S \cong GL(m, F) \times Sp(n - m, F)$$

and

$$M_{(m)}^G \cong GL(m, F) \times GSp(n - m, F)$$

in a natural way.

Let π be an admissible representation of finite length of $GL(m, F)$ and let σ be a similar representation of $Sp(n - m, F)$ (resp. $GSp(n - m, F)$). Define

$$\pi \rtimes \sigma = \mathrm{Ind}_{P_{(m)}^S}^{Sp(n,F)}(\pi \otimes \sigma)$$

$$\left(\text{resp. } \pi \rtimes \sigma = \mathrm{Ind}_{P_{(m)}^G}^{GSp(n,F)}(\pi \otimes \sigma) \right).$$

Then

$$(\pi_1 \times \pi_2) \rtimes \sigma \cong \pi_1 \rtimes (\pi_2 \rtimes \sigma),$$

since the parabolic induction does not depend on the stages of induction. Also

$$(\pi \rtimes \sigma)^\sim \cong \tilde{\pi} \rtimes \tilde{\sigma}.$$

If σ is a representation of a GSp-group, then

$$\chi(\pi \rtimes \sigma) = \pi \rtimes (\chi\sigma)$$

for any character χ of $F^\times$.

Denote by

$$R_n(S) = R[Sp(n, F)],$$

$$R_n(G) = R[GSp(n, F)],$$

$$R(S) = \bigoplus_{n \in \mathbb{Z}_+} R_n(S)$$

and

$$R(G) = \bigoplus_{n \in \mathbb{Z}_+} R_n(G).$$

Then one lifts $\rtimes$ to a biadditive mappings

$$R \times R(S) \to R(S)$$

and

$$R \times R(G) \to R(G)$$

in a similar way as we lifted $\times$ to an operation on R. One can factor these mappings through

$$\mu : R \otimes R(S) \to R(S)$$

and

$$\mu : R \otimes R(G) \to R(G).$$

In this way $R(S)$ and $R(G)$ become modules over R.

If $\pi \in R$ and $\sigma \in R(S)$, then

$$\pi \rtimes \sigma = \tilde{\pi} \rtimes \sigma.$$

This follows from the fact about the Jordan-Hölder series of the parabolically induced representations from the associate pairs.

Suppose that π is an irreducible representation of some GL-group and σ a similar representation of some GSp-group. Then we have also in $R(G)$

$$\pi \rtimes \sigma = \tilde{\pi} \rtimes (\omega_\pi \sigma). \tag{10.1}$$

For an irreducible representation σ of $Sp(n, F)$ (resp. $GSp(n, F)$) set

$$\mu^*(\sigma) = \sum_{m=0}^{n} \text{s.s.} \left(r_{M_{(m)}^S}^{Sp(n,F)}(\sigma) \right)$$

$$\left(\text{resp. } \mu^*(\sigma) = \sum_{m=0}^{n} \text{s.s.} \left(r_{M_{(m)}^G}^{GSp(n,F)}(\sigma) \right) \right).$$

Here (0) denotes the empty partition of 0, i.e. $M_{(0)}^S = Sp(n, F)$ and $M_{(0)}^G = GSp(n, F)$. Then we can consider $\mu^*(\sigma) \in R \otimes R(S)$ (resp.$R \otimes R(G)$). Lift μ^* to an additive mapping

$$\mu^* : R(S) \to R \otimes R(S)$$

$$(\text{resp. } \mu^* : R(G) \to R(G)).$$

Then μ^* is coassociative, i.e.

$$(1 \otimes \mu^*) \circ \mu^* = (m^* \otimes 1) \circ \mu^*.$$

To have some understanding of the Jacquet modules of the parabolically induced representations in this setting, one should know what is $\mu^* \circ \mu$, i.e. what is $\mu^*(\pi \rtimes \sigma)$. For an irreducible admissible representations π_1, π_2, π_3 and π_4 of some general linear groups and σ a similar representation of some GSp-group set

$$(\pi_1 \otimes \pi_2 \otimes \pi_3) \tilde{\rtimes} (\pi_4 \otimes \sigma) = \tilde{\pi}_1 \times \pi_2 \times \pi_4 \otimes \pi_3 \rtimes \omega_{\pi_1} \sigma.$$

Let

$$s : R \otimes R \to R \otimes R$$

be the homomorphism $s \left(\sum r_i \otimes s_i \right) = \sum s_i \otimes r_i$. Set

$$M^* = (1 \otimes m^*) \circ s \circ m^*.$$

Then for an irreducible admissible representation π of $GL(n, F)$ and σ a similar representation of $GSp(m, F)$, we have the following theorem ([Td15], Theorem 5.2.).

10.1. Theorem. $\mu^*(\pi \rtimes \sigma) = M^*(\pi) \tilde{\rtimes} \mu^*(\sigma)$.

A similar formula holds for Sp-groups ([Td15]).

As we could already see, one very important problem of the representation theory and the harmonic analysis, is a construction of the square integrable representations of G. If we exclude the case of the unramified irreducible admissible representations (these are irreducible subquotients of $\mathrm{Ind}_{P_{\min}}^{G}(\chi)$ where χ is a character of $M_{\min}$ which is trivial on the maximal compact subgroup of $M_{\min}$), and the case of $GL(n)$, then in general, very little is known about construction of the essentially square integrable representations of reductive groups over local non-archimedean fields (see also [R1]). Let us recall that the main interest of constructing of the square integrable representations comes, among others, from the Langlands classification and the Plancherel formula. We shall use now the formula for $\mu^* \circ \mu$ to construct in a pretty simple way some new essentially square integrable representations of GSp-groups.

Recall that $GL(2, F) \cong GSp(1, F)$. Now the fact that $\nu \times 1_{F^\times}$ is reducible means in the new notation that $\nu \rtimes 1_{F^\times}$ is reducible. In the same way as for GL-groups one can construct recursively representations $\delta([\nu, \nu^n], \chi)$ where χ is a character of $F^\times$. They are unique subquotients of $(\nu^n \times \nu^{n-1} \times \cdots \times \nu) \rtimes \chi$ which satisfy

$$\mu^* \left(\delta([\nu, \nu^n], \chi) \right) = 1 \otimes \delta([\nu, \nu^n], \chi) + \nu^n \otimes \delta([\nu, \nu^{n-1}], \chi) +$$

$$\delta([\nu^{n-1}, \nu^n]) \otimes \delta([\nu, \nu^{n-2}], \chi) + \cdots + \delta([\nu, \nu^n]) \otimes \chi.$$

These representations are essentially square integrable. They are a special examples of a more general family of square integrable representations as we shall see soon.

Let ρ be a cuspidal representation of $GL(n, F)$ and let σ be a cuspidal representation of $GSp(m, F)$. Write

$$\rho = \nu^\alpha \rho_o,$$

where $\alpha \in \mathbb{R}$ and where ρ_o is a unitarizable representation. In the construction of the square integrable representations we are interested when $\rho \rtimes \sigma$ reduces because of the Proposition 8.14. Let φ be a character of $F^\times$. Then $\rho \rtimes \sigma$ reduces if and only if $\varphi(\rho \rtimes \sigma) \cong \rho \rtimes \varphi\sigma$ reduces. Therefore, without a lost of generality we may suppose that σ is unitarizable. We have now

$$M^*(\rho) = 1 \otimes 1 \otimes \rho + 1 \otimes \rho \otimes 1 + \rho \otimes 1 \otimes 1$$

and

$$\mu^*(\sigma) = 1 \otimes \sigma.$$

Thus

$$\mu^*(\rho \rtimes \sigma) = 1 \otimes \rho \rtimes \sigma + \rho \otimes \sigma + \tilde{\rho} \otimes \omega_\rho \sigma.$$

Suppose that $\nu^\alpha \rho_o \rtimes \sigma$ reduces for some $\alpha \in \mathbb{R}$. If $\alpha = 0$, then the Frobenius reciprocity gives

$$\rho_o \cong \tilde{\rho}_o \,(\cong \overline{\rho}_o)$$

and

$$\sigma \cong \omega_{\rho_o} \sigma.$$

Suppose that $\alpha \neq 0$. Take a positive-valued character χ such that $\chi(\rho \rtimes \sigma) \cong \rho \rtimes (\chi\sigma)$ has a unitary central character. Now $\rho \rtimes (\chi\sigma)$ has a square integrable subquotient by the criterion mentioned in the fourth section. Thus, $\rho \rtimes (\chi\sigma)$ and $\tilde{\rho} \rtimes (\chi^{-1}\sigma)$ have an irreducible subquotient in common. Since $\rho \rtimes (\chi\sigma)$ has no non-trivial cuspidal subquotients, looking at the Jacquet modules one obtains

$$\overline{\rho} \otimes \chi^{-1}\sigma = \rho \otimes \sigma \ \text{ or } \ \overline{\rho} \otimes \chi^{-1}\sigma = \tilde{\rho} \otimes \omega_\rho \sigma.$$

If $\overline{\rho} = \rho$, then ρ is unitarizable, i.e. $\alpha = 0$. Thus

$$\rho \cong \overline{\rho} \ \text{ and } \ \chi^{-1}\sigma \cong \omega_\rho \sigma.$$

This implies $\rho_o \cong \tilde{\rho}_o$ and $\sigma \cong \omega_{\rho_o}\sigma$.

If $\nu^{\alpha_o}\rho_o \rtimes \sigma$ reduces with $\alpha_o \neq 0$, then $\nu^{\alpha_o}\rho \rtimes \sigma$ contains a unique essentially square integrable subquotient. Note that then also $\nu^{-\alpha_o}\rho_o \rtimes \sigma$ reduces. Therefore we can take $\alpha_o > 0$. Denote that subquotient by $\delta(\nu^{\alpha_o}\rho, \sigma)$. We have seen up to now that such situation appears for $\alpha_o = 1$. Suppose therefore that $\alpha_o = 1$ (a similar treatment holds for any $\alpha_o > 0$). One can define recursively representations $\delta([\nu\rho, \nu^n\rho], \sigma)$ as subquotients of $(\nu^n\rho \times \nu^{n-1}\rho \times \cdots \times \nu\rho) \rtimes \sigma$ which satisfy

$$\mu^*(\delta([\nu\rho, \nu^n\rho], \sigma)) = 1 \otimes \delta([\nu\rho, \nu^n\rho], \sigma) + \nu^n\rho \otimes \delta([\nu\rho, \nu^{n-1}\rho], \sigma) +$$

$$\delta([\nu^{n-1}\rho, \nu^n\rho]) \otimes \delta([\nu\rho, \nu^{n-2}\rho], \sigma) + \cdots + \delta([\nu\rho, \nu^n\rho]) \otimes \sigma.$$

These representations are essentially square integrable. They will be called an *essentially square integrable representations of the Steinberg type*.

It is interesting to note that even if $\rho \rtimes \sigma = \nu^\alpha\rho_o \rtimes \sigma$ is irreducible for any $\alpha \in \mathbb{R}$, in same cases it is possible to attach also to these representations a series of essentially square integrable representations. We shall explain it now.

Suppose that $\rho_o \cong \tilde{\rho}_o$ and $\omega_{\rho_o}\sigma \not\cong \sigma$. This provides that $\rho \rtimes \sigma = (\nu^\alpha\rho_o) \rtimes \sigma$ is irreducible for any $\alpha \in \mathbb{R}$. Consider the representation

$$\nu\rho_o \times \rho_o \rtimes \sigma.$$

We have already seen that

$$M^*(\nu\rho_o) = 1 \otimes 1 \otimes \nu\rho_o + 1 \otimes \nu\rho_o \otimes 1 + \nu\rho_o \otimes 1 \otimes 1$$

and

$$\mu^*(\rho_o \rtimes \sigma) = 1 \otimes \rho_o \rtimes \sigma + \rho_o \otimes \sigma + \tilde{\rho}_o \otimes \omega_{\rho_o}\sigma.$$

Thus,

$$\mu^*(\nu\rho_o \times \rho_o \rtimes \sigma) = 1 \otimes \nu\rho_o \times \rho_o \rtimes \sigma$$

$$+ [\rho_o \otimes \nu\rho_o \rtimes \sigma + \tilde{\rho}_o \otimes \nu\rho_o \rtimes \omega_{\rho_o}\sigma$$

$$+ \nu\rho_o \otimes \rho_o \rtimes \sigma + \nu^{-1}\tilde{\rho}_o \otimes \omega_{\nu\rho_o}(\rho_o \rtimes \sigma)]$$

$$+ [\nu\rho_o \times \rho_o \otimes \sigma + \nu\rho_o \times \tilde{\rho}_o \otimes \omega_{\rho_o}\sigma$$

$$+ \nu^{-1}\tilde{\rho}_o \times \tilde{\rho}_o \otimes \omega_{\nu\rho_o}\omega_{\rho_o}\sigma + \nu^{-1}\tilde{\rho}_o \times \rho_o \otimes \omega_{\nu\rho_o}\sigma]$$

(recall that $\tilde{\rho}_o \cong \rho_o$ and $\omega_{\rho_o}^2 = 1_{F^\times}$). To make things precise, suppose that ρ_o is a representation of $GL(m, F)$ and that σ is a representation of $GL(k, F)$. Then

$$\text{s.s.} \left(r_{P^G_{(m,m)}}^{GSp(2m+k,F)}(\nu\rho_o \times \rho_o \rtimes \sigma) \right) =$$

$$\nu\rho_o \otimes \rho_o \otimes \sigma + \rho_o \otimes \nu\rho_o \otimes \sigma +$$

$$\nu\rho_o \otimes \tilde{\rho}_o \otimes \omega_{\rho_o}\sigma + \tilde{\rho}_o \otimes \nu\rho_o \otimes \omega_{\rho_o}\sigma +$$

$$\nu^{-1}\tilde{\rho}_o \otimes \tilde{\rho}_o \otimes \omega_{\nu\rho_o}\omega_{\rho_o}\sigma + \tilde{\rho}_o \otimes \nu^{-1}\tilde{\rho}_o \otimes \omega_{\nu\rho_o}\omega_{\rho_o}\sigma +$$

$$\nu^{-1}\tilde{\rho}_o \otimes \rho_o \otimes \omega_{\nu\rho_o}\sigma + \rho_o \otimes \nu^{-1}\tilde{\rho}_o \otimes \omega_{\nu\rho_o}\sigma.$$

Note that this is a multiplicity one representation. The above formula implies also that for each irreducible subquotient τ we have

$$r^{GSp(2m+k,F)}_{P^G_{(m,m)}}(\tau) \neq 0.$$

We know

$$m^*(\delta([\rho_o, \nu\rho_o])) = 1 \otimes \delta([\rho_o, \nu\rho_o]) + \nu\rho_o \otimes \rho_o + \delta([\rho_o, \nu\rho_o]) \otimes 1.$$

Thus

$$M^*\left(\delta([\rho_o, \nu\rho_o])\right) =$$

$$(1 \otimes m^*)\left(\delta([\rho_o, \nu\rho_o]) \otimes 1 + \rho_o \otimes \nu\rho_o + 1 \otimes \delta([\rho_o, \nu\rho_o])\right)$$

$$= \delta([\rho_o, \nu\rho_o]) \otimes 1 \otimes 1 + \rho_o \otimes 1 \otimes \nu\rho_o + \rho_o \otimes \nu\rho_o \otimes 1$$

$$+ 1 \otimes 1 \otimes \delta([\rho_o, \nu\rho_o]) + 1 \otimes \nu\rho_o \otimes \rho_o + 1 \otimes \delta([\rho_o, \nu\rho_o]) \otimes 1.$$

Now

$$\mu^*\left(\delta([\rho_o, \nu\rho_o]) \rtimes \sigma\right) = 1 \otimes \delta([\rho_o, \nu\rho_o]) \rtimes \sigma +$$

$$\nu\rho_o \otimes \rho_o \rtimes \sigma + \tilde{\rho}_o \otimes \nu\rho_o \rtimes \omega_{\rho_o}\sigma +$$

$$\delta([\nu^{-1}\rho_o, \rho_o]) \otimes \omega_{\rho_o}\omega_{\nu\rho_o}\sigma + \rho_o \times \nu\rho_o \otimes \omega_{\rho_o}\sigma + \delta([\rho_o, \nu\rho_o]) \otimes \sigma.$$

Therefore,

$$r^{GSp(2m+k,F)}_{P^G_{(m,m)}}(\delta([\rho_o, \nu\rho_o]) \rtimes \sigma) =$$

$$\rho_o \otimes \nu^{-1}\rho_o \otimes \omega_{\rho_o}\omega_{\nu\rho_o}\sigma + \rho_o \otimes \nu\rho_o \otimes \omega_{\rho_o}\sigma +$$

$$\nu\rho_o \otimes \rho_o \otimes \omega_{\rho_o}\sigma + \nu\rho_o \otimes \rho_o \otimes \sigma.$$

By (10.1) we have in $R(G)$

$$\nu\rho_o \times \rho_o \rtimes \sigma = \nu\rho_o \times \rho_o \rtimes \omega_{\rho_o}\sigma.$$

Thus we have in $R(G)$

$$\delta([\rho_o, \nu\rho_o]) \rtimes \omega_{\rho_o}\sigma \leqslant \nu\rho_o \times \rho_o \rtimes \sigma.$$

The same calculation gives

$$r^{GSp(2m+k,F)}_{P^G_{(m,m)}}(\delta([\rho_o, \nu\rho_o]) \rtimes \omega_{\rho_o}\sigma) =$$

$$\rho_o \otimes \nu^{-1}\rho_o \otimes \omega_{\nu\rho_o}\sigma + \rho_o \otimes \nu\rho_o \otimes \sigma +$$

$$\nu\rho_o \otimes \rho_o \otimes \sigma + \nu\rho_o \otimes \rho_o \otimes \omega_{\rho_o}\sigma.$$

We can now conclude that there exist subquotients

$$\tau_1, \ldots, \tau_p \in GSp(2m+k, F)\tilde{\ }$$

of $\delta([\rho_o, \nu\rho_o]) \rtimes \sigma$ and $\delta([\rho_o, \nu\rho_o]) \rtimes \omega_{\rho_o}\sigma$ such that

$$r^{GSp(2m+k,F)}_{P^G_{(m,m)}}(\tau_1 + \cdots + \tau_p) = \nu\rho_o \otimes \rho_o \otimes \sigma + \nu\rho_o \otimes \rho_o \otimes \omega_{\rho_o}\sigma.$$

Clearly, $p \leqslant 2$. Without a lost of generality we can suppose that $\nu\rho_o \otimes \rho_o \otimes \sigma$ is a quotient of

$$r^{GSp(2m+k,F)}_{P^G_{(m,m)}}(\tau_1).$$

Otherwise, $\nu\rho_o \otimes \rho_o \otimes \omega_{\rho_o}\sigma$ is a quotient, and one proceeds in the same way as we shall do now. The Frobenius reciprocity implies

$$\tau_1 \hookrightarrow \nu\rho_o \times \rho_o \rtimes \sigma.$$

Since $\rho_o \rtimes \sigma$ is irreducible, we have $\rho_o \rtimes \sigma \cong \rho_o \rtimes \omega_{\rho_o}\sigma$. Thus

$$\nu\rho_o \times \rho_o \rtimes \sigma \cong \nu\rho_o \times \rho_o \rtimes \omega_{\rho_o}\sigma.$$

Therefore $\tau_1 \hookrightarrow \nu\rho_o \times \rho_o \rtimes \omega_{\rho_o}\sigma$. Now the Frobenius reciprocity implies that $\nu\rho_o \otimes \rho_o \otimes \omega_{\rho_o}\sigma$ is also a quotient of

$$r^{GSp(2m+k,F)}_{P^G_{(m,m)}}(\tau_1).$$

Therefore, $p = 1$.

Denote

$$\tau_1 = \delta([\rho_o, \nu\rho_o], \sigma).$$

It is now easy to get from the formula for $\mu^*(\nu\rho_o \times \rho_o \rtimes \sigma)$ that

$$\mu^*\left(\delta([\rho_o, \nu\rho_o], \sigma)\right) = 1 \otimes \delta([\rho_o, \nu\rho_o], \sigma)$$

$$+\nu\rho_o \otimes \rho_o \rtimes \sigma + \delta([\rho_o, \nu\rho_o]) \otimes (\sigma + \omega_{\rho_o}\sigma).$$

This representation is essentially square integrable by the square integrability criterion.

In a similar way as above, one constructs representations $\delta([\rho_o, \nu^n\rho_o], \sigma)$ which are subquotients of $\nu^n\rho_o \times \nu^{n-1}\rho_o \times \cdots \times \nu\rho_o \times \rho_o \rtimes \sigma$ which satisfy

$$\mu^*\left(\delta([\rho_o, \nu^n\rho_o], \sigma)\right) = 1 \otimes \delta([\rho_o, \nu^n\rho_o], \sigma) +$$

$$\nu^n\rho_o \otimes \delta\left([\rho_o, \nu^{n-1}\rho_o], \sigma\right) +$$

$$\delta\left([\nu^{n-1}\rho_o, \nu^n\rho_o]\right) \otimes \delta\left([\rho_o, \nu^{n-2}\rho_o], \sigma\right) +$$

$$\vdots$$

$$+\delta\left([\nu^2\rho_o, \nu^n\rho_o]\right) \otimes \delta([\nu\rho_o, \rho_o], \sigma)+$$

$$\delta([\nu\rho_o, \nu^n\rho_o]) \otimes \rho_o \rtimes \sigma+$$

$$\delta([\rho_o, \nu^n\rho_o]) \otimes (\sigma + \omega_{\rho_o}\sigma).$$

These representations appear as common irreducible subquotients of

$$\nu^n\rho_o \rtimes \delta\left([\rho_o, \nu^{n-1}\rho_o], \sigma\right)$$

and

$$\delta\left([\nu^{n-1}\rho_o, \nu^n\rho_o]\right) \rtimes \delta\left([\rho_o, \nu^{n-2}\rho_o], \sigma\right).$$

Here we denote

$$\delta(\rho_o, \sigma) = \rho_o \rtimes \sigma.$$

With this notation we have

$$\mu^*\left(\delta([\rho_o, \nu^n\rho_o], \sigma)\right) =$$

$$\sum_{k=0}^{n} \delta\left([\nu^{k+1}\rho_o, \nu^n\rho_o]\right) \otimes \delta\left([\rho_o, \nu^k\rho_o], \sigma\right)$$

$$+\delta\left([\rho_o, \nu^n\rho_o]\right) \otimes (\sigma + \omega_{\rho_o}\sigma).$$

Representations

$$\delta([\rho_o, \nu^n\rho_o], \sigma), \quad n \geqslant 1,$$

are essentially square integrable.

Suppose for a moment that σ is a character of $GSp(0, F)$. Then

$$\delta([\rho_o, \nu^n\rho_o], \sigma)|Sp(nm, F)$$

is a sum of two square integrable representations which are not equivalent. This follows easily from the Clifford theory for p-adic groups which we have already mentioned ([GbKn]).

The simplest example of the above representations $\delta([\rho_o, \nu^n\rho_o], \sigma)$ is the case when ρ_o and σ are characters of $F^\times$, ρ_o is of order 2 and $n = 1$. These representations were pointed out by F. Rodier in [R1]. Because of that, we shall call these representations *essentially square integrable representations of the Rodier type*.

Now one can consider "mixed" case. New essentially square integrable representations are constructed using several Steinberg and Rodier type essentially square integrable representations (see [Td13] for an explicit description of that representations).

11. ON THE REDUCIBILITY OF PARABOLIC INDUCTION.

In this section we shall see how the formulas for $\mu^*(\pi \rtimes \sigma)$ can be used in the study of the reducibility of the parabolically induced representations. Not too many general methods exist for this purpose. There is a very good technology for this problem for the general linear groups developed by J. Bernstein and A.V. Zelevinsky ([Ze1]). W. Casselman introduced in [Cs1] a method which was after that used in various cases of parabolically induced representations by one-dimensional characters (usually unramified). We have outlined a part of that method in the sixth section.

We shall present here one method which works pretty well in different situations. For this method it does not matter if the inducing representation is one-dimensional or not. One very simple application of this method will be explained now. More sophisticated applications are announced in [Td13]. C. Jantzen used this method in his thesis [Jn].

We shall illustrate the method on the following example. Let ρ_o be a cuspidal unitarizable representation of $GL(m, F)$, where $m \geqslant 2$, such that

$$\rho_o \cong \tilde{\rho}_o \text{ and } \omega_{\rho_o} \neq 1_{F^\times}.$$

Let σ be a character of $F^\times$. Then we have defined

$$\delta([\rho_o, \nu^n \rho_o], \sigma).$$

We shall prove

11.1. Proposition. *Let χ be a character of $F^\times$ different from ω_{ρ_o}. Then*

$$\chi \rtimes \delta([\rho_o, \nu^n \rho_o], \sigma)$$

is reducible if and only if $\chi = \nu^{\pm 1}$. If we have reducibility, then we get a multiplicity one representation of length two.

We shall first show that for $\chi \neq \omega_{\rho_o}$ we have the following

11.2. Lemma. *The representation $\chi \times \rho_o \rtimes \sigma$ is irreducible if $\chi \neq \nu^{\pm 1}$. If $\chi = \nu^{\pm 1}$, then we have a multiplicity one representation of length two.*

Proof. The reducibility for $\chi = \nu^{\pm 1}$ is clear. Suppose that χ is a non-unitary character different from $\nu^{\pm 1}$. Since

$$\chi \times (\rho_o \rtimes \sigma) \cong \chi \rtimes (\rho_o \rtimes \omega_{\rho_o} \sigma) \cong$$

$$(\chi \times \rho_o) \rtimes \omega_{\rho_o} \sigma \cong (\rho_o \times \chi) \rtimes \omega_{\rho_o} \sigma \cong$$

$$\rho_o \times (\chi \rtimes \omega_{\rho_o} \sigma) \cong \rho_o \times \chi^{-1} \rtimes \chi \omega_{\rho_o} \sigma \cong \chi^{-1} \times \rho_o \rtimes \chi \omega_{\rho_o} \sigma,$$

the elementary properties of the Langlands classification imply that $\chi \times \rho_o \rtimes \sigma$ is irreducible (the long intertwining operator is an isomorphism).

Further we consider

$$\mu^*(\chi \times \rho_o \rtimes \sigma) =$$

$$[(\chi \otimes 1 \otimes 1 + 1 \otimes \chi \otimes 1) + (1 \otimes 1 \otimes \chi)] \,\tilde{\rtimes}$$

$$[1 \otimes \rho_o \rtimes \sigma + (\rho_o \otimes \sigma + \rho_o \otimes \omega_{\rho_o} \sigma)] =$$

$$+1 \otimes \chi \times \rho_o \rtimes \sigma +$$

$$\left[\chi^{-1} \otimes \rho_o \rtimes \chi \sigma + \chi \otimes \rho_o \rtimes \sigma \right] +$$

$$\left[\rho_o \otimes \chi \rtimes \sigma + \rho_o \otimes \chi \rtimes \omega_{\rho_o} \sigma \right] + \tag{11.1}$$

$$\left[\chi^{-1} \times \rho_o \otimes \chi \sigma + \chi^{-1} \times \rho_o \otimes \chi \omega_{\rho_o} \sigma + \chi \times \rho_o \otimes \sigma + \chi \times \rho_o \otimes \omega_{\rho_o} \sigma \right].$$

In the line (11.1) both representations are irreducible by the seventh, or by the last section. Note that

$$\chi \rtimes \sigma \not\cong \chi \rtimes \omega_{\rho_o} \sigma$$

since $\chi \neq \omega_{\rho_o}$ (look at the Jacquet modules). Thus

$$\rho_o \otimes \chi \rtimes \sigma \not\cong \rho_o \otimes \chi \rtimes \omega_{\rho_o} \sigma. \tag{11.2}$$

Looking at the Jacquet module for $P^G_{(m)}$ (this is the line (11.1)), we see that $\chi \times \rho_o \rtimes \sigma$ is a representation of the length $\leqslant 2$ and that it is a multiplicity one representation (see also the section eight).

Suppose now that χ is a unitary character (such that $\chi \neq \omega_{\rho_o}$). Consider the Frobenius reciprocity for

$$\rho_o \rtimes (\chi \rtimes \sigma).$$

Since the representations in the line (11.1) are not isomorphic by (11.2), we get that the commutator of the representation $\chi \times \rho_o \rtimes \sigma$ consists of the scalar operators only, since the commutator is one-dimensional by the Frobenius reciprocity applied to the subgroup $P^G_{(m)}$. Now the unitarizability of the representation implies the irreducibility. $\qquad\square$

Proof of Proposition 11.1. We have directly

$$\mu^*\left(\chi \rtimes \delta([\rho_o, \nu^n \rho_o], \sigma)\right) =$$

$$1 \otimes \chi \rtimes \delta([\rho_o, \nu^n \rho_o], \sigma) +$$

$$\chi \otimes \delta([\rho_o, \nu^n \rho_o], \sigma) + \chi^{-1} \otimes \chi \delta([\rho_o, \nu^n \rho_o], \sigma) +$$

$$\nu^n \rho_o \otimes \chi \rtimes \delta([\rho_o, \nu^{n-1}\rho_o], \sigma)+$$

$$\vdots$$

$$+\chi \times \delta([\nu^2 \rho_o, \nu^n \rho_o]) \otimes \delta([\rho_o, \nu \rho_o], \sigma) + \chi^{-1} \times \delta([\nu^2 \chi_o, \nu^n \rho_o]) \otimes \chi \delta([\rho_o, \nu \rho_o], \sigma)+$$

$$\boxed{\delta([\nu \rho_o, \nu^n \rho_o]) \otimes \chi \times \rho_o \rtimes \sigma}+$$

$$\chi \times \delta([\nu \rho_o, \nu^n \rho_o]) \otimes \rho_o \rtimes \sigma + \chi^{-1} \times \delta([\nu \rho_o, \nu^n \rho_o]) \otimes \chi^{-1}(\rho_o \rtimes \sigma)+$$

$$\delta([\rho_o, \nu^n \rho_o]) \otimes \chi \rtimes (\sigma + \omega_{\rho_o} \sigma)+$$

$$+\chi \times \delta([\rho_o, \nu^n \rho_o]) \otimes (\sigma + \omega_{\rho_o} \sigma) + \chi^{-1} \rtimes \delta([\rho_o, \nu^n \rho_o]) \otimes \chi(\sigma + \omega_{\rho_o} \sigma).$$

Applying the Lemma 8.12. to the boxed member, we get the irreducibility.

For the reducibility of $\nu \rtimes \delta([\rho_o, \nu^n \rho_o], \sigma)$ one considers the representation

$$\delta([\rho_o, \nu^n \rho_o]) \rtimes \delta(\nu, \sigma)$$

and shows that these two representations have a common irreducible subquotient. It must be a proper subquotient of $\nu \rtimes \delta([\rho_o, \nu^n \rho_o], \sigma)$. This proves the reducibility for $\chi = \nu^{\pm 1}$. $\qquad\square$

11.3. Remark.

Considering the restriction to the symplectic group, one can prove that

$$\omega_{\rho_o} \rtimes \delta([\rho_o, \nu^n \rho_o], \sigma)$$

reduces.

REFERENCES

[Ar1]   Arthur, J., *Automorphic representations and number theory*, Canadian Mathematical Society Conference Proceedings, Vol. 1, Providence, Rhode Island, 1981, pp. 3-51.

[Ar2]   Arthur, J., *On some problems suggested by the trace formula*, Lie Group Representations II, Proceedings, University of Maryland 1982-83, Lecture Notes in Math. 1041, Springer-Verlag, Berlin, 1984, pp. 1-49.

[Ar3]   Arthur, J., *Unipotent automorphic representations: conjectures*, Astérisque **171-172** (1989), 13-71.

[Au]   Aubert, A.-M., *Description de la correspondance de Howe en terms de classification de Kazhdan-Lusztig*, Invent. Math. **103** (1991), 379-415.

[Ba]   Barbasch, D., *The unitary dual for complex classical groups*, Invent. Math. **96** (1989), 103-176.

[BaMo]   Barbasch, D., Moy, A., *A unitarity criterion for p-adic groups*, Invent. Math. **98** (1989), 19-37.

[Bg]   Bargmann V., *Irreducible unitary representations of the Lorentz group*, Ann. of Math. **48** (1947), 568-640.

[Be1]   Bernstein, J.N., *All reductive p-adic groups are tame*, Functional Anal. Appl. **8** (1974), 91-93.

[Be2]   Bernstein, J., *P-invariant distributions on $GL(N)$ and the classification of unitary representations of $GL(N)$ (non-archimedean case)*, Lie Group Representations II, Proceedings, University of Maryland 1982-83, Lecture Notes in Math. 1041, Springer-Verlag, Berlin, 1984, pp. 50-102.

[BeDe]   Bernstein, J., rédigé par Deligne, P., *Le "centre" de Bernstein*, Représentations des Groupes Réductifs sur un Corps Local, by Bernstein, J.-N., Deligne, P., Kazhdan, D. and Vignéras, M.-F., Hermann, Paris, 1984.

[BeDeKz] Bernstein, J., Deligne, P., Kazhdan, D., *Trace Paley-Wiener theorem for reductive p-adic groups*, J. Analyse Math **42** (1986), 180-192.

[BeZe1] Bernstein, I. N., Zelevinsky, A.V., *Representations of the group $GL(n, F)$, where F is a local non-Archimedean field*, Uspekhi Mat. Nauk. **31** (1976), 5-70.

[BeZe2] Bernstein, J., Zelevinsky, A.V., *Induced representations of reductive p-adic groups I*, Ann. Sci. École Norm Sup. **10** (1977), 441-472.

[Bl1]   Borel, A., *Linear algebraic groups*, Benjamin, New York, 1969.

[Bl2]   Borel, A., *Automorphic L-functions*, Symp. Pure Math. 33, part 2, Amer. Math. Soc., Providence, Rhode Island, 1979, pp. 27-61.

[BlJc]   Borel, A., Jacquet, H., *Automorphic forms and automorphic representations*, Symp. Pure Math. 33, part 1, Amer. Math. Soc., Providence, Rhode Island, 1979, pp. 189-202.

[BlTi1] Borel, A., Tits, J., *Groupes réductifs*, Publ. Math. I.H.E.S. **27** (1965), 55-150.

[BlTi2] Borel, A., Tits, J., *Complements à l'article, Groupes réductifs*, Publ. Math. I.H.E.S. **41** (1972), 253-276.

[BlWh]   Borel, A., Wallach, N., *Continuous cohomology, discrete subgroups, and representations of reductive groups*, Princeton University Press, Princeton, 1980.

[Bb1]   Bourbaki, N., *Groupes et algèbres de Lie (chapitres 4, 5 et 6)*, Hermann, Paris, 1968.

[Bb2]   Bourbaki, N., *Groupes et algèbres de Lie, Groupes de Lie réels compacts (chapitre 9)*, Masson, Paris, 1982.

[Bb3]   Bourbaki, N., *Intégration*, Hermann, Paris.

[BhTi]   Bruhat, F., Tits, J., *Groupes réductifs sur un corp local*, Publ. Math. I.H.E.S. **41** (1972), 5-251.

[BuKu]   Bushnell, C.J., Kutzko, P., *The admissible dual of $GL(N)$ via compact open subgroups* (1993), Princeton University Press, Princeton.

[Cy]   Carayol, H., *Représentations cuspidales du group linéaire*, Ann. Sci. École Norm. Sup **17** (1984), 191-226.

[Ct]   Cartier, P., *Representations of p-adic groups; a survey*, Symp. Pure Math. 33, part 1, Amer. Math. Soc., Providence, Rhode Island, 1979, pp. 111-155.

[Cs1]   Casselman, W., *Introduction to the theory of admissible representations of p-adic reductive groups*, preprint.

[Cs2]   Casselman, W., *Characters and Jacquet modules*, Math. Ann. **230** (1977), 101-105.

[Cs3]   Casselman, W., *Jacquet modules for real groups*, Proceedings of the International Congress of Mathematics, Helsinki, 1978.

[Cs4]   Casselman, W., *The unramified principal series of ℘-adic groups I. The spherical function*, Compositio Math. **40,** Fasc. 3 (1980), 387-406.

[CsMi]   Casselman, W., Miličić, D., *Asymptotic behavior of matrix coefficients of admissible representations*, Duke Math. J. **49,** n. 4 (1982), 869-930.

[Cl1]   Clozel, L., *Sur une conjecture de Howe-I*, Compositio Math. **56** (1985), 87-110.

[Cl2]   Clozel, L., *Orbital integrals on p-adic groups: A proof of the Howe conjecture*, Ann. Math. **129** (1989), 237-251.

[Co]   Corwin, L., *A construction of the supercuspidal representations of $GL_n(F)$, F p-adic*, Tran. Amer. Math. Soc. **337** (1993), 1-58.

[Cu]   Curtis, C.W., *Representations of finite groups of Lie type*, Bulletin Amer. Math. Soc. **1** (1979), 721-757.

[DeLu]   Deligne, P., Lusztig, G., *Representations of reductive groups over finite fields*, Ann. of Math. **103** (1976), 103-161.

[DeKzVg] Deligne, P., Kazhdan, D., Vignéras, M.-F., *Représentations des algèbres centrales simples p-adiques*, Représentations des Groupes Réductifs sur un Corps Local, by Bernstein, J.-N., Deligne, P., Kazhdan, D. and Vignéras, M.-F., Hermann, Paris, 1984.

[Dj]   Dijk, G. van, *Computation of certain induced characters of ℘-adic groups*, Math. Ann. **199** (1972), 229-240.

[Dx]   Dixmier, J., *C*-algebras et leurs Représentations*, Gauthiers-Villars, Paris, 1969.

[Du]   Duflo, M., *Représentations unitaires irréductibles des groups simples complexes de rang deux*, Bull. Soc. Math. France **107** (1979), 55-96.

[Fa]   Faddeev, D.K, *On multiplication of representations of classical groups over finite field with representations of the full linear group (in Russian)*, Vestnik Leningradskogo Universiteta **13** (1976), 35-40.

[Fe]   Fell J.M.G., *Non-unitary dual space of groups*, Acta Math. **114** (1965), 267-310.

[Gb1]   Gelbart, S., *Automorphic forms on adele groups*, Ann. of Math. Stud. 83, Princeton University Press, Princeton, 1975.

[Gb2]   Gelbart, S.S., *Elliptic curves and automorphic representations*, Advan. in Math. **21** (1976), 235-292.

[Gb3]   Gelbart, S.S., *An elementary introduction to the Langlands program*, Bulletin Amer. Math. Soc. **10** (1984), 177-219.

[GbJc]   Gelbart, S. and Jacquet, H., *Forms on GL(2) from the analytic point of view*, Symp. Pure Math. 33, part 1, Amer. Math. Soc., Providence, Rhode Island, 1979, pp. 213-251.

[GbKn]   Gelbart, S.S., Knapp, A. W., *L-indistinguishability and R groups for the special linear group*, Advan. in Math. **43** (1982), 101-121.

[GbPSRa] Gelbart, S., Piatetski-Shapiro, I., Rallis, S., *Explicit constructions of automorphic L-functions*, Lecture Notes in Math 1254, Springer-Verlag, Berlin, 1987.

[GbSh]   Gelbart, S., Shahidi, F., *Analytic properties of automorphic L-functions*, Academic Press, Boston, 1988.

[GlN1]   Gelfand I.M., Naimark M.A., *Unitary representations of the Lorentz group*, Izvestiya Akad. Nauk SSSR, Ser. Mat. (Russian) **11** (1947), 411-504.

[GlN2]   Gelfand, I.M., Naimark, M.A., *Unitare Derstellungen der Klassischen Gruppen (German translation of Russian publication from 1950)*, Akademie Verlag, Berlin, 1957.

[GlGrPS] Gelfand, I. M., Graev, M., Piatetski-Shapiro, *Representation theory and automorphic functions*, Saunders, Philadelphia, 1969.

[GlKz]   Gelfand, I.M., Kazhdan, D.A., *Representations of $GL(n,k)$*, Lie groups and their Representations, Halstead Press, Budapest, 1974, pp. 95-118.

[Gr]   Gross, H.B., *Some applications of Gelfand pairs to number theory*, Bulletin Amer. Math. Soc. **24, n.2** (1991), 277-301.

[Gu]   Gustafson, R., *The degenerate principal series for $Sp(2n)$*, Mem. of the Amer. Math. Society **248** (1981), 1-81.

[HC1]   Harish-Chandra, *Harmonic analysis on reductive p-adic groups*, Symp. Pure Math. 26, Amer. Math. Soc., Providence, Rhode Island, 1973, pp. 167-192.

[HC2]   Harish-Chandra, *Collected papers*, Springer-Verlag, Berlin, 1983.

[He1]   Henniart, G., *On the local Langlands conjecture for $GL(n)$: the cyclic case*, Ann. of Math. **123** (1986), 145-203.

[He2]   Henniart, G., *Représentations des groupes réductifs p-adiques*, Séminaire Bo urbaki n. *736 (1991)*, Astérisque **201-202-203** (1981), 193-219.

[Ho1]   Howe, R., *Tamely ramified supercuspidal representations of GL_n*, Pacific J. Math. **73** (1977), 437-460.

[Ho2]   Howe, R., *Some qualitative results on the representation theory of Gl_n over a p-adic field*, Pacific J. Math. **73** (1977), 479-538.

[Ho3]   Howe, R., *θ-series and automorphic forms*, Symp. Pure Math. 33, part 1, Amer. Math. Soc., Providence, Rhode Island, 1979, pp. 275-286.

[HoMe]   Howe, R., Moore, C.C., *Asymptotic properties of unitary representations*, J. Functional Analysis **32, n.1** (1979), 72-96.

[HoMo]   Howe, R. with the collaboration of Moy, A., *Harish-Chandra homomorphism for p-adic groups*, CBMS Regional Conference Series 59, Providence, 1985.

[Hu]   Humpreys, J.E., *Linear algebraic groups*, Springer-Verlag, New York, 1975.

[Jc1]   Jacquet, H, *Représentations des groupes linéaires p-adiques*, Theory of Group Representations and Fourier Analysis (Proceedings of a conference at Montecatini, 1970), C.I.M.E., Edizioni Cremonese, Roma.

[Jc2]   Jacquet, H, *Generic representations*, Non-Commutative Harmonic Analysis, Lecture Notes in Math. 587, Spring Verlag, Berlin, 1977, pp. 91-101.

[Jc3]   Jacquet, H., *Principal L-functions of the linear group*, Symp. Pure Math. 33, part 2, Amer. Math. Soc., Providence, Rhone Island, 1979, pp. 63-86.

[Jc4]   Jacquet, H., *On the residual spectrum of $GL(n)$*, Lie Group Representations II, Proceedings, University of Maryland 1982-83, Lecture Notes in Math. 1041, Springer-Verlag, Berlin, 1984, pp. 185-208.

[JaLs]   Jacquet, H, Langlands, R.P., *Automorphic Forms on $GL(2)$*, Lecture Notes in Math 114, Springer-Verlag, Berlin, 1970.

[Jn]   Jantzen, C., *Degenerate principal series for symplectic groups* (1990), Ph.D. thesis, University of Chicago.

[Kz1]   Kazhdan, D.A., *Connection of the dual space of a group with the structure of its closed subgroups*, Functional Anal. Appl. 1 (1967), 63-65.

[Kz3]   Kazhdan, D., *On lifting*, Lie Group Representations II, Proceedings, University of Maryland 1982-83, Lecture Notes in Math. 1041, Springer-Verlag, Berlin, 1984, pp. 209-249.

[Kz4]   Kazhdan, D., *Cuspidal geometry of p-adic groups*, J. Analyse Math. **47** (1986), 1-36.

[KzLu]   Kazhdan, D., Lusztig, G., *Proof of the Deligne-Langlands conjecture for Hecke algebras*, Invent. Math. **87** (1987), 153-215.

[Ke]	Keys, D., *On the decomposition of reducible principal series representations of p-adic Chevalley groups*, Pacific J. Math **101** (1982), 351-388.
[Ki]	Kirillov, A.A., *Elements of the Theory of Representations*, Nauka, Moskva, 1978.
[Kn]	Knapp, A.W., *Representation Theory of Semisimple Groups*, Princeton University Press, Princeton, 1986.
[KnZu]	Knapp A.W., Zuckerman G.J., *Classification of irreducible tempered representations of semisimple Lie groups*, Proc. Nat. Acad. Sci. USA **73** (1976), 2178-2180.
[Ku1]	Kutzko, P., *The Langlands conjecture for GL(2) of a local field*, Ann. of Math. **112** (1980), 381-412.
[Ku2]	Kutzko, P.C., *On the supercuspidal representations of GL_N and other p-adic groups*, Proceedings of the International Congress of Mathematics, Berkeley, 1986.
[KuMo]	Kutzko, P., Moy, A., *On the local Langlands conjecture in prime dimension*, Ann. of Math. **121** (1985), 495-517.
[Lb]	Labesse, J.-P., *Le formule des traces d'Arthur-Selberg*, Séminaire Bourbaki n. 636 (1984-85), Astérisque.
[Lang]	Lang, S., $SL_2(\mathbb{R})$, Addison-Wesley, Reading, 1975.
[Ls1]	Langlands, R.P., *Problems in the theory of automorphic forms*, Lecture Notes in Math 170, Springer-Verlag, Berlin, 1970.
[Ls2]	Langlands, R.P., *On the Functional Equation Satisfied by Einsenstein Series*, Lecture Notes in Math 544, Springer-Verlag, Berlin, 1976.
[Ls3]	Langlands, R.P., *On the classification of irreducible representations of real algebraic groups*, Representation Theory and Harmonic Analysis on Semisimple Lie Groups, P.J. Sally, Jr. and D. A. Vogan, Jr. editors, Amer. Math. Soc., Providence, 1989.
[Md]	Macdonald, I. G., *Spherical functions on groups of p-adic type*, Univ. of Madras Publ., 1971.
[Mt1]	Mautner, F., *Spherical functions over p-adic fields I*, Amer. J. Math. **80** (1958), 441-457.
[Mt2]	Mautner, F., *Spherical functions over p-adic fields II*, Amer. J. Math. **86** (1964), 171-200.
[Mi]	Miličić, D., *On C^*-algebras with bounded trace*, Glasnik Mat. **8(28)** (1973), 7-21.
[MgVgWd]	Mœglin, C., Vignéras, M.-F. and Waldspurger J.-L., *Correspondance de Howe sur un corps p-adique*, Lecture Notes in Math 1291, Springer-Verlag, Berlin, 1987.
[MgWd1]	Mœglin, C., Waldspurger, J.-L., *Le spectre résiduel de GL(n)*, Ann. Sci. École Norm. Sup **22** (1989), 605-674.
[MgWd2]	Mœglin, C., Waldspurger, J.-L., *Décomposition spectrale et series d'Eisenstein*, preprint.
[Mr]	Morris, *P-cuspidal representations*, Proc. London Math. Soc. **(3)57** (1988), 329-356.
[Mo1]	Moy, A., *Local constants and the tame Langlands correspondence*, Amer. J. Math. **108** (1986), 863-930.
[Mo2]	Moy, A., *Representations of GSp(4) over a p-adic field: parts 1 and 2*, Compositio Math. **66** (1988), 237-328.
[PS]	Piatetski-Shapiro, *Complex representations of GL(2, K) for finite fields F*, Contemporary mathematics **16** (1983), 1-71.
[Ri]	Ritter, J., editor, *Representation Theory and Number Theory in connection with the Local Langlands Conjecture*, Proceedings of a Conference held at the University of Augsburg, Germany, December 8-14, 1985.
[Ro1]	Rodier, F., *Décomposition de la série principale des groupes réductifs p-adiques*, Non-Commutative Harmonic Analysis, Lecture Notes in Math. 880, Springer-Verlag, Berlin, 1981.
[Ro2]	Rodier, F., *Représentations de GL(n, k) où k est un corps p-adique*, Séminaire Bourbaki n. 587 (1982), Astérisque **92-93** (1982), 201-218.

[Ro3]   Rodier, F., *Sur les représentations non ramifiées des groupes réductifs p-adiques; l'example de GSp(4)*, Bull. Soc. Math. France **116** (1988), 15-42.

[Ro4]   Rodier, F., *Intégrabilité locale des caractères du groupe $GL(n,k)$ où k est un corps de nombres de caractéristiques positive*, Duke Math. **52** (1985), 771-792.

[Sah]   Sahi S., *On Kirillov's conjecture for archimedean fields*, Compos. Math. **72** (1989), 67-86.

[Sa]   Sally, J.P.,Jr., *Some remarks on discrete series characters for reductive p-adic groups*, Representations of Lie Groups, Kyoto, Hiroshima, 1986, 1988, pp. 337-348.

[SaSl] Sally, P.J., Shalika, A., *The Fourier transform of orbital integrals on SL_2 over a p-adic field*, Lie Group Representations II, Proceedings, University of Maryland 1982-83, Lecture Notes in Math. 1041, Springer-Verlag, Berlin, 1984, pp. 303-338.

[SaTd] Sally, P.J., Tadić, M., *Induced representations and classifications for $GSp(2,F)$ and $Sp(2,F)$*, Mémoires Soc. Math. France **52** (1993), 75-133.

[Se]   Serre, J.-P., *Linear Representations of Finite Groups*, Springer-Verlag, New York, 1977.

[Sh1]   Shahidi, F., *Fourier Transforms of intertwining operators and Plancherel measure for $GL(n)$*, Amer. J. Math. **106** (1984), 67-111.

[Sh2]   Shahidi, F., *A proof of Langlands conjecture on Plancherel measures; complementary series for p-adic groups*, Ann. of Math. **132** (1990), 273-330.

[Sh3]   Shahidi, F., *L-functions and representation theory of p-adic groups*, p-adic Methods and Applications, Oxford University Press.

[Sh4]   Shahidi, F., *Twisted endoscopy and reducibility of induced representations for p-adic groups*, Duke Math. J. **66** (1992), 1-41.

[Sl2]   Shalika, J.A., *A theorem on semisimple p-adic groups*, Ann. of Math. **95** (1972), 226-242.

[Sl2]   Shalika, J.A., *The multiplicity one theorem for GL_n*, Ann. of Math. **100** (1974), 171-193.

[Si1]   Silberger, A., *The Langlands quotient theorem for p-adic groups*, Math. Ann. **236** (1978), 95-104.

[Si2]   Silberger, A., *Introduction to harmonic analysis on reductive p-adic groups*, Princeton University Press, Princeton, 1979.

[Si3]   Silberger, A., *Isogeny restrictions of irreducible representations are finite direct sums of irreducible admissible representations*, Proceedings Amer. Math. Soc. **73** (1979), 263-264.

[Si4]   Silberger, A., *Discrete series and classifications for p-adic groups I*, Amer. J. Math. **103** (1981), 1231 - 1321.

[Sp]   Speh B., *Unitary representations of $GL(n,\mathbb{R})$ with non-trivial $(\mathfrak{g},K)$-cohomology*, Invent. Math. **71** (1983), 443-465.

[Sr]   Springer, T.A., *Reductive Groups*, Symp. Pure Math. 33, part 1, Amer. Math. Soc., Providence, Rhode Island, 1979, pp. 3-27.

[St]   Steinberg, R., *Lectures on Chevalley groups*, Yale University, 1968.

[Td1]   Tadić, M., *Harmonic analysis of spherical functions on reductive groups over p-adic fields*, Pacific J. Math. (1983), 215-235.

[Td2]   Tadić, M., *Unitary dual of p-adic $GL(n)$, Proof of Bernstein Conjectures*, Bulletin Amer. Math. Soc. **13** (1985), 39-42.

[Td3]   Tadić, M., *Proof of a conjecture of Bernstein*, Math. Ann. **272** (1985), 11-16.

[Td4]   Tadić, M., *Unitary representations of general linear group over real and complex field*, preprint MPI/SFB 85-22 Bonn (1985).

[Td5]   Tadić, M., *Spherical unitary dual of general linear group over non-archimedean local field*, Ann. Inst. Fourier, **36** (1986), 47-55, n. 2.

[Td6]	Tadić, M., *Classification of unitary representations in irreducible representations of general linear group (non-archimedean case)*, Ann. Sci. École Norm. Sup **19** (1986), 335-382.
[Td7]	Tadić, M., *Topology of unitary dual of non-archimedean GL(n)*, Duke Math. J. **55** (1987), 385-422.
[Td8]	Tadić, M., *On limits of characters of irreducible unitary representations*, Glasnik Mat. **23 (43)** (1988), 15-25.
[Td9]	Tadić, M., *Unitary representations of GL(n), derivatives in the non-archimedean case*, Berichte Math. Stat. Sekt., Forschungsgesellschaft Joaneum, Graz **281** (1987), 281/1-281/19.
[Td10]	Tadić, M., *Geometry of dual spaces of reductive groups (non-archimedean case)*, J. Analyse Math. **51** (1988), 139-181.
[Td11]	Tadić, M., *Notes on representations of non-archimedean SL(n)*, Pacific J. Math. **152** (1992), 375-396.
[Td12]	Tadić, M., *Induced representations of GL(n, A) for p-adic division algebras A*, J. reine angew. Math. **405** (1990), 48-77.
[Td13]	Tadić, M., *On Jacquet modules of induced representations of p-adic symplectic groups*, Harmonic Analysis on Reductive Groups, Proceedings, Bowdoin College 1989, Progress in Mathematics 101, Birkhäuser, Boston, 1991, pp. 305-314.
[Td14]	Tadić, M., *Representations of p-adic symplectic groups*, Compositio Math. (to appear).
[Td15]	Tadić, M., *A structure arising from induction and restriction of representations of classical p-adic groups*, preprint.
[Td16]	Tadić, M., *An external approach to unitary representations*, Bulletin Amer. Math. Soc. **28** (1993), no. 2, 215-252.
[Tt]	Tate, J., *Number theoretic background*, Symp. Pure Math. 33, part 2, Amer. Math. Soc., Providence, Rhode Island, 1979, pp. 3-26.
[Ti]	Tits, J., *Reductive groups over local fields*, Symp. Pure Math. 33, part 1, Amer. Math. Soc., Providence, Rhode Island, 1979, pp. 26-69.
[Vo1]	Vogan, D.A., *The unitary dual of GL(n) over an archimedean field*, Invent. Math. **82** (1986), 449-505.
[Vo2]	Vogan, D.A., *Representations of real reductive groups*, Birkhaüser, Boston, 1981.
[Wd1]	Waldspurger, J.-L., *Un exercice sur GSp(4, F) et les représentations de Weil*, Bull. Soc. Math. France **115** (1987), 35-69.
[Wd2]	Waldspurger, J.-L., *Représentation métaplectique et conjectures de Howe*, Séminaire Bourbaki n. 674 (1986-87), Astérisque **92-93** (1982), 201-218.
[Wr]	Warner, G., *Harmonic analysis on Semi-Simple Lie Groups I, II*, Springer-Verlag, Berlin, 1972.
[We1]	Weil, A., *Sur certains groupes d'operateurs unitaires*, Acta Math. **111** (1964), 143-211.
[We2]	Weil, A., *Basic number theory*, Springer-Verlag, New York, 1974.
[Wn]	Winarsky, N., *Reducibility of principal series representations of p-adic Chevalley groups*, Amer. J. Math. **100** (1978), 941-956.
[Ze1]	Zelevinsky, A.V., *Induced representations of reductive p-adic groups II, On irreducible representations of GL(n)*, Ann. Sci École Norm Sup. **13** (1980), 165-210.
[Ze2]	Zelevinsky, A. V., *Representations of Finite Classical Groups, A Hopf Algebra Approach*, Lecture Notes in Math 869, Springer-Verlag, Berlin, 1981.

Universität Goettingen, Matematisches Institut Geometrie und Analysis, Sonderforschungsbereich 170, Busenstrasse 3-5, D-3400 Goettingen, Germany.

e–mail `tadic@cfgauss.uni-math.gwdg.de`

C∞ VECTORS

N. R. WALLACH

In these lectures we will study the space of C^∞ vectors of a (strongly continuous) representation, (π, H), of a Lie Group, G on a Hilbert space. That is, the space

$$H^\infty = \{v \in H \mid g \mapsto \pi(g)v \text{ is of class } C^\infty\}.$$

This space has a natural structure of a Fréchet space such that the action of G is C^∞. The emphasis will be on the case when G is reductive (e.g. $G = \mathrm{GL}(n,\mathbb{R})$, $\mathrm{O}(p,q)$) although examples will be given for other classes of groups (for example Heisenberg groups). A development will be given of the lecturer's theory (with Casselman) which in particular implies that for so-called admissible representations the space of C^∞ vectors depends only on the underlying algebraic structure of the representation (the corresponding $(\mathfrak{g}, K)$-module). The relationship with asymptotic expansions and the growth of matrix coefficients will be emphasized. The theory is intimately connected with the condition of "moderate growth" and thereby gives an explanation of why the condition arises in such diverse areas as the theory of regular singularities for systems of differential equations and in the theory of automorphic forms. Applications will be given to boundary value problems (e.g. vector valued Poisson integral representations), automorphic forms and Whittaker functions (related to the Fourier expansions of automorphic forms).

The lectures will be aimed at an audience that is familiar with basic algebra, analysis and Lie groups. For preparatory reading we suggest:

M. Reed and B. Simon, *Functional Analysis* I, Academic Press, New York, 1972. [RS]

N. Wallach, *Harmonic analysis on homogeneous spaces*, Marcel Dekker, 1972. [W1]

More advanced students could consult:

N. Wallach, *Real reductive groups I,II,* Academic Press, Boston, 1988,1992. [RRGI,II]

INTRODUCTION.

In the last twenty years there have been major advances in our understanding of the representation theory of real reductive groups. The main thrust of this research has been in the theory of $(\mathfrak{g}, K)$-modules (see [V],[BW]). This theory is algebraic in nature. With the help of powerful tools of algebraic geometry (e.g. the Weil conjectures) this theory has answered many of the problems that were originally framed in analytic terms (e.g. Vogan's determination of the composition series of principal series representations). For this reason, many students and young researchers are (for the most part) ignorant of the underpinnings of the subject that come from analysis. This is unfortunate since a significant number of the main applications of the theory have been in the direction of number theory (i.e. the Langlands Program to develop a non-abelian class field theory). The $(\mathfrak{g}, K)$-modules play an important role, but the smooth models of representations are becoming the more important versions of the representations to number theorists. The obvious reasons are twofold. The first is that the group itself does not act on $(\mathfrak{g}, K)$-modules. The second is that the key objects in the applications are such things as Eisenstein series and Whittaker vectors which turn out to be continuous functionals on smooth models of the representations. Simply put, the theory of $(\mathfrak{g}, K)$-modules is the study of Taylor series in a neighborhood of a maximal compact subgroup whereas the applications involve the study of neighborhoods of infinity in the group.

The purpose of these lectures is to give a fairly easily accessible introduction to the smooth models of representations, that is, the theory of C^∞-vectors. The lectures aim at an exposition of a theorem of Casselman and the speaker that shows that for the class of admissible, finitely generated, $(\mathfrak{g}, K)$-module there is (up to isomorphism) precisely one smooth model satisfying an appropriate growth condition at infinity. The crux of the matter is the growth condition which we will describe in the case of $G = GL(2, \mathbb{R})$ (2×2 matrices of determinant one). If $g \in G$ then we set $\|g\|^2 = \text{tr}(g^T g) + \text{tr}((g^{-1})^T g^{-1})$ (g^T the usual transpose). We will say that a C^∞ function, f, on G is of (uniform) moderate growth if there exist $d \geqslant 0$ such that

$$|X_1 \cdots X_n f(g)| \leqslant C\|g\|^d$$

where C depends on f and $X_1, ..., X_n$ which are 2×2 matrices and if X is a 2×2 matrix then

$$Xf(g) = \frac{d}{dt} f(ge^{tX})_{|t=0}.$$

The key functions associated with representations in number theory have moderate growth. For example, in the definition of an automorphic form the condition of moderate growth is a critical (but perhaps mysterious) ingredient. Harish-Chandra once said "Without the condition [moderate growth] there is no theory of automorphic forms." It should also be pointed out that the condition of moderate growth also arises naturally in the theory of regular singular differential equations (e.g. the singularities of solutions are no worse than poles). To emphasize the condition of moderate growth, our construction of the initial completions of Harish-Chandra modules is in terms of functions of moderate growth which is analogous to (and inspired by) a construction of Schmid [Sch] in the case of analytic vectors.

The theorem alluded to above has a proof that evolved over years of false starts and blunders. Both Casselman and the lecturer have no doubt that there will be a much more intuitive proof that the one that is sketched here. In fact, Joseph Bernstein has recently announced an elegant proof of a special case of the theorem that in addition takes into account dependence on parameters. However, even though the material presented here will not be the final word, many of the techniques and asymptotic results leading to the theorem have useful independent applications.

At the end of the main body of this article there is an appendix covering a substantial amount of the background needed for these lectures and a few of the more technical generalities that will be used. We hope that it will be used only for reference purposes.

The last lecture is devoted to applications. We close this introduction with an example related to the Poisson integral representation of harmonic functions. Recall that if $D = \{z \in \mathbb{C} : |z| < 1\}$ and if $S = \{z \in \mathbb{C} : |z| = 1\}$ then the usual Poisson kernel is given by $P(z, b) = \frac{1-|z|^2}{|z-b|^2}$, $z \in D$, $b \in S$. We say that a formal sum $T(b) = \sum a_n b^n$ is a hyperfunction on S if for each $0 < r < 1$, $\sup_n |a_n| r^{|n|} < \infty$. What this implies is that if $f(b)$ is a real analytic function on S then if $f(b) = \sum_n f_n b^n$ is its Fourier series then we may write

$$\int_S T(b) f(b) db = \sum_n a_n f_{-n}$$

with the series converging absolutely. This is because $\sum_n f_n b^n$ is the Fourier series of an analytic function if and only if there is some $0 < r < 1$ such that $|f_n| \leqslant C r^{|n|}$. A theorem of Koebe says that if f is a harmonic function on D then there exists a hyperfunction T on S such that

$$f(z) = \int_S T(b) P(z, b) db.$$

We call T the boundary value of f. On the other hand there are the distributions on S. These are the series $T(b) = \sum a_n b^n$ such that there is some $d \geqslant 0$ such that $|a_n| \leqslant C(1 + |n|)^d$. These can be "integrated" against C^∞ functions on S since $\sum f_n b^n$ is the Fourier series of a C^∞ function if and only if $\lim_{|n| \to \infty} |n|^d |f_n| = 0$ for all $d \geqslant 0$. The necessary and sufficient condition that harmonic function, f, have a distribution as its boundary value is that there exists $d \geqslant 0$ and $C > 0$ such that

$$|f(z)| \leqslant C(1 - |z|^2)^{-d}, z \in D.$$

If we set

$$G = SU(1,1) = \left\{ \begin{bmatrix} a & b \\ \bar{b} & \bar{a} \end{bmatrix} \Big| |a|^2 - |b|^2 = 1 \right\}.$$

Then G acts transitively on D by linear fractional transformations

$$z \mapsto \frac{az + b}{\bar{b}z + \bar{a}} = \begin{bmatrix} a & b \\ \bar{b} & \bar{a} \end{bmatrix} \cdot z.$$

The growth condition above then translates precisely to the condition that $g \mapsto f(g \cdot 0)$ be of moderate growth.

1. THE NOTION OF A C^∞ VECTOR.

1.1. Let G denote a Lie group with Lie algebra $\mathfrak{g}$. If V is a Fréchet space (A.1) then we denote by $GL(V)$ the group of all continuous linear maps of V to V that are bijective (hence, by the open mapping theorem) with continuous inverse. If π is a homomorphism of G into $GL(V)$ then (π, V) is said to be a (Fréchet) *representation* of G if the map

$$g, v \mapsto \pi(g)v$$

is continuous from $G \times V$ to V. If V is a Banach (resp. Hilbert) space then (π, V) is called a Banach (resp. Hilbert) representation of G. If (π, V) is a Hilbert representation of G and if $\pi(g)$ is a unitary operator for $g \in G$ then we say that (π, V) is a *unitary* representation of G.

Examples. In these examples G will be $\mathbb{R}$ under addition.

1. $V = L^1(\mathbb{R})$ and $\pi(t)f(x) = f(x-t)$. We must show that the map $t, f \mapsto f(\cdot - t)$ is continuous from $\mathbb{R} \times L^1(\mathbb{R})$ into $L^1(\mathbb{R})$. We note that $\|\pi(t)f\| = \|f\|$ for all t, f. Thus $\|\pi(t)f - \pi(s)g\| \leqslant \|f - g\| + \|\pi(t-s)g - g\|$. It is therefore enough to show that $\lim_{h \to 0} \|\pi(h)g - g\| = 0$ for all $g \in L^1(\mathbb{R})$. For this we observe that the space of continuous compactly supported functions on $\mathbb{R}$, $C_c(\mathbb{R})$ is dense in $L^1(\mathbb{R})$. Let $\epsilon > 0$ be given. Let $u \in C_c(\mathbb{R})$ be such that $\|g - u\| < \epsilon$. We assume that $u(x) = 0$ if $|x| > R$ and that $|h| < 1$. Then uniform continuity of u implies that there exists $\delta > 0$ such that if $|h| < \delta$ then $|u(x - h) - u(x)| < \epsilon \chi(x)/(2R + 2)$ with χ the characteristic function of $[-R - 1, R + 1]$. Thus

$$\|\pi(h)u - u\| \leqslant \frac{\epsilon}{2R + 2} \int_{-\infty}^{\infty} \chi(x)dx = \epsilon.$$

Thus if $|h| < \epsilon$ (and $|h| < 1$) then

$$\|\pi(h)g - g\| = \|\pi(h)g - \pi(h)u + \pi(h)u - u + u - g\| \leqslant 2\|g - u\| + \|\pi(h)u - u\| < 3\epsilon.$$

We therefore see that $(\pi, L^1(\mathbb{R}))$ is indeed a Banach representation of $\mathbb{R}$.

2. $V = L^2(\mathbb{R})$ and $\pi(t)f(x) = f(x - t)$. If we argue in exactly the same way we did in the previous example it is enough to show that

$$\lim_{h \to 0} \|\pi(h)f - f\| = 0$$

for all $f \in L^2(\mathbb{R})$. Fix $\epsilon > 0$. This time observe that $C_c(\mathbb{R})$ is dense in $L^2(\mathbb{R})$ and choose $u \in C_c(\mathbb{R})$ such that $\|f - u\| < \epsilon$. If u vanishes outside of $[-R, R]$, $|h| < 1$ and $\delta > 0$ is so small that $|u(x - h) - u(x)| < \epsilon \chi(x)/(2R + 2)^{1/2}$ for $|h| < \delta$ then the obvious modification of the above argument implies that $(\pi, L^2(\mathbb{R}))$ is a unitary representation of $\mathbb{R}$.

3. $V = \mathcal{S}(\mathbb{R})$ and $\pi(t)f(x) = f(x - t)$. Here $\mathcal{S}(\mathbb{R})$ is the space of all infinitely differentiable functions on $\mathbb{R}$ such that

$$\nu_{p,q}(f) = \sup_{x \in \mathbb{R}} |x|^p |f^{(q)}(x)| < \infty$$

for all $p, q \geq 0$, $p, q \in \mathbb{Z}$ (the integers). We endow $\mathcal{S}(\mathbb{R})$ with the topology induced by the semi-norms $\nu_{p,q}$. We note that if we take $a(x) = 1 + x^2$ and $m = 1$, $X_1 = \frac{d}{dx}$ then the observations in A.7 imply that $\mathcal{S}(\mathbb{R})$ is a Fréchet space. In this case we prove that if $f \in \mathcal{S}(\mathbb{R})$ then the map $t \mapsto \pi(t)f$ is C^∞ from $\mathbb{R}$ to $\mathcal{S}(\mathbb{R})$. Fix $f \in \mathcal{S}(\mathbb{R})$. Set $\Phi(t) = \pi(t)f$. If we show that Φ is differentiable with values in $\mathcal{S}(\mathbb{R})$ and that $\Phi'(t) = \pi(t)f'$ then clearly we would have Φ' is differentiable with derivative $\pi(t)f''$, etc. So Φ is of class C^∞. Now

$$\nu_{p,q}\left(\frac{\pi(t+h)f - \pi(t)f}{h} - \pi(t)f'\right) =$$

$$\sup_{x} |x|^p \left| \frac{f^{(q)}(x - t - h) - f^{(q)}(x - t)}{h} - f^{(q+1)}(x - t) \right| =$$

$$\sup_{x} |x + t|^p \left| \frac{f^{(q)}(x - h) - f^{(q)}(x)}{h} - f^{(q+1)}(x) \right|.$$

Since t is fixed, the last expression is dominated by

$$\sup C\zeta(x) \left| \frac{f^{(q)}(x - h) - f^{(q)}(x)}{h} - f^{(q+1)}(x) \right|$$

with $\zeta(x) = 1 + |x| + \ldots + |x|^p$ and C depends on t. Taylor's theorem implies that

$$f^{(q)}(x - h) - f^{(q)}(x) = h f^{(q+1)}(x) + \frac{h^2}{2} f^{(q+2)}(\theta)$$

with θ between x and $x - h$. Thus

$$\nu_{p,q}\left(\frac{\pi(t+h)f - \pi(t)f}{h} - \pi(t)f'\right) = \sup C\zeta(x) \frac{|h|}{2} |f^{(q+2)}(\theta)| \leq C'|h|.$$

This implies our contention. From this it is an easy matter to see that $(\pi, \mathcal{S}(\mathbb{R}))$ is a Fréchet representation of $\mathbb{R}$. Obviously, it is more than that.

We say that a Fréchet representation, (π, V), of G is *smooth* if the map $g \mapsto \pi(g)f$ is is a C^∞ map of G into V for all $f \in V$. Thus example 3 above is a smooth Fréchet represntation. We now come to the central concept in these lectures. If (π, V) is a representation of G on a Fréchet space then we say that $v \in V$ is a C^∞ vector if the map $g \mapsto \pi(g)v$ is of class C^∞.

1.3. If (π, H) is a Hilbert representation of G then we set $\check{H}$ equal to the space of all $\lambda : H \to \mathbb{C}$ continuous that are $\mathbb{R}$ linear and $\lambda(cv) = \bar{c}\lambda(v)$. Then the map $v \mapsto \lambda_v = (w \mapsto \langle v, w \rangle)$ defines a linear bijection of H onto $\check{H}$. If $g \in G$ then we set $\check{\pi}(g)\lambda(v) = \lambda(\pi(g)^{-1}v)$. If we use the above identification of $\check{H}$ with H then the action is just $\check{\pi}(g) = \pi(g^{-1})^*$ (the usual adjoint of $\pi(g^{-1})$). We note that if $\dim H = \infty$ the map $T \mapsto T^*$ is not continuous in the strong operator topology. Thus one must actually give a proof of the following result.

Lemma. *If (π, H) is a Hilbert representation of G then $(\check{\pi}, H)$ is also a Hilbert representation of G.*

We will defer the proof of this result to the end of the next number.

1.4. Let (π, H) be a Banach representation of G. Set H^∞ equal the space of all C^∞ vectors of H.

Theorem. *(Gårding) H^∞ is dense in H.*

To prove this result we will need two lemmas.

Lemma 1. *If $f \in C_c^\infty(G)$ and $v \in H$ then we set $\pi(f)v = \int_G f(g)\pi(g)v\,dg$ (see A.6). We have*

    1. $\pi(f)$ is a bounded operator on H.
    2. $\pi(f)v \in H^\infty$ for all $v \in H$.

Proof. The principle of uniform boundedness implies that if $\omega \subset G$ is a compact subset then there exists a constant $C_\omega < \infty$ such that

$$\|\pi(g)\| \leqslant C_\omega, \, g \in \omega.$$

(Here $\|...\|$ denotes the usual operator norm.) Thus if $f \in C_c^\infty(G)$ and if $\mathrm{supp} f \subset U$ and open subset with compact closure ω then

$$\|\pi(f)v\| \leqslant \int_U |f(g)| \|\pi(g)\| dg \|v\| \leqslant C_\omega \|f\|_{1,U} \|v\|.$$

Here $\|f\|_{1,U}$ is the L^1-norm of f as a function on U. This proves 1. We note that we have also shown that $f \mapsto \pi(f)$ extends to a bounded operator from $L^1(U, dx)$ to the space of all bounded operators on H, $End(H)$.

We may assume that if $x \in U$ then $x^{-1} \in U$. Set $V = UU$. Then V has compact closure. We note

    I. If $f \in C_c^\infty(G)$ and $\mathrm{supp} f \subset U$ then $(g \mapsto l_g f)$ is a C^∞ map from U to $L^1(V, dg)$.

Assuming I, we prove 2. We note that

$$\pi(g)\pi(f)v = \int_G f(x)\pi(gx)vdx = \int_G f(g^{-1}x)\pi(x)vdx = \pi(l_g f)v.$$

Thus $g \mapsto \pi(g)\pi(f)v$ is the compostion of a linear map $(f \mapsto \pi(f)v)$ and a C^∞ map $x \mapsto l_x f$.

We now prove I. If $f \in C^\infty(G)$ and if $X \in \mathfrak{g} = Lie(G)$ then set

$$\check{X}f(g) = \frac{d}{dt}f(\exp(-tX)g)_{|t=0}.$$

Assume that f has support in U. $l_{g \exp tX} f(x) = f(\exp(-tX)g^{-1}x)$. Taylor's theorem implies that $f(\exp(-tX)y) = f(y) + t\check{X}f(y) + t^2 E(t,y)$ and there exists $\epsilon > 0$ such that if $|t| < \epsilon$ then $|E(t,y)| \leqslant C\chi(y)$ where χ is the characteristic function of U. Thus

$$\left\| \frac{l_{g \exp tX} f - l_g f}{t} - l_g \check{X} f \right\|_1 = |t|C.$$

This implies that $F = g \mapsto l_g f$ is of class C^1 from U to $L^1(V)$ with $XF = F_1$ and $F_1(g) = l_g \check{X} f$. If we apply the same argument to $\check{X}f$ we find that F_1 is C^1, etc.

The second lemma involves some more notation and concepts. Let $\{\varphi_j\}$ be a sequence of elements in $C_c^\infty(G)$ such that

Delta 1. $\varphi_j(x) \geqslant 0$ and $\varphi_j(1) > 0$ for all j.

Delta 2. $\mathrm{supp}\varphi_j \supset \mathrm{supp}\varphi_{j+1}$ and $\cap_j \mathrm{supp}\varphi_j = \{1\}$.

Delta 3. $\|\varphi_j\|_1 = 1$ all j.

We will call $\{\varphi_j\}$ a *delta sequence* or *approximate identity*. To see that such a sequence exists let (U, Ψ) be a chart for G (A.2) such that $\Psi(U) = \{x \in \mathbb{R}^n | \sum x_i^2 < 2\}$ and $\Psi(1) = 0$. Let $\alpha \in C_c^\infty(\mathbb{R})$ be non-negative with support contained in $\{-1, 1\}$ and $\alpha(0) = 1$. Set for $x \in \Psi(U)$, $\beta_j(x) = \alpha(j^2\|x\|^2)$. Then $\beta_j(x) = 0$ if $\|x\| \geqslant 1/j$. Define $\psi_j(g) = \beta_j(\Psi(g))$ if $g \in U$ and $\psi_j(g) = 0$ otherwise. Set $\varphi_j = \psi_j/\|\psi_j\|$. Then $\{\varphi_j\}$ satisfies Delta 1-3.

Lemma 2. *If $\{\varphi_j\}$ is a delta sequence then $\lim_{j \to \infty} \pi(\varphi_j)v = v$ for all $v \in H$.*

Proof. Fix $v \in H$. Let $\epsilon > 0$ be given then there exists, U, an open neighborhood of 1 such that $\|\pi(g)v - v\| < \epsilon$ for $g \in U$. Thus there exists j_o such that if $j \geqslant j_o$ then $\mathrm{supp}\varphi_j \subset U$. Thus if $w \in H$ and $j \geqslant j_o$ then we have

$$\|\pi(\varphi_j)v - v\| = \left\| \int_U \varphi_j(g)(\pi(g)v - v)dg \right\| \leqslant \epsilon \int_U \varphi_j(g)dg = \epsilon.$$

Here we gave used $\int_G \varphi_j(g)vdg = v$.

The above theorem is a direct concequence of Lemmas 1 and 2.

We will now prove Lemma 1.3. If $\varphi \in C_c^\infty(G)$ then we set $\check{\varphi}(g) = \delta(g^{-1})\overline{\varphi(g^{-1})}$. If $v, w \in H$ then

$$\langle v, \pi(\check{\varphi})w \rangle = \int_G \varphi(g^{-1})\delta(g)\langle v, \pi(g)w\rangle dg = \int_G \varphi(g^{-1})\delta(g^{-1})\langle \pi(g)^*v, w\rangle dg =$$

$$\int_G \varphi(g)\delta(g)\delta(g)^{-1}\langle \pi(g^{-1})^*v, w\rangle dg = \int_G \varphi(g)\langle \check{\pi}(g)v, w\rangle dg.$$

We therefore define $\check{\pi}(\varphi) = \pi(\check{\varphi})^*$. Then $\check{\pi}(g)\check{\pi}(\varphi) = \check{\pi}(l_g\varphi)$. If supp$\varphi \subset \omega$ a compact subset of G then one checks that $\|\check{\pi}(\varphi)\| \leqslant C_\omega \|\varphi\|_1$. The argument above now implies that if $v \in H$ then $g \mapsto \check{\pi}(g)\check{\pi}(\varphi)v$ is continuous. We also note that the arguments above also imply that the space spanned by the $\check{\pi}(\varphi)v$ is dense in H. Now let $v \in H$ and let $\epsilon > 0$ be given. Then there exists $v_o \in H$ such that the map $g \mapsto \check{\pi}(g)v_o$ is continuous and $\|v - v_o\| < \epsilon$. We have

$$\|\check{\pi}(g)v - v\| = \|\check{\pi}(g)v - \check{\pi}(g)v_o + \check{\pi}(g)v_o - v_o + v_o - v\| \leqslant$$

$$\|\check{\pi}(g)(v - v_o)\| + \|\check{\pi}(g)v_o - v_o\| + \epsilon.$$

Let U be an open neighborhood of 1 such that if $x \in U$ then $x^{-1} \in U$ and so small that $\|\pi(g)\| < 2$ for all $g \in U$. Since $\|T^*\| = \|T\|$ we see that $\|\check{\pi}(g)\| \leqslant 2$ for $g \in U$. Thus

$$\|\check{\pi}(g)v - v\| \leqslant 3\epsilon + \|\check{\pi}(g)v_o - v_o\|.$$

This implies that $\lim_{g \to 1} \check{\pi}(g)v = v$. From this it is an easy matter to complete the proof of Lemma 1.3.

1.5. Let (π, H) be a Banach representation of G. If $v \in H^\infty$ then for each $X \in \mathfrak{g}$ the map $g, t \mapsto \pi(g \exp tX)v$ is jointly C^∞. Thus if we set $\pi(X)v = \frac{d}{dt}\pi(\exp tX)v_{|t=0}$ then $\pi(X)v \in H^\infty$. Thus $\pi(X)$ defines a linear operator on H^∞. If we apply Lemma A.11 we see that $\pi(aX+bY) = a\pi(X)+b\pi(Y)$ and $\pi([X,Y]) = \pi(X)\pi(Y)-\pi(Y)\pi(X)$ for $a, b \in \mathbb{R}$ and $X, Y \in \mathfrak{g}$. Thus we have a representation of $\mathfrak{g}$ on H^∞. Theorem A.9 implies that for each $x \in U(\mathfrak{g}_C)$ there is an operator $\pi(x)$ such that $x \mapsto \pi(x)$ is a homomorphism of $U(\mathfrak{g}_C)$ into the endomorphisms of H^∞. If $x \in U(\mathfrak{g}_C)$ we define the semi-norm $q_x(v) = \|\pi(x)v\|$. If $X_1, ..., X_n$ is a basis of $\mathfrak{g}_C$ then Theorem A.9 implies that the elements $X_{i_1}X_{i_2}\cdots X_{i_k}$ with $1 \leqslant i_1 \leqslant i_2 \leqslant ... \leqslant i_k \leqslant n$ form a basis of $U(\mathfrak{g}_C)$. Thus if we endow H^∞ with the topology induced by the semi-norms $q_{X_{i_1}\cdots X_{i_k}}$ then each q_x is continuous with respect to this topology. We will use this topology throughout in referring to H^∞. We note that if $v \in H^\infty$, $x \in G$ then $g \mapsto \pi(g)\pi(x)v$ is also of class C^∞. Thus $\pi(g)H^\infty \subset H^\infty$ for all $g \in G$.

Theorem. H^∞ is a Fréchet space and (π, H^∞) is a smooth Fréchet representation of G. Also the space spanned by the elements $\pi(\alpha)v$, $\alpha \in C_c^\infty(G)$ is dense in H^∞.

Proof. Let $\{v_j\}$ be a Cauchy sequence in H^∞. This means that if $x \in U(\mathfrak{g}_C)$ then $\{\pi(x)v_j\}$ is a Cauchy sequence in H. Since H is complete, for each $x \in U(\mathfrak{g}_C)$ there exists $u_x \in H$ such that

$$\lim_{j \to \infty} \pi(x)v_j = u_x$$

in H. To show that $\lim_{j \to \infty} v_j = u_1$ in H^∞ we must show that $u_x = \pi(x)u_1$. We prove this for $x \in U^j(\mathfrak{g}_C)$ by induction on j. If $j = 0$ then $x = c1$ so the result is clear. Assume for j. If $x \in U^{j+1}(\mathfrak{g}_C)$ then we can write $x = \sum_i X_i y_i + y_0$ with $X_i \in \mathfrak{g}$ and $y_i \in U^j(\mathfrak{g}_C)$. Hence it is enough to show that if $\{w_j\}$ is a Cauchy sequence in H^∞ with $\lim_{j \to \infty} w_j = z$ then for each $X \in \mathfrak{g}$, $y = \lim_{j \to \infty} \pi(X)w_j = \pi(X)z$. To prove this, we set $F_j(t) = \pi(\exp tX)w_j$. F_j is a C^∞ map of $\mathbb{R}$ into H. Thus

$$F_j(t) = w_j + \int_0^t F_j'(s)ds. \tag{1}$$

We have seen that $F_j'(t) = \pi(\exp tX)\pi(X)w_j$. Then taking the limit as $j \to \infty$ in (1) we see that

$$\pi(\exp tX)z = z + \int_0^\infty \pi(\exp sX)yds.$$

Thus $\frac{d}{dt}\pi(\exp tX)z = \pi(\exp tX)y$. So $y = \pi(X)z$ as desired.

To prove the second assertion we note that if $U^j(\mathfrak{g}_C)$ is the subspace of $U(\mathfrak{g}_C)$ spanned by j or less products of elements of $\mathfrak{g}$ then $\dim U^j(\mathfrak{g}_C) < \infty$. We also note that the map $U^j(\mathfrak{g}_C) \otimes H^\infty \to H^\infty$, $x \otimes v \mapsto \pi(x)v$ is of class C^∞. If $x \in U(\mathfrak{g}_C)$ and if $g \in G$ then $\pi(g)\pi(x)v = \pi(\mathrm{Ad}(g)x)\pi(g)v$. Thus

$$\|\pi(x)(\pi(g)v - v)\| = \|\pi(g)(\pi(\mathrm{Ad}(g^{-1})x - x)v)\|.$$

Thus (π, H^∞) is a Fréchet representation of G. We note that the integrals in the first part of the proof are all convergent in H^∞. From this it is an exercise to show that if $v \in H^\infty$, $X \in \mathfrak{g}$ then

$$\lim_{t \to \infty} \frac{\pi(\exp tX)v - v}{t} = \pi(X)v$$

in H^∞.

The last assertion is also easy and left to the reader.

Note. It is a theorem of Dixmier and Malliavin that the space spanned by the vectors $\pi(\alpha)v$, $\alpha \in C_c^\infty(G)$, $v \in H$ is exactly H^∞.

1.6. Example. Let $H = L^2(G)$ with $\pi(g)f(x) = f(xg)$. One can argue as in Example 2 in 1.1 to see that (π, H) is a unitary representation of G. Let V be as in

A.7 with $X_1, ..., X_n$ any basis of $\mathfrak{g}$. If $\alpha \in C_c^\infty(G)$ $f \in L^2(G)$ then $\pi(\alpha)f \in H^\infty$ and $\pi(\alpha)f \in V$. We note that $V \subset H^\infty$ and the topology on V is the subspace topology. Since V is dense, and complete this implies that $V = H^\infty$. In other words, $L^2(G)^\infty$ is the space of all $f \in C^\infty(G)$ such that $xf \in L^2(G)$ for all $x \in U(\mathfrak{g}_\mathbb{C})$.

1.7. Example. In this example we take $G = GL(n, \mathbb{R})$. On $\mathbb{R}^n$ we take the standard inner product $(x, y) = \sum x_i y_i$ and the standard norm $\|x\| = \sqrt{(x, x)}$. We also write $\{e_i\}$ for the standard basis. If $\nu \in \mathbb{C}$ then we define V^ν to be the space of all $f \in C^\infty(\mathbb{R}^n - (0))$ such that $f(tx) = t^\nu f(x)$ for $t > 0$, $x \in \mathbb{R}^n - (0)$. We set $\mathbb{R}^+ = (0, \infty)$ and $S^{n-1} = \{x \in \mathbb{R}^n | \|x\| = 1\}$ then the map $(t, x) \mapsto tx$ defines a diffeomorphism of $\mathbb{R}^+ \times S^{n-1}$ onto $\mathbb{R}^n$. Thus if $f \in V^\nu$ then f is completely determined by its restriction to S^{n-1} and $V^\nu_{|S^{n-1}} = C^\infty(S^{n-1})$. On S^{n-1} we define vector fields $(X_i)_x = e_i - x_i x$ (here we have identified $T(S^{n-1})_x$ with $\{v \in \mathbb{R}^n | (v, x) = 0\}$). We define a topology on $C^\infty(S^{n-1})$ as in the end of A.7 using $X_1, ..., X_n$ and the function $a \equiv 1$. This is the usual C^∞ topology. We pull back this topology to V^ν. If $g \in G$, $f \in V^\nu$ then we set $\pi_\nu(g)f(x) = f(g^{-1}x)$.

Lemma. (π_ν, V^ν) is a smooth Fréchet representation of G.

Proof. We first define the topology on V^ν more intrinsicly. If ω is a compact subset of $\mathbb{R}^n - (0)$ and if I is a multiindex and if $f \in C^\infty(\mathbb{R}^n - (0))$ then we set $p_{\omega, I}(f) = \sup_{x \in \omega} |\partial^I f(x)|$ (see A.7 for notation). If $N \in \mathbb{N}^+ = \{1, 2, ...\}$ then we set $C_N = \{x \in \mathbb{R}^n | \frac{1}{N} \leqslant \|x\| \leqslant N\}$. If ω is compact then there exists $N \in \mathbb{N}^+$ such that $\omega \subset C_N$. We may thus use the $p_{C_N, I}$ to define a topology on $C^\infty(\mathbb{R}^n - (0))$ which makes it a Fréchet space. Now V^ν is the subspace of $f \in C^\infty(\mathbb{R}^n - (0))$ satisfying

$$\sum_i x_i \frac{\partial f}{\partial x_i} = \nu f.$$

Thus V^ν is a Fréchet space. If $\varphi \in C^\infty(S^{n-1})$ then set $\varphi(tx) = t^\nu \varphi(x)$ for $t > 0$, $x \in S^{n-1}$. Then it is easily checked that the map $\varphi \mapsto \varphi_\nu$ is continuous from $C^\infty(S^{n-1})$ to V^ν. The open mapping theorem implies that this map defines an isomorphism of Fréchet spaces. We define an action of G on $W = C^\infty(\mathbb{R}^n \setminus (0))$ by $\sigma(g)f(x) = f(g^{-1}x)$. Thus to prove the lemma we need only show

1) (σ, W) is a smooth Fréchet representation of G.

If $g \in G$ then set $g^{-1} = [g^{ij}]$. Then $\partial_i \sigma(g)f(x) = \sum_k g^{ki}(\partial^k f)(g^{-1}x)$. We therefore see that

$$\partial^I \sigma(g)f = \sum_{|J|=|I|} \alpha_{J,I}([g^{ij}])\sigma(g)\partial^J f$$

with $\alpha_{I,J}$ a polynomial on $M_n(\mathbb{C})$ homogenous of degree $|I|$ and $\alpha_{I,J}(I) = \delta_{I,J}$. Cramer's rule implies that

$$|\alpha_{I,J}([g^{ij}])| \leqslant C|\det(g)|^{-|I|}\|g\|^{(n-1)}.$$

Hence
$$p_{\omega,I}(\sigma(g)f) \leqslant C|\det(g)|^{-|I|}\|g\|^{(n-1)} \sum_{|J|=|I|} p_{g^{-1}\omega,J}(f).$$

Also
$$\partial^I \sigma(g)f - \partial^I f = \sum_{|J|=|I|} \alpha_{J,I}([g^{ij}])\sigma(g)\partial^J f - \partial^I f =$$

$$\sum_{|J|=|I|} (\alpha_{J,I}(g^{ij}) - \alpha_{J,I}(I))\sigma(g)\partial^J f + \sigma(g)\partial^I f - \partial^I f.$$

If $\epsilon > 0$ is given and if $\|g - I\| < \delta$ for $\delta > 0$ sufficiently small then
$$|\alpha_{J,I}(g^{ij}) - \alpha_{J,I}(I)| < \epsilon.$$

If $\omega \subset \mathbb{R}^n - (0)$ is compact then for $\delta > 0$ possibly smaller
$$|\sigma(g)\partial^I f(x) - \partial^I f(x)| < \epsilon$$

for $x \in \omega$ and $\|g - I\| < \delta$. This implies that if $\|g - I\| < \delta$ then
$$p_{\omega,I}(\sigma(g)f - f) \leqslant C\epsilon$$

with C depending only on f, ω and δ. Thus $\lim_{g \to I} \sigma(g)f = f$ in W. We note that if $h \in G$ then

$$p_{I,\omega}(\sigma(gh)f - \sigma(g)f) \leqslant C|\det(g)|^{-|I|}\|g\|^{(n-1)} \sum_{|J|=|I|} p_{g^{-1}\omega,J}(\sigma(h)f - f).$$

Thus (σ, W) is indeed a Fréchet representation.

To prove that it is smooth we define for $A = [a_{ij}] \in M_n(\mathbb{R})$ the vector field

$$(X_A)_x f = \frac{d}{dt} f(e^{-tX}x)_{|t=0} = \frac{d}{dt} f(x - tAx)_{|t=0} = -\sum_{ij} a_{ij} x_j \frac{\partial}{\partial x_i} f(x).$$

Now
$$\sup_{x \in \omega} \left| \partial^I \left(\frac{\sigma(e^{tA})f(g^{-1}x) - f(g^{-1}x)}{t} - X_A f(g^{-1}x) \right) \right| \leqslant$$

$$C \sup_{x \in g^{-1}\omega} \left| \partial^I \left(\frac{\sigma(e^{tA})f(x) - f(x)}{t} - X_A f(x) \right) \right|$$

with $C > 0$ depending on g. Now we can write

$$\partial^I \left(\frac{\sigma(e^{tA})f(x) - f(x)}{t} - X_A f(x) \right) = \frac{\sigma(e^{tA})\sigma(e^{-tA})\partial^I \sigma(e^{tA})f(x) - \partial^I f(x)}{t} -$$

$$- \partial^I X_A f(x).$$

215

If we take the limit as $t \to 0$ we get

$$X_A \partial^I f - [X_A, \partial^I] f - \partial^I X_A f(x) = 0.$$

Now our standard argument using Taylor's theorem implies that

$$\lim_{t \to 0} \frac{\sigma(g e^{tA}) f - \sigma(g) f}{t} = \sigma(g) X_A f$$

in W. Thus the map $g \mapsto \sigma(g) f$ is of class C^1. The same argument with f replaced by $X_A f$ implies C^2, etc.

1.8. We now look at a variant of this representation that is a Hilbert representation. On S^{n-1} we use the volume form $\omega_x(v_1, ..., v_{n-1}) = dx(x, v_1, ..., v_{n-1})$ (here dx is the standard volume form on $\mathbb{R}^n$. We note that with this definition we have

$$\int_{\mathbb{R}^n} f(x) dx = \int_{S^{n-1}} \left(\int_0^\infty r^{n-1} f(r \cdot) dr \right) \omega.$$

If $\varphi \in C^\infty(S^{n-1})$ then there exists $f \in C_c^\infty(\mathbb{R}^{n-1} - (0))$ such that

$$\int_0^\infty r^{n-1} f(rx) dr = \varphi(x), \ x \in S^{n-1}.$$

If $g \in G$ then we set $\tau(g) \varphi(x) = \varphi \left(\frac{g^{-1} x}{\|g^{-1} x\|} \right)$. Let f be as above for φ then

$$|\det(g)| \int_{S^{n-1}} \varphi \omega = |\det(g)| \int_{\mathbb{R}^n} f(x) dx = \int_{\mathbb{R}^n} f(g^{-1} x) dx$$

$$= \int_{S^{n-1}} \int_0^\infty r^{n-1} f(g^{-1} rx) dr \omega =$$

$$\int_{S^{n-1}} \int_0^\infty r^{n-1} f(r \|g^{-1} x\| (g^{-1} x / \|g^{-1} x\|)) dr \omega =$$

$$\int_{S^{n-1}} \|g^{-1} x\|^{n-1} \int_0^\infty r^{n-1} f(r(g^{-1} x / \|g^{-1} x\|)) dr \omega = \int_{S^{n-1}} \|g^{-1} \cdot \|^{-n} \tau(g) \varphi \omega.$$

On V^ν we put the pre-Hilbert space structure

$$\langle f, g \rangle = \int_{S^{n-1}} f \bar{g} \omega.$$

216

We note that $\pi_\nu(g)\varphi_\nu(x) = \varphi_\nu(g^{-1}x) = \|g^{-1}x\|^\nu \varphi(g^{-1}x/\|g^{-1}x\|)$. We set

$$\tau_\nu(g)\varphi(x) = \|g^{-1}x\|^\nu \varphi(g^{-1}x/\|g^{-1}x\|).$$

We have already shown that for fixed ν, $\lim_{g \to I} \tau_\nu(g)f = f$ uniformly on compacta. Thus if we set H^ν equal to the Hilbert space completion of V^ν relative to the above inner product then (π_ν, H^ν) is a Hilbert representation of G. We also note that the map $\varphi \mapsto \varphi_\nu$ (which we denote by T) is a unitary bijection of $L^2(S^{n-1})$ onto H^ν such that $T\pi_\nu(g) = \tau_\nu(g)T$. Let $SL^\pm(n, \mathbb{R}) = {}^oG = \{g \in G | \, |\det(g)| = 1\}$. If $g \in {}^oG$ and $f, h \in L^2(S^{n-1})$ then

$$\langle \tau_\nu(g)f, \tau_{-n-\bar\nu}(g)h \rangle = \langle f, h \rangle$$

for all $\nu \in \mathbb{C}$ (that is the point of the calculation above). From this we see that the conjugate dual representation of (π_ν, H^ν) is equivalent with $(\pi_{-\bar\nu-n}, H^{-\bar\nu-n})$. Furthermore, (π_ν, H^ν) is unitary if $\mathrm{Re}\,\nu = -n/2$.

We leave it as an exercise to the reader to prove that $(\pi_\nu, (H^\nu)^\infty)$ is just (π_ν, V^ν) as a smooth Fréchet representation.

1.9. We close this section with one more notion. Let (π, V) be a Fréchet representation of G. We say that $v \in V^\infty$ is a (weakly) *analytic vector* if the map $g \mapsto \lambda(\pi(g)v)$ is real analytic for all $\lambda \in V'$. We set V^ω equal to the space of all analytic vectors in V. We note that if $v \in V^\omega$ then $\pi(g)\pi(x)v \in V^\omega$ for all $g \in G$, $x \in U(\mathfrak{g}_\mathbb{C})$.

Theorem. *Assume that G is connected. If $W \subset V^\omega$ is a subspace such that $\pi(U(\mathfrak{g}_\mathbb{C}))W \subset W$ and if $Cl(W)$ is the closure if W in V then $Cl(W)$ is $\pi(g)$ invariant for all $g \in G$.*

Proof. We note that if $W^\perp = \{\lambda \in V' | \lambda(W) = 0\}$ then $Cl(W) = (W^\perp)^\perp$ (cf. [[Tr, Corollary 1,p.186]]. Let $w \in W$ and $\lambda \in W^\perp$ then if $X \in \mathfrak{g}$ and t is sufficiently small

$$\lambda(\pi(\exp tX)w) = \sum_{n=0}^\infty \frac{t^n}{n!}\lambda(\pi(X)^n w) = 0.$$

Thus $\lambda(\pi(\exp tX)w) = 0$ for all $t \in \mathbb{R}$ (here we are using real analyticity again). Thus $\pi(\exp X)w \in (W^\perp)^\perp$ for all $X \in \mathfrak{g}$. Since $\exp(\mathfrak{g})$ generates $\mathfrak{g}$ we see that $\pi(g)W \subset Cl(W)$ for all $g \in G$.

2. The condition of moderate growth.

2.1. Let G be a real reductive group (see A.14). Let G_1 and G_o be as in A.14 and let $\nu : G \to G_o$ be as in A.14. Let $K = \nu^{-1}(K_o)$ with $K_o = G_o \cap O(n)$. Then K_o is a maximal compact subgroup of G (see A.15). Then $G_1 \subset GL(n, \mathbb{R})$ we consider the map $\mu : G \to GL(n, \mathbb{R}) \times GL(n, \mathbb{R}) \subset GL(2n, \mathbb{R})$ (in block diagonal form) given by $\mu(g) = (\varphi(g), \varphi(g^{-1})^T)$. Set $\|g\| = \|\mu(g)\|$(operator norm). We have

1. $\|xy\| \leqslant \|x\|\,\|y\|$ for $x, y \in G$.
2. If $k \in K$ then $\|kg\| = \|gk\| = \|g\|$, $g \in G$.
3. $\|g\| = \|g^{-1}\|$.
4. If $X \in \mathfrak{p}$ (see A.14) then $\|\exp tX\| = \|\exp X\|^t$ for $t \geqslant 0$.
5. If $0 < r < \infty$ then the set $\{g \in G|\,\|g\| \leqslant r\}$ is compact.

Lemma 1. *Let (π, V) be a Banach representation of G then there exist constants $C > 0$ and $d \geqslant 0$ such that $\|\pi(g)\| \leqslant C\|g\|^d$ (here we are using the operator norm on End(V)).*

Proof. We note that $\|g\| \geqslant 1$ for all $g \in G$ $(1 = \|gg^{-1}\| \leqslant \|g\|\,\|g^{-1}\| = \|g\|^2)$. Set $\sigma(g) = \log\|g\|$. We also set $\varphi(g) = \log\|\pi(g)\|$. Then $\sigma(xy) \leqslant \sigma(x) + \sigma(y)$ and $0 \leqslant \sigma(x) = \sigma(x^{-1})$, $\varphi(xy) \leqslant \varphi(x) + \varphi(y)$, $x, y \in G$. Set $B_r = \{g \in G|\,\sigma(g) \leqslant r\}$. 5. above implies that B_r is compact for all $0 < r < \infty$. The principle of uniform boundedness implies that there exists $C > 0$ such that $\varphi(g) \leqslant C$ for $g \in B_1$.

We note that $K \subset B_1$. Thus if $k \in K$, $x \in G$ then $-C + \varphi(x) \leqslant \varphi(kx) \leqslant C + \varphi(x)$. Indeed, $\varphi(x) = \varphi(k^{-1}kx) \leqslant \varphi(k^{-1}) + \varphi(kx) \leqslant C + \varphi(kx)$. Now let $X \in \mathfrak{p}$ then if $t \geqslant 0$ we have $\sigma(\exp tX) = t\sigma(\exp X)$ for $t \geqslant 0$. Let $j \geqslant 0$ be an integer such that $j \leqslant \sigma(\exp X) < j + 1$. Then $\sigma(\exp(X/(j+1))) < 1$. Hence

$$\varphi(\exp X) \leqslant (j+1)C \leqslant C(1 + \sigma(\exp X)).$$

Thus if $k \in K$, $X \in \mathfrak{p}$ then

$$\varphi(k \exp X) \leqslant C + C(1 + \sigma(\exp X)) = C(2 + \sigma(\exp X)).$$

Hence

$$\|\pi(k \exp X)\| \leqslant e^{2C}\|k \exp X\|^C.$$

Since every element of G is of the form $k \exp X$, $k \in K$, $X \in \mathfrak{p}$ (see A.15) the lemma is proved.

We note that the same argument proves that if $\alpha : G \to (0, \infty)$ has the properties

I. $\alpha(xy) \leqslant \alpha(x)\alpha(y)$, $x, y \in G$.

II. If $\Omega \subset G$ is compact then there exists $C_\Omega < \infty$ such that $\alpha(x) \leqslant C_\Omega$ for all $x \in \Omega$. Then there exist constants $A > 0$, $b \geqslant 0$ such that $\alpha(x) \leqslant A\|x\|^b$.

We say that a function $\alpha : G \to (0, \infty)$ is a *norm* on G if it satisfies

Norm 1. $\alpha(xy) \leqslant \alpha(x)\alpha(y)$, $x, y \in G$.
Norm 2. $\alpha(x) = \alpha(x^{-1})$, $x \in G$.
Norm 3. There exist $C > 0$, $d > 0$ such that $\|x\| \leqslant C\alpha(x)^d$.
α is said to be K-invariant if $\alpha(kx) = \alpha(xk)$, $x \in G$, $k \in K$.

We note that if we have two realizations of G as a real reductive group then each gives a norm on G. We also note that if we set $\alpha(g) = \mathrm{tr}(\mu(g)\mu(g)^T)$ then $\|g\| \leqslant \alpha(g)^{\frac{1}{2}}$. Thus α is a norm. Thus there exist C^∞ norms. We will allow ourselves use the notation $\|...\|$ for whatever norm on G that we define for a particular application.

2.2. Lemma. *If* $\|\ldots\|$ *is a continuous norm on* G *then there exists* $d_o > 0$ *such that*

$$\int_G \|g\|^{-d_o} dg < \infty$$

(here we have fixed an invariant measure on G.)

Proof. We first note that we may assume that $G = {}^\circ G$. Indeed if A is a split component of G then if we set $\alpha(ag) = \|a\|\,\|g\|$, $a \in A$ and $g \in {}^\circ G$ then it is easily seen that α satisfies 1.-5. above (replacing $\|\ldots\|$ with α) thus the proof of Lemma 1.1 implies that α is a norm. We show that there exists $d > 0$ such that

$$\int_A \|a\|^{-d} da < \infty.$$

For this it is enough to show that if $H_1, \ldots, H_r$ are symmetric $n \times n$ matrices such that $[H_i, H_j] = 0$ and $H_1, \ldots, H_r$ are linearly independent and if we set H_i' equal to the block diagonal $2n \times 2n$ matrix with diagonal blocks H_i and $-H_i$ then

$$\int_{\mathbb{R}^r} \|e^{\sum x_i H_i'}\|^{-d} dx < \infty$$

for some $d > 0$. For this we may use $\beta(e^{\sum x_i H_i'}) = \prod_i \|e^{x_i H_i'}\|$ by arguing as above. Now $\|e^{x_i H_i'}\| = \|e^{H_i'}\|^{|x_i|}$ and $\|e^{H_i}\| = e^{a_i}$ with $a_i > 0$. Thus we must show that

$$\int_{\mathbb{R}^r} e^{-\sum |x_i| a_i} dx < \infty$$

which is obvious. We may thus assume that $G = {}^\circ G$.

Fix (Q, A) a minimal standard p-pair. Let $\alpha_1, \ldots, \alpha_l$ be the simple roots of $\Phi(Q, A)$ (A.18). Let $H_1, \ldots, H_l$ be defined by $\alpha_i(H_j) = \delta_{ij}$. Set $A^+ = \{a \in A \mid a^\alpha > 1, \alpha \in \Phi(Q, A)\}$. Then $A^+ = \{\exp(\sum x_i H_i) \mid x_i > 0\}$. We have seen that there exist $a_i > 0$ such that if $x_i \geqslant 0$ then

$$\|\exp(\sum x_i H_i)\| \geqslant e^{\sum a_i x_i}$$

Thus if $d > 0$, $d \in \mathbb{Z}$ and $da_i > 2\rho(H_i) + 1$, $i = 1, \ldots, l$ then if $x_i \geqslant 0$

$$\|\exp(\sum x_i H_i)\|^{-d} \leqslant e^{-\sum x_i} e^{-2\rho(\sum x_i H_i)}$$

Thus if we set $\lambda(\sum x_i H_i) = \sum x_i$ we have (see A.18)

$$\int_G \|g\|^{-d} dg = \int_{A^+} \gamma(a) a^{-2\rho} a^{-\lambda} da.$$

Since $\gamma(a) \leqslant a^{2\rho}$ if $a \in A^+$ and $\int_{A^+} a^{-\lambda} da < \infty$ the result follows.

2.3. Fix $\|...\|$ a norm on G. Let (π, H) be a Banach representation of G. Then there exist $C > 0$, $d \geqslant 0$ such that $\|\pi(g)\| \leqslant C\|g\|^d$. Let $x \in U(\mathfrak{g}_C)$ and let q_x be the corresponding continuous semi-norm on H^∞ ($q_x(v) = \|\pi(x)v\|$). Assume that $x \in U^k(\mathfrak{g}_C)$. Let $x_1, ..., x_d$ be a basis of $U^k(\mathfrak{g}_C)$ then $\mathrm{Ad}(g)x_i = \sum_j \phi_{ji}(g)x_j$ and if we set $\varphi(g) = \max_{ij} |\varphi_{ij}(g)|$ then φ satisfies I,II in 2.1. Thus there exist $C_1 > 0$, $d_1 \geqslant 0$ such that $\varphi(g) \leqslant C_1\|g\|^{d_1}$. Now $x = \sum b_i x_i$ thus

$$q_x(\pi(g)v) = \|\pi(x)\pi(g)v\| \leqslant \sum_i |b_i| \|\pi(x_i)\pi(g)v\| =$$

$$\sum_i |b_i| \|\pi(g)\pi(\mathrm{Ad}(g^{-1})x_i)v\| \leqslant \sum_i |b_i| C\|g\|^d \| \sum \varphi_{ji}(g^{-1})\pi(x_j)v\| \leqslant$$

$$\left(\sum_i |b_i| C C_1 \|g\|^{d+d_1}\right) \sum_j \|\pi(x_j)v\|.$$

We have proved

Lemma. *If $x \in U(\mathfrak{g}_C)$ then there exist $A > 0, b \geqslant 0$ and $x_1, ..., x_d \in U(\mathfrak{g}_C)$ such that*

$$q_x(\pi(g)v) \leqslant A\|g\|^b \sum_i q_{x_i}(v), \quad v \in H^\infty.$$

Let (π, V) be a smooth Fréchet representation of G. Let $S = \{q_j\}$ be the set of semi-norms used to define the topology of V. Then we say that (π, V) is of moderate growth if for each i there exist constants $A_i > 0$, $b_i \geqslant 0$ and $i_1, ..., i_k$ such that

$$q_i(\pi(g)v) \leqslant A\|g\|^b \sum_j q_{i_j}(v), \quad v \in V, g \in G.$$

Example. Let $G = GL(n, \mathbb{R})$ and let (π_ν, V^ν) be as in 1.7. Then in the course of the proof of Lemma 1.7 we have shown that this representation has moderate growth.

The above lemma says that if (π, H) is a Hilbert representation of G then (π, H^∞) is a smooth Fréchet representation of moderate growth.

2.4. We denote by $\mathcal{F}_{mod}(G)$ the category of all smooth Fréchet representations of moderate growth with morphisms $T : (\pi, V) \to (\sigma, W)$ with $T : V \to W$ continuous such that $T(V)$ is a topological summand of W and $T \circ \pi(g) = \sigma(g) \circ T$. If T satisfies all of these conditions but $T(V)$ is not necessarily a topological summand of W then we say that T is a *continuous G-homomorphism*.

2.5. If $x \in U(\mathfrak{g}_C)$, $d \geqslant 0$ then we set

$$q_{x,d}(f) = \sup_{g \in G} \|g\|^d |xf(g)|$$

for $f \in C^\infty(G)$. Set $\mathcal{S}(G) = \{f \in C^\infty(G) | q_{x,d}(f) < 0, x \in U(\mathfrak{g}_\mathbb{C}), d \geqslant 0\}$. We note that we could define $\mathcal{S}(G)$ using a basis (x_j) of $U(\mathfrak{g}_\mathbb{C})$ and $d \in \mathbb{Z}, d \geqslant 0$. We endow $\mathcal{S}(G)$ with the topology given by these semi-norms. If we use $\sigma(g) = \|g\|$ and if $X_1, ..., X_n$ is a basis of $\mathfrak{g}$ then $\mathcal{S}(G)$ is the space W as in A.7. Thus $\mathcal{S}(G)$ is a Fréchet space.

If $f \in C^\infty(G)$ then we set $\pi_L(g)f(x) = f(g^{-1}x)$ and $\pi_R(g)f(x) = f(xg)$.

Lemma. $C_c^\infty(G)$ *is dense in* $\mathcal{S}(G)$. *Both* $(\pi_L, \mathcal{S}(G))$ *and* $(\pi_R, \mathcal{S}(G))$ *are smooth Fréchet representations of* G *of moderate growth.*

Proof. In this proof we will use $\alpha(g) = \mathrm{tr}(\mu(g)\mu(g)^T)$ for the norm. If $X \in \mathfrak{g}$ then $X^k\alpha(g) = \sum_j \binom{k}{j}\mathrm{tr}(\mu(g)\mu(X)^j(\mu(X)^T)^{k-j}\mu(g)^T)$. So $|X^k\alpha(g)| \leqslant C_{k,X}\alpha(g)$. If $x \in U(\mathfrak{g}_\mathbb{C})$ then x can be written as a linear combination of terms of the form X^k with $X \in \mathfrak{g}$. We therefore see that if $x \in U(\mathfrak{g}_\mathbb{C})$ then $|x\alpha(g)| \leqslant C_x\alpha(g)$.

Let $h \in C^\infty(\mathbb{R})$ be such that $h(x) = 1$ for $|x| \leqslant 1$ and $h(x) = 0$ for $|x| > 2$. Set $u_r(g) = h(\alpha(g)/r)$ for $r > 0$. If $X \in \mathfrak{g}$ then $Xu_r(g) = h'(\alpha(g)/r)X\alpha(g)/r$. If we apply the observations above we find that if $k > 0$ then

$$|X^k(u_r - 1)(g)| \leqslant D_{k,X}\alpha(g)^{d_{k,X}}/r.$$

If also $u_r(g) - 1 = h(\alpha(g)/r) - h(0) = (\alpha(g)/r)h'(\theta)$ with $0 < \theta < \alpha(g)/r$. Thus the above inequality is also true for $k = 0$. Thus if $x \in U^j(\mathfrak{g}_\mathbb{C})$ and if $x_1, ..., x_d$ is a basis of $U^j(\mathfrak{g}_\mathbb{C})$ then there exist $d_i \geqslant 0$ and $E > 0$ such that

$$q_{x,d}(u_rf - f) \leqslant \frac{E}{r}\sum_i q_{x_i,d_i}(f).$$

This implies that $\lim_{r\to\infty} u_rf = f$.

We now use any norm to define the topology on $\mathcal{S}(G)$. We note that if $x \in U(\mathfrak{g}_\mathbb{C})$, $g \in G$ and $f \in \mathcal{S}(G)$ then $\sup_{y\in G}\|y\|^r|xf(g^{-1}y)| = \sup_{y\in G}\|gy\|^r|xf(y)|$. Thus

$$q_{x,r}(\pi_L(g)f) \leqslant \|g\|^r q_{x,r}(f).$$

We now show that the map $g \mapsto \pi_L(g)f$ is C^∞ from G to $\mathcal{S}(G)$. Let $X \in \mathfrak{g}$. Let $x \in U(\mathfrak{g}_\mathbb{C})$ then

$$xf(\exp(-tX)g^{-1}y) = xf(g^{-1}y) + t\check{X}xf(g^{-1}y) + \frac{t^2}{2}\check{X}^2xf(\exp(-\theta X)g^{-1}y)$$

with

$$\check{X}f(y) = \frac{d}{dt}f(\exp(-tX)y)_{|t=0}$$

and θ is between 0 and t. Now $\check{X}f(y) = -(\mathrm{Ad}(y)^{-1}X)f(y)$. If $0 < |t| < 1$

$$\sup_{y\in G}\|y\|^r\left|\frac{xf(\exp(-tX)g^{-1}y) - xf(g^{-1}y)}{t} - x\check{X}f(g^{-1}y)\right| \leqslant$$

$$\frac{|t|}{2} \sup_{\substack{y \in G \\ |\theta| < 1}} \|y\|^r |\check{X}^2 f(\exp(\theta X)g^{-1}x)|.$$

Using the above observation it is now an easy matter to see that

$$\lim_{t \to 0} \frac{\pi_L(g \exp tX)f - \pi_L(g)f}{t} = \pi_L(g)\check{X}f$$

in $\mathcal{S}(G)$. Thus the map is of class C^1 and since $\check{X}f \in \mathcal{S}(G)$ the argument above implies that the map is of class C^2 etc. To prove the result for π_R we note that $[\check{X},\check{Y}] = [X,Y]\check{}$. Thus we have a map of $U(\mathfrak{g}_{\mathbb{C}})$ into the differential operators on $G : X \mapsto \check{X}$. The image is the algebra of right invariant differential operators. Using the properties of norms it is easily seen that the topology on $\mathcal{S}(G)$ is given by the semi-norms

$$p_{x,r}(f) = \sup_{g \in G} \|g\|^r |\check{x}f(g)|.$$

The arguments for π_R are essentially the same as those for π_L using this family of seminorms.

2.6. Let (π, H) be a Banach representation of G. Let $f \in \mathcal{S}(G)$. Let $d > 0$ be such that

$$\int_G \|g\|^{-d} dg < \infty.$$

Let $d_o \geqslant 0$ be such that there exists $C > 0$ with $\|\pi(g)\| \leqslant C\|g\|^{d_o}$ then

$$\|f(g)\pi(g)v\| \leqslant C\|g\|^{-d} q_{1,d+d_o}(f)\|v\|.$$

This implies that the integral

$$\int_G f(g)\pi(g)v dg$$

converges aboslutely and defines a bounded operator $\pi(f) : H \to H$. Furthermore the map $\mathcal{S}(G) \times H \to H$, $f, v \mapsto \pi(f)v$ is continuous.

If (π, V) is a smooth Fréchet representation of G and $\{q_j\}$ are as usual. Then if $v \in V$ and A_i, b_i and $i_1, ..., i_k$ are as in 2.3 then

$$q_i(f(g)\pi(g)v) \leqslant A_i q_{1,b_i+d}(f)\|g\|^{-d} \sum_j q_{i_j}(v).$$

Thus the integral

$$\int_G f(g)\pi(g)v dg$$

converges and defines a continuous operator $\pi(f) : V \to V$. Furthermore the map $\mathcal{S}(G) \times V \to V$, $f, v \mapsto \pi(f)v$ is continuous.

We apply this to $(\pi_L, \mathcal{S}(G))$ and $(\pi_R, \mathcal{S}(G))$. We note that

$$\pi_L(f)\phi(x) = \int_G f(g)\phi(g^{-1}x)dg = \int_G f(xg)\phi(g^{-1})dg = \pi_R(\tilde{\phi})f(x)$$

with $\tilde{\phi}(g) = \phi(g^{-1})$. We set $f \star \phi = \pi_L(f)\phi$ then one can check that $\star$ defines an associative multiplication on $\mathcal{S}(G)$ which makes $\mathcal{S}(G)$ into a Fréchet algebra (i.e. the map $f, \phi \mapsto f \star \phi$ is continuous.

Lemma. *If (π, H) is a Banach representation of G and if $f \in \mathcal{S}(G)$ then $\pi(f)H \subset H^\infty$.*

Proof. Let $v \in H$, $f \in \mathcal{S}(G)$ then $g \mapsto \pi(g)\pi(f)v$ factors into

$$g \mapsto \pi_L(g)f \mapsto (\pi_L(g)f)v$$

so the map is C^∞ by Lemma 2.5.

2.7. We now introduce another class of natural Fréchet spaces of functions on G. Fix $d > 0$. Let $\mathcal{A}_d(G)$ denote the space of all $f \in C^\infty(G)$ such that if $x \in U(\mathfrak{g}_{\mathbb{C}})$

$$p_{d,x}(f) = \sup_{g \in G} \|g\|^{-d}|xf(g)| < \infty.$$

We endow $\mathcal{A}_d(G)$ with the topology induced by the seminorms, $p_{d,x}$, $x \in U(\mathfrak{g}_{\mathbb{C}})$. If $f \in \mathcal{A}_d(G)$ and $g \in G$ then we set $\sigma_d(g)f(x) = f(xg)$.

Lemma. *$\mathcal{A}_d(G)$ is a Fréchet space and $(\sigma_d, \mathcal{A}_d(G))$ is a smooth Fréchet representation of moderate growth.*

Proof. We note that if $\{x_j\}$ is a basis of $U(\mathfrak{g}_{\mathbb{C}})$ then the seminorms p_{d,x_i} define the topology on $\mathcal{A}_d(G)$. We must therefore show that $\mathcal{A}_d(G)$ is sequentially complete. Let $\{f_j\}$ be a Cauchy sequence in $\mathcal{A}_d(G)$. Then if $U \subset G$ is open with compact closure the material in A.7 implies that there exists $g_U \in C^\infty(U)$ such that $xf_{j|U} \to xg_U$ uniformly on U. We note that if $V \subset G$ is open with compact closure then $g_U(x) = g_V(x)$ for $x \in U \cap V$. Thus there exists $g \in C^\infty(G)$ such that $g_{|U} = g_U$ for all such U. We next note that $g \in \mathcal{A}_d(G)$. Fix $x \in U(\mathfrak{g}_{\mathbb{C}})$. There exists j such that if $i \geqslant j$ then $p_{x,d}(f_i - f_j) < 1$. If $y \in G$ there exists $i > j$ such that $|xg(y) - xf_i(y)| < 1$. Thus

$$|xg(y)| = |xg(y) - xf_i(y) + xf_i(y) - xf_j(y) + xf_j(y)| \leqslant$$
$$1 + \|y\|^d + \|y\|^d p_{x,d}(f_j).$$

Since $\|y\| \geqslant 1$ we see that

$$|xg(y)| \leqslant (2 + p_{x,d}(f_j))\|y\|^d.$$

223

Hence, $p_{x,d}(g) \leqslant 2 + p_{x,d}(f_j)$. We modify this argument to show that $f_j \to g$ in $\mathcal{A}_d(G)$. Let $\epsilon > 0$ be given. Fix $x \in U(\mathfrak{g}_\mathbb{C})$. Let N be such that if $i, j \geqslant N$ then $p_x(f_i - f_j) < \epsilon$. Let $j \geqslant N$ and let $y \in G$. Then there exists $i \geqslant j$ such that $|xg(y) - xf_i(y)| < \epsilon$. Thus

$$|xg(y) - xf_j(y)| = |xg(y) - xf_i(y) + xf_i(y) - xf_j(y)| \leqslant$$

$$|xg(y) - xf_i(y)| + |xf_i(y) - xf_j(y)| \leqslant \epsilon + \epsilon\|y\|^d.$$

Thus as before

$$\|y\|^d |xg(y) - xf_j(y)| \leqslant 2\epsilon$$

for all $y \in G$. This completes the proof that $\mathcal{A}_d(G)$ is a Fréchet space.

Let $x \in U^j(\mathfrak{g}_\mathbb{C})$ and let $x_1, ..., x_m$ be a basis of $U^j(\mathfrak{g}_\mathbb{C})$ then $\mathrm{Ad}(g)x = \sum_i \varphi_i(g)x_i$. There exists a constants $C > 0$, $a > 0$ such that

$$|\varphi_i(g)| \leqslant C\|g\|^a, \ g \in G.$$

We now estimate $(\sigma = \sigma_d)$

$$p_{x,d}(\sigma(y)f) = \sup_{g \in G} \|g\|^d |x\sigma(y)f(g)| = \sup_{g \in G} \|g\|^d |(\mathrm{Ad}(y)^{-1}x)f(gy)| \leqslant$$

$$C\|y^{-1}\|^a \sum_i \sup_{g \in G} \|gy^{-1}\|^d |x_i f(g)| \leqslant C\|y^{-1}\|^{a+d} \sum_i \sup_{g \in G} \|g\|^d |x_i f(g)| =$$

$$C\|y\|^{a+d} \sum_i p_{x_i,d}(f).$$

This implies that $(\sigma_d, \mathcal{A}_d(G))$ is a Fréchet representation of G. If we show that the representation is smooth then we will have shown that it is of moderate growth. We now prove the smoothness. We first observe that if $y \in U(\mathfrak{g}_\mathbb{C})$ then

$$p_{x,d}(yf) = p_{xy,d}(f).$$

This combined with our usual Taylor series argument (see 2.5) can be used to complete the proof.

3. $(\mathfrak{g}, K)$ MODULES.

3.1. Let G be a real reductive group with Lie algebra $\mathfrak{g}$. Let K be as in A.14. Let V be a module for $\mathfrak{g}$ that is also a module for K such that

1. If $v \in V$ then there is a finite dimensional space W_v such that $Kv \subset W_v$ and the map $k \mapsto kv$ is of class C^∞ from K to W_v.

2. If $X \in \mathfrak{g}$, $k \in K$ and $v \in V$ then $kXv = \mathrm{Ad}(k)Xkv$.

3. If $Y \in \mathfrak{k} = Lie(K) \subset G$ then $\frac{d}{dt} \exp(tY)v_{|t=0} = Yv$ for $v \in V$.

Let V be a $(\mathfrak{g}, K)$–module. If $v \in V$ then span$\{Kv\}$ is a finite dimensional subspace of V on which K acts smoothly (see 1.) we will use the notation W_v for this space. Thus as a representation of K, W_v is a finite dierect sum of irreducible K submodules, $W_v = W_1 \oplus ... \oplus W_d$. Let $\hat{K}$ denote the set of all equivalence classes of irreducible, finite dimensional, continuous (hence smooth) K-modules. If $\gamma \in \hat{K}$ then we set $V(\gamma)$ equal to the sum of all irreducible K-submodules of V in the class of γ. Then $V = \oplus_{\gamma \in \hat{K}} V(\gamma)$. $V(\gamma)$ is called the γ–isotypic component of V.

Example. Let (π, V) be a smooth Fréchet representation of G. Let $\mathfrak{g}$ act by $\pi(X)$, $X \in \mathfrak{g}$ and K act by $\pi(k)$, $k \in K$. Then this action satisfies 2. and 3. but not necessarily 1. We set $V_K = \{v \in V | \dim \text{span}\{\pi(K)v\} < \infty\}$ then V_K satisfies 1. and 3. We must check that V_K is invariant by the action of $\mathfrak{g}$. For this we note that since 2. is satisfied in V, $\pi(K)\pi(\mathfrak{g})v \subset \pi(\text{Ad}(K)\mathfrak{g})\pi(K)v$ thus

$$\dim \text{span}\{\pi(K)\pi(\mathfrak{g})v\} \leqslant \dim \mathfrak{g} \cdot \dim \text{span}\{\pi(K)v\}.$$

So V_K is indeed an invariant subspace of V under both K and $\mathfrak{g}$. We note that it is almost never G-invariant. We will call V_K the *underlying* $(\mathfrak{g}, K)$-module of V.

Theorem. If (π, V) is a smooth Fréchet representation of G then V_K is a dense subspace of V.

Example. Let $G = GL(n, \mathbb{R})$ and let (π_ν, V^ν) be as in 1.7. $K = O(n)$. We note that under the map $\varphi \to \varphi_\nu$ we may identify V^ν with $C^\infty(S^{n-1})$. Under this identification $\pi_\nu(k)$ becomes $\sigma(k)\varphi(x) = \varphi(k^{-1}x)$. Let $\mathcal{P}^j$ denote the space of all $f \in C^\infty(S^{n-1})$ such that f is the restriction of a polynomial function on $\mathbb{R}^n$ of degree at most j. Then $\sigma(k)\mathcal{P}^j \subset \mathcal{P}^j$ for $k \in K$. Thus if $\mathcal{P} = \cup_j \mathcal{P}^j$ then $\mathcal{P}_\nu \subset (V^\nu)_K$. Since S^{n-1} is compact, the Stone-Weierstrauss theorem implies that $\mathcal{P}$ is dense in $C(K)$ relative to the topology of uniform convergence. We will use the material in the next number to show that $((V^\nu)_K)_{|S^{n-1}} = \mathcal{P}$.

3.2. The proof of Theorem 3.1 involves some preparation. If $\gamma \in \hat{K}$ then we fix $V_\gamma \in \gamma$. If $(\ ,\ )$ is an inner product on V_γ then we define

$$\langle v, w \rangle = \int_K (kv, kw)dk.$$

Here dk is normalized (total volume 1) invariant measure on K. Then with respect to $\langle ..., ... \rangle$, V_γ is a unitary representation of K.

1. If $b(\ ,\ )$ is a sesquilinear form on V_γ such that $b(kv, kw) = b(v, w)$, $v, w \in V_\gamma$, $k \in K$, then there is $c \in \mathbb{C}$ such that $b = c\langle ..., ... \rangle$.

Indeed, there exists $T \in End(V_\gamma)$ such that $b(v, w) = \langle Tv, w \rangle$. The invariance hypothesis on b implies that $Tkv = kTv$, $v \in V_\gamma$, $k \in K$. Schur's Lemma implies that $T = cI$ for some $c \in \mathbb{C}$.

We define a linear map $\Psi_\gamma : V_\gamma^* \otimes V_\gamma \to C^\infty(K)$ by $\Psi_\gamma(\lambda \otimes v)(k) = \lambda(kv)$. On V_γ^* we put the *contragredient representation*, $k\lambda(v) = \lambda(k^{-1}v)$. On V_γ^* we put the inner product dual to $\langle ..., ...\rangle$. That is, if $\lambda \in V_\gamma^*$ then there exists $\lambda^b \in V_\gamma$ such that $\lambda(v) = \langle v, \lambda^b\rangle$. We set $\langle \lambda, \mu\rangle = \langle \mu^b, \lambda^b\rangle$.

We note that

$$l_k \Psi_\gamma(\lambda \otimes v) = \Psi_\gamma(k\lambda \otimes v), \quad r_k \Psi_\gamma(\lambda \otimes v) = \Psi_\gamma(\lambda \otimes kv).$$

We set $d(\gamma) = \dim V_\gamma$. The following result is the classical Schur orthogonality relations.

2. Let $\gamma, \gamma' \in \hat{K}$. Then $\int_K \Psi_\gamma(\lambda \otimes v)(k)\overline{\Psi_{\gamma'}(\lambda' \otimes v')(k)}dk = d(\gamma)^{-1}\delta_{\gamma,\gamma'} \langle v, v'\rangle \langle \lambda, \lambda'\rangle$.

For fixed λ, λ' the above formula gives a sesquilinear pairing of $V_\gamma, V_{\gamma'}$ which is K-invariant. As in the proof of 1., this implies that there is a linear map $T(\lambda, \lambda') = T : V_{\gamma'} \to V_\gamma$ such that $Tkv = kTv$ for $v \in V_{\gamma'}$, $k \in K$. Thus if $\gamma \neq \gamma'$ we have $T = 0$ and if $\gamma = \gamma'$ then $T = c(\lambda, \lambda')I$. Clearly, $c(,)$ is sesquilinear and K−invariant. Hence, $c = C\langle , \rangle$, $C \in \mathbb{C}$. We must show that $C = d(\gamma)^{-1}$. Let $v_1, ..., v_d$ be an orthonormal basis of V_γ and let $\lambda_1, ..., \lambda_d$ be the dual basis of V_γ^* then $\sum_{ij}\Psi_\gamma(\lambda_i \otimes v_j)(k)\overline{\Psi_\gamma(\lambda_i \otimes v_j)(k)}$ is constant on K. Evaluating at $1 \in K$ we find that the constant value is d. Thus applying what we already know we find that $d = Cd^2$. Hence $C = d^{-1}$ as was to be shown.

If $\gamma \in \hat{K}$ we define $\chi_\gamma(k)$ to be the trace of the action of k on V_γ. Then in the notation above, $\chi_\gamma = \sum_i \Psi_\gamma(\lambda_i \otimes v_i)$. We set $\alpha_\gamma(k) = d(\gamma)\chi_\gamma(k)$. If $f, g \in C(K)$ define $f \star g(k) = \int_K f(x)g(x^{-1}k)dx$.

3. If $\gamma, \gamma' \in \hat{K}$ then $\alpha_\gamma \star \Psi_{\gamma'}(\xi) = \delta_{\gamma,\gamma'}\Psi_{\gamma'}(\xi)$ for $\xi \in V_{\gamma'}^* \otimes V_{\gamma'}$.

We leave this calculation (using 2. above) as an exercise to the reader.

We now come to the crux of the matter.

Theorem.

1. If $f \in C^\infty(K), \gamma \in \hat{K}$ then $\alpha_\gamma \star f \in \Psi_\gamma(V_{\gamma'}^* \otimes V_{\gamma'})$.

2. For $i = 1, 2, ...,$ let $F_i \subset \hat{K}$ be a finite subset such that $F_i \subset F_{i+1}$ and $\cup_i F_i = \hat{K}$. If $f \in C^\infty(K)$ then

$$\lim_{i \to \infty} \sum_{\gamma \in F_i} \alpha_\gamma \star f = f$$

in $C^\infty(K)$.

We will not give a complete proof of this result. However, we will indicate how it reduces to the elliptic regularity theorem. If $(,)$ is any inner product on $\mathfrak{k} = Lie(K)$ then we set $Q(X, Y) = \int_K (\mathrm{Ad}(k)X, \mathrm{Ad}(k)Y)dk$. Then Q is an inner product on $\mathfrak{k}$ and $Q(\mathrm{Ad}(k)X, \mathrm{Ad}(k)Y) = Q(X, Y)$ for $X, Y \in \mathfrak{k}$, $k \in K$.

If $X_1, ..., X_n$ is an orthonormal basis of $\mathfrak{k}$ then we set $C = -\sum X_i^2 \in U(\mathfrak{k}_\mathbb{C})$. Then one checks that C is an elliptic second order differential opertor on K(cf.

[F.Warner]). The elliptic regularity theory (same reference) implies that if $\lambda \in \mathbb{C}$ and if we set $C^\infty(K)_\lambda = \{f \in C^\infty(K)|\,Cf = \lambda f\}$ then $\dim C^\infty(K)_\lambda < \infty$. Also, is an easy matter to check that $l_k C = C l_k$ and $r_k C = C r_k$ thus $C^\infty(K)_\lambda$ is a finite dimensional subspace of $C^\infty(K)$ that is invariant under both the left and right regular representation of K. Using the L^2−inner product on $C^\infty(K)_\lambda$ it is an easy matter to see that the space splits into irreducible invariant subspaces under the action $\rho(x,y) = l_x r_y$ of $K \times K$ on $C^\infty(K)$. It is also standard that such an irreducible subspace must be of the form $\Psi_\gamma(V_\gamma^* \otimes V_\gamma)$ for some $\gamma \in \hat{K}$.

Since K is compact, one can see that if $C^\infty(K)_\lambda \neq 0$ then $\lambda \in \mathbb{R}$ and $\lambda \geqslant 0$. Let

$$0 = \lambda_0 < \lambda_1 < \ldots$$

be the set of all λ such that $C^\infty(K)_\lambda \neq 0$. Applying elliptic regularity again one finds that if $f \in C^\infty(K)$ then there exist $f_i \in C^\infty(K)_{\lambda_i}$ such that $\sum f_i$ converges to f in $C^\infty(K)$. Since $\langle Cf, g \rangle = \langle f, Cg \rangle$ (here we are using the L^2 inner product) then the spaces $C^\infty(K)_\lambda$ are mutually perpendicular. Set $S_i = \{\gamma \in \hat{K}|\, \Psi_\gamma(V_\gamma^* \otimes V_\gamma) \subset C^\infty(K)_{\lambda_i}\}$. Then the above remarks easily imply that $f_i = \sum_{\gamma \in S_i} \alpha_\gamma \star f$. If we set $F_i = \cup_{j \leqslant i} S_j$ the theorem follows.

3.3. We will now prove Theorem 3.1. We use the notation of the previous number. Let $\gamma \in \hat{K}$. If $v \in V$ then we set

$$E_\gamma v = \int_K \overline{\alpha_\gamma(k)}\pi(k)v\,dk.$$

If $\lambda \in V'$ (continuous functional on V) then $k \mapsto \lambda(\pi(k)v) = c_{\lambda,v} \in C^\infty(K)$. We note that $\alpha_\gamma \star c_{\lambda,v} = c_{\lambda,E_\gamma v}$. Since $\sum_{\gamma \in \hat{K}} \alpha_\gamma \star c_{\lambda,v} = c_{\lambda,v}$, it follows that if $\lambda(E_\gamma V) = 0$ for all $\gamma \in \hat{K}$, $\lambda = 0$. We now show that $E_\gamma V \subset V_K$. So fix $\gamma \in \hat{K}$, $v \in V$ and set $w = E_\gamma v$. Then $\alpha_\gamma \star r_k c_{\lambda,w} = r_k c_{\lambda,w}$. Hence, $\{r_k c_{\lambda,w}|\lambda \in V', k \in K\} \subset \Psi_\gamma(V_\gamma^* \otimes V_\gamma)$. This implies that $V'_{|\mathrm{span}\{\pi(K)w\}}$ is a finite dimensional space. The Hahn-Banach theorem now implies that $\mathrm{span}(\pi(K)w)$ is finite dimensional. Thus $w \in V_K$. We have therefore shown that if $\lambda \in V'$ and if $\lambda(V_K) = 0$ then $\lambda = 0$. Thus another application of the Hahn-Banach theorem implies that V_K is dense in V.

Example 3.1 (continued). Let $\Delta = \sum_i \partial_i^2$. Set $\mathcal{H}^j$ equal to the space of all polynomials f on $\mathbb{R}^n$ that are homogeneous of degree j and satisfy $\Delta f = 0$. We assert that the space $\mathcal{H}^j$ defines an irreducible representation of $K(= O(n))$ under the action $\tau(k)f(x) = f(k^{-1}x)$. We first note that $\tau(k)\Delta = \Delta\tau(k)$ for $k \in K$. Thus $\tau(K)\mathcal{H}^j \subset \mathcal{H}^j$. So $(\tau, \mathcal{H}^j)$ is indeed a representation of K.

Set P^j equal to the space of all polynomials on $\mathbb{R}^n$ homogeneous of degree j. Set $r_n^2 = r^2 = \sum_{i=1}^n x_i^2$.

1. If $f \in P^j$ and $\tau(k)f = f$ for all $k \in K$ then j must be even and there is a constant c such that $f = cr^j$.

Indeed, since $-I \in K$ and $\tau(-I)f = (-1)^j f$ we see that j must be even. Since $Ke_n = S^{n-1}$, f is constant on S^{n-1} thus $r^{-j}f$ is constant on $\mathbb{R}^n - (0)$. The assertion follows.

Let $M = \{k \in K \,|\, ke_n = e_n\}$ then M is isomorphic with $O(n-1)$.

2. If W is a non-zero K-invariant subspace of $\mathcal{H}^j$ then there exists a non-zero $w \in W$ such that $\tau(m)w = w$ for all $m \in M$.

Indeed, if for all $g \in W$, $g(e_n) = 0$ then for all $k \in K$, $g \in W$, $\tau(k)g(e_n) = 0$. Thus for all $g \in W$, $g_{|S^{n-1}} = 0$ so $g = 0$. On $\mathcal{H}^j$ we have the K-inner product

$$\langle f, g \rangle = \int_{S^{n-1}} f\bar{g}_{|S^{n-1}} \omega.$$

Let $\lambda(f) = f(e_n)$ then $\lambda^b \in W$ (defined by $\lambda(f) = \langle f, \lambda^b \rangle$) satisfies $\tau(m)\lambda^b = \lambda^b$.

3. There is a unique (up to scalar multiple) $\varphi \in \mathcal{H}^j$ such that $\tau(m)\varphi = \varphi$ for all $m \in M$. Furthermore, φ can be chosen such that $\langle f, \varphi \rangle = f(e_n)$ for all $f \in \mathcal{H}^j$.

To prove this assertion we need only prove the uniqueness statement. The second part is implied by the argument in 2. Let $f \in \mathcal{H}^j$ be invariant under M. We expand f in terms of powers of x_n. Thus $f = \sum f_k x_n^{j-k}$ with f_k a polynomial in $x_1, ..., x_{n-1}$. Since $\tau(m)f = f$ each f_k is invariant under $O(n-1)$. Thus $f_k = 0$ if k is odd and if k is even then $f_k = c_k r_{n-1}^k$. Since $r_n = r_{n-1} + x_n^2$, we see that there exist constants b_k such that

$$f = \sum b_k r^{2k} x_n^{j-2k}.$$

We note that $0 = \Delta f = \sum_{k>0} d_k r^{2k-2} x_n^{j-2k}$ with $d_k \neq 0$ for $1 \leqslant k \leqslant j/2$. Thus if $b_0 = 0$ then $f = 0$. This implies the uniqueness assertion.

We can now prove the irreducibilty. Suppose that $W \subset \mathcal{H}^j$ is K-invariant and non-zero. Then $\varphi \in W$. If $f \in W^\perp$ then $f(e_n) = \langle f, \varphi \rangle = 0$. Thus since $W^\perp$ is also K invariant, $W^\perp = 0$ (see the proof of 1.). Hence $W = \mathcal{H}^j$.

We also note that a variant of the argument used to prove 3. implies

4. $P^j = \sum_{k=0}^{j/2} r^{2k} \mathcal{H}^{j-2k}$.

We now complete the discussion of example 3.3. We note that the above discussion implies that $\mathcal{P} = \oplus_{j \geqslant 0} \mathcal{H}^j_{|S^{n-1}}$ is a decomposition into irreducible spaces. Since V_K^ν is dense in V^ν and $C^\infty(K)$ is dense in $C(K)$ we see that $(V_K^\nu)_{|S^{n-1}}$ is dense in $C(K)$. If $\gamma \in \hat{K}$ then $E_\gamma \mathcal{P} \subset \mathcal{P}$. Thus for each $\gamma \in \hat{K}$, $V^\nu(\gamma)_{|S^{n-1}} \cap \mathcal{P}$ is dense in $V^\nu(\gamma)$ and this implies that $V^\nu(\gamma) = \mathcal{H}^j_{|S^{n-1}}$ for some j.

If (π, H) is a Hilbert representation of G then we set $H_K = (H^\infty)_K$ and we call H_K the *underlying* $(\mathfrak{g}, K)$-*module* of H. We note that since H^∞ is dense in H, Theorem 3.1 implies that H_K is dense in H.

3.4. If V is a $(\mathfrak{g}, K)$-module then we say that V is *admissible* if $\dim V(\gamma) < \infty$ for all $\gamma \in \hat{K}$. We will now give a construction of a large class of admissible $(\mathfrak{g}, K)$ modules. We say that V is *finitely generated* if there exists a finite dimensional subspace W of V such that $V = U(\mathfrak{g}_C)W$.

We say that a $(\mathfrak{g}, K)$-module V is *irreducible* if the only subspaces W of V such that $\mathfrak{g}W \subset W$ and $KW \subset W$ are $W = \{0\}$ and $W = V$. We need a variant of Schur's lemma which is based on the following observation of Dixmier.

Lemma. *Let V be a countable dimensional vector space over $\mathbb{C}$. If T is an endomorphism of V then there exists $\lambda \in \mathbb{C}$ such that $T - \lambda I$ is not bijective.*

Proof. Suppose note then $T - \lambda I$ is bijective for all $\lambda \in \mathbb{C}$. Let R denote the algebra of all rational functions on $\mathbb{C}$. If $p/q \in R$ then define $\tau(p/q)$ as follows: Let $q(x) = a \prod(x - a_i)$ then $\tau(p/q) = ap(T) \prod(T - a_i I)^{-1}$. Then τ defines an algebra homomorphism of R into $End(V)$. Let $v \in V$ be non-zero. Suppose that $\tau(p/q)v = 0$. Then $p(T)v = 0$. Now $p = c \prod(x - c_i)$ thus if $p \neq 0$ then $\prod(T - c_i I)v = 0$. But this clearly would imply that $v = 0$. Hence the map $R \to V$, $r \mapsto \tau(r)v$ is injective. Now this implies that $\dim_\mathbb{C} V \geqslant \dim_\mathbb{C} R$. We note that the set $\{\frac{1}{x-c} | c \in \mathbb{C}\}$ is linearly independent in R. Thus $\dim R \geqslant \text{Card}(\mathbb{C})$ which is uncountable. This is a contradiction so the lemma follows.

3.5. We define $\mathcal{C}(\mathfrak{g}, K)$ to be the category of all $(\mathfrak{g}, K)$-modules with morphisms the linear maps $T : V \to W$ with $Txv = xTv$ for $x \in \mathfrak{g}$ or $x \in K$, $V, W \in \mathcal{C}(\mathfrak{g}, K)$. We denote $\text{Hom}_{\mathcal{C}(\mathfrak{g},K)}$ by $\text{Hom}_{\mathfrak{g},K}$.

Lemma. *(Schur's Lemma for $\mathcal{C}(\mathfrak{g}, K)$). If $V \in \mathcal{C}(\mathfrak{g}, K)$ is irreducible then*

$$\text{Hom}_{\mathfrak{g},K}(V, V) = \mathbb{C}I.$$

Proof. We note that if $W \subset V$ is a K-invariant subspace then $U(\mathfrak{g}_C)W$ is both K and $\mathfrak{g}$-invariant. This implies that V is finitely generated and in particular V is countable dimensional. Thus if $T \in \text{Hom}_{\mathfrak{g},K}(V, V)$ then there exists $\lambda \in \mathbb{C}$ such that $T - \lambda I$ is either not injective or not surjective. If $T - \lambda I$ is not injective then $\ker(T - \lambda I) = V$ so $T = \lambda I$. If $(T - \lambda I)V \neq V$ then $(T - \lambda I)V = 0$ so $T = \lambda I$.

3.6. We will now recall a basic theorem of Harish-Chandra. Let $Z(\mathfrak{g}_C)$ denote the center of $U(\mathfrak{g}_C)$. We say that G is of *inner type* if $\text{Ad}(G)$ is contained in the identity component of the group of automorphisms of $\mathfrak{g}_C$. Assume that G is of inner type and if $\gamma \in \hat{K}$ then since $\text{Ad}(g)z = z$ for $z \in Z(\mathfrak{g}_C)$, $Z(\mathfrak{g}_C)V(\gamma) \subset V(\gamma)$ for all $V \in \mathcal{C}(\mathfrak{g}, K)$.

Theorem. *If G is of inner type and if V is a finitely generated $(\mathfrak{g}, K)$-module then $V(\gamma)$ is finitely generated as a $Z(\mathfrak{g}_C)$-module.*

For a proof cf. [RRG I,3.4].
This result combined with Lemma 3.4 (and its proof) implies

Corollary. *If V is an irreducible $(\mathfrak{g}, K)$-module then V is finitely generated and admissible.*

Proof. In the proof of Lemma 3.4 we showed that V is finitely generated. If $z \in Z(\mathfrak{g}_C)$ then the action of z on V is in $\mathrm{Hom}_{\mathfrak{g}.K}(V, V)$. Thus if $z \in Z(\mathfrak{g}_C)$ there is $\chi(z) \in \mathbb{C}$ such that z acts on V by $\chi(z)I$. Let $\gamma \in \hat{K}$ and let $u_1, ..., u_d$ $(d < \infty)$ be such that $V(\gamma) = \sum Z(\mathfrak{g}_C)u_i$. Then $V(\gamma) = \sum \mathbb{C}u_i$. Thus $\dim V(\gamma) < \infty$.

3.7. If $X, Y \in \mathfrak{g}$ then define $B(X,Y) = \mathrm{tr}XY$ (note that $\mathfrak{g}$ is identified with $Lie(G_o) \subset M_n(\mathbb{R})$). Since $X \in \mathfrak{g}$ implies that $X^T \in \mathfrak{g}$ and since $\mathrm{tr}XX^T > 0$ if $X \neq 0$ we see that B is non-degenerate. We also note that if $X \in \mathfrak{k}$ then $X^T = -X$ so $B(X,X) < 0$ if $X \neq 0$. Let $X_1, ..., X_k$ be a basis of $\mathfrak{k}$ such that $B(X_i, X_j) = -\delta_{ij}$. We note that $\mathfrak{k}^\perp$ in $\mathfrak{g}$ (relative to B) is the space of all $X \in \mathfrak{g}$ such that $X^T = X$. Let $Y_1, ..., Y_n$ be an orthonormal basis of $\mathfrak{k}^\perp$. Set $C = -\sum X_i^2 + \sum Y_i^2 \in U(\mathfrak{g}_C)$. If Z_i is a basis of $\mathfrak{g}$ and if Z^i are defined by $B(Z_i . Z^j) = \delta_{ij}$ then $C = \sum Z_i Z^i$. From this it is easily seen that $\mathrm{Ad}(g)C = C$ for $g \in G$. We set $\Delta = C - 2C_K$ with $C_K = -\sum X_i^2$. Then Δ defines an elliptic operator on G with analytic coefficients.

Theorem. *Let (π, H) be an admissible Hilbert representation or an admissible smooth Fréchet represetation of G then $H_K \subset H^\omega$ (see 1.9).*

Proof. There exists λ such that if $v \in H(\gamma)(= (H_K)(\gamma))$ such that $C_K v = \lambda v$. Also since $CH(\gamma) \subset H(\gamma)$ there exists a nonconstant monic polynomial $p(x)$ such that $p(C)H(\gamma) = 0$. Thus $p(\Delta + 2C_K)H(\gamma) = 0$. Set $q(x) = p(x + 2\lambda)$ then $q(x)$ is a monic nonconstant polynomial and $q(\Delta)H(\gamma) = 0$. This implies that if $\xi \in H'$, $v \in H(\gamma)$ and if $f(g) = \xi(\pi(g)v)$ then $q(\Delta)f = 0$. Analytic elliptic regularity ([Hö]) now implies that f is real analytic.

Corollary. *If (π, H) is an admissible Hilbert representation of G then (π, H) is irreducible if and only if H_K is an irreducible $(\mathfrak{g}, K)$-module.*

Proof. If H_K is reducible then there is a proper $(\mathfrak{g}, K)$-submodule $V \neq 0$ of H_K. We note that $H_K \cap Cl(V) = V$. Thus $Cl(V) \neq H$. Let G^o be the identity component of G. Then Theorem 1.9 implies that $Cl(V)$ is invariant under G^o. Since $G = G^o K$ and $Cl(V)$ is obviously invariant under K we see that H is reducible. Suppose that H is reducible. Then there exists a closed non-zero invariant subspace, W, of H. If $W \cap H_K = H_K$ then $Cl(W) = H$. Thus $W \cap H_K \neq H_K$ and since $W \neq 0$, $W_K \neq 0$. Clearly $W \cap H_K$ is $(\mathfrak{g}, K)$-nvariant. Hence H_K is reducible.

Example. Let (π_ν, H^ν) be as in 1.8, $G = GL(n, \mathbb{R})$. Then $(H^\nu)^\infty = V^\nu$. So the material in Example 3.1 (and its continuation) implies that H^ν is admissible. If $f \in H^\nu$ define $Sf(x) = f(-x)$. Then S defines a unitary operator on H^ν, $S\pi_\nu(g) = \pi_\nu(g)S$ and $S^2 = I$. Let $H^\nu_\pm = \{f \in H^\nu | Sf = \pm f\}$.

Lemma. (π_ν, H_+^ν) is irreducible if and only if $\nu \notin 2\mathbb{N} \cup (-n - 2\mathbb{N})$. (π_ν, H_-^ν) is irreducible if and only if $\nu \notin (1 + 2\mathbb{N}_+) \cup (-n - 1 - 2\mathbb{N}_+)$.

Proof. We show that if ν is not equal to $0,2,4,\dots$ then $Z_\nu = \pi_\nu(U(\mathfrak{g}_C))1_\nu = \oplus_{j \geqslant 0} P^{2j} 1_{\nu - 2j}$. Indeed, let σ be as in 1.7. Then $\sigma(\mathfrak{g})$ is the span of the operators $x_i \partial_j$. Also $1_\nu(x) = \|x\|^\nu = (x_1^2 + \dots + x_n^2)^{\frac{\nu}{2}}$. Thus $x_i \partial_j 1_\nu = \nu x_i x_j (x_1^2 + \dots + x_n^2)^{\frac{\nu}{2} - 1}$. Thus if $\nu \neq 0$ then $Z_\nu \supset P^0 1_\nu \oplus P^2 1_{\nu - 2}$. Suppose $f \in P^2$ then

$$x_i \partial_j (f 1_{\nu - 2}) = x_i (\partial_j f) 1_{\nu - 2} + (\nu - 2) x_i x_j f 1_{\nu - 4}.$$

Thus if $\nu \neq 0, 2$ then $Z_\nu \supset P^4 1_{\nu - 4}$. Continuing this argument we see that if $\nu \neq 0, 2, 4, \dots$ then $Z_\nu \supset P^{2j} 1_{\nu - 2j}$ for $j \in \mathbb{N}$. Suppose that $\nu \notin 2\mathbb{N} \cup (-n - 2\mathbb{N})$ and that $X \subset (H_+^\nu)_K = Y_+^\nu$ is a $(\mathfrak{g}, K)$-invariant subspace. If $1_\nu \in X$ then $X = V$ by the above observations. If $1_\nu \notin X$ then $\langle X, 1_{-n-\bar\nu} \rangle = 0$. Since X is $(\mathfrak{g}, K)$-invariant we see that $\langle X, Z_{-n-\bar\nu} \rangle = 0$. But as above, under our conditions $Z_{-n-\bar\nu} = Y_+^{-n-\bar\nu}$. So $X = 0$. Thus Y^ν is irreducible under the hypothesis for H_+^ν. Thus the condition is sufficient for irreducibility (we are using the above corollary). If $\nu = 2j$ with $j \in \mathbb{N}$ then it is easily seen that $Z_\nu = P^0 1_\nu \oplus \dots \oplus P^{2j} 1_\nu$. Thus H_+^ν is reducible. Also $Z_\nu^\perp$ in $H_+^{-n-\bar\nu}$ is proper, non-zero and invariant. Thus the condition is indeed necessary and sufficient.

The result for H_-^ν is proved similarly. We show that if $Z_\nu^- = \pi_\nu(P^1 1_{\nu-1})$ then $Z_\nu^- = (H_-^\nu)_K = Y_-^\nu$ if $\nu \notin 2\mathbb{N} + 1$. As before, if $f \in P^1$ then

$$x_i \partial_j (f 1_{\nu - 1}) = x_i (\partial_j f) 1_{\nu - 1} + (\nu - 1) x_i x_j 1_{\nu - 3}.$$

Now proceed as above. Since all the $\mathcal{H}^j$ have different dimensions they are inequivalent. Assume that $\nu \notin (2\mathbb{N} + 1) \cup (-n - 1 - 2\mathbb{N})$. Let $X \subset Y_-^\nu$ be a $(\mathfrak{g}, K)$-invariant subspace. If $P^1 1_{\nu-1} \subset X$ then $X = Y^\nu$. Otherwise, $X \subset (P^1 1_{-n-2-\bar\nu})^\perp$. Thus as before $X = 0$. If $\nu = 2j + 1$, $j \in \mathbb{N}$ then $Z_\nu^- = P^1 1_{\nu-1} \oplus \dots \oplus P^{2j+1}$. Thus the necessity is also proved as above.

Note. If $n \geqslant 3$ then the $\mathcal{H}^j$ are already irreducible under the action of $SO(n)$ (exercise). Thus the above result applies to $SL(n, \mathbb{R})$ if $n \geqslant 3$. If $n = 2$ then $\mathcal{H}^j = \mathbb{C}(x + iy)^j \oplus \mathbb{C}(x - iy)^j$ is a decomposition into invariant subspaces for $SO(2)$. We leave it as an exercise to the reader to show that for $SL(2, \mathbb{R})$ the above result applies *except* that if $\nu = -1$ then the subspaces $\oplus_{j \geqslant 0} \mathbb{C}(x + iy)^{2j+1} 1_{-2j-2}$ and $\oplus_{j \geqslant 0} \mathbb{C}(x - iy)^{2j+1} 1_{-2j-2}$ are invariant subspaces of H_-^{-1} as $(Lie(SL(2, \mathbb{R})), SO(2))$-modules.

3.8. Let (Q, A) be a p-pair. Let $\nu \in \mathfrak{a}_C^*$ and let (σ, W) be a Hilbert representation of ${}^\circ M$ which is unitary as a representation of K. We define a Hilbert representation $(\pi_{Q,\sigma,\nu}, H^{Q,\sigma,\nu})$ as follows. Let $H_o = H_o^{Q,\sigma,\nu}$ be the space of continuous functions on G with values in W such that $f(nmag) = a^{\rho_Q + \nu} \sigma(m) f(g)$. If $f, g \in H_o$ then set

$$\langle f, g \rangle = \int_K \langle f(k), g(k) \rangle dk.$$

This defines a pre-Hilbert space structure on H_o since $G = QK$. We set $\pi_{Q,\sigma,\nu}(g)f(x) = f(xg)$. Set $H^{Q,\sigma,\nu}$ equal to the Hilbert space completion of H_o.

Lemma. *If $g \in G$ then $\pi_{Q,\sigma,\nu}(g)$ extends to a bounded operator on $H^{Q,\sigma,\nu}$ and $(\pi_{Q,\sigma,\nu}, H^{Q,\sigma,\nu})$ is a Hilbert representation of G. If $(\check{\sigma}, \check{W})$ is the conjugate dual representation of (σ, W) then the conjugate dual represention of $(\pi_{Q,\sigma,\nu}, H^{Q,\sigma,\nu})$ is equivalent to $(\pi_{Q,\check{\sigma},-\bar{\nu}}, H^{Q,\check{\sigma},-\bar{\nu}})$ where $\bar{\nu}(h) = \overline{\nu(h)}$ for $h \in \mathfrak{a}$. Furthermore, if (σ, W) is a unitary representation and if $\nu \in i\mathfrak{a}^*$ then $(\pi_{Q,\sigma,\nu}, H^{Q,\sigma,\nu})$ is a unitary representation of G.*

Proof. Set $\pi = \pi_{Q,\sigma,\nu}$. Let $f \in H_o$ then

$$\langle \pi(g)f, \pi(g)f \rangle = \int_K \langle f(kg), f(kg) \rangle dk.$$

If $g \in G$ then we can write $g = nmak$ with $n \in N$, $m \in M$, $a \in A$ and $k \in K$. In this decomposition the only ambiguity is that we can replace m, k with $mu, u^{-1}k$ with $u \in K \cap M$. Also $a(g) = a$ defines a C^∞ map from G to A. We also write $m(g) = m$, $k(g) = k$. We have

$$\langle \pi(g)f, \pi(g)f \rangle = \int_K a(kg)^{2\rho+2\mathrm{Re}\nu} \langle \sigma(m(kg))f(k(kg)), \sigma(m(kg))f(k(kg)) \rangle dk.$$

We now make a few observations that will also be used later.

1. Let ω be a compact subset of $\mathfrak{a}_{\mathbb{C}}^*$ then there exist constants $C_\omega > 0$, $d_\omega > 0$ such that
$$a(g)^{2\rho+2\mathrm{Re}\nu} \leqslant C_\omega \|g\|^{d_\omega}, \quad \nu \in \omega.$$

2. There exist constants $B_\sigma > 0$, $c_\sigma > 0$ such that
$$\|\sigma(m(g))\| \leqslant B_\sigma \|g\|^{c_\sigma}.$$

We will discuss 1. and 2. shortly. We note that they imply
3. If $g \in G$ and if $\nu \in \omega$ then

$$\|\pi_{Q,\sigma,\nu}(g)f\|^2 \leqslant C_\omega B_\sigma \|g\|^{d_\omega + c_\sigma} \|f\|^2.$$

Thus, in particular, we see that $\pi(g)$ extends to a bounded operator in $H = H^{Q,\sigma,\nu}$. We can argue as in 1.1 example 2. to prove that (π, H) is a representation of G (the argument uses only uniform continuity on compact sets).

We are left with 1. and 2. We will only prove them in the case when Q is minimal and $G = {}^oG$ (since that is the only case that will be used seriously). Then M is compact so 2. is obvious. We now prove 1. Recall that we have $\varphi : G \to G_o \subset$

$GL(n, \mathbb{R})$. There is an orthonormal basis, $v_1, ..., v_n$ of $\mathbb{R}^n$ such that if $a \in A$ and $n \in N$ then the matrix of $\varphi(na)$ is of the form

$$\begin{bmatrix} \lambda_1(a) & * & * & * & & * \\ & \cdot & * & * & & * \\ & & \cdot & * & & * \\ & & & \cdot & & * \\ & & & & & \lambda_n(a) \end{bmatrix}$$

and $\varphi(a)$ is diagonal with diagonal entries $\lambda_i(a)$. Thus if we use the version of $\|...\|$ given as α in 2.1 then we see that $\|nak\| \geq \|a\|$. Thus $\|g\| \geq \|a(g)\|$. Since $a \mapsto a^{2\rho + 2Re\nu}$ is a representation of A it follows that if $\nu \in \omega$ then there exists $d = d_\omega \geq 0$ such that $a^{2\rho + 2Re\nu} \leq C\|a\|^d$. Hence $a(g)^{2\rho + 2Re\nu} \leq C\|a(g)\|^d \leq C\|g\|^d$ if $\nu \in \omega$.

With the above notation in place we will use the additional notation $I_{Q,\sigma,\nu} = (H^{Q,\sigma,\nu})_K$. We will show that if $W_{K \cap M}$ is an admissible, finitely generated $(^o\mathfrak{m}, K \cap M)$-module then $I_{Q,\sigma,\nu}$ is admissible.

Example. As usual, let $G = GL(n, \mathbb{R})$. Let $P = \{g \in G | g\mathbb{R}e_1 = \mathbb{R}e_1\}$. Then

$$P = \left\{ \begin{bmatrix} c & x \\ 0 & g \end{bmatrix} \,\middle|\, g \in GL(n-1, \mathbb{R}), c \in \mathbb{R}^\times, x \in \mathbb{R}^{n-1} \right\}.$$

Show that the representation $(\pi_\nu, H_\pm^\nu)$ is equivalent with $(\pi_{P,\sigma,\mu}, H^{P,\sigma,\mu})$ for appropriate σ, μ (find the parameters!).

4. The Subrepresentation Theorem.

4.1. Let G be a real reductive group and let (Q, A) be a minimal standard p-pair in G. Then $Q = MAN$ and $M = Q \cap K$. Let (σ, H_σ) be an irreducible unitary representation of M (necessarily finite dimensional). We will use the notation of the previous section.

Proposition. *Let $\nu \in \mathfrak{a}_C^*$ and let $\gamma \in \hat{K}$. Then*

$$\dim Hom_K(V_\gamma, H^{Q,\sigma,\nu}) = \dim Hom_M(V_\gamma, H_\sigma).$$

In particular this implies that $I_{Q,\sigma,\nu}$ is admissible. The Proposition is an application of Frobenius reciprocity. To see this we note that if $f \in H_o$ then $f_{|K}$ determines f and if H_o^σ is the space of all continuous maps, φ, of K to H_σ such that $\varphi(mk) = \sigma(m)\varphi(k)$ then $H_{o\,|K} = H_o^\sigma$. We set $\pi_\sigma(k)\varphi(x) = \varphi(xk)$. On H_o^σ we put the inner product

$$\langle f, g \rangle = \int_K \langle f(k), g(k) \rangle dk.$$

Let H^σ be the Hilbert space completion of H^σ_o. Then $\pi_\sigma(k)$ extends to a unitary orperator on H^σ and as in 3., (π_σ, H^σ) is a unitary representation of K. The map $H^{Q,\sigma,\nu}_o \to H^\sigma_o$, $f \mapsto f_{|K}$ extends to a unitary map, T, of $H^{Q,\sigma,\nu}$ to H^σ. Clearly, $T\pi_{Q,\sigma,\nu}(k) = \pi_\sigma(k)T$. Thus we need only prove the above formula for H^σ. But in this case it is precisely Frobenius reciprocity (see [W1]).

4.2. We now introduce another variant of Frobenius reciprocity. If $V \in \mathcal{C}(\mathfrak{g}, K)$ then we set $\mathfrak{n}V = \text{span}\{Xv \mid X \in \mathfrak{n}, v \in V\}$.

Proposition. *If V is admissible and finitely generated then V is finitely generated as a $U(\mathfrak{n}_C)$-module. In particular, $V/\mathfrak{n}V$ is finite dimensional. Furthermore, if W_ν is the $(\mathfrak{m} + \mathfrak{a}, M)$-module corresponding to the representation of MA on H_σ given by $\sigma_\nu(ma) = a^{\rho+\nu}\sigma(m)$ then*

$$\text{Hom}_{\mathfrak{g},K}(V, I_{Q,\sigma,\nu}) = \text{Hom}_{\mathfrak{m}+\mathfrak{a},M}(V/\mathfrak{n}V, W_\nu).$$

Here the equality really means that there is a natural isomorphism between the two spaces. We will now describe that isomorphism. If $T \in \text{Hom}_{\mathfrak{g},K}(V, I_{Q,\sigma,\nu})$ then set $\check{T}(v) = T(v)(1)$. Then it is clear that $\check{T} \in \text{Hom}_{\mathfrak{m}+\mathfrak{a},M}(V/\mathfrak{n}V, W_\nu)$. Let p denote the natural projection of V onto $V/\mathfrak{n}V$. If $T \in \text{Hom}_{\mathfrak{m}+\mathfrak{a},M}(V/\mathfrak{n}V, W_\nu)$ then set $\hat{T}(v)(namk) = a^{\nu+\rho}T(p(kv))$. Then one checks that $\hat{T} \in \text{Hom}_{\mathfrak{g},K}(V, I_{Q,\sigma,\nu})$. Since $\check{\hat{T}} = T$ and $\hat{\check{T}} = T$ the formula in the proposition follows. The difficult part of the result is the finiteness assertion. Since this will take us far afield we will just give a reference to this result of Osborne.

4.3. Let V be an irreducible $(\mathfrak{g}, K)$-module. If $W = V/\mathfrak{n}V \neq 0$ then W is a finite dimensional $(\mathfrak{q}, M)$-module (see Corollary 3.5). Thus W has a non-zero irreducible quotient X. Let $H \in \mathfrak{a}$ be such that $\lambda(H) > 0$ for all $\lambda \in \Phi(Q, A)$ (see A.14). Since X is finite dimensional the Jordan normal form of the action of H in X implies that $X = \oplus X_\mu$ with $HX_\mu \subset X_\mu$ and $H - \mu I$ is nilpotent as a transformation of X_μ. Now $\mathfrak{n}_\lambda X_\mu \subset X_{\mu+\lambda(H)}$. Thus if μ_o is the maximum of the μ such that $X_\mu \neq 0$ then $\mathfrak{n}X_{\mu_o} = 0$. Since $[\mathfrak{q}, \mathfrak{n}] \subset \mathfrak{n}$ and $\text{Ad}(M)\mathfrak{n} \subset \mathfrak{n}$, this imples that X_{μ_o} is a non-zero invariant subspace of X. Hence $X_{\mu_o} = X$. Now X is irreducible as a $(\mathfrak{m} + \mathfrak{a}, M)$-module. Since A is central in MA, Schur's lemma implies that there exists $\nu \in \mathfrak{a}^*_C$ such that $hv = (\nu + \rho)(h)v$ for $v \in X$. We therefore see that X is irreducible as an M-module. Since M is compact we can put an M invariant inner product on X. Denote the corresponding irreducible unitary representation of M by (σ, H_σ). Then in the notation of the previous number $X = W_\nu$ and the composition of natural projections $V \to V/\mathfrak{n}V \to X$ defines an element of $\text{Hom}_{\mathfrak{m}+\mathfrak{a},M}(V/\mathfrak{n}V, W_\nu)$. Let T be the corresponding $(\mathfrak{g}, K)$ homomorphism of V into $I_{Q,\sigma,\nu}$. Then $T \neq 0$ so T is injective. We have therefore shown

Lemma. *If V is an irreducible $(\mathfrak{g}, K)$-module and if $V \neq \mathfrak{n}V$ then there exists an irreducible unitary representation (σ, H_σ) of M and $\nu \in \mathfrak{a}_C^*$ such that V is equivalent with a subrepresentation of $I_{Q,\sigma,\nu}$.*

4.4. We will now show that if V is admissible and finitely generated and $V \neq 0$ then $V \neq \mathfrak{n}V$. This in particular will imply that in the notation of the above lemma every irreducible $(\mathfrak{g}, K)$-module is equivalent with a subrepresentation of some $I_{Q,\sigma,\nu}$. This is the Casselman subrepresentation theorem. Our proof will involve the following theorem of [H-C1], [Lep], Rader (unpublished).

Theorem. *Assume that G is connected. If V is an irreducible $(\mathfrak{g}, K)$-module then there exists an irreducible unitary representation (σ, H_σ) of M and $\nu \in \mathfrak{a}_C^*$ such that V is equivalent with a subquotient of $I_{Q,\sigma,\nu}$.*

Notice that this theorem seems to be the same as the above lemma without the hypothesis $V \neq \mathfrak{n}V$. The difference has to do with the fact that a subrepresentation is a *very* special case of a subquotient. As we shall see that the condition $V \neq \mathfrak{n}V$ is important and (for irreducible V) is implied by the subrepresentation theorem (Proposition 4.2) but *not* by the subquotient theorem.

4.5. Theorem. *If V is a non-zero finitely generated $(\mathfrak{g}, K)$-module then $V \neq \mathfrak{n}V$.*

The proof will take up quite a bit of space. We prove the result by induction on $\dim \mathfrak{g}$. If $\dim \mathfrak{g} = 1$ then $\mathfrak{n} = 0$ so the result is obvious. We assume the result for all $(\mathfrak{g}, K)$ with $1 \leqslant \dim \mathfrak{g} < r$. We now prove it for $\dim \mathfrak{g} = r$.

We first note that if K^o is the identity component of K then V is finitely generated as a $(\mathfrak{g}, K^o)$-module. We may therefore assume that G is connected. We now assume that $V = \mathfrak{n}V$ we will be trying to prove that $V = 0$. If not then V has a non-zero irreducible quotient W and since $\mathfrak{n}V = V$ we must have $\mathfrak{n}W = W$. We may thus assume that V is irreducible.

Theorem 4.4 implies that there exist $(\sigma, H_\sigma) \in \hat{M}$, $\mu \in \mathfrak{a}_C^*$ and $X \subset Y \subset I_{Q,\sigma,\mu}$ invariant subspaces such that $V \cong X/Y$. Since $I_{Q,\sigma,\nu}$ consists of analytic vectors in $H = H^{Q,\sigma,\mu}$ we see that $H_1 = Cl(X) \subset Cl(Y) = H_2$ are G-invariant subspaces of H. Set U equal to the quotient Hilbert space (isomorphic with the orthogonal complement of H_1 in H_2). Let π denote the induced action of G on U. Then (π, U) is an irreducible representation of G and $U_K \cong V$.

We now apply 3.8 3. which implies

1. There exists $C > 0$, $d > 0$ such that if $v, w \in U$ then $|\langle \pi(g)v, w \rangle| \leqslant C\|g\|^d \|v\|\|w\|$ for all $g \in G$.

We now come to the crux of the matter. We assume that $V = U_K$.

2. If $v, w \in V$ and if $n > 0$ then there exists $C_{v,w,n} > 0$ such that

$$|\langle \pi(g)v, w \rangle| \leqslant C_{v,w,n}\|g\|^{-n}, \ g \in G.$$

Before we prove this we will derive the theorem from the assertion.

We realize $(\pi_{Q,\sigma,\mu}, H^{Q,\sigma,\nu})$ as follows. We have seen that as a Hilbert space $H^{Q,\sigma,\nu} \cong H^\sigma$. If $f \in H^\sigma_\sigma$ set $f_\nu(nak) = a^{\nu+\rho}f(k)$. Then we set $\pi_\nu(g)f(k) = f_\nu(kg)$. Hence we $(\pi_{Q,\sigma,\nu}, H^{Q,\sigma,\nu}) \cong (\pi_\nu, H^\sigma)$. Notice that the Hilbert space is now fixed. Fix $v, w \in V$. Set $f(g) = \langle \pi(g)v, w \rangle$ for $g \in G$. 3.6 3. implies that if $f_1, f_2 \in H^\sigma$ then

$$ \nu \mapsto \int_G \overline{f(g)} \langle \pi_\nu(g)f_1, f_2 \rangle dg $$

is a holomorphic function on $\mathfrak{a}^*_C$ which we will denote as $\Psi(v, w, f_1, f_2) = \Psi$.

We assert that $\Psi(\nu) \equiv 0$. Since Ψ is holomorphic it is enough to show that $\Psi(i\nu) \equiv 0$ for $\nu \in \mathfrak{a}^*$. Suppose then $\Psi(i\nu) \neq 0$ for some $\nu \in \mathfrak{a}^*$. Then

$$ \Psi(u, w, g, f_2)(i\nu) = \langle g, T(v) \rangle $$

with $T(v) \in H^\sigma$ for all $g \in H^\sigma_K i$. If $k \in K$ then $\Psi(ku, w, g, f_2)(i\nu) = \Psi(u, w, kg, f_2)(i\nu)$ and if $X \in \mathfrak{g}$ then $\Psi(Xu, w, g, f_2)(i\nu) = -\Psi(u, w, \pi_{i\nu}(X)g, f_2)(i\nu)$. Thus $T(ku) = kT(u)$ for $k \in K$ and $u \in V$ also $TXu = \pi_{i\nu}(X)T(u)$ (here we are using the fact that $\pi_{i\nu}$ is unitary). Thus we have $T : V \to I_{Q,\sigma,i\nu}$ is a $(\mathfrak{g}, K)$-module homomorphism. Now Proposition 4.2 implies a contradiction.

But if $\Psi(\nu) \equiv 0$ for all f_1, f_2 then since (π, U) is a subquotient of π_μ, $f = \langle \pi_\mu(g)f_1, f_2 \rangle$ for appropriate f_1, f_2. Thus $f = 0$. Since $v, w \in V$ were arbitrary this implies that $V = 0$. We are left with the proof of 1.

Let $A^- = \{a \in A \mid a^\alpha \leq 1, \alpha \in \Phi(Q, A)\}$ then $G = KA^-K$. Let Δ be the set of all simple roots of $\Phi(Q, A)$. If $\alpha \in \Delta$ then let $H \in \mathfrak{a}$ be such that $\alpha(H) = 1$ and $\beta(H) = 0$ for $\beta \in \Delta$ and $\beta \neq \alpha$. Let $\mathfrak{n}^\alpha$ be the sum of the eigenspaces for $\text{ad}(H)$ with strictly positive eigenvalues. Then $\mathfrak{n}^\alpha = \oplus_{\mu \in \Sigma_\alpha} \mathfrak{n}_\mu$ with $\mu = \sum_{\beta \in \Delta} n_\beta \beta$ and, $n_\beta \in \mathbb{N}$, $n_\alpha > 0$. This implies that $\mathfrak{n}^\alpha$ is an ideal in $\mathfrak{n}$. Now V is finitely generated as a $U(\mathfrak{n}_C)$-module. Thus $V/\mathfrak{n}^\alpha V$ is finitely generated as a $U(\mathfrak{n}/\mathfrak{n}^\alpha)$-module. Hence as a module for $\mathfrak{n}^*$ with $\mathfrak{n}^* = \ker \text{ad}(H)_{|\mathfrak{n}}$. We note that if $M^\alpha = \{g \in G \mid \text{Ad}(g)H = H\}$ then M^α is a real reductive group and thus $Z = V/\mathfrak{n}^\alpha V$ is a finitely generated $(\mathfrak{m}^\alpha, K \cap M^\alpha)$-module. Now our assumption $\mathfrak{n}V = V$ implies that $\mathfrak{n}^*Z = Z$. The inductive hypothesis implies that $Z = 0$ ($1 \leq \dim M^\alpha < \dim G$). Hence $\mathfrak{n}^\alpha V = V$ for all α. If $\beta \in \Sigma_\alpha$ then $a^\beta \leq a^\alpha$. Let $X_1, ..., X_r$ be a basis of $\mathfrak{n}^\alpha$ such that $X_i \in \mathfrak{n}_{\mu_i}$. Let $y \in V$ then $y \in \mathfrak{n}^\alpha V$, $y = \sum X_i y_i$. Thus if $z \in V$ and $a \in A^-$ then

$$ \langle \pi(a)y, z \rangle = \sum \langle \pi(a)X_i y_i, z \rangle = \sum a^{\mu_i} \langle \pi(X_i)\pi(a)y_i, z \rangle = $$

$$ -\sum a^{\mu_i} \langle \pi(a)y_i, \check{\pi}(X_i)z \rangle. $$

Now there exist $C > 0$, $m > 0$ such that $\|\pi(g)\| \leq C\|g\|^m$. Now this implies that

$$ |\langle \pi(a)y, z \rangle| \leq C' \|a\|^m a^\alpha. $$

Let $l = |\Delta|$. Then taking the l-fold product of this inequality and then taking the l-th root we have that if $y, z \in V$ then

$$|\langle \pi(a)y, z \rangle| \leqslant C''\|a\|^m a^{\frac{1}{l}\sum_{\alpha \in \Delta} \alpha}.$$

Now there exists $B > 0$ and $s > 0$ such that

$$a^{\frac{1}{l}\sum_{\alpha \in \Delta} \alpha} \leqslant B\|a\|^{-s}, \ a \in A^-.$$

Let $y_1, ..., y_n, z_1, ..., z_k$ be respectively bases of $\mathrm{span}Ky$ and $\mathrm{span}Kz$. Then there exists a constant $D > 0$ such that

$$|\langle \pi(a)y_i, z_j \rangle| \leqslant D\|a\|^{m-s}.$$

Now $k_1 y = \sum \varphi_i(k_1)y_i$ and $k_2^{-1}z = \sum \psi_i(k_2)z_i$ with φ_i, ψ_j continuous functions on K. We therefore see that

$$|\langle \pi(ak_1)y, \pi(k_2)^{-1}z \rangle| \leqslant D'\|a\|^{m-s}, \ a \in A^-, k_1, k_2 \in K.$$

This implies that if $y, z \in V$ then there exists $C_{y,z,1} > 0$ such that

$$|\langle \pi(g)y, z \rangle| \leqslant C_{y,z,1}\|g\|^{m-s}, \ g \in G.$$

We now use exactly the same argument to show that there exists $C_{y,z,2} > 0$ such that

$$|\langle \pi(g)y, z \rangle| \leqslant C_{y,z,2}\|g\|^{m-2s}, \ g \in G$$

etc.

4.6. Theorem 4.5 is the basis of a theory developed by Casselman and the author that we call the theory of the *real Jacquet module*. There are two implications of the theory that we will be using in these lectures. Since the methods of proof involve the development of more algebraic formalism, we will just state the results and give appropriate references.

Theorem. *Let $V \in \mathcal{C}(\mathfrak{g}, K)$ be admissible and finitely generated then V has finite length as a $\mathfrak{g}$-module. That is if $V_1 \subset V_2 \subset ... \subset V_k \subset ...$ or $V_1 \supset V_2 \supset ... \supset V_k \supset ...$ with V_k a $\mathfrak{g}$-submodule of V then there exists j such that $V_j = V_k$ for $k \geqslant j$.*

We note that the ascending chain condition follows from the fact that $U(\mathfrak{g}_C)$ is a Noetherian algebra over $\mathbb{C}$ (cf.[RRGI]). The descending chain condition is usually proved using Harish-Chandra's theory of characters. However, the theory gives a more "elementary" proof (cf. [RRGI, 4.2.1]).

4.7. Theorem. *Let $V \in \mathcal{C}(\mathfrak{g}, K)$ be admissible and finitely generated then there exists, (π, H), a Hilbert representation of G such that V is $(\mathfrak{g}, K)$-isomorphic with H_K and such that the space of $K - C^\infty$ vectors of (π, H) is equal to the space of $G - C^\infty$ vectors.*

This result of Casselman was initially proved using the theory of regular singular systems of differential equations. For a proof using the theory of the real Jacquet module see [RRGI,4.2.4]. We note that this result implies that there exists a smooth Fréchet representation (π, H^∞) of moderate growth such that $(H^\infty)_K \cong V$. If (π, W) is a smooth Fréchet representation of G of moderate growth such that $W_K \cong V$ then we will call (π, W) a *realization* of V. If (π, H) is a Hilbert representation of G such that $H_K \cong V$ then we will call (π, H) a *Hilbert realization* of V.

4.8. Theorem. *Let $V \in \mathcal{C}(\mathfrak{g}, K)$ then $V \in \mathcal{H}$ if and only if V is finitely generated as a $U(\mathfrak{n}_C)$-module.*

This result is due to Stafford and the author and makes serious use of the theory of the real Jacquet module. See [RRGI,4.2.6] for a proof of this theorem.

4.9. We denote by $\mathcal{H}$ the full subcategory of all $V \in \mathcal{C}(\mathfrak{g}, K)$ that are admissible and finitely generated. We denote by $\mathcal{FH}$ the full subcategory of all $V \in \mathcal{F}_{mod}(G)$ such that V_K is admissible and finitely generated. In the next section we will see that there exists a functor $V \mapsto \bar{V}$ from $\mathcal{H}$ to $\mathcal{FH}$ with $\bar{V}_K = V_K$ that defines an equivalence of categories. A critical step in the proof of this theorem (due to Casselman and the author) is an estimate on matrix coefficients that we will now discuss.

Let $(\pi, V) \in \mathcal{FH}$. We set (as usual) V' equal to the space of all continuous linear functionals on V. If $\lambda \in V'$ and $g \in G$ then we define $\pi^T(g)\lambda$ by $\pi^T(g)\lambda(v) = \lambda(\pi(g)^{-1}v)$. We set $V'_K = \{\lambda \in V' | \pi^T(K)\lambda$ spans a finite dimensional space$\}$. If $V \in \mathcal{C}(\mathfrak{g}, K)$ is admissible then we define actions of $\mathfrak{g}$ and K on V^* (the algebraic dual space of V) by $X\lambda(v) = -\lambda(Xv)$ and $k\lambda(v) = \lambda(k^{-1}v)$ for $\lambda \in V^*$, $X \in \mathfrak{g}$, $k \in K$. We set $\tilde{V} = \{\lambda \in V^* | K\lambda$ spans a finite dimensional space$\}$. Then $\mathfrak{g}\tilde{V} \subset \tilde{V}$ ($\text{span}K\mathfrak{g}\lambda \subset \mathfrak{g}\text{span}K\lambda\}$. If $\lambda \in V(\gamma)^*$ then we extend λ to V by setting $\lambda(V(\gamma')) = 0$ for $\gamma' \neq \gamma$. Then it is easy to see that $\tilde{V} = \oplus_{\gamma \in \hat{K}} V(\gamma)^*$. Thus $\tilde{V} \in \mathcal{C}(\mathfrak{g}, K)$ and is admissible.

Proposition. *If $V \in \mathcal{H}$ then $\tilde{V} \in \mathcal{H}$. If $V \in \mathcal{FH}$ then $(V'_K)_{|V_K} = (V_K)\tilde{}$.*

Proof. We must show that if $V \in \mathcal{H}$ then $\tilde{V}$ is finitely generated. That is it satisfies the ascending chain condition. Let $X_1 \subset X_2 \subset \dots$ with X_i a $(\mathfrak{g}, K)$-submodule of $\tilde{V}$. Set $X_i^\perp = \{v \in V | X_i(v) = 0\}$. Then $X_1^\perp \supset X_2^\perp \supset \dots$ is a descending chain of $(\mathfrak{g}, K)$- submodules of V. Hence there exists j such that $X_j^\perp = X_k^\perp$ for all $k \geq j$. The admissibility implies that if X is a $(\mathfrak{g}, K)$-submodule of $\tilde{V}$ then $(X^\perp)^\perp = X$. Thus $X_j = X_k$ for $k \geq j$. As for the second assertion we note that $(V'_K)_{|V_K} \subset (V_K)\tilde{}$.

If $\gamma \in \hat{K}$ then set E_γ equal to the projection of V onto $V(\gamma)$ that corresponds to the direct sum decomposition $V = Cl(\oplus V(\gamma))$. Then E_γ is continuous (it is given by the integral formula in 3.3) and thus if $\lambda \in V({}^\cdot\gamma)^*$ then $\lambda \circ E_\gamma \in V'_K$ thus $(V'_K)_{|V_K} \supset (V_K)^{\cdot}$.

4.10. We are now ready to prove the basic estimate that will be used in the next chapter.

Theorem. *Let* $V \in \mathcal{H}$. *There exists* $d > 0$ *(depending only on* V*) such that if* $(\pi, W) \in \mathcal{FH}$ *is such that* $W_K \cong V$ *and if* $\lambda \in W'_K$ *then there exists a continuous seminorm,* q_λ, *on* W *such that* $|\lambda(\pi(g)v)| \leqslant \|g\|^d q_\lambda(v)$.

The point of this theorem is that the d depends only on V. If we cared only about dependence on W the result would follow directly from Lemma 4.8 and the definition of moderate growth. Our proof of this lemma actually proves more than the assertion.

We first observe that if $G \neq {}^0G$ then $G \cong S \times {}^0G$. Let $\mathfrak{s} = Lie(S)$. Then $\mathfrak{s}V(\gamma) \subset V(\gamma)$ for all $\gamma \in \hat{K}$. This implies that $V = \oplus_\mu V_\mu$ with the sum over $\mu \in \mathfrak{s}^*_C$ and $V_\mu = \{v \in V | (h - \mu(h))^r v = 0, h \in \mathfrak{s}$ for some $r\}$. Clearly each V_μ is a $(\mathfrak{g}, K)$-submodule of V. Since V is of finite length, it follows that $\Sigma = \{\mu \in \mathfrak{s}^*_C | V_\mu \neq 0\}$ is finite. Set $V_{\mu,r} = \{v \in V_\mu | (h - \mu(h))^r v = 0, h \in \mathfrak{s}\}$. Then $V_{\mu,r} \subset V_{\mu,r+1}$ this since V has finite length we see that there exists r depending only on V such that $V = \oplus_\mu V_{\mu,r}$. Let $(\pi, W) \in \mathcal{FH}$ be such that $W_K \cong V$. Set $W_\mu = Cl((W_K)_{\mu,r})$ then $W = \oplus_\mu W_\mu$ and $W_\mu = \{v \in W | (h - \mu(h))^r v = 0, h \in \mathfrak{s}\}$.

Scholium. *Let* $f \in C^\infty(\mathbb{R}^n)$ *and assume that there are complex numbers* $c_1, ..., c_n$ *and an integer* r *such that* $(\partial_i - c_i)^r f = 0$ $i = 1, ..., n$. *Then there exists a polynomial* $p(x)$ *of degree at most* r *such that* $f(x) = p(x)e^{\sum c_i x_i}$ *for* $x \in \mathbb{R}^n$.

This is pretty obvious. Set $g(x) = e^{-\sum c_i x_i} f(x)$ then $\partial_i^r g = 0$ for $i = 1, ..., n$.

Let $h_1, ..., h_m$ be a basis of $\mathfrak{s}$. Let $\mu \in \Sigma$. Set $c_i(\mu) = \mu(h_i)$. If $v \in W$ then $v = \sum v_\mu$ and thus if $\lambda \in W'_K$ then $\lambda(\pi(sg)v) = \sum_\mu \lambda(\pi(s)\pi(g)v_\mu)$. The map $\mathbb{R}^m \to S$, $x \mapsto \exp(\sum x_i h_i)$ defines a diffeomorphism of $\mathbb{R}^n$ onto S. Hence if $s = \exp(\sum x_i h_i)$ then

$$\lambda(\pi(s)\pi(g)v_\mu) = s^\mu \sum_{|I| \leqslant r} x^I \lambda_I(\pi(g)v_\mu)$$

with $\lambda_I \in (W_K)^{\cdot}$. Clearly, there exists $p \geqslant 0$ such that $|x^I| \leqslant C\|s\|^p$ for all $|I| \leqslant r$. Also, since $s \mapsto s^\mu$ is a representation of S if we take C, p sufficiently large then we have

$$|\lambda(\pi(s)\pi(g)v_\mu)| \leqslant C\|s\|^p \sum_{|I| \leqslant r} \lambda_I(\pi(g)v_\mu).$$

Since $sg \mapsto \|s\|\|g\|$, $s \in S$, $g \in {}^0G$ defines a norm on G, this implies that the theorem will be proved if we can prove it in the case when $G = {}^0G$. We now make that assumption and begin the proof in the next number.

4.11. Let $\Delta = \{\alpha_1, ..., \alpha_l\}$ let $h_1, ..., h_l \in \mathfrak{a}$ be such that $\alpha_i(h_j) = \delta_{ij}$. Let $Z = \tilde{V}/\mathfrak{n}\tilde{V}$. Then Z is a finite dimensional $(\mathfrak{m} + \mathfrak{a}, M)$-module. Thus as above $Z = \oplus Z_\mu$ with $\mu \in \mathfrak{a}_{\mathbb{C}}^*$. Let $\Sigma = \{\mu| Z_\mu \neq 0\}$. Define $\Lambda \in \mathfrak{a}^*$ by $\Lambda(h_i) = \max\{\text{-Re}\mu(h_i)| \mu \in \Sigma\}$. We will show

1. There exist $p \geqslant 0$ such that if $\nu \in (W_K')$ then there exists a continuous semi-norm σ_ν on W such that

$$|\nu(\pi(a)y)| \leqslant (1 + \log \|a\|)^p a^\Lambda \sigma_\nu(y), \ y \in W, \ a \in A^+.$$

Here $A^+ = \{a \in A| a^\alpha \geqslant 1, \alpha \in \Phi(Q, A)\}$.

We first show how 1. implies the desired result. As above there exists $s \geqslant 0$ and $C > 0$ such that

$$(1 + \log \|a\|)^p a^\Lambda \leqslant C\|a\|^s, \ a \in A^+.$$

Let $\{\nu_j\}$ be a basis for the finite dimensional space span$K\nu$. If $k \in K$ then $k\nu = \sum \varphi_i(k)\nu_i$ with φ_i continuous. Thus if $a \in A^+$ then

$$|\nu(\pi(k_1 a k_2)y)| = |k_1^{-1}\nu(\pi(a)\pi(k_2)y)| \leqslant \sum_i |\varphi_i(k_1^{-1})||\nu_i(\pi(a)\pi(k_2)y)| \leqslant$$

$$\|a\|^s C' \sum \sigma_{\nu_i}(\pi(k_2)y).$$

Since the representation of K on W is continuous an K is compact the principle of uniform boundedness implies that there exists a continuous seminorm q_ν on W such that $C' \sum \sigma_{\nu_i}(\pi(k_2)y) \leqslant q_\nu(y)$ for all $k \in K$ and $y \in W$. Hence if $g = k_1 a k_2$, $k_i \in K$, $a \in A^+$ then

$$|\nu(\pi(g)y)| \leqslant \|a\|^s q_\nu(y) = \|g\|^s q_\nu(y)$$

for all $y \in W$. This clearly implies the theorem.

We are therefore left with the proof of 1. We first note that the definition of moderate growth combined with the argument in the proof of Lemma 2.2 implies that there exists $\delta \in \mathfrak{a}^*$ such that if $\nu \in (W_K')$ then there exists a continuous seminorm p_ν on W such that

$$|\nu(\pi(a)y)| \leqslant a^\delta p_\nu(y), \ y \in W, \ a \in A^+. \tag{1}$$

Let $\alpha \in \Delta$ and let M^α and $\mathfrak{n}^\alpha$ be as in 4.5. Let $H \in \mathfrak{a}$ be such that $\beta(H) = 0$ for $\beta \in \Delta - \{\alpha\}$ and $\alpha(H) = 1$. Then $\mathfrak{m}^\alpha \oplus \mathbb{R}H = \{X \in \mathfrak{g}| [H, X] = 0\}$. Now $\tilde{V}/\mathfrak{n}^\alpha\tilde{V}$ is finitely generated as a $U(\mathfrak{n}/\mathfrak{n}^\alpha)$-module. This implies that $\tilde{V}/\mathfrak{n}^\alpha\tilde{V}$ is an element of the category $\mathcal{H}$ for $(\mathfrak{m}^\alpha \oplus \mathbb{R}H, K \cap M^\alpha)$. Let q be the natural projection of $\tilde{V}$ onto $\tilde{V}/\mathfrak{n}^\alpha\tilde{V}$. $\circ$

i) If $q(\nu) = 0$ then there exists a continuous seminorm, p_ν', on W such that

$$|\nu(\pi(a)y)| \leqslant a^{\delta-\alpha} p_\nu'(y), \ y \in W, \ a \in A^+.$$

This is proved by the main argument used in 4.5.

We now suppose that $q(\nu) \neq 0$. We set $Z = \tilde{V}/\mathfrak{n}^\alpha \tilde{V}$. Then since

$$\mathrm{Ad}(m)H = H$$

for $m \in M^\alpha$ we see that $Z = \oplus Z_\mu$ with $Z_\mu = \{z \in Z \mid (H - \mu I)^d z = 0, \text{ for some } d\}$. We note that since Z is finitely generated there exists d such that $(H - \mu)^d Z_\mu = 0$ and the set $\Sigma^\alpha = \{\mu \mid Z_\mu \neq 0\}$ is finite. We also note that if $^*\mathfrak{n} = \mathfrak{m}^\alpha \cap \mathfrak{n}$ then $^*\mathfrak{n} \oplus \mathfrak{n}^\alpha = \mathfrak{n}$ and $\mathfrak{n}^\alpha$ is an ideal in $\mathfrak{n}$. Theorem 4.5 implies that $Z_\mu/^*\mathfrak{n}Z_\mu \neq 0$. Thus for each $\mu \in \Sigma^\alpha$ there exists $\zeta \in \Sigma$ such that $\zeta(H) = \mu$. Let P_μ be the projection of Z onto Z_μ relative to the above direct sum decomposition. For each μ we choose $\nu_\mu \in \tilde{V}$ such that $P_\mu q(\nu) = q(\nu_\mu)$. Then $\nu - \sum \nu_\mu \in \mathfrak{n}^\alpha \tilde{V}$.

We now estimate $|\nu_\mu(\pi(a)y)|$. So set $\zeta = \nu_\mu$ and assume $q(\zeta) \neq 0$. Let $\bar{\zeta}_1, ..., \bar{\zeta}_r$ be a basis of $\sum \mathbb{C}H^j q(\zeta)$ with $\bar{\zeta}_1 = q(\zeta)$. Set $\zeta_1 = \zeta$ and choose $\zeta_i \in \tilde{V}$ such that $q(\zeta_i) = \bar{\zeta}_i$. Then

$$H\bar{\zeta}_i = \sum b_{in}\bar{\zeta}_n$$

and if $B = [b_{ij}]$ then

$$(B - \mu I)^r = 0. \tag{2}$$

We also note that

$$H\zeta_i - \sum b_{in}\zeta_n = \omega_i \in \mathfrak{n}^\alpha \tilde{V}.$$

If $a \in A^+$ then we write $a = a'a_t$ with $a' \in A^+$, $(a')^\alpha = 1$ and $a_t = \exp tH$, $t \geqslant 0$. We set

$$F(t, a'; y) = \begin{bmatrix} \zeta_1(\pi(a_t a')y) \\ \zeta_2(\pi(a_t a')y) \\ \vdots \\ \zeta_r(\pi(a_t a')y) \end{bmatrix}$$

and

$$G(t, a'; y) = \begin{bmatrix} \omega_1(\pi(a_t a')y) \\ \omega_2(\pi(a_t a')y) \\ \vdots \\ \omega_r(\pi(a_t a')y) \end{bmatrix}.$$

Then

$$\frac{d}{dt}F(t, a'; y) = -BF(t, a'; y) - G(t, a'; y).$$

This implies that

$$F(t, a'; y) = e^{-tB} F(0, a'; y) - e^{-tB} \int_0^t e^{sB} G(s, a'; y) ds. \tag{3}$$

241

(1) above implies that

$$\|F(0, a'; y)\| \leqslant (a')^\delta \beta(y) \tag{4}$$

with β a continuous seminorm on W. (i) above implies that

$$\|G(s, a'; y)\| \leqslant e^{s(\delta(H)-1)}(a')^\delta \beta'(y), s \geqslant 0.$$

with β' a continuous seminorm on W. (2) above implies that

$$\|e^{sB}\| \leqslant C(1 + |s|)^r e^{Re\,\mu}, s \in \mathbb{R}. \tag{5}$$

We replace β with $\beta + \beta'$ then our estimates (1),(2),(4),(5) to (3) we find that if $t \geqslant 0$ then

$$\|F(t, a'; y)\| \leqslant C(1+t)^r e^{-tRe\,\mu}(a')^\delta (1 + C \int_0^t (1+s)^r e^{s(\delta(H)+Re\,\mu-1)}ds)\beta(y).$$

We note that if $s \geqslant 0$ then $(1+s)^r e^{-s/3} \leqslant C' < \infty$. Thus if $t \geqslant 0$ then we have

$$\|F(t, a'; y)\| \leqslant C(1+t)^r e^{-tRe\,\mu}(a')^\delta \beta(y) + C'C(1+t)^r e^{t(\delta(H)-2/3)}(a')^\delta \beta(y). \tag{6}$$

Now our definition of Λ implies that $-Re\,\mu \leqslant \Lambda(H)$ so (6) implies that

$$|\zeta(\pi(a_t a')y)| \leqslant C(1+t)^r e^{t\Lambda(H)}(a')^\delta \beta(y) + C'C(1+t)^r e^{t(\delta(H)-2/3)}(a')^\delta \beta(y) \tag{7}$$

for $t \geqslant 0$. If $\delta(H) - 2/3 \leqslant \Lambda(H)$ then if we put (7) together with (i) and the various definitions we have shown that we can replace δ by δ' with $\delta'(H) = \Lambda(H)$ and $\delta'(h) = \delta(h)$ if $\alpha(h) = 0$ at the cost of putting a factor $(1+t)^r$ in (1). If $\delta(H) - 2/3 > \Lambda(H)$ then we can replace δ with $\delta - \frac{1}{2}\alpha$ in (1). We can then run through the entire argument again. And eventually we will have

$$|\nu(\pi(a_t a')y)| \leqslant (1+t)^r e^{t\Lambda(H)}(a')^\delta \beta(y)$$

for $t \geqslant 0$. We now run this procedure through all the simple roots and 1. follows.

5. The functor from $\mathcal{H}$ to $\mathcal{FH}$.

5.1. Let $V \in \mathcal{H}$ and let (π, W) be a realization of V of moderate growth. We assume that $V = W_K$. Proposition 4.9 says that $(W'_K)_{|V} = \tilde{V}$ and $\tilde{V} \in \mathcal{H}$. Thus there exist $\lambda_1, ..., \lambda_p \in W'_K$ such that $\sum U(\mathfrak{g}_C)\lambda_i = W'_K$. Theorem 4.10 implies that there exists $m > 0$ depending only on V and a continuous seminorm q on W such that if $v \in W$, $g \in G$ then

$$|\lambda_i(\pi(g)v)| \leqslant \|g\|^m q(v).$$

If $r \geqslant m$ then define $T : W \to \times^p \mathcal{A}_r(G)$ by $T(v) = (\lambda_1(\pi(\cdot)v), ..., \lambda_p(\pi(\cdot)v))$. On $\times^p \mathcal{A}_r(G)$ we put the product topology (i.e. use the seminorms $q_{r,x}(f_1, ..., f_p) = \sum q_{r,x}(f_i)$) with $q_{r,x}(f) = \sup_{g \in G} \|g\|^{-r} |x f(g)|$. We let G act on $\times^p \mathcal{A}_r(G)$ by

$$\sigma_r(g)(f_1, ..., f_p) \doteq (\sigma_r(g)f_1, ..., \sigma_r(g)f_p).$$

$(\sigma_r(g)f(x) = f(xg).)$ Also, $U(\mathfrak{g}_C)$ acts on $\times^p \mathcal{A}_r(G)$ by $x(f_1, ..., f_p) = (x f_1, ..., x f_p)$ and this action corresponds to the differential of σ_r. We note that

$$T(\pi(g)v) = \sigma_r(g)T(v), \quad g \in G, \tag{1}$$

and

$$T(\pi(x)v) = \dot{x}T(v), \quad x \in U(\mathfrak{g}_C). \tag{2}$$

Thus $q_{r,x}(T(v)) \leqslant pq(\pi(x)v)$ for $x \in U(\mathfrak{g}_C)$. This implies that T defines a continuous linear map from W to $\times^p \mathcal{A}_r(G)$. Set $W_r = Cl(T(V))$ the closure taken in $\mathcal{A}_r(G)$. Then (1) and (2) combined with Theorem 3.7 imply that $T(V) \subset (\times^p \mathcal{A}_r(G))^\omega$ and thus W_r is a G-invariant subspace of $\times^p \mathcal{A}_r(G)$ by Theorem 1.9. Let $\pi_r(g) = \sigma_r(g)_{|W_r}$. Then $(W_r)_K = T(V)$ and $T\pi(g) = \pi_r(g)T$.

We now show that T is injective on W. Indeed, let $Z = \ker T_{|V}$. Then $\lambda_i(Z) = 0$ for $i = 1, ..., p$. Thus since Z is $U(\mathfrak{g}_C)$-invariant, $\sum U(\mathfrak{g}_C)\lambda_i(Z) = 0$. Hence $\check{V}(Z) = 0$. So $Z = 0$. This implies that T is injective on V. Let $X = \ker T$ then X is a closed G-invariant subspace of W. Hence X_K is dense in X. Clearly, $X_K \subset \ker T_{|V} = 0$. Thus $X = 0$.

This discussion leads us to the following result.

Theorem. *Let* $V \in \mathcal{H}$. *Then there exists a smooth Fréchet representation, of moderate growth,* $(\pi, \bar{V})$, *of* G *such that* $\bar{V}_K = V$ *and such that if* (σ, U) *is a smooth Fréchet representation of moderate growth such that* $U_K \in \mathcal{H}$ *and if* $S : U_K \to V$ *is an injective* $(\mathfrak{g}, K)$-*homorphism then* S *extends to a continuous* G-*homomorphism (i.e.* $S \circ \sigma(g) = \pi(g) \circ S, g \in G)$ *of* U *into* $\bar{V}$. *Furthermore,* $\bar{V}$ *is uniquely characterized by these properties.*

Proof. We first observe that the space W_r and the action π_r depend only on V. Indeed, if (σ, Z) is another smooth Fréchet representation of moderate growth of G with $Z_K = V$ then let $\lambda'_i \in Z'_K$ be such that $\lambda'_{i|K} = \lambda_{i|K}$. Then $\lambda'_i(xkv) = \lambda_i(xkv)$ for $x \in U(\mathfrak{g}_C)$, $k \in K$, $v \in V$. Since the functions $g \mapsto \lambda_i(\pi(g)v)$ and $g \mapsto \lambda'_i(\sigma(g)v)$ are real analytic we see that they agree on $G^\circ K = G$. Thus the space of functions $T(V)$ is independent of the realization. But this implies that T extends to a continuous map $T' : Z \to W_r$ such that $T'\sigma(g) = \pi_r(g)T$ for $g \in G$. If $s \geqslant m$ then (π_s, W_s) has the same property. Thus there are G-homomorphisms $T_{r,s} : W_s \to W_r$ and $T_{s,r} : W_r \to W_s$ such that $T_{s,r}T_{r,s}$ is the identity on W_s and $T_{r,s}T_{s,r}$ is the identity map on W_r. Thus W_r doesn't depend on $r \geqslant m$.

We set $S = \{\lambda_1, ..., \lambda_p\}$ and write $T = T_{S,r}$, $W_r = W_{S,r}$, and $\pi_r = \pi_{S,r}$. If S' is a subset of $\tilde{V}$ generating $\tilde{V}$ as a $U(\mathfrak{g}_C)$-module then we assert that the isomorphism $L = T_{S',r}(T_{S,r})^{-1} : (W_{S,r})_K \to (W_{S',r})_K$ extends to a G-module isomorphism of $W_{S,r}$ onto $W_{S',r}$. Indeed, we can use S' rather than S in the above construction and $W_{S,r}$ rather than W. Then the corresponding map "T" as above when restricted to $(W_{S,r})_K$ is L. Thus L has a continuous extension that is a G-module homomorphism. If we interchange the roles of S and S' then we see that T is an isomorphism.

We identify V with its image under $T_{S,r}$ for some generating set S of $\tilde{V}$ and some $r \geqslant m$. We set $\bar{V} = W_{S,r}$. If X is a $(\mathfrak{g}, K)$-submodule of V then the closure of X in $\bar{V}$ is just the representation constructed as above for X using $S_{|X}$ and r (notice that the same m works for X). Thus the closure of X in $\bar{V}$ is just $\bar{X}$. The uniqueness of $\bar{V}$ is also clear.

5.2. We now show that if $V, W \in \mathcal{H}$ and if $A \in \mathrm{Hom}_{\mathcal{H}}(V, W)$ then A extends to a continuous G-homomorphism if $\bar{V}$ to $\bar{W}$. Let $r \geqslant 0$ be such that r is greater than or equal to the "m" for both V and W. Let $S \subset \tilde{W}$ and $F \subset \tilde{V}$ be respectively generating sets. If $\lambda \in \tilde{V}$ set $\tilde{A}(\lambda) = \lambda \circ A$. Put $S' = \tilde{A}(S) \cup F$. Let $p = |S|$, $q = |F|$. We label S as $\lambda_1, ..., \lambda_p$, and S' as $\tilde{A}(\lambda_1), ..., \tilde{A}(\lambda_p), \mu_{p+1}, ..., \mu_{p+q}$. Then $T_{S,r}(W) \subset \times^p \mathcal{A}_r(G)$ and $T_{S',r}(W) \subset \times^{p+q} \mathcal{A}_r(G)$. Let p_1 be the projection of $\times^{p+q} \mathcal{A}_r(G)$ onto $\times^p \mathcal{A}_r(G)$ given bu $p_1(f_1, ..., f_{p+q}) = (f_1, ..., f_p)$. Then $p_1(Cl(T_{S',r}(V))) \subset Cl(T_{S,r}(W))$. If we identify V and W respectively with their images in $Cl(T_{S',r}(V))$ and $Cl(T_{S,r}(W))$ then p_1 induces a continuous G-homomorphism of $\bar{V}$ to $\bar{W}$ extending A. We we will write $\bar{A}$ for this extension.

To see that $\bar{A} \in \mathrm{Hom}_{\mathcal{FH}}(\bar{V}, \bar{W})$ we must show that $\bar{A}(\bar{V})$ is a (closed) topological direct summand of $\bar{W}$. At this time, there is no direct proof of this fact. This fact will appear as a concequence of our main theorem. (It is actually equivalent to the main theorem.) We will now discuss this result.

5.3. If $V \in \mathcal{H}$ then we define V^*_{mod} to be the set of all $\lambda \in V^*$ (algebraic dual space) such that there exists $r \geqslant 0$ and for each $v \in V$ there exists a real analytic function $f_{\lambda,v}$ on G such that

1. $x f_{\lambda,v}(k) = \lambda(kxv)$ for $x \in U(\mathfrak{g}_C)$, $k \in K$.
2. If $v \in V$ then there exists $C(v) > 0$ such that $|f_{\lambda,v}(g)| \leqslant C(v)\|g\|^r$, $g \in G$.

We note that V^*_{mod} is a subspace of V^* (take $f_{a\lambda+b\mu,v} = a f_{\lambda,v} + b f_{\mu,v}$).

Lemma. Let $V \in \mathcal{H}$. If $W \in \mathcal{FH}$ is such that $W_K = V$ then $W'_{|K} \subset V^*_{mod}$.

Proof. Let $\{q_j\}$ define the topology of W. Then if $\lambda \in W'$ then there exist n amd $C > 0$ such that

$$|\lambda(w)| \leqslant C \sum_{i \leqslant n} q_i(w), \quad w \in W.$$

Since $V \subset W^\omega$, the lemma now follows from the definition of $\mathcal{F}_{mod}$ (see 2.3).

5.4. We now come to one of the main results of Casselman and the author.

Theorem. $V \in \mathcal{H}$. Then $\bar{V}'_{|K} = V^*_{mod}$.

The only proofs (at this time) of this result are very complicated. Before we discuss the proof we will give some immediate consequences.

We first recall a theorem of Banach (cf. [Tr; Theorem 37.2,p. 382].

5.5. Theorem. *Let E and F be Fréchet spaces and let A be a continuous linear map of E into F. Define $A^T : F' \to E'$ by $A^T(\lambda) = \lambda \circ A$. Then $A(E) = F$ if and only if the following two conditions are satisfied*

1. A^T *is injective.*
2. $A^T(F')$ *is weakly closed in E'.*

Here the weak topology on E' is the topology induced by the seminorms $q_v(\lambda) = |\lambda(v)|$, $v \in E$. This topology is not a Fréchet topology in general thus convergence must be given in terms of nets. The theorem has a simply stated implication.

Corollary. *Let E and F be Fréchet spaces and let A be a continuous linear map of E into F. Then A is bijective, hence an isomorphism (A^{-1} is also continuous)if and only if A^T is bijective.*

Proof. If A^T is bijective then both conditions of the Theorem are clearly satisfied. Thus A is surjective. If $v \in E$ and if $A(v) = 0$ then $A^T F'(v) = 0$. Thus $E'(v) = 0$. So $v = 0$ by the Hahn-Banach theorem.

5.6. We now apply this material to our constructs. The first implication is perhaps the most spectacular. The result says that there is only one way to "integrate" an admissible finitely generated $(\mathfrak{g}, K)$−module to a Fréchet representation of G of moderate growth.

Theorem. *Let $V \in \mathcal{FH}$ then the identity map of V_K to V_K extends to an isomorphism of V with $\overline{(V_K)}$.*

Proof. Theorem 5.1 implies that there is a continuous G-homomorphism, A, of V into $\overline{(V_K)}$ with restriction to V_K the identity map. Now $\overline{(V_K)}'_{|V_K} = (V_K)^*_{mod}$. Thus $A^T(\overline{(V_K)}')_{|V_K} = V^*_{mod}$. Thus Lemma 5.3 implies that $A^T(\overline{(V_K)}')_{|V_K} = V'_{|V_K}$. Since V_K is dense in V we see that $A^T(\overline{(V_K)}') = V'$. If $A^T(\lambda) = 0$ then $\lambda(V_K) = 0$ thus $\lambda = 0$. Thus A^T is bijective. Corollary 5.5 implies the result.

5.7. Theorem. *Let $V \in \mathcal{H}$ and let W be a $(\mathfrak{g}, K)$-submodule of V. Then the closure of W, $Cl(W)$ in $\bar{V}$ is isomorphic with $\bar{W}$. Furthermore, $Cl(W)$ is a topological summand of $\bar{V}$.*

Proof. Let (π, H) be a Hilbert realization of V such that the space of $K - C^\infty$ vectors of H is equal to the space of $G - C^\infty$ vectors and such that the action of K is unitary.

Then $H^\infty \in \mathcal{FH}$. Thus we may assume that $H^\infty = \bar{V}$. Let H_1 be the closure of W in H. Then (π, H_1) is a Hilbert realization of W and it is clear that the $K-C^\infty$ vectors of H_1 is equal to the space of $G-C^\infty$ vectors. Since $(\pi_{|K}, H)$ is unitary, $H_1^\perp$ is K-invariant. Thus $H = H_1^{\infty_K} \oplus (H_1^\perp)^{\infty_K}$ (topological direct sum ∞_K indicating $K - C^\infty$ vectors). But $H_1^\infty = H_1^{\infty_K}$ and W is dense in H_1^∞ thus $H_1^\infty = Cl(W)$ in $\bar{V}$. The result now follows from Theorem 5.6.

5.8. We continuing in this "magic kingdom" we have

Theorem. *Let* $(\pi, V) \in \mathcal{FH}$ *and let* $(\sigma, W) \in \mathcal{F}_{mod}(G)$. *Let* $A : W \to V$ *be continuous and such that* $A \circ \sigma(g) = \pi(g) \circ A$ *for* $g \in G$. *Then* $A(W)$ *is a (closed) topological summand of* V.

Proof. Let $Y = Cl(A(W))$ in V. Then the above material implies that $Y \cong \overline{(Y_K)}$. Thus Y is a topological summand of V. Since $A(W)$ is dense in Y it is clear that $A^T : Y' \to W'$ is injective. The content of the theorem (in light of the previous results) is that $A(W) = Y$. That is, we must show that $A^T(Y')$ is weakly closed in W'. Let λ_α be a net in Y' such that $A^T(\lambda_\alpha)$ converges to λ in W'. We must show that $\lambda \in A^T(Y)$. Let $Z = \ker A$. Then clearly, $\lambda(Z) = 0$. We note that $W_K = \sum_{\gamma \in \hat{K}} \sigma_{|K}(\bar{\alpha}_\gamma)W$. Thus $A(W_K) = Y_K$ and $\ker A_{|W_K} = Z \cap W_K$. Thus λ defines an element $\xi \in (Y_K)^*$. We assert that $\xi \in (Y_K)^*_{mod}$. Indeed, let $v \in Y_K$ and let $w \in W_K$ be such that $A(w) = v$. Set $f_{\xi,v}(g) = \lambda(\sigma(g)w)$ (notice that this is well defined). We note that since $Y_K \in \mathcal{H}$ there exists a monic polynomial $p(x)$ such that $p(\Delta)v = 0$ (see the proof of Theorem 3.7). Thus $p(\Delta)f_{\xi,v} = 0$. Hence $f_{\xi,v}$ is real analytic. It is also easily seen that $f_{\xi,v}$ satisfies 1. in 5.3 and since $W \in \mathcal{F}_{mod}(G)$ it satisfies 2. Thus Theorems 5.4,5.7 with imply that there exists $\mu \in Y'$ such that $\mu_{|Y_K} = \xi$. Now $(A^T\mu - \lambda)_{|W_K} = 0$. Thus since W_K is dense in W, $A^T(\mu) = \lambda$. This completes the proof.

We are finally ready to complete the functorial properties of $V \to \bar{V}$.

Corollary. *If* $V, W \in \mathcal{H}$ *and if* $A \in Hom_\mathcal{H}(V, W)$ *then* $\bar{A} \in Hom_{\mathcal{FH}}(\bar{V}, \bar{W})$. *Furthermore, the functor* $V \to \bar{V}$ *defines an equivalence of categories between* $\mathcal{H}$ *and* $\mathcal{FH}$.

Proof. The first assertion is an immediate consequence of the above theorem. Suppose that

$$0 \to V \xrightarrow{\alpha} W \xrightarrow{\beta} Z \to 0$$

is an exact sequence in $\mathcal{H}$. Then $\bar{\beta}(\bar{W})$ is a topological summand of $\bar{Z}$ since $\beta(W) = Z$ we see that $\bar{\beta}(\bar{W}) = \bar{Z}$. It is clear that $\bar{\alpha}$ is injective and $\bar{\alpha}(\bar{V})$ is dense in $\ker \bar{\beta}$. Thus since $\operatorname{Im} \bar{\alpha}$ is closed in $\bar{W}$, $\ker \bar{\beta} = \operatorname{Im} \bar{\alpha}$. Thus

$$0 \to \bar{V} \xrightarrow{\bar{\alpha}} \bar{W} \xrightarrow{\bar{\beta}} \bar{Z} \to 0$$

is exact in $\mathcal{FH}$. We define an inverse functor from $\mathcal{FH}$ to $\mathcal{H}$ by $V \to V_K$.

5.9. We now come to the critical role of $\mathcal{S}(G)$ in this theory.

Theorem. *Let* $(\pi, V) \in \mathcal{FH}$. *Let* $v_1, ..., v_r \in V$ *then* $\sum_i \pi(\mathcal{S}(G))v_i$ *is a* $G-$ *invariant topological summand of* V.

Note. The critical aspect of this result is that there are no closures in the statement (since if there were it would be trivial).

Proof. We note that if $\times^r \mathcal{S}(G)$ is defined in the same way as $\times^r \mathcal{A}_m(G)$ above and if we use the action $l_g(f_1, ..., f_r) = (l_g f_1, ..., l_g f_r)$ then $(l, \times^r \mathcal{S}(G)) \in \mathcal{F}_{mod}(G)$. We define $T : \times^r \mathcal{S}(G) \to V$ by $T(f_1, ..., f_r) = \sum \pi(f_i)v_i$. Then $T \circ l_g = \pi(g) \circ T$. Thus Theorem 5.8 implies the theorem.

Corollary. *Let* $V \in \mathcal{FH}$ *then* V_K *is irreducible if and only if* V *is algebraically irreducible as an* $\mathcal{S}(G)-$ *module (i.e. the only, not necessarily closed,* $\pi(\mathcal{S}(G))$*-invariant subspaces of* V *are 0 and* V.

Proof. We have seen that V is irreducible as a Fréchet representation of G if and only if V_K is irreducible as a $(\mathfrak{g}, K)$-module. If V is irreducible and if $v \in V$ is non-zero then the above theorem implies that $\pi(\mathcal{S}(G))v$ is a closed G-invariant subspace. Hence $\pi(\mathcal{S}(G))v = V$. This implies that V is indeed algebraically irreducible as an $\mathcal{S}(G)$-module. If V is algebraically irreducible as an $\mathcal{S}(G)$-module and if W is a closed G-invariant subspace of V then $\pi(\mathcal{S}(G))W \subset W$. Thus $W = 0$ or $W = V$.

5.10. We will now discuss the steps involved in proving Theorem 5.4. We first give yet another realization of $\bar{V}$ for $V \in \mathcal{H}$. We revert to the notation of 5.1. Except that we fix $\|...\|$ to be $\text{tr}(\varphi(g)^T \varphi(g)) + \text{tr}(\varphi(g^{-1})^T \varphi(g^{-1}))$ (A.14). Let $d > 0$ be such that

$$\int_G \|g\|^{-d} dg < \infty.$$

For the sake of simplicity we normalize dg such that the above integral is 1. We define an inner product on $\times^p \mathcal{A}_r(g)$ as follows: If

$$f = (f_1, ..., f_p), g = (h_1, ..., h_p) \in \times^p \mathcal{A}_r(G)$$

then set

$$\langle f, h \rangle = \sum_{i=1}^p \int_G \|g\|^{-2r-d} \langle f_i(g), h_i(g) \rangle dg.$$

Then

$$\langle \sigma_r(x)f, \sigma_r(x)f \rangle = \sum_{i=1}^p \int_G \|g\|^{-2r-d} \langle f_i(gx), f_i(gx) \rangle dg =$$

$$\sum_{i=1}^{p} \int_G \|gx^{-1}\|^{-2r-d} \langle f_i(g), f_i(g) \rangle dg \leq \|x\|^{-2r-d} \|f\|^2 \qquad (1)$$

since $\|gx^{-1}\| \geq \|g\|\|x\|^{-1}$. We also note that if $f \in \times^P \mathcal{A}_r(G)$ then

$$\|f\| \leq q_{r,1}(f). \qquad (2)$$

Set $\mathcal{H}_{p,r}$ equal to the Hilbert space completion of $\times^P \mathcal{A}_r(G)$ relative to this inner product. Then (1) combined with our usual argument using uniform continuity on compacta imply that the operators $\sigma_r(g)$ extend to bounded operators of $\mathcal{H}_{p,r}$ and that $(\sigma_r, \mathcal{H}_{p,r})$ is a Hilbert representation of G. (2) implies that the injection $f \mapsto f$ is a continuous operator from $\times^P \mathcal{A}_r(G)$ into $(\mathcal{H}_{p,r})^\infty$.

Now let T_r be as in 5.1. Let H_r be the closure of $T_r(V)$ in $\mathcal{H}_{p,r}$. Then since $T_r(V)$ consists of analytic vectors H_r is G-invariant. Let (π_r, H_r) be the corresponding Hilbert representation of G. By the above we see that the canonical injection of $\bar{V}$ into H_r^∞ is a continuous G-homomorphism. Since $(H_r)_K = \bar{V}_K$ the identity map from $(H_r)_K$ to V extends to a continuous G-homomorphism from H_r^∞ to $\bar{V}$ (Theorem 5.1). Thus $H_r^\infty \cong \bar{V}$. It is also not hard to show that the space of $K-C^\infty$ vectors of H_r is equal to the space of $G-C^\infty$ vectors.

We have sketched the proof of

Proposition. *If $V \in \mathcal{H}$ then there exists a Hilbert realization (π, H) of V such that the space of $K-C^\infty$ of H equals the space of $K-C^\infty$ vectors and $H^\infty \cong \bar{V}$.*

5.11. The next result is complicated but not terribly hard (see *[RRGII;11.6.2]*)

Proposition. *Let (π, H) be a Hilbert representation of G such that the space of $K-C^\infty$ of H equals the space of $G-C^\infty$ vectors with $H_K \in \mathcal{H}$ and $H^\infty \cong \overline{H_K}$. If $(\check{\pi}, \check{H})$ is the conjugate dual representation of π then $(\check{H}^\infty)'_{|\check{H}_K} = (\check{H}_K)^*_{mod}$.*

If we take H_r corresponding to $\check{V}$ then we set $(\check{H}_r)^\infty = \bar{\bar{V}}$. We thus have a smooth Fréchet representation of moderate growth, $(\pi, \bar{\bar{V}})$, of G with $\bar{\bar{V}}_K = V$. The content of Theorem 5.4 is that $\bar{V} \cong \bar{\bar{V}}$. As usual, our proof of this fact is very complicated involving Casselman's automatic continuity theorem for conical vectors and Langlands' classifiction of irreducible objects in $\mathcal{H}$. We will scketch the key ideas in the case of $SL(2, \mathbb{R})$.

5.12. We assume for the remainder of this section that $G = SL(2, \mathbb{R})$. Let

$$h = \begin{bmatrix} 1 & 0 \\ 0 & -1 \end{bmatrix}, \; e = \begin{bmatrix} 0 & 1 \\ 0 & 0 \end{bmatrix}, \; f = \begin{bmatrix} 0 & 0 \\ 1 & 0 \end{bmatrix}, \; j = \begin{bmatrix} 0 & 1 \\ -1 & 0 \end{bmatrix}$$

Then $\mathfrak{g} = \mathbb{R}e \oplus \mathbb{R}h \oplus \mathbb{R}f$. We set $C = \frac{1}{2}h^2 + ef + fe \in U(\mathfrak{g}_\mathbb{C})$ we note that $\mathrm{Ad}(g)C = C$ $g \in G$. $K = SO(2)$. Let $V \in \mathcal{H}$ be irreducible hence $\check{V}$ is irreducible. Then there exists $c \in \mathbb{C}$ such that $C = cI$ on V (hence on $\tilde{V}$). We rewrite $C = \frac{1}{2}h^2 - h + 2ef$ ($[f, e] = -h$). Let $p : \tilde{V} \to \tilde{V}/e\tilde{V}$ be the canonical projection. If $\lambda \in \tilde{V}$ then $\lambda p(\lambda) = p(C\lambda) = p(\frac{1}{2}h^2\lambda - h\lambda)$. Since $[h, e] = 2e$, h acts naturally on $\tilde{V}/e\tilde{V}$. We therefore have

$$(h - 1)^2 u = (2c + 1)u, \quad u \in \tilde{V}/e\tilde{V}. \tag{1}$$

We write the elements of $SO(2)$ as

$$k(\theta) = \begin{bmatrix} \cos\theta & \sin\theta \\ -\sin\theta & \cos\theta \end{bmatrix} = e^{\theta j}.$$

We put $\gamma_n(k(\theta)) = e^{in\theta}$. Assuming that $V \neq 0$ (as we should) there exists n such that $\tilde{V}(\gamma_n) \neq 0$. Fix $\lambda \in \tilde{V}(\gamma_n) - (0)$. Then since e, h, j is a basis of $\mathfrak{g}$, and $j\lambda = in\lambda$ the elements

$$h^l e^m p(\lambda)$$

span $\tilde{V}/e\tilde{V}$. Since $ep(\lambda) = 0$ we see that $\tilde{V}/e\tilde{V} = \mathbb{C}p(\lambda) + \mathbb{C}hp(\lambda)$. This implies that $\dim \tilde{V}/e\tilde{V} \leqslant 2$. For simplicity we assume that $\lambda \neq -\frac{1}{2}$. Then the eigenvalues of h on $\tilde{V}/e\tilde{V}$ are contained in the set $\{1 \pm \sqrt{2c+1}\}$. We will assume that both eigenvaues actually appear. The argument is a bit simpler if there is only one. Let μ_1, μ_2 be the eigenvalues. We assume that $Re\,\mu_1 \leqslant Re\,\mu_2$. If $\lambda \in \tilde{V}$ then $p(\lambda) = \bar{\lambda}_1 + \bar{\lambda}_2$ with $H\bar{\lambda}_i = \mu_i\bar{\lambda}_i$, $i = 1, 2$. Let $\lambda_i \in \tilde{V}$ be such that $p(\lambda_i) = \bar{\lambda}_i$. Then

$$\lambda = \lambda_1 + \lambda_2 + e\xi \tag{2}$$

with $\xi \in \tilde{V}$. We also note that

$$H\lambda_i = \mu_i\lambda_i + e\xi_i \tag{3}$$

with $\xi_i \in \tilde{V}$.

Let (π, W) be a smooth Fréchet representation of moderate growth of G such that $W_K = V$. Set $a_t = \exp th$. We have seen in 4.10 1. that there exists $d \geqslant 0$ such if $\nu \in \tilde{V}$ then then if $t \geqslant 0$ then

$$|\nu(\pi(a_t)w)| \leqslant (1 + t)^d e^{-Re\,\mu_1 t} q_\nu(w)$$

with q_ν a continuous seminorm on W. Set $\varphi_i(w, t) = \lambda_i(\pi(a_t)w)$, $\psi_i(w, t) = e\xi_i(\pi(a_t)w)$. Then

$$\frac{d}{dt}\varphi_i = -\mu_i\varphi_i - \psi_i.$$

Thus

$$\varphi_i(t) = e^{-\mu_i t}\varphi_i(0) + \int_0^t e^{-\mu_i(t-s)}\psi_i(s)ds. \tag{4}$$

Now

$$\psi_i(t) = e\xi_i(\pi(a_t)w) = -\xi_i(\pi(e)\pi(a_t)w) = -\xi_i(\pi(a_t)\pi(\mathrm{Ad}(a_t^{-1})e)w)$$

$$= e^{-2t}\xi_i(\pi(a_t)\pi(e)w).$$

This implies that if $t \geqslant 0$ then

$$|\psi_i(t)| \leqslant e^{-2t}(1+t)^d e^{-\mathrm{Re}\,\mu_1 t}q_{\xi_i}(\pi(e)w). \tag{5}$$

Also if $\psi(t) = e\xi(\pi(a_t)w)$ then ψ satisfies a the same estimate with q_ξ rather that q_{ξ_1}. Hence

$$|e^{\mu_1 t}\psi_1(t)| \leqslant e^{-2t}(1+t)^d q_{\xi_1}(\pi(e)w).$$

This

$$\int_0^\infty |e^{\mu_1 s)}\psi_1(s)|ds \leqslant C q_{\xi_1}(\pi(e)w).$$

Set

$$\lambda_1^o(w) = \varphi_1(w,0) + \int_0^\infty e^{\mu_1 s}\psi_1(w,s)ds.$$

Then since $\lambda_1 \in W'$ there exists a continuous seminorm q on W such that

$$|\lambda_1^o(w)| \leqslant q_1(w).$$

In other words $\lambda_1^o \in W'$. We also note that

$$e^{\mu_1 t}\varphi_1(w,t) - \lambda_1^o(w) = \int_t^\infty e^{s\mu_1}\psi_1(w,s)ds \tag{6}$$

for $t \geqslant 0$. We have

$$\lim_{t\to+\infty} e^{\mu t}\lambda(\pi(a_t)w) = \lambda^o(w), \quad w \in W.$$

Also

$$\left|\int_t^\infty e^{s\mu_1}\psi_1(w,s)ds\right| \leqslant Ce^{-t}e^{-\mathrm{Re}\,\mu_1 t}q_{\xi_1}(\pi(e)w). \tag{7}$$

If $\lambda_1^o = 0$ for all $\lambda \in \tilde{V}$ then there would exist for each $\lambda \in \tilde{V}$ a continuous seminorm, p_λ, on W such that

$$|\lambda_1(\pi(a_t)w)| \leqslant e^{-\mathrm{Re}\,\mu_1 t}e^{-t}p_{\lambda,1}(w).$$

If we look at λ_2 and use (4) again we find that

$$|\lambda_2(\pi(a_t)w)| \leqslant C_1 e^{-\mathrm{Re}\,\mu_2 t}|\lambda(w)| + C_2 e^{-t}e^{-\mathrm{Re}\,\mu_1}q_{\xi_2}(\pi(e)w).$$

If we "plug" this estimate into (2) and use (5) (and the remark after (5) we find that if $\lambda \in \tilde{V}$ there is a continuous seminorm on W, $p_{\lambda,2}$ such that

$$|\lambda(\pi(a_t)w)| \leqslant \max(e^{-\operatorname{Re}\mu_1 t}e^{-t}, e^{-\operatorname{Re}\mu_2 t})p_{\lambda,2}(w), t \geqslant 0.$$

If we now put this estimate into our machine then after a finite number of stages we have

$$|\lambda(\pi(a_t)w)| \leqslant e^{-\operatorname{Re}\mu_2 t}p_{\lambda,2}(w), t \geqslant 0.$$

We can now use (5) and the argument thereafter to define λ_2^o and have

$$e^{\mu_2 t}\varphi_2(w,t) - \lambda_2^o(w) = \int_t^\infty e^{s\mu_2}\psi_2(w,s)ds$$

Assume that $\lambda_2^o = 0$ for all $\lambda \in \tilde{V}$. Then we have

$$\lambda(\pi(a_t)w) = e\xi(\pi(a_t)w) + \int_t^\infty e^{s\mu_1}e\xi_1(\pi(a_s)w)ds + \int_t^\infty e^{s\mu_2}e\xi_2(\pi(a_s)w)ds. \quad (8)$$

Thus if $\lambda_i^o = 0$, $i = 1, 2$ then for all $\lambda \in \tilde{V}$ we have

$$|\lambda(\pi(a_t)w)| \leqslant e^{-t}e^{-\operatorname{Re}\mu_2 t}\sigma_{1,\lambda}(w), t \geqslant 0.$$

With $\sigma_{1,\lambda}$ a continuous seminorm on W. Using this inequality in (8) we find that

$$|\lambda(\pi(a_t)w)| \leqslant e^{-2t}e^{-\operatorname{Re}\mu_2 t}\sigma_{2,\lambda}(w), t \geqslant 0$$

etc. Thus for each $n \geqslant 0$ and $\lambda \in \tilde{V}$ there is a continuous seminorm $\sigma_{n,\lambda}$ on W such that

$$|\lambda(\pi(a_t)w)| \leqslant e^{-\operatorname{Re}\mu t}e^{-nt}\sigma_{\lambda,n}(w), t \geqslant 0.$$

This combined with the argument in the proof of 4.5 leads to the contradiction $V = 0$. Thus there exists λ such that $\lambda_1^o \neq 0$ or if not then there exists λ such that $\lambda_2^o \neq 0$. In the first case set $\mu = \mu_1$ and $\lambda^o = \lambda_1^o$ in the second set $\mu = \mu_2$ and $\lambda^o = \lambda_2^o$. Then

$$\lim_{t \to +\infty} e^{\mu t}\lambda(\pi(a_t)v) = \lambda^o(v), v \in W.$$

Now $\lambda(\pi(a_t)\pi(f)w) = -e^{-2t}(f\lambda)(\pi(a_t)w)$. Hence

$$\lim_{t \to +\infty} e^{\mu t}\lambda(\pi(a_t)\pi(f)w) = 0, \lambda \in \tilde{V}, w \in W.$$

We therefore see that $\lambda^o(\pi(f)W) = 0$ for all $\lambda \in \tilde{V}$. Let

$$s = \begin{bmatrix} 0 & 1 \\ -1 & 0 \end{bmatrix} \in K.$$

Then $\mathrm{Ad}(s)e = -f$, $\mathrm{Ad}(s)f = -e$. Thus of we set $\bar{\lambda}(w) = \lambda^\circ(\pi(s)w)$ then $\bar{\lambda} \in W'$ and $\bar{\lambda}(\pi(e)W) = 0$. Furthermore, it is clear that if $\bar{\lambda} = 0$ then $\lambda^\circ = 0$.

Let $M = \{\pm I\} \subset K$, $A = \{a_t | t \in \mathbb{R}\}$, $N = \{\exp xe | x \in \mathbb{R}\}$. Then $Q = MAN$ is a proper parabolic subgroup of G. Since M is central in G, M either acts on V by the trivial character or the signum character. Set χ equal to that character. Fix λ such that $\delta = \bar{\lambda} \neq 0$. Let $\nu \in \mathfrak{a}_C^*$ be such that $\nu(h) = \mu - 1$. Define $T : W \to C^\infty(G)$ by

$$T(w)(g) = \delta(\pi(g)w).$$

Then $\delta(\pi(n)\pi(g)w) = \delta(\pi(g)w)$, $w \in W$, $g \in G$, $n \in N$. $\delta(\pi(mg)w) = \chi(m)\delta(\pi(g)w)$ and $\delta(\pi(a_t g)w) = e^{\mu t}\delta(\pi(g)w)$. Thus $T : W \to (H^{Q,\chi,\nu})^\infty$ is a continuous non-zero G-homomorphism.

For the sake of simplicity we will now assume that $\pi_{Q,\chi,\nu}$ is irreducible (this is not so serious in this case but it is the cause of a significant number of the difficulties in the proof in the general case). The above implies that $(H^{Q,\chi,\nu})^\infty = \overline{I_{Q,\chi,\nu}}$. Since $H^{Q,\chi,\nu} \cong (H^{Q,\chi,-\bar{\nu}})'$ it follows that $(H^{Q,\chi,\nu})^\infty = \overline{I_{Q,\chi,\nu}}$. This completes the proof in this special case.

6. Some examples.

6.1. In this number we will say a few words about automorphic forms. Let G be a real reductive group of inner type with $G = {}^\circ G$ (i.e. the center of G is compact). Let Γ be a discrete subgroup of G such that, with respect to the measure on $\Gamma\backslash G$ induced by dg, $\Gamma\backslash G$ has finite volume. Then a C^∞ function, f, on G is said to be an *automorphic form* if it satisfies the following four conditions:

1. $f(\gamma g) = f(g)$, $g \in G$, $\gamma \in \Gamma$.
2. $\dim \mathrm{span}\ r_K f < \infty$.
3. $\dim Z(\mathfrak{g}_C)f < \infty$. Here $Z(\mathfrak{g}_C)$ is as usual the center of $U(\mathfrak{g}_C)$.
4. There exists $d \geqslant 0$ such that if $x \in U(\mathfrak{g}_C)$ then

$$|xf(g)| \leqslant C_x \|g\|^d.$$

The condition 4. is just that $f \in \mathcal{A}_d(G)$. We set $\mathcal{A}(\Gamma\backslash G)$ equal to the space of all automorphic forms on G.

If $f \in \mathcal{A}(\Gamma\backslash G)$ then we set $V_f = \mathrm{span}\ U(\mathfrak{g}_C)r_K f$. $\mathfrak{g}$ acts on V_f by the usual action of $\mathfrak{g}$ as left invariant vector fields and $k \in K$ acts by r_k. Then V_f is a finitely generated $(\mathfrak{g}, K)$-module. Theorem 3.6 implies that V_f is admissible. Hence we see that $V_f \in \mathcal{H}$. We note that 4. implies that $V_f \subset \mathcal{A}_d(G)$. Our results now imply that the closure $Cl(V_f)$ in $\mathcal{A}_d(G)$ is a closed G-invariant subspace isomorphic as a smooth Fréchet module with our canonical completion $\overline{V_f}$. We note that $\overline{V_f} = \sigma_r(\mathcal{S}(G))f = f \star \mathcal{S}(G)$. With this in mind we are led to a more flexible definition of automorphic form.

We say that $f \in C^\infty(G)$ is a *smooth automorphic form* if f satisfies the conditions 1.,3.,4. above. We set $\mathcal{A}^\infty(\Gamma\backslash G)$ equal to the space of all smooth automorphic forms. If $f \in \mathcal{A}^\infty(\Gamma\backslash G)$ and if $\gamma \in K$ then $f_\gamma = r(\bar{\alpha}_\gamma)f \in \mathcal{A}(\Gamma\backslash G)$. Furthermore, $\sum_{\gamma \in \hat{K}} f_\gamma$ converges to f in $\mathcal{A}_d(G)$. Now $V_{f_\gamma} \in \mathcal{H}$ for each $\gamma \in \hat{K}$. Set $Z = Z(\mathfrak{g}_C)f$. If $\chi : Z(\mathfrak{g}_C) \to \mathbb{C}$ is an algebra homomorphism then we set $Z_\chi = \{g \in Z | (z - \chi(z))^r g = 0$ for some r and for all $z \in Z(\mathfrak{g}_C)\}$. Since $\dim Z < \infty$ we see that $Z = \oplus Z_\chi$. Let $\Sigma_f = \{\chi | Z_\chi \neq 0\}$. Then Σ_f is a finite set and there exists $r_f = r > 0$ such that $(z - \chi(z))^r Z_\chi = 0$. If $g \in U(\mathfrak{g}_C)\mathrm{span} r_K f = X_f$ then $g \in \mathcal{A}^\infty(\Gamma\backslash G)$ and $\Sigma_g \subset \Sigma_f$, $r_g \leqslant r_f$. Similarly, for $g \in U(\mathfrak{g}_C)\mathrm{span} r_K f_\gamma$. A basic theorem of Harish-Chandra [H-C2] states that if $\chi, r \geqslant 1$ and γ are fixed then

$$\dim\{f \in \mathcal{A}(\Gamma\backslash G)| (z - \chi(z))^r f = 0, z \in Z(\mathfrak{g}_C), f = f_\gamma\} < \infty.$$

This implies that $\dim\{g| g = h_\gamma, h \in X_f\} < \infty$. Hence $\sum_{\gamma \in \hat{K}} V_{f_\gamma}$ (algebraic sum) is an admissible $(\mathfrak{g}, K)$-module. We now apply the key result involved in the proof of the above mentioned theorem of Harish-Chandra (beyond Theorem 3.6 and a result of Langlands on cusp forms). It says that if χ is fixed then there are up to isomorphism only a finite number of irreducible $(\mathfrak{g}, K)$-modules on which $Z(\mathfrak{g}_C)$ acts by χ. We therefore see that $Y_f = \sum_{\gamma \in \hat{K}} V_{f_\gamma} \in \mathcal{H}$. Clearly, $f \in Cl(Y_f)$ in $\mathcal{A}_d(G)$.

On the other hand we may consider $Cl(\mathrm{span} r_G f)$ in $\mathcal{A}_d(G)$ which is a smooth Fréchet representation of G containing f and hence f_γ for all $\gamma \in \hat{K}$. The upshot is the following result.

Proposition. *If $f \in \mathcal{A}^\infty(\Gamma\backslash G)$ and if $d \geqslant 0$ is such that $f \in \mathcal{A}_d(G)$ then the closure of $\mathrm{span} r_G f$ in $\mathcal{A}_d(G)$ is a subspace, A_f, of $\mathcal{A}^\infty(\Gamma\backslash G)$ independent of the choice of d. The corresponding smooth, Fréchet representation is in $\mathcal{FH}$. Furthermore, $A_f = \sigma_d(\mathcal{S}(G))f = f \star \mathcal{S}(G)$.*

We note that such a result followed from a finiteness assertion and otherwise used nothing special about $\Gamma\backslash G$. We will have a bit more to say about automorphic forms later in this chapter.

6.2. Let $U^K = U(\mathfrak{g}_C)^K = \{x \in U(\mathfrak{g}_C)| \mathrm{Ad}(k)x = x, k \in K\}$. If $\chi : U^K \to \mathbb{C}$ is an algebra homomorphism then we set $\mathcal{A}_\chi(G/K)$ equal to the space of all functions $f \in C^\infty(G)$ such that

1. $f(gk) = f(g)$, $g \in G$, $k \in K$.
2. $uf = \chi(u)f$, $u \in U^K$.
3. There exists $d \geqslant 0$ such that if $x \in U(\mathfrak{g}_C)$, $|xf(g)| \leqslant C_x\|g\|^d$, $g \in G$.

We note that $Z(\mathfrak{g}_C) \subset U^K$. Thus as in 6.1, $V_f = U(\mathfrak{g}_C)f \subset \mathcal{A}_d(G)$ and $\sigma_d(\mathcal{S}(G))f = Cl(V_f)$ in $\mathcal{A}_d(G)$. Also $Cl(V_f) \cong \overline{V_f}$. We now form an auxiliary $(\mathfrak{g}, K)$-module,

$$Y^\chi = U(\mathfrak{g}_C)/(U(\mathfrak{g}_C)\mathfrak{k} + \sum_{u \in U^K} U(\mathfrak{g}_C)(u - \chi(u))).$$

253

Here, K acts by $kx = \mathrm{Ad}(k)x$ and $\mathfrak{g}$ acts by left multiplication. Theorem 3.6 implies that $Y^\chi \in \mathcal{H}$. The map $x \mapsto xf$ of $U(\mathfrak{g}_C)$ to V_f induces a $(\mathfrak{g}, K)$-module homomorphism of Y^χ onto V_f thus our theory implies that this homomorphism extends to a continuous G-homomorphism of $\overline{Y^\chi}$ onto $\overline{V_f}$. We now give a simpler discription of Y^χ which involves the Harish-Chandra classification of all such χ.

Fix $Q = {}^o MAN$ a minimal parabolic subgroup of G. Let σ_o denote the trivial one dimensional representation of ${}^o M$ (i.e. $\sigma_o(m) = 1$ all $m \in {}^o M$). If $\nu \in \mathfrak{a}^*_C$ then we set $I_\nu = I_{Q,\sigma_o,\nu}$. We set $(\pi_\nu, V^\nu) = (\pi_{Q,\sigma_o,\nu}, (H^{Q,\sigma_o,\nu})^\infty)$. We note that if $f \in V^\nu$ then $f_{|K} \in C^\infty({}^o M \backslash K)$. Furthermore, if $\varphi \in C^\infty({}^o M \backslash K)$ and if we set $\varphi_\nu(nak) = a^{\nu+\rho}\varphi(k)$ then $\varphi_\nu \in V^\nu$. We may thus look upon (π_ν, V^ν) as $(\pi_\nu, C^\infty({}^o M \backslash K))$ with $\pi_\nu(g)\varphi(k) = \varphi_\nu(kg)$. We note that if γ_o is the class of the trivial one dimensional representation of K then $C^\infty({}^o M \backslash K)(\gamma_o) = \mathbb{C}1$ where 1 denotes the function on K that takes the constant value 1. This implies that $V^\nu(\gamma_o)$ is one dimensional. We therefore see that there exists a homomorphism $\chi_\nu : U^K \to \mathbb{C}$ such that $\pi_\nu(u)1 = \chi_\nu(u)1$ for $u \in U^K$.

Let $W = \{s \in GL(\mathfrak{a}) | \text{there exists } k \in K \text{ such that } \mathrm{Ad}(k)_{|\mathfrak{a}} = s\}$. W is sometimes called the *small Weyl group*. Harish-Chandra has shown (cf. [RRGI,3.6.6])

1. If $\chi : U^K \to \mathbb{C}$ is an algebra homomorphism with $\chi(1) = 1$ then there exists $\nu \in \mathfrak{a}^*_C$ such that $\chi = \chi_\nu$. Furthermore, $\chi_\nu = \chi_{\nu'}$ if and only if there exists $s \in W$ such that $\nu' = \nu \circ s^{-1} = s\nu$.

We note that

2. If $\nu \in \mathfrak{a}^*_C$ then there exists $s \in W$ such that $Re(s\nu, \alpha) \geqslant 0$ for all $\alpha \in \Phi(Q, A)$. We write $(\mathfrak{a}^*_C)^+$ for the set of all $\nu \in \mathfrak{a}^*_C$ satisfying this condition.

A theorem of Kostant [K] implies that if ν satisfies the dominance condition in 2. then $I_\nu = U(\mathfrak{g}_C)1_\nu$. From this one can prove (cf. [RRGII;11.3])

Proposition. *If $\chi : U^K \to \mathbb{C}$ is an algebra homomorphism such that $\chi(1) = 1$ then there exists $\nu \in (\mathfrak{a}^*_C)^+$ such that $\chi_\nu = \chi$ and such that the map $Y^\chi \to I_\nu$ induced by $x \mapsto x1_\nu$ is a $(\mathfrak{g}, K)$-isomorphism.*

We now return to the situation at the beginning of this number. Then our main theorem now implies that if ν is related to χ as above then there is a continuous surjective linear map $T : C^\infty({}^o M \backslash K) \to V_f$ such that $T\pi_\nu(g) = \sigma_d(g)T$. Let $\delta(g) = g(1)$ for $g \in C^\infty(G)$ then δ defines a continuous functional on all of the spaces $\mathcal{A}_r(G)$. We also note that if $g \in \mathcal{A}_r(G)$ then $g(x) = \delta(\sigma_r(x)g)$. Set $\lambda = \delta \circ T$ then $\lambda \in (C^\infty({}^o M \backslash K))'$ and $\lambda(\pi_\nu(g)1) = f(g)$, $g \in G$. We have proved the following Theorem of [O-S], [W2].

Theorem. *Let $\chi : U^K \to \mathbb{C}$ be an algebra homomorphism and let $\nu \in \mathfrak{a}^+_C$ be as in the proposition above. If $f \in \mathcal{A}_\chi(G/K)$ then there exists $\lambda \in C^\infty({}^o M \backslash K)'$ such that $f(g) = \lambda(\pi_\nu(g)1)$, $g \in G$.*

6.3. We will now show how Theorem 6.2 relates to the example of the introduction.

Let $G = SU(1,1)$. Then we take

$$a_t = \begin{bmatrix} \cosh t & \sinh t \\ \sinh t & \cosh t \end{bmatrix}, t \in \mathbb{R}.$$

We note that if we set

$$n_x = \begin{bmatrix} 1 + ix & -ix \\ ix & 1 - ix \end{bmatrix}, x \in \mathbb{R}$$

then if we put ${}^{\circ}M = \{\pm I\}$, $A = \{a_t | t \in \mathbb{R}\}$ and $N = \{n_x | x \in \mathbb{R}\}$ then $Q = {}^{\circ}MAN$ is a minimal parabolic subgroup of G. We also note that we can take

$$K = \left\{ k(\theta) = \begin{bmatrix} e^{i\theta} & 0 \\ 0 & e^{-i\theta} \end{bmatrix} \middle| \theta \in \mathbb{R} \right\}.$$

Set $h = \begin{bmatrix} 0 & 1 \\ 1 & 0 \end{bmatrix}$. Then $a_t = e^{th}$. Let G act on $\mathbb{C}^2$ in the natural way by 2×2 matrices. We note that $(a_t)^{\rho} = e^t$. Set $v = (1,1) \in \mathbb{C}^2$. Then $n_x v = v$, $x \in \mathbb{R}$ and $av = a^{\rho}v$ for $a \in A$. This implies that

$$\|g^{-1}v\| = \sqrt{2}a(g)^{-\rho}, \, g \in G.$$

On the other hand G acts on $D = \{z \in \mathbb{C} | |z| < 1\}$ by linear fractional transformations. (See the introduction.) Thus if

$$g = \begin{bmatrix} a & b \\ \bar{b} & \bar{a} \end{bmatrix}, |a|^2 - |b|^2 = 1$$

then $z = g \cdot 0 = \frac{b}{\bar{a}}$. Also $\sqrt{2}a(gk(\theta))^{-\rho} = \|(\bar{a}e^{i\theta} - be^{-i\theta}, -\bar{b}e^{i\theta} + ae^{-i\theta})\| = \sqrt{2}|a||z - e^{2i\theta}|$. Now $1 - |z|^2 = 1 - |\frac{b}{\bar{a}}|^2$. Thus $1 - |z|^2 = |a|^{-2}$. We therefore see that

$$a(gk(\theta))^{2\rho} = \frac{1 - |z|^2}{|z - e^{2i\theta}|^2} = P(z, e^{2i\theta}).$$

One can show that $f \in C^{\infty}(D)$ is harmonic if and only if $u(g \mapsto f(g \cdot 0)) = \chi_{\rho}(u)(g \mapsto f(g \cdot 0))$. Also $\|g\|^2 = 2(|a|^2 + |b|^2) = 2(2|a|^2 - 1)$. Thus $\|g\|^2 \leq 4|a|^2$. Clearly, $2|a|^2 \leq \|g\|^2$. Thus the condition of moderate growth on $g \mapsto f(g \cdot 0)$ translates to $|f(z)| \leq C(1 - |z|^2)^{-d}$ for some $d \geq 0$. Thus Theorem 6.2 does indeed imply the result in the introduction.

6.4. We now combine examples 6.1 and 6.2. Let $\mathcal{A}_\chi(\Gamma \backslash G / K) = \mathcal{A}_\chi(G/K) \cap \mathcal{A}(\Gamma \backslash G)$. Let $\chi = \chi_\nu$ as in 7.2 with $\nu \in (\mathfrak{a}_\mathbb{C}^*)^+$. If $f \in \mathcal{A}_\chi(\Gamma \backslash G / K)$ then there exists $\lambda \in C^\infty({}^{\circ}M \backslash K)'$ such that

$$f(g) = \lambda(\pi_\nu(g)1), \, g \in G.$$

This observation implies a form of Frobenius reciprocity for automorphic forms (generalizing results of [G-G-PS] for the case when $\Gamma \backslash G$ is compact). We first need a bit of notation.

If H is a closed subgroup of G and if (π, V) is a smooth, Fréchet representation of G then we set $(V')^H = \{\lambda \in V' | \lambda(\pi(h)v) = \lambda(v), h \in H, v \in V\}$. Then we have

Theorem. *Define* $T_\nu : ((V^\nu)')^\Gamma \to \mathcal{A}_\chi(\Gamma\backslash G/K)$ *by* $T_\nu(\lambda)(g) = \lambda(\pi_\nu(g)1)$. *Then* T_ν *is bijective.*

As indicated at the end of 6.1 we have not really used anything special about Γ thus there are analogous results for any closed subgroup. We emphasize this example because there is no obvious way to ascertain Γ-invariance on $(\mathfrak{g}, K)$-modules.

6.5. Let H be a closed subgroup of G such that $\det(\mathrm{Ad}(h)) = \det(\mathrm{Ad}(h)_{|Lie(H)}) = 1$ for $h \in H$. Then there exists a volume form ω on the homogeneous space $X = H\backslash G$ that is invariant under the right action of G on X $((Hg \cdot x = H(gx))$. We can thus form the corresponding unitary representation $(\pi_H, L^2(H\backslash G))$ given by $\pi_H(g)f(x) = f(xg)$. Suppose that W is a closed irreducible invariant subspace of $L^2(X)$. Let $\pi(g) = \pi(g)_{|W}$. Harish-Chandra [cf. [RRGI;3.4.9] has proved that $W_K \in \mathcal{H}$. We note that W^∞ is the space defined in A.7 using $X_i = \pi_H(x_i)$ with $x_1, ..., x_n$ a basis of $\mathfrak{g}$. This implies that $\delta : W^\infty \to \mathbb{C}$, $\delta(f) = \check{f}(1)$ defines an element of $(W^\infty)'$. If $f \in W^\infty$ then $f(g) = \delta(\pi(g)f)$. Thus we see that there exists $d \geqslant 0$ such that $f \in \mathcal{A}_d(G)$. Fix (Q, A) a minimal p-pair in G. If we apply the subrepresentation theorem $(4.4, 4.5)$ then we find that there exists an irreducible finite dimensional unitary representation of ${}^\circ M$, σ and $\nu \in \mathfrak{a}_{\mathbb{C}}^*$ such that W_K is isomorphic with a subrepresentation of $I_{Q,\sigma,-\bar\nu}$. Thus since $(\pi, W) \cong (\check{\pi}, W)$ we see that there is a $(\mathfrak{g}, K)$-homomorphism of $I_{Q,\sigma,\nu}$ onto W_K. Our main theorem implies that $W^\infty = \overline{W_K}$. We therefore have a surjective continuous G-homomorphism $T : (H^{Q,\sigma,\nu})^\infty \to W^\infty$. As before $\delta \circ T \in (((H^{Q,\sigma,\nu})^\infty)')^H$. We therefore have a boundary value map in this case in terms of distributions.

6.6. The reader who is interested in further examples of applications should consult ([RRG II;11.9], [W2], [Ca]).

APPENDIX.

In this appendix we collect some results in functional analysis and Lie theory that will be used in the body of the lectures. We begin with the functional analysis.

A.1. Let V be a vector space over $\mathbb{C}$. Then a *seminorm* on V is a function $p : V \to [0, \infty) = \mathbb{R}^+$ such that $p(v + w) \leqslant p(v) + p(w)$ and $p(xv) = |x|p(v)$ for $v, w \in V$ and $x \in \mathbb{C}$. Let S be a countable set of seminorms on V such that if $v \in V$ is such that $p(v) = 0$ for all $p \in S$ then $v = 0$. We say that $\lim_{n \to \infty} v_n = v$ if $\lim_{n \to \infty} p(v_n - v) = 0$ for all $p \in S$. With this notion of convergence we have a topological vector space. Let $\tilde{S}$ be the set of seminorms of the form $\sum_p x_p p$ with $x_p \geqslant 0$, $x_p \in \mathbb{Q}$ and $x_p = 0$ for all but a finite number of p. If $p \in \tilde{S}$ set $B_p = \{v \in V | p(v) < 1\}$ then a subset U in V is open if for each $v \in U$ there exists $p \in \tilde{S}$ such that $v + B_p \subset U$.

Lemma. *If p is a continous seminorm on V then there exists $q \in \tilde{S}$ such that $p(v) \leqslant q(v)$ for all $v \in V$.*

Proof. Let B_p is open. Hence there exists $q \in \tilde{S}$ such that $B_q \subset B_p$. Thus if $q(v) \leqslant 1$ then $p(v) \leqslant 1$. Thus if $q(v) = 0$ then $p(tv) \leqslant 1$ for all $t \in \mathbb{R}$. Thus $p(v) \leqslant \frac{1}{t}$ for all $t > 0$. So $q(v) = 0$ implies $p(v) = 0$. If $v \in V$ and $q(v) \neq 0$ then $q(\frac{v}{q(v)}) = 1$ hence $p(\frac{v}{q(v)}) \leqslant 1$ so $p(v) \leqslant q(v)$.

We say that a sequence $\{v_n\}$ in V is *Cauchy* if for each $p \in S$ given $\epsilon > 0$ there exists N such that if $n, m \geqslant N$ then $p(v_n - v_m) < \epsilon$. We say that (V, S) is *complete* if every Cauchy sequence converges in this case V endowed with this topology is called a *Fréchet space*. If $S = \{\|...\|\}$ (i.e. $\|...\|$ is a *norm*) and (V, S) is complete then V is called a *Banach space*. If (V, S) is not complete then we can use the usual method to complete the space. That is let $\bar{V}$ denote the set of equivalence classes of Cauchy sequences $\{v_n\}$ in V under the relation $\{v_n\} \equiv \{w_n\}$ if $\lim_{n \to \infty} p(v_n - w_n) = 0$ for all $p \in S$. If $\{v_n\}$ is a Cauchy sequence then so is $p(v_n)$ ($|p(v) - p(w)| \leqslant p(v - w)$). Thus $\lim_{n \to \infty} p(v_n)$ exists and depends only on the class of $\{v_n\}$. We denote this limit by $p\{v_n\}$ and note that it defines a seminorm on $\bar{V}$ (the vector space structure is given by componentwise addition and scalar multiplication. We identify V with the set of equivalence classes of constant sequences.

A *pre-Hilbert* space structure on V is an inner product, $\langle \ , \ \rangle$ on V. If $\langle \ , \ \rangle$ is a pre-Hilbert space structure on V then we set $\|v\| = \sqrt{\langle v, v \rangle}$. If $(V, \| \ \|)$ is a Banach space then we call V a *Hilbert space*. As above if V is a pre-Hilbert space and we complete it as a Banach space then the inner product extends to $\bar{V}$ using the obvious limit.

A.2. Before we give some examples of these concepts we recall how one integrates on manifolds. Our basic reference will be [FWar]. Let M be an n-dimensional C^∞ ·manifold. That is, M is a Hausdorf topological space and the is an open covering, $\{U_j\}_{j=1}^N$, $N \leqslant \infty$, with $\Phi_j : U_j \to W_j$ a homeomorphism of U_j onto an open subset W_j of $\mathbb{R}^n$ and for all i, j,

$$\Phi_j \circ \Phi_i^{-1} : \Phi_i(U_i \cap U_j) \to \Phi_j(U_i \cap U_j)$$

is of class C^∞. The set $\{(U_j, \Phi_j)\}$ is called a C^∞ *atlas* for M. U is an open subset of M and $\Phi : U \to W$ is a homeomorphism of U onto W an open subset of $\mathbb{R}^n$ then we say that (U, Φ) is a *chart* for M if

$$\Phi \circ \Phi_j^{-1} : \Phi_j(U \cap U_j) \to \Phi(U \cap U_j)$$

is C^∞ for all j. Any countable number of charts that cover M will be called an atlas for M.

Examples. 1. $\mathbb{R}^n$, $N = 1$ and $U_1 = \mathbb{R}^n$, Φ_1 the identity map.

2. $S^n = \{x \in \mathbb{R}^{n+1} | \sum x_i^2 = 1\}$. $U_i = \{x \in S^n | x_i > 0\}$, $\Phi_i(x) = (x_1, ..., x_{i-1}, x_{i+1}, ..., x_n)$.

3. If M, N are C^∞ manifolds with atlases $\{(U_i, \Phi_i)\}$ and $\{(V_j, \Psi_j)\}$ respectively then on $M \times N$ we take the atlas $\{(U_i \times V_j, \Phi_i \times \Psi_j)\}$ with $\Phi_i \times \Psi_j(x, y) = (\Phi_i(x), \Psi_j(y))$. This puts the structure of a C^∞ manifold on $M \times N$.

A.3. We say that $f : M \to \mathbb{R}$ or $f : M \to \mathbb{C}(= \mathbb{R}^2)$ is C^∞ if for each chart of M, (U, Φ), $f \circ \Phi^{-1}$ is C^∞ on $\Phi(U)$. The set of real valued (resp. complex valued) C^∞ functions on M, $C^\infty(M, \mathbb{R})$ (resp. $C^\infty(M)$) forms a commutative algebra over $\mathbb{R}$ (resp. $\mathbb{C}$). We define the *support* of $f \in C^\infty(M)$ (supp (f)) to be the closure of $\{p \in M | f(p) \neq 0\}$. We denote by $C_c^\infty(M)$ the space of all compactly supported elements of $C^\infty(M)$. If $\{V_j\}$ is an open covering of M then a *partition of unity subordinate to* $\{W_j\}$ is a collection $\{\varphi_i\}$ of real valued elements of $C_c^\infty(M)$ such that

PU1. $0 \leqslant \varphi_i(x) \leqslant 1$ for all $x \in M$.

PU2. If $x \in M$ then there is an open neighborhood, U, of x in M such that $\{i | \varphi_i |_U \neq 0\}$ is finite and and $\sum_i \varphi_i(x) = 1$.

PU3. For each i there exists j such that $\varphi_i(M - U_j) = \{0\}$ (i.e supp $\varphi_i \subset U_j$). There exists a partition of unity subordinate to any covering of M.

A.4. If $x \in M$ then we define $T(M)_x$ to be the space of all linear maps $L : C^\infty(M, \mathbb{R}) \to \mathbb{R}$ such that $L(fg) = f(x)Lg + g(x)Lf$. If (U, Φ) is a chart for M with $x \in U$ we set $\Phi(y) = (x_1(y), ..., x_n(y))$. Then $\{x_1, ..., x_n\}$ is called a *system of local coordinates* on U. If $y \in U$ and $f \in C^\infty(M, \mathbb{R})$ then we write

$$\frac{\partial}{\partial x_i}_y f = \frac{\partial(f \circ \Phi^{-1})(\Phi(y))}{\partial x_i}.$$

Then it can be shown that $\left\{ \frac{\partial}{\partial x_1}_y, ..., \frac{\partial}{\partial x_n}_y \right\}$ is a basis if $T(M)_x$. A C^∞ curve on M is a continuous map $\sigma : (a, b) \to M$ such that if $t \in (a, b)$ there exists $\epsilon > 0$ and a chart (U, Ψ) for M with $\sigma((t - \epsilon, t + \epsilon)) \subset U$ and $\Psi \circ \sigma$ is of class C^∞ on $(t - \epsilon, t + \epsilon)$. It is also well known that if $v \in T(M)_x$ then there exists a C^∞ curve $\sigma : (-a, a) \to M$ such that $\sigma(0) = x$ and $vf = \frac{d}{dt} f(\sigma(t))_{|t=0}$. More generally, if σ is a C^∞ curve in M we define $\sigma'(t)f = \frac{d}{dt} f(\sigma(t))$. Then $\sigma'(t) \in T_{\sigma(t)}(M)$.

A *vector field* on M is an assignment $x \mapsto X_x$ with $X_x \in T(M)_x$ and such that $Xf = (x \mapsto X_x f) \in C^\infty(M, \mathbb{R})$ for all $f \in C^\infty(M, \mathbb{R})$. We note that $X : C^\infty(M, \mathbb{R}) \to C^\infty(M, \mathbb{R})$ is a derivation of the algebra ($X(fg) = fXg + gXf$) and every derivation is of this form. We denote by $Vect(M)$ the $C^\infty(M, \mathbb{R})$-module of all vector fields in M. If $X, Y \in Vect(M)$ then we note that $XY - YX$ defines a derivation of $C^\infty(M, \mathbb{R})$ hence $[X, Y] = XY - YX \in Vect(M)$.

If N is another C^∞ manifold then a continuous map $f : M \to N$ is said to be C^∞ if for each $p \in N$ and (U, Φ) a chart for N with $p \in U$ there is a chart (V, Ψ) for M such that $f(V) \subset U$ and $\Phi \circ f \circ \Psi^{-1} : \Psi(V) \to \Phi(U)$ is C^∞. If f is a C^∞ map of M to N and if $p \in M$ we define $df_p : T(M)_p \to T(N)_{f(p)}$ by $df_p(L)g = L(g \circ f)$.

A.5. We define $\Omega^n(M)$ to be the space of all assignments $x \mapsto \omega_x$ with ω_x an alternating n-multilinear map of $\times^n T(M)_x$ to $\mathbb{R}$ such that $\omega(X_1, ..., X_n) = (x \mapsto \omega_x(X_{1_x}, ..., X_{n_x})) \in C^\infty(M, \mathbb{R})$ for all $X \in Vect(M)$. If N is another n dimensional manifold and if $f : M \to N$ is C^∞ then we define $f^* : \Omega^n(N) \to \Omega^n(M)$ by

$$(f^*\omega)_p(v_1, ..., v_n) = \omega_{f(p)}(df_p(v_1), ..., df_p(v_n))$$

for $v_1, ..., v_n \in T(M)_p$.

If U is open in M and if $x_1, ..., x_n$ is a system of local coordinates on U then $\omega_y \left(\frac{\partial}{\partial x_1}_y, ..., \frac{\partial}{\partial x_n}_y \right) = a(y)$ with $a \in C^\infty(U, \mathbb{R})$. We say that ω is a *volume form* on M if $\omega_x \neq 0$ for all $x \in M$. Assume that ω is a volume form on M. By shrinking U and possibly replacing x_1 with $-x_1$ we may (and do) assume that $a(y) > 0$ for $y \in U$. If $f \in C_c(M)$ (compactly supported continuous function) and $f(M - U) = 0$ then we define

$$\int_M f\omega = \int_{\Phi(U)} f(\Phi^{-1}(x))a(\Phi^{-1}(x))dx$$

where $\Phi(y) = (x_1(y), ..., x_n(y))$. The change of variables theorem implies that this formula makes sense. That is, it is independent of the choice of coordinate system (subject to the positivity condition). Let $\{(U_j, \Phi_j)\}$ be an atlas for M such that the corresponding local coordinates satisfy the above positivity condition. Let $\{\varphi_i\}$ be a partition of unity subordinate to $\{U_j\}$ then we set

$$\int_M f\omega = \sum_i \int_M \varphi_i f\omega$$

for $f \in C(M)$ (complex valued continuous functions) provided that the series on the right converges with f replaced by $|f|$. The terms in the right hand side have been defined and it is easily seen that the sum is also independent of all choices. Notice that if $M = \mathbb{R}^n$ and ω is defined by $\omega_y(\frac{\partial}{\partial x_1}_y, ..., \frac{\partial}{\partial x_n}_y) = 1$ for $x_1, ..., x_n$ the usual linear coordinates and $y \in \mathbb{R}^n$ then this definition gives the ordinary Riemann integral. The corresponding volume form on $\mathbb{R}^n$ will be denoted dx.

The standard change of variables theorem of advanced calculus can be reformulated as follows. Let $f : M \to M$ be C^∞ then if $\omega \in \Omega^n(M)$, $f^*\omega = F\omega$. The function F depends only on f and not on ω. One has

$$\int_M (\varphi \circ f)f^*\omega = \int_M \varphi\omega$$

if f is bijective and f^{-1} is C^∞ (i.e. f is a *diffeomorphism*).

A.6. Let V be a Fréchet space. If f is a continuous function from $[a, b]$ to V with $-\infty < a < b < \infty$ then we can define $\int_a^b f(x)dx$ in the usual way one defines Riemann integrals in first year calculas (partitions and all that). We can define an integral of a V valued continuous function on a rectangle $\{(x_1, ..., x_n)|\, a_i \leqslant x_i \leqslant b_i\}$ as an iterated integral. If M is a C^∞-manifold and if f is a continuous function with values in V such that

$$\int_M (p \circ f)\omega < \infty$$

for all continuous semi-norms on V then we can define

$$\int_M f\omega \in V$$

using a partition of unity as above.

A.7. Fix ω a volume form on M if $f, g \in C_c(M)$ then we set

$$\langle f, g \rangle = \int_M f\bar{g}\omega.$$

Then $\langle \, , \, \rangle$ defines a pre-Hilbert space structure on $C_c(M)$. Let $L^2(M, \omega)$ denote the Hilbert space completion of this pre-Hilbert space. We write $\|f\|_2$ for the norm associated with this inner product. We also define

$$\|f\|_1 = \int_M |f|\omega$$

for $f \in C_c(M)$ and $L^1(M, \omega)$ to be the completion of $C_c(M)$ with respect to $\|...\|_1$.

We now give an important collection of Fréchet spaces. Let M be a C^∞ manifold with volume form ω. Let $X_1, ..., X_m \in Vect(M)$. We assume that if $x \in M$ then $X_{1_x}, ..., X_{m_x}$ span $T(M)_x$. If $i_1, i_2, ..., i_k \in \{1, ..., m\}$ and if $f \in C_c^\infty(M)$ (compactly supported, complex valued, C^∞ function on M) then we set

$$p_{i_1...i_n}(f) = \|X_{i_1} \cdots X_{i_k} f\|_2$$

if it exists. Let V denote the space of all $f \in C^\infty(M)$ such that $p_{i_1...i_k}(f)$ exists for all $i_1, ..., i_k$ endowed with the topology induced by these semi-norms. We assert that V is a Fréchet space. To see this we must show that it is complete. We first note that if $\varphi \in C_c^\infty(M)$ then the map $L_\varphi(f) = \varphi f$ is continuous on V. Indeed, $X_{i_1} \cdots X_{i_k}(\varphi f)$ $= \sum a_{j_1...j_p, l_1, ..., l_{k-p}}(X_{j_1} \cdots X_{j_p}\varphi)(X_{l_1} \cdots X_{l_{k-p}}f)$ with the coefficients independent of φ and f. This implies that

$$p_{i_1...i_n}(\varphi f) \leqslant C \sum_{0 \leqslant r \leqslant k} \sum_{1 \leqslant l_i \leqslant m} p_{l_1...l_r}(f)$$

with $C = m^k(\max |a_{j,l}|) \max_{0 \leqslant r \leqslant k} \max_{1 \leqslant j_i \leqslant m} \sup_{x \in M} |X_{j_1} \cdots X_{j_p}\varphi(x)|$. Let $\{f_n\}$ be a Cauchy sequence in V. If $\varphi \in C_c^\infty(M)$ then $\{\varphi f_n\}$ is also a Cauchy sequence. We choose an atlas $\{(U_j, \Phi_j)\}$ for M such that $\Phi_j(U_j) = R = \{x \in \mathbb{R}^n | \, |x_i| < 1, 1 \leqslant i \leqslant n\}$ and such that there exist $i_1, .., i_n$ such that $X_{i_1}, ..., X_{i_n}$ are linear independent at each $p \in U_j$. Let $\{\varphi_i\}$ be a partition of unity subordinate to $\{U_j\}$. We relabel the chart (possibly giving the same element a multiple count) so that $\mathrm{supp}\varphi_i \subset U_i$.

On $\mathbb{R}^n$ we will use standard multi-index notation. If $I = (i_1, ..., i_n) \in \mathbb{N}^n$ then

$$\partial^I = \frac{\partial^{|I|}}{\partial x_1^{i_1} \cdots \partial x_n^{i_n}} \, , |I| = i_1 + \cdots + i_n.$$

We set $p_I(\alpha)^2 = \int_R |\partial^I \alpha(x)|^2 dx$. If $\alpha_n = (\varphi_i f_n) \circ \Phi_i^{-1}$ on R then α_n has compact support in R and $\{\alpha_n\}$ is a Cauchy sequence with respect to the seminorms p_I

(here one must invert the appropriate minor and observe that $\omega_{|U_i} = a\Phi_i^* dx$ with a bounded above and below on the union of the supports of the α_n). Note that supp$\alpha_n \subset \Phi_i(\text{supp}\varphi_i)$. We observe that the Schwarz inequality implies that

$$\int_R |\partial^I \alpha(x)| dx \leqslant 2^{\frac{n}{2}} p_I(f).$$

Also if $J = (1,...,1)$ and if $\alpha \in C^\infty(R)$ extends to a continuous function on the closure of R vanishing on the boundary then

$$|\alpha(x)| = \left| \int_{-1}^{x_1} \cdots \int_{-1}^{x_n} \partial^J \alpha(x) dx \right| \leqslant 2^{\frac{n}{2}} p_J(\alpha).$$

Thus if $\nu_I(\alpha) = \sup_{x \in R} |\partial^I \alpha(x)|$ then for such α

$$\nu_I(\alpha) \leqslant 2^{\frac{n}{2}} p_{I+J}(\alpha).$$

Thus $\{\alpha_n\}$ is a Cauchy sequence with respect to the topology of uniform convergence with all derivatives and there exists a fixed compact subset of R containing the supports of all of the α_n. This implies that there exists $\alpha \in C_c^\infty(R)$ such that $\lim_{n \to \infty} \partial^I \alpha_n(x) = \partial^I \alpha(x)$ uniformly on R. Let $g_i(p) = \alpha(\Phi_i(p))$, $p \in U_i$ and $g_i(p) = 0$ if $p \notin U_i$. Then supp$g_i \subset \text{supp}\varphi_i$. Thus $f = \sum_i g_i$ defines a C^∞ function on M (see PU2 in A.3). It is an easy matter (using another partition of unity argument to prove that $\lim_{n \to \infty} f_n = f$ in V.

Let $a \in C^\infty(M; \mathbb{R})$ have strictly positive values and such that $\{p \in M \,|\, a(p) \leqslant r\}$ is compact for all $0 \leqslant r < \infty$. We also assume that there exists $d_o > 0$ such that

$$\int_M a^{-d_o} \omega < \infty.$$

We define $q_{d,i_1...i_k}(f) = \sup_{p \in M} a(p)^r |X_{i_1} \cdots X_{i_k} f(p)|$. Let W be the space of all $f \in C^\infty(M)$ such that $q_{d,i_1...i_k}(f) < \infty$ for all $d \geqslant 0$, $1 \leqslant i_1, ..., i_k \leqslant m$. We assert that W endowed with the topology induced by these semi-norms is complete. Indeed, we note that $q_{i_1...i_k}(f) \leqslant q_{d_o/2, i_1...i_k}(f)$ with C independent of f. Thus if $\{f_n\}$ is a Cauchy sequence in W it is Cauchy in V. Hence there exists $f \in V$ such that $\lim_{n \to \infty} f_n = f$ in V. Also we note that we have also shown that $X_{i_1} \cdots X_{i_k} f_n \to X_{i_1} \cdots X_{i_k} f$ uniformly on compacta.

Set $U_r = \{p \in M \,|\, a(p) > r\}$. Then $M - U_r$ is compact. Let $d > 0$ be fixed. There exists N such that if $n \geqslant N$ then $q_{d,i_1...i_k}(f_n - f_N) < 1$. Thus, if $p \in M$, $n \geqslant N$ then

$$a(p)^d |X_{i_1} \cdots X_{i_k} f_n(p)| \leqslant 1 + q_{d,i_1...i_k}(f_N).$$

Thus $|X_{i_1} \cdots X_{i_k} f_n(p)| \leqslant a(p)^{-d} C$ for $n \geqslant N$ with C depending only on N. Now given $\epsilon > 0$ there exists $N_1 \geqslant N$ such that if $n \geqslant N_1$ then

$$|X_{i_1} \cdots X_{i_k}(f - f_n)(p)| < \epsilon$$

for all $p \in M$. Hence $|X_{i_1} \cdots X_{i_k} f(p)| < \epsilon + Ca(p)^{-d}$ for all $\epsilon > 0$ and $p \in M$. So $f \in W$. We have also shown that there esists a constant D such that

$$|X_{i_1} \cdots X_{i_k}(f - f_n)(p)| \leqslant Da(p)^{-d}$$

for $n \geqslant N_1$. Let $0 < s < d$. Let $\epsilon > 0$ and let $r > 0$ be so large that if $p \in U_r$ then $Da(p)^{-d+s} < \epsilon$. Let N_2 be so large that $|X_{i_1} \cdots X_{i_k}(f - f_n)(p)| \leqslant \epsilon r^{-s}$ for $n \geqslant N_2$ and $p \in M - U_r$. Then $a(p)^s |X_{i_1} \cdots X_{i_k}(f - f_n)(p)| < \epsilon$ for $p \in M - U_r$ and $a(p)^s |X_{i_1} \cdots X_{i_k}(f - f_n)(p)| \leqslant Da(p)^{-d+s} < \epsilon$ for $p \in U_r$. Thus $q_{s,i_1...i_k}(f - f_n) < \epsilon$ for $p \in M$. This completes the proof of completeness of W.

A.8. A *Lie group* is a group G that is also a C^∞ manifold such that the operations $G \times G \to G$, $x, y \mapsto xy$ and $G \to G$, $x \mapsto x^{-1}$ are C^∞. If G is a Lie group we define two families of C^∞ maps of G to G. $L(g)(x) = gx$ and $R(g)(x) = xg$ for $x \in G, g \in G$.

Example. $G = GL(n, \mathbb{R})$ the group of all invertible transformations of $\mathbb{R}^n$. G is open in $M_n(\mathbb{R})$ the space of all $n \times n$ matrices. We look upon $M_n(\mathbb{R})$ as $\mathbb{R}^{n^2}$ using the matrix entries as coordinates. Then martix multiplication is a polynomial map of $M_n(\mathbb{R}) \times M_n(\mathbb{R}) \to M_n(\mathbb{R})$ and Cramer's rule implies that the map $g \mapsto g^{-1}$ is also C^∞.

If G is a Lie group then we define the Lie algebra of G to be the space of all $X \in Vect(G)$ such that $dL(g)_x X_x = X_{gx}$ (i.e. X is left invariant). We denote this space by $Lie(G)$. We note that if $X, Y \in Lie(G)$ then $[X, Y] \in Lie(G)$. It is easily seen that if $\mathfrak{g} = Lie(G)$ then

Lie 1. $[X, Y] = -[Y, X]$, $X, Y \in \mathfrak{g}$.

Lie 2. $[X, [Y, Z]] = [[X, Y], Z] + [Y, [X, Z]]$, $X, Y, Z \in \mathfrak{g}$.

In general, a vector space over a field k (characteristic not 2) with a bilinear map $x, y \mapsto [x, y]$ satisfying Lie1 and Lie2 is called a *Lie algebra* over k. So $Lie(G)$ is a Lie algebra over $\mathbb{R}$. Note that for a C^∞ manifold M, $Vect(M)$ is a Lie algebra under $[\ ,\ ]$.

Let G be a Lie group with identity element 1. Set $\mathfrak{g} = Lie(G)$. If $X \in \mathfrak{g}$ and $X_1 = 0$ then $0 = dL(g)_1 X_1 = X_g$ thus $X = 0$. If $v \in T(G)_1$ let $\sigma : (-a, a) \to G$ be a C^∞ curve such that $\sigma(0) = 1$ and $\sigma'(0) = v$. We consider the C^∞ map $\alpha : G \times (-a, a) \to G, \alpha(g, t) = g\sigma(t)$. Then $X = (g \mapsto d\alpha_{g,0}(0, \frac{d}{dt}))$ is a left invariant vector field on G. Clearly, $X_1 = v$. Thus dim $\mathfrak{g} = n = \dim G$.

In the example of $G = GL(n, \mathbb{R})$, let $X_{ij} \in Lie(G)$ be such that $(X_{ij})_I = \frac{\partial}{\partial x_{ij}}\big|_I$. Then $X_{ij} f(g) = \frac{d}{dt}\big|_{t=0} f(g(I + tE_{ij}))$ with E_{ij} the matrix with a 1 in the i, j position and zeros everywhere else. Thus $X_{ij} = \sum_r x_{ri} \frac{\partial}{\partial x_{rj}}$. If $X \in Lie(G)$ then $X = \sum_{i,j} a_{ij} X_{ij}$ with $[a_{ij}] \in M_n(\mathbb{R})$. If we set $\Psi(X) = A$ then $\Psi([X, Y]) = \Psi(X)\Psi(Y) - \Psi(Y)\Psi(X)$. Hence $Lie(G)$ is isomorphic with $M_n(\mathbb{R})$ under matrix commutator. We will therefore identify $Lie(GL(n, \mathbb{R}))$ with $M_n(\mathbb{R})$.

A.9. If M is a C^∞ manifold of dimension n then a differential operator on M of order at most k is a linear map $D : C^\infty(M) \to C^\infty(M)$ such that if $p \in M$ and (U, Ψ) is a chart for M with $p \in U$ and if $\{x_1, ..., x_n\}$ is the corresponding system of local coordinates on U and $a_I \in C^\infty(U)$, $I \in \mathbb{N}^n$, $|I| \leqslant k$ such that $Df(y) = \sum_{|I| \leqslant k} a_I(y)\partial^I f(y)$ for $y \in U$. Thus a vector field is a differential operator of order at most 1. We define $\mathrm{ord}(D)$ to be the smallest integer for which D has order at most k. We note that if D, E are differential operators of order k, l respectively then $D \circ E$ is a differential operator of order at most $k + l$. Thus the space of all differential operators, Diff (M), is an algebra over $\mathbb{C}$. We set $\mathrm{Diff}^k(M)$ equal to the space of all differential operators of order at most k. Then Diff (M) is a filtered algebra with respect to order (i.e. $\mathrm{Diff}^k(M) \, \mathrm{Diff}^l(M) \subset \mathrm{Diff}^{k+l}(M)$).

If G is a Lie group then we set $l_g f(x) = f(g^{-1}x)$ and $r_g f(x) = f(xg)$ for $x, g \in G$ and $f \in C(G)$. We set $U(G)$ equal to the space of all differential operators D on G such that $D \circ l_g = l_g \circ D$ for all $g \in G$ (i.e. the left invariant differential operators). The following basic result contains the Poincarè-Birkhoff-Witt thoerem and a theorem of L. Schwartz.

Theorem. $U(G)$ is generated by $\mathfrak{g}$ as an algebra. If $X_1, ..., X_n$ is a basis of $\mathfrak{g}$ then the elements $X_{i_1} X_{i_2} \cdots X_{i_k}$ with $1 \leqslant i_1 \leqslant i_2 \leqslant ... \leqslant i_k \leqslant n$ form a basis of $U(G)$. If $\mathcal{A}$ is an associative algebra over $\mathbb{C}$ with unit 1 and if $\rho : \mathfrak{g} \to \mathcal{A}$ is such that $\rho([X, Y]) = \rho(X)\rho(Y) - \rho(Y)\rho(X)$ for $X, Y \in \mathfrak{g}$ then there exists a unique homomorphism of $U(G)$ into $\mathcal{A}$, which we denote by $\breve{\rho}$ such that $\breve{\rho}(I) = 1$ and $\breve{\rho}(X) = \rho(X)$ for $X \in \mathfrak{g}$.

Since $X_1, ..., X_n$ is a basis of $\mathfrak{g}$, $(X_1)_1, ..., (X_n)_1$ is a basis of $T(G)_1$. We can thus find a chart, (U, Ψ) for G such that $\Psi(1) = 0$ and if $\{x_1, ..., x_n\}$ is the corresponding system of local coordinates on U then $X_{i\,|U} = \sum_j a_{ji} \frac{\partial}{\partial x_i}$ with $a_{ij} \in C^\infty(U)$ and $a_{ij}(1) = \delta_{ij}$. This implies that if $I \in \mathbb{N}^n$ then

$$X_1^{i_1} \cdots X_n^{i_n} f(1) = \partial^I f(1)$$

if $\partial^J f(1) = 0$ for $|J| < |I|$. Thus in particular, if we set $x^I = x_1^{i_1} \cdots x_n^{i_n}$ then

$$X_1^{i_1} \cdots X_n^{i_n} x^J(1) = \delta_{IJ} i_1! \cdots i_n!$$

for all $|J| = |I|$. This implies that the elements $X_1^{i_1} \cdots X_n^{i_n}$, $|I| = k$ are linearly independent modulo $U(G) \cap \mathrm{Diff}^{k-1}(G)$. Also if $D \in U(G)$ then $Df(g) = D(l_g f)(1)$. Thus if $D = \sum_{|I| \leqslant k} a_I \partial^I$ then $D - \sum_{|I|=k} a_I(1) X_1^{i_1} \cdots X_n^{i_n}$ is in $U(G)$ and is of order at most $k - 1$. We have thus proved the first two assertions of the theorem.

We now prove the last assertion. For this let $T(\mathfrak{g})$ denote the tensor algebra on the complexification of $\mathfrak{g}$. That is if $X_1, ..., X_n$ is a basis of $\mathfrak{g}$ over $\mathbb{R}$ then $T(\mathfrak{g})$ is the space with basis all symbols $X_{i_1} X_{i_2} \cdots X_{i_k}$ and multiplication given by juxtaposition $(X_{i_1} X_{i_2} \cdots X_{i_k} \cdot X_{j_1} X_{j_2} \cdots X_{j_l} = X_{i_1} X_{i_2} \cdots X_{i_k} X_{j_1} X_{j_2} \cdots X_{j_l})$. We form the two sided ideal $I(\mathfrak{g}) = \sum_{X,Y \in \mathfrak{g}} T(\mathfrak{g})(XY - YX - [X,Y])T(\mathfrak{g})$ in $T(\mathfrak{g})$. We denote by $U(\mathfrak{g})$ the algebra $T(\mathfrak{g})/I(\mathfrak{g})$. We note that $U(\mathfrak{g})$ has the mapping property asserted for $U(G)$. Using Lie1 in A.8 it is easily seen that the terms $X_1^{i_1} \cdots X_n^{i_n}, I \in \mathbb{N}^n$ span $U(\mathfrak{g})$. But the universal mapping property of $U(\mathfrak{g})$ implies that there is a sujective algebra homomorphism of $U(\mathfrak{g})$ onto $U(G)$. Hence the monomials $X_1^{i_1} \cdots X_n^{i_n}, I \in \mathbb{N}^n$ form a basis of $U(\mathfrak{g})$ and $U(\mathfrak{g})$ is isomorphic with $U(G)$. This completes the proof.

Note. The algebra $U(\mathfrak{g})$ will sometimes be denoted $U(\mathfrak{g}_{\mathbb{C}})$. Also we will drop the notation $U(G)$ and identify it with $U(\mathfrak{g})$.

A.10. If G and H are Lie groups then a *Lie homomorphism* of H to G is a C^∞ map that is a group homomorphism. If $\varphi : G \to H$ is a Lie homomorphism we define $d\varphi : Lie(H) \to Lie(G)$ by $d\varphi(X)_1 = d\varphi_1(X_1)$ for $X \in Lie(H)$. It can be shown that $d\varphi$ defines a Lie algebra homomorphism ([FWar]). If $H = \mathbb{R}$ under addition then a Lie homomorphism of $\mathbb{R}$ into G is called a *one parameter subgroup*. If φ is a one parameter subgroup of G then $d\varphi(\frac{d}{dt}) = X \in Lie(G)$. We will call X the *generator* of φ. The following summerizes the basics of one parameter subgroups.

Theorem. *Let G be a Lie group with Lie algebra $\mathfrak{g}$. Then there exists a C^∞ map* $\exp : \mathfrak{g} \to G$ *such that if φ is a one parameter subgroup of G with generator X then* $\varphi(t) = \exp(tX), t \in \mathbb{R}$. *Furthermore, if $X \in \mathfrak{g}$ then $t \mapsto \exp(tX)$ is a one parameter subgroup of G generated by X.*

For a proof see [FWar]. If we choose a basis of $\mathfrak{g}$ then we can identify $\mathfrak{g}$ with $\mathbb{R}^n$. We also identify $T(\mathbb{R}^n)_x$ with $\mathbb{R}^n$ using the basis $\frac{\partial}{\partial x_1}, ..., \frac{\partial}{\partial x_n}$. Thus we have the identification $T(\mathfrak{g})_X = \mathfrak{g}$. With this understanding, $d\exp_0(X) = X$ for all $X \in \mathfrak{g}$. The inverse function theorem of advanced calculus now implies that there is an open neighborhood of 0, U_0, in $\mathfrak{g}$ and an open neighborhood of 1 in G such that

$$\exp : U_0 \to U_1$$

is a homeomorphism with C^∞ inverse $\log : U_1 \to U_0$. By shrinking U_0 we may assume that it is convex and if $X \in U_0$ then $tX \in U_0$ for $|t| \leqslant 1$ (i.e. starlike).

Lemma. *If $X, Y \in \mathfrak{g}$ and if $t \in \mathbb{R}$ is so small that $\exp tX \exp tY \in U_1$ then*

$$\log(\exp tX \exp tY) = tX + tY + \frac{t^2}{2}[X,Y] + O(t^3).$$

For a proof see e.g. [FWar].

A.11. If G is a Lie group with Lie algebra $\mathfrak{g}$ then we define for $g \in G$, $\text{Int}(g)x = gxg^{-1}$. We set $di\,\text{Int}(g) = \text{Ad}(g)$. Then $\text{Ad} : G \to GL(\mathfrak{g})$ is a Lie homomorphism which is called the *adjoint representation* of $\mathfrak{g}$. We note that the above Lemma implies that

$$d\text{Ad}(X) = \text{ad}(X)$$

with $\mathrm{ad}(X)Y = [X,Y]$.

We also not that if we define an atlas for G, (U_g, Ψ_g) with $U_g = gU_1$ and $\Psi_g(x) = \log(g^{-1}x)$, $x \in U_g$ then on $\Psi_g(U_g \cap U_x)$, $\Psi_x \circ \Psi_g^{-1}$ is real analytic. We therefore see that a Lie group has the structure of a real analytic manifold (defined the same way as a C^∞ manifold with C^∞ replaced by real analytic. We also have the obvious notions of real analytic functions and maps.

A.12. If H is a subgroup of G that has the structure of a Lie group and if the canonical injection of H into G ($i(h) = h$) is a Lie homomorphism such that $di : Lie(H) \rightarrow Lie(G)$ is injective then we say that H is a *Lie subgroup* of G. We note that if H is a Lie subgroup of G then we can identify $di(Lie(H))$ with a subalgebra, $\mathfrak{h}$, of $\mathfrak{g}$. Let $\exp_H$ denote the exponential map of H. Then $t \mapsto i \circ \exp_H(tX)$ is a one parameter subgroup of G generated by X. Hence $\exp_H X = \exp X$ for $X \in \mathfrak{h}$. Also $\mathfrak{h} = \{X \in \mathfrak{g}| \exp(tX) \in H, t \in \mathbb{R}\}$.

More generally, if H is a Lie group and if $\varphi : H \rightarrow G$ is a Lie homomorphism then $\varphi(\exp_H(X)) = \exp(d\varphi(X))$ for $X \in Lie(H)$.

Theorem. *Let H be a closed subgroup of G. Then H has a unique structure of a Lie group such that H is a Lie subgroup of G.*

The basic idea is that the Lemma in A.**10** implies that $\{X \in \mathfrak{g}| \exp tX \in H, t \in \mathbb{R}\}$ is a Lie subalgebra of $\mathfrak{g}$ (cf.[FWar] for a proof of the theorem).

A.13. Let G be a Lie group with Lie algebra $\mathfrak{g}$. If $X_1, ..., X_n$ is a basis of $\mathfrak{g}$ we define $\omega \in \Omega^n(G)$ by $\omega_g(X_{1_g}, ..., X_{n_g}) = 1$ for $g \in G$. Then $L(g)^*\omega = \omega$ for $g \in G$. In general $\omega \in \Omega^n(G)$ is said to be left invariant if $L(g)^*\omega = \omega$ for all $g \in G$. The above construction implies that left invariant n-forms exist. It is also clear that the space of all left invariant n-forms is one dimensional.

Fix, ω, a non-zero left invariant n-form on G. If $f \in C_c(G)$ then we write

$$\int_G f(g)dg = \int_G f\omega.$$

The left invariance of ω implies that

$$\int_G l_g f(x)dx = \int_G f(x)dx.$$

That is,

$$\int_G f(g^{-1}x)dx = \int_G f(x)dx.$$

We will call dg a choice of *left invariant measure* on G.

We note that if $X \in \mathfrak{g}$ then

$$dR(g)_x X_x = dR(g)_x dL(g^{-1})_{gx} X_{gx} = d\,\mathrm{Int}(g^{-1})_{gx} X_{gx} = (\mathrm{Ad}(g^{-1})X)_{gx}.$$

This implies that if ω is a left invariant n-form on G then

$$R(g)^*\omega = \det(\mathrm{Ad}(g))^{-1}\omega.$$

Hence

$$\int_G f(xg)dx = |\det(\mathrm{Ad}(g))|^{-1} \int_G f(x)dx.$$

We say that G is *unimodular* if $|\det(\mathrm{Ad}(g))| = 1$ for all $g \in G$. If G is unimodular then

$$\int_G f(gx)dx = \int_G f(xg)dx = \int_G f(x)dx, \ g \in G, \ f \in C_c(G).$$

Obviously, we have a similar situation with right invariance. We leave it to the reader to set up the analogous notion of right invariant measure.

If G is compact then we note that since $|\mathrm{Ad}(G)|$ is a compact subgroup of $(0,\infty)$, $|\mathrm{Ad}(G)| = \{1\}$ so G is unimodular if G is compact.

A.14. We say that G is a *linear real algebraic group* if G is a subgroup of $GL(n,\mathbb{R})$ and there exist real valued polynomial functions on $M_n(\mathbb{R})$, $f_1, ..., f_d$ such that

$$G = \{g \in GL(n,\mathbb{R})|f_i(g) = 0, \ i = 1, ..., d\}.$$

Notice that in particular G is a closed subgroup of $GL(n,\mathbb{R})$ thus G is a Lie subgroup. We will say that a linear real algebraic group is *symmetric* if whenever $g \in G$, $g^T \in G$ (here g^T is the usual transpose of g).

Examples.
   1. $GL(n,\mathbb{R})$ here take the $f_i = 0$.
   2. $SL(n,\mathbb{R}) = \{g \in GL(n,\mathbb{R})| \det(g) = 1\}$. Take $d = 1$ and $f_1 = \det -1$.
   3. $O(n) = \{g \in GL(n,\mathbb{R})| g^Tg = I\}$. Here we take $f_{ij}(g) = x_{ij}(g^Tg - I)$. Note that this group is compact.
   4. $SO(n) = \{g \in O(n)| \det(g) = 1\}$. In addition to the f_{ij} in 3. use $\det -1$.

We say that a Lie group, G, is a *real reductive group* if there exists a symmetric linear real algebraic group, G_1, an open subgroup, G_o, of G_1 and a surjective Lie homomorphism $\varphi : G \to G_o$ such that $d\varphi$ is bijective and φ has finite kernel.

If G is a real reductive group then we can use the φ in the definition to identify $Lie(G)$ with $Lie(G_o) = Lie(G_1)$. $Lie(G_1)$ can in turn be identified with a Lie subalgebra of $Lie(GL(n,\mathbb{R}))$ which we have identified with $M_n(\mathbb{R})$. This algebra can be described using the polynomials f_i as the $X \in M_n(\mathbb{R})$ such that $df_{i\ I}(X) = 0$ for all i.

Examples.
   1. $Lie(GL(n,\mathbb{R})) = M_n(\mathbb{R})$.
   2. $Lie(SL(n,\mathbb{R})) = \{X \in M_n(\mathbb{R})| \mathrm{tr}X = 0\}$ ($d\det_I = \mathrm{tr}$).
   3. $Lie(O(n)) = \{X \in M_n(\mathbb{R})| X^T + X = 0\}$.

4. $Lie(SO(n)) = Lie(O(n))$.

A.15. If G is a real reductive group with $Lie(G) = \mathfrak{g}$. We then have φ, G_1, G_o. We thus we can define $\theta : \mathfrak{g} \to \mathfrak{g}$ by $\theta(X) = -X^T$. We will call θ a *Cartan involution* of $\mathfrak{g}$. We also set $B(X, Y) = \text{tr}(XY)$ for $X, Y \in \mathfrak{g}$. We note that B is symmetric, non-degenerate and

$$B(\text{Ad}(g)X, \text{Ad}(g)Y) = B(X, Y)$$

for $X, Y \in \mathfrak{g}$ and $g \in G$.

Let $\mathfrak{z}(\mathfrak{g}) = \{X \in \mathfrak{g} |\, [X, \mathfrak{g}] = 0\}$. We set $G^\dagger = \{g \in G | \text{Ad}(g)|_{\mathfrak{z}(\mathfrak{g})} = I\}$ then $G^\dagger$ is a real reductive subgroup of G.

Let $\mathfrak{p} = \{X \in \mathfrak{g} |\, \theta(X) = -X\}$. Set $K_o = O(n) \cap G_o$ and $K = \varphi^{-1}(K_o)$. Then $Lie(K) = \{X \in \mathfrak{g} |\, \theta(X) = X\}$. We note that K is a compact subgroup of G. We observe that if $X \in \mathfrak{p}$ then $X^T = X$ thus X is diagonalizable over $\mathbb{R}$. The polar decomposition of elements of $GL(n, \mathbb{R})$ implies

Proposition. *The map* $K \times \mathfrak{p} \to G$, $k, X \mapsto k \exp X$ *is a diffeomorphism onto* G.

We note that this implies that K is a maximal compact subgroup of G. Indeed suppose that K' is a subgroup of G with $K' \supset K$, $K' \neq K$. Then there exists $k' = k \exp X \in K - K'$, $k \in K$, $X \in \mathfrak{p}$. Thus $\exp X \in K'$ with $X \neq 0$. Since $X^T = X$, the group $e^{\mathbb{Z}X}$ is unbounded. Hence K' is not compact.

If $\mathfrak{a} \subset \mathfrak{p}$ is a real subspace such that if $X, Y \in \mathfrak{a}$ then $[X, Y] = 0$ we will call $\mathfrak{a}$ an *abelian subspace*. If $\mathfrak{a}$ is an abelian subspace of $\mathfrak{p}$ then we note that the elements of $\mathfrak{a}$ are simultaneously diagonalizable. If $\lambda \in \mathfrak{a}^*$ then we set

$$\mathfrak{g}_\lambda = \{X \in \mathfrak{g} |\, [H, X] = \lambda(H)X, H \in \mathfrak{a}\}.$$

We note that if $X \in M_n(\mathbb{R})$ is diagonalizable over $\mathbb{R}$ then $\text{ad}(X)$ is diagonalizable as an endomorphism of $M_n(\mathbb{R})$. Set $\Phi(\mathfrak{g}, \mathfrak{a}) = \{\lambda \neq 0 |\, \mathfrak{g}_\lambda \neq 0\}$. Then

$$\mathfrak{g} = \mathfrak{g}_0 \oplus \oplus_{\lambda \in \Phi(\mathfrak{g}, \mathfrak{a})} \mathfrak{g}_\lambda.$$

We first concetrate on the case $\mathfrak{a} = \mathfrak{s}(\mathfrak{g}) = \{X \in \mathfrak{p} |\, \text{ad}(X) = 0\}$ (i.e. $\mathfrak{g} = \mathfrak{g}_0$ and $\mathfrak{a}$ is maximal subject to this property).

Theorem. *We assume that* $G = G^\dagger$. *Let* $\mathfrak{a} = \mathfrak{s}(\mathfrak{g})$. *Set* $X(G) = \{\chi : G \to \mathbb{R}^+ = (0, \infty) |\, \chi$ *a Lie homomorphism*$\}$. *Put* $^\circ G = \cap_{\chi \in X(G)} \ker \chi$. *Then*

1. $^\circ G$ *is a real reductive subgoup of* G.
2. $A = \exp \mathfrak{a}$ *is a closed subgroup of* G *and* $\exp : \mathfrak{a} \to A$ *is a Lie isomorphism of* $\mathfrak{a}$ *onto* A *(here* $\mathfrak{a}$ *is looked upon as a Lie group under addition).*
3. *The map* $A \times {}^\circ G \to G$, $a, g \mapsto ag$ *is a Lie isomorphism of* $A \times {}^\circ G$ *onto* G.

For a proof cf.[RRG I,2.2.2].

Example. $G = GL(n, \mathbb{R})$. Then $\mathfrak{z}(\mathfrak{g}) = \mathbb{R}I$ so $G = G^\dagger$. $A = \{\lambda I | \lambda > 0\}$. $^\circ G = \{g \in G | \det(g) = \pm 1\}$.

A.16. We now look at $\mathfrak{a} \subset \mathfrak{p}$ an abelian subspace. Let $M = \{g \in G | \operatorname{Ad}(g)_{|\mathfrak{a}} = I\}$. Then it is easily seen that M is a real reductive subgroup of G. It can also be shown that if $G = G^\dagger$ then $M = M^\dagger$. Let $\mathfrak{m} = Lie(M) \subset \mathfrak{g}$. We note that $\mathfrak{s}(\mathfrak{m}) \supset \mathfrak{a}$. We will say that $\mathfrak{a}$ is a *special abelian subspace* of $\mathfrak{p}$ if $\mathfrak{a} = \mathfrak{s}(\mathfrak{m})$. We will assume that $\mathfrak{a}$ is special. Fix a basis $\{H_1, ..., H_l\}$ of $\mathfrak{a}$. If $\lambda \in \mathfrak{a}^*$ then we say that $\lambda > 0$ if $\lambda(H_1) > 0$ or if $\lambda(H_1) = 0$ then $\lambda(H_2) > 0$, etc. Set $\Phi^+ = \{\lambda \in \Phi(\mathfrak{g}, \mathfrak{a}) | \lambda > 0\}$. Put $\mathfrak{n} = \oplus_{\lambda \in \Phi^+} \mathfrak{g}_\lambda$. Then $\mathfrak{n}$ is a Lie subalgebra of $\mathfrak{g}$ and if $X \in \mathfrak{n}$ then $\operatorname{ad}(X)_{|\mathfrak{n}}$ is a nilpotent transformation of $\mathfrak{n}$. It is an easy matter to see that if $N = \exp(\mathfrak{n})$ then N is a closed Lie subgroup of G and $\exp : \mathfrak{n} \to N$ is a diffeomorphism.

Set $\mathfrak{q} = \mathfrak{m} \oplus \mathfrak{n}$. Then $\mathfrak{q}$ is called a *(standard) parabolic subalgebra* of $\mathfrak{g}$. The pair $(\mathfrak{q}, \mathfrak{a})$ is called a *(standard) parabolic pair* (p-pair for short) of $\mathfrak{g}$.

Theorem. *Set $Q = MN$. Then the map $^\circ M \times A \times N \to Q$ is a Lie isomorphism onto Q. Furthermore, $G = QK$.*

For a proof cf. [RRG I,2.2.7].

Q will be called a *(standard) parabolic subgroup* of G. The pair (Q, A) will be called a *(standard) p-pair* in G. The decomposition $Q = {}^\circ MAN$ is called a *(standard) Langlands decomposition.*

A.17. If $\mathfrak{a}$ is a special abelian subalgebra of $\mathfrak{p}$ let (Q, A) be as above. If $\lambda \in \mathfrak{a}_C^*$ (complex valued linear functionals) we define $a^\lambda = e^{\lambda(H)}$ if $a = \exp(H), H \in \mathfrak{a}$. We define $\rho_Q = \rho$ by $\rho(H) = \frac{1}{2} \operatorname{tr}(\operatorname{ad}(H)_{|\mathfrak{n}})$ for $H \in \mathfrak{a}$.

On K we denote by dk the normalized invariant measure ($\int_K dk = 1$). On A we use the measure da such that $\exp^* da$ is the measure dx corresponding to an orthonormal basis of $\mathfrak{a}$ with respect to B ($B_{|\mathfrak{a}}$ is positive definite). On $\mathfrak{n}$ we put the inner product $\langle X, Y \rangle = -B(X, \theta Y)$ and use the left invariant measure dn on N such that $\exp^* dn$ is dx with respect to an orthonormal basis of $\mathfrak{n}$ with respect to $\langle \, , \, \rangle$. (We note that if $n \in N$ then $\operatorname{Ad}(n) - I$ is nilpotent on $\mathfrak{g}$. Thus N is unimodular.

Lemma. *G is unimodular (as is $^\circ M$). We can choose invariant measure dg on G and invariant measure dm on $^\circ M$ such that*

$$\int_G f(g)dg = \int_{N \times A \times {}^\circ M \times K} a^{-2\rho} f(namk) dn \, da \, dm \, dk.$$

For a proof cf. [RRG I,2.2.4].

A.18. We now assume that $Q = {}^\circ MAN$ is a minimal parabolic subgroup of G. Then $^\circ M \subset K$. Hence we have the *Iwasawa decomposition of G*

1. $G = NAK$ furthermore the map $N \times A \times K \to G$, $n, a, k \mapsto nak$ is a diffeomorphism. We note that since NA is connected, this implies that $G = G^\circ K = KG^\circ$.

268

We write $g = n(g)a(g)k(g)$ and note that $n : G \to N$, $a : G \to A$, $k : G \to K$ are C^∞ maps.

Using Lemma A.16 one can show (cf.[RRGI,2.4.1])

1. $\int_K f(k)dk = \int_K a(kg)^{2\rho} f(k(kg))dk$ for (say) $f \in C(K)$. Here we are using normalized invariant measure on K.

A.19. Let (Q, A) be a p-pair. Then relative to the action under ad of $\mathfrak{a}$ we have $\mathfrak{n} = \oplus_{\lambda \in \mathfrak{a}^*} \mathfrak{n}_\lambda$ ($\mathfrak{n}_\lambda = \{X \in \mathfrak{n}|\, [h, X] = \lambda(h)X, h \in \mathfrak{a}\}$. Set $\Phi(Q, A) = \{\lambda \in \mathfrak{a}^*|\, \mathfrak{n}_\lambda \neq 0\}$. Let Δ be the set of all $\lambda \in \Phi(Q, A)$ such that λ cannot be written in the form $\lambda_1 + ... + \lambda_k$ with $\lambda_i \in \Phi(Q, A)$ and $k > 1$. Such λ will be called *simple*.

1. Let $r = \dim \mathfrak{a} \cap [\mathfrak{g}, \mathfrak{g}]$ then $|\Delta| = r$ and if $\lambda \in \Phi(Q, A)$ then there are non-negative integers n_α, $\alpha \in \Delta$ such that $\lambda = \sum_{\alpha \in \Delta} n_\alpha \alpha$.

We now assume that Q is minimal. Let $A^+ = \{a \in A|\, a^\alpha \geqslant 1, \alpha \in \Phi(Q, A)\}$.

2. $G = KA^+K$. Futhermore there exist normalizations of invariant measures dg, da on G and A respectively such that

$$\int_G f(g)dg = \int_{K \times A^+ \times K} \gamma(a)f(k_1 a k_2)dk_1 da dk_2$$

with $\gamma(a) = \prod_{\alpha \in \Phi(Q,A)}(a^\alpha - a^{-\alpha})^{\dim \mathfrak{n}_\alpha}$.

REFERENCES

[B-W]   A. Borel, N. Wallach, *Continuous cohomology, discrete subgroups, and representation of reductive groups*, Princeton University Press, Princeton, 1980.

[Ca]   W. Casselman, *Canonical extensions of harish-Chandra modules*, Can. J. Math. **41** (1989), 385-438.

[D-M]   J. Dixmier, P. Malliavan, *Factorisations de fonctions et de vecteurs indéfiniment différentiables*, Bull. Sci. Math. **102** (1978), France, 307-330.

[G-G-PS]   I.M. Gelfand, M.I. Graev, I.I. Piatetski-Shapiro, *Rapresentation theory and automorphic forms*, W.B. Saunder Co., Philadelphia, 1969.

[H-C1]   Harish-Chandra, *Representations of semisimple Lie groups III*, Trans. Amer. Math. Soc. **76** (1954), 234-253.

[H-C2]   Harish-Chandra, *Automorphic forms on semisimple Lie groups*, vol. 62, Springer-Verlag, Berlin, 1968.

[Hö]   L. Hörmander, *Linear differential operators*, Springer-Verlag, Berlin, 1979.

[K]   B. Kostant, *On the existence and irreducibility of certain series of representations*, Bull. Amer. Math. Soc. **75** (1969), 627-642.

[Lep]   J. Lepowsky, *Algebraic results on representations of semisimple Lie groups*, Trans Amer. Math. Soc. **176** (1973), 1-44.

[O-S]   T. Oshima, J. Sekiguchi, *Eigenspaces of invariant differential operators in an affine symmetric space*, Invetiones math. **57** (1980), 1-81.

[R-S]   M. Reed, B. Simon, *Functional Analysis I*, Academic Press, New York, 1972.

[Sch]   W. Schmid, *Boundary value problems for group invariant differential equations*, Astérisque (1985), 311-321.

[Tr]   F. Treves, *Topological vector spaces, distributions and kernels*, Academic Press, New York, 1967.

[V]   D. Vogan, *Representations of real reductive groups*, vol. 15, Birkhauser, Boston, 1981.

[W1]   N. Wallach, *Harmonic analysis on homogenous spaces*, Marcel Dekker, New York, 1972.

[W2]   N. Wallach, *Asymptotic expansions of generalized matrix entries of representations of real reductive groups*, vol. 1204, Springer-Verlag, 1983.

[RRG I-II] N. Wallach, *Real reductive groups*, vol. I, II, Academic Press, Boston, 1988, 1992.

[FWar]   F. Warmer, *Foundation of differential geometry and Lie groups*, vol. II, Scott, Foresman and Co., 1971.

UCSD, Department of Mathematics, La Jolla CA 92093, U.S.A.

e–mail nwallach@euclid.ucsd.edu

270

OLD AND NEW ABOUT KAZHDAN'S PROPERTY (T)

ALAIN VALETTE

CONTENTS

0. INTRODUCTION

Property (T) is a representation-theoretic property that a locally compact group may have or not. It may be viewed as a weak form of rigidity (as such, it was used e.g. in Margulis' proof of the structure theorem for normal subgroups of lattices in higher rank groups).

As the title of Kazhdan's original paper [Kaz] rightfully suggests, this property provides a connection between the structure of the dual space of a locally compact group G (namely, isolation of the trivial representation) and the algebraic structure of some closed subgroups of G (namely, those subgroups with finite co-volume). If G has property (T) and H is a subgroup with finite co-volume, then:

- H is compactly generated: in particular, if H is discrete, then H is finitely generated;
- the quotient $H/cl\,H'$ is compact (here $cl\,H'$ denotes the closure of the commutator subgroup H' of H);
- if H is discrete, then for any fixed integer $d > 0$, there are only finitely many unitary irreducible representations of H in degree d.

Property (T) also proved useful in the solution of a number of problems ranging from ergodic theory to graph theory. Some of these applications are discussed in Chapters 3, 4 and especially 5 of these Notes.

The monograph [HaV], by Pierre de la Harpe and the author was entirely devoted to property (T) and its applications. This set of notes may be viewed as an updating of [HaV].

1. MAIN DEFINITIONS AND RESULTS

1.a. Some representation theory

All the way, we shall deal with *locally compact groups*, i.e. topological groups whose underlying topological space is locally compact, σ-compact. We have in mind 3 basic classes of examples:

- discrete countable groups;
- real or complex Lie groups;
- algebraic groups over local fields (a local field $\mathbb{F}$ may be archimedean, in which case $\mathbb{F} = \mathbb{R}$ or $\mathbb{C}$; by a non-archimedean local field we mean, in characteristic 0, a finite extension of a p-adic field $\mathbb{Q}_p$; in positive characteristic, a field of formal power series over a finite field).

Up to a few exceptions (always explicitly mentioned), representations will always be strongly continuous unitary representations on complex Hilbert spaces. Formally, a *representation* of the locally compact group G is a pair $(\pi, \mathcal{H}_\pi)$ where $\mathcal{H}_\pi$ is a complex Hilbert space and $\pi : G \to U(\mathcal{H}_\pi)$ is a homomorphism from G into the unitary group $U(\mathcal{H}_\pi)$ of $\mathcal{H}_\pi$; moreover π is *strongly continuous* in the sense that the map

$$G \times \mathcal{H}_\pi \to \mathcal{H}_\pi : (g, \xi) \to \pi(g)\xi$$

is continuous.

We shall denote by $\mathcal{H}_\pi^1$ the unit sphere of $\mathcal{H}_\pi$

1.1. Definitions. Let π be a representation of G.

a) Given $\varepsilon > 0$ and a compact subset $K \subseteq G$, we say that a vector $\xi \in \mathcal{H}_\pi^1$ is (ε, K)-*invariant* if $\max_{k \in K} \|\pi(k)\xi - \xi\| < \varepsilon$.

b) We say that π *almost has invariant vectors* (or that π *weakly contains the trivial representation*) if, for any (ε, K), there exists an (ε, K)-invariant vector in $\mathcal{H}_\pi$.

Here is now our main subject of concern.

1.2. Definition. We say that G has *property (T)*, or is a *Kazhdan group*, if any representation of G that almost has invariant vectors has non-zero invariant vectors.

A stupid but useful representation of G is the *trivial representation* 1_G, i.e. the constant representation of G on the 1-dimensional Hilbert space $\mathbb{C}$. We may then rephrase 1.2 by saying that G has property (T) if and only if whenever a representation of G weakly contains 1_G, it really contains 1_G as a subrepresentation.

1.3. Examples. a) Here is the only simple class of examples of Kazhdan groups: any compact group G has property (T). Indeed, let π be a representation of G

that almost has invariant vectors. For $\varepsilon > 0$, let ξ_ε be an (ε, G)-invariant vector. Averaging over G with respect to a normalized Haar measure dg, we get

$$\eta_\varepsilon = \int_G \pi(g)\xi_\varepsilon dg,$$

clearly an invariant vector; and for ε small enough η_ε is non-zero.

b) The *left regular representation* of a locally compact group G is the representation on $L^2(G)$ defined by $(\lambda(g)\xi)(s) = \xi(g^{-1}s)$ $(\xi \in L^2(G), g, s \in G)$. The group G is compact if and only if λ has non-zero invariant vectors, and G is amenable if and only if λ almost has invariant vectors ([Zim], Theorem 7.1.8). As a consequence, we see that a Kazhdan group is amenable if and only if it is compact.

A good deal of the theory of Kazhdan groups rests on general, usually fairly easy, representation-theoretic results. The truly hard results are the ones that provide explicit examples. We begin with the "soft" part of the theory.

The first result deals with quotients of Kazhdan groups.

1.4. Proposition. *Let G be a Kazhdan group.*

i) Let H be a locally compact group. If there exists a continuous homomorphism $\varphi : G \to H$ with dense range, then H has property (T);

ii) any quotient of G has property (T);

iii) any homomorphism from G into an amenable locally compact group has relatively compact range;

iv) G is unimodular;

v) if G' denotes the commutator subgroup of G and $cl\, G'$ its closure, then the group $G/cl\, G'$ is compact. (In particular, if G is discrete, then G/G' is finite).

Proof. i) Let π be representation of H that almost has invariant vectors. Then $\pi \circ \varphi$ is a representation of G with the same property; so we find $\xi \in \mathcal{H}_\pi^1$ which is $\pi(\varphi(G))$-invariant. By density of $\varphi(G)$ in H, the vector ξ is also $\pi(H)$-invariant.

ii) is a special case of i).

iii) is a consequence of i) above and example 1.3.(b).

iv) Since the only compact subgroup of $\mathbb{R}_+^\times$ is $\{1\}$ and $\mathbb{R}_+^\times$ is amenable, the assertion follows from iii).

v) is a consequence of ii) and iii).

Here is another basic result.

1.5. Proposition. *A Kazhdan group is compactly generated. (In particular a discrete Kazhdan group is finitely generated.)*

Proof. Let G be a Kazhdan group. Since G is σ-compact we find an increasing family $(K_n)_{n \geqslant 1}$ of compact neighbourhoods of 1 in G, such that $G = \bigcup_{n=1}^\infty K_n$. Let H_n be the subgroup generated by K_n; since K_n is a neighbourhood of 1, the subgroup

H_n is open; we denote by π_n the permutation representation of G on $l^2(G/H_n)$, and set $\pi = \bigoplus_{n=1}^{\infty} \pi_n$.

We claim that π almost has invariant vectors. Indeed, let K be a compact subset in G; let n be such that $K \subseteq K_n$, and let ξ_n be the characteristic function of the base-point of G/H_n. Then $\xi_n \in \mathcal{H}_\pi^1$ and ξ_n is $\pi(K)$-invariant, a fortiori (ε, K)-invariant for all $\varepsilon > 0$.

By property (T), π has non-zero invariant vectors; therefore some π_m has non-zero invariant vectors. This means that $l^2(G/H_m)$ contains non-zero constant functions, i.e. G/H_m is finite. Since $G = \bigcup_{n=1}^{\infty} H_n$, we have $G = H_k$ for some $k \geqslant m$. $\qquad \square$

1.6. Proposition. *i) Let $1 \to N \to G \to G/N \to 1$ be a short exact sequence of locally compact groups. If both N and G/N are Kazhdan, then so is G.*

ii) Let G_1, G_2 be locally compact groups. $G_1 \times G_2$ is Kazhdan if and only if G_1 and G_2 are.

Proof. (i) Let π be a representation of G that almost has invariant vectors. Let $\mathcal{H}_\pi^N$ be the subspace of $\pi(N)$-invariant vectors in $\mathcal{H}_\pi$; because N has property (T) we have $\mathcal{H}_\pi^N \neq \{0\}$; on the other hand $\mathcal{H}_\pi^N$ is $\pi(G)$-invariant, because N is normal in G. The restriction of π to the orthogonal $(\mathcal{H}_\pi^N)^\perp$ does not almost have invariant vectors (if it had, it would have non-zero $\pi(N)$-invariant vectors, contradicting $\mathcal{H}_\pi^N \cap (\mathcal{H}_\pi^N)^\perp = \{0\}$). So the restriction of π to $\mathcal{H}_\pi^N$ almost has invariant vectors; but this restriction factors through G/N, which has property (T). So there must be non-zero $\pi(G)$-invariant vectors in $\mathcal{H}_\pi^N$.

(ii) Follows from (i) and 1.4.(ii). $\qquad \square$

1.7. Definitions. Let G be a locally compact group, and let H be a closed subgroup.

a) H is *co-compact*, or *uniform*, in G if the homogeneous space G/H is compact;

b) H has *finite co-volume* in G if G/H carries a G−invariant probability measure;

c) H is a *lattice* in G if H is a discrete subgroup with finite co-volume in G.

Examples of lattices

1) For $n \geqslant 2$, the group $SL_n(\mathbb{Z})$ is a lattice in $SL_n(\mathbb{R})$; similarly, $Sp_{2n}(\mathbb{Z})$ is a lattice in $Sp_{2n}(\mathbb{R})$.

2) Let S be a compact Riemann surface of genus at least 2. It follows from uniformization theory that the fundamental group Γ of S is a co-compact lattice in $PSL_2(\mathbb{R})$.

3) More generally, let M be a n-dimensional compact *hyperbolic* manifold (i.e., a Riemannian manifold with constant sectional curvature -1). Then the fundamental group Γ of M is a co-compact lattice in the Lorentz group $SO_0(n, 1)$.

For a nice elementary introduction to the theory of lattices, see the recent paper [GrP] by Gromov and Pansu.

The fact that property (T) is inherited by lattices is a crucial fact:

1.8. Proposition. *Let H be a closed subgroup with finite co-volume in G. The following are equivalent:*

(i) *G has property (T);*

(ii) *H has property (T).*

Proof. (i)$\Longrightarrow$(ii) We shall use freely the properties of weak containment and induction of representations (see [Fel]). We just recall that, if σ is a representation of H, the induced representation $\mathrm{Ind}_H^G \sigma$ is the representation of G by left translations on the Hilbert space

$$\{ \xi : G \to \mathcal{H}_\sigma : \xi(gh) = \sigma(h^{-1})\xi(g) \text{ for } g \in G, h \in H; \int_{G/H} \|\xi(\dot{g})\|^2 d\dot{g} < \infty \}$$

Now, let π be a representation of H that weakly contains the trivial representation 1_H. Inducing up from H to G, the representation $\mathrm{Ind}_H^G \pi$ weakly contains $\mathrm{Ind}_H^G 1_H$. But the latter is nothing but the permutation representation of G on $L^2(G/H)$. Since H has finite co-volume, $\mathrm{Ind}_H^G 1_H$ contains the trivial representation 1_G. By transitivity of weak containment, $\mathrm{Ind}_H^G \pi$ weakly contains 1_G so that, by property (T), $\mathrm{Ind}_H^G \pi$ contains 1_G. This means that we can find a non-zero function $\eta : G \to \mathcal{H}_\pi$ such that, for $g, s \in G, h \in H$:

$$\begin{cases} \eta(gh) = \pi(h^{-1})\eta(g) \\ \eta(s^{-1}g) = \eta(g) \end{cases}$$

Set $\varsigma = \eta(1)$. Then $\varsigma \neq 0$ and $\pi(h)\varsigma = \varsigma$ for $h \in H$.

(ii)$\Longrightarrow$(i) See Corollaire 19, Chapitre 4 in [HaV]. $\qquad\square$

The following result makes possible a quantitative approach of property (T); we shall elaborate on that in chapter 4.

1.9. Proposition. *Let G be a compactly generated locally compact group; let K be a compact generating subset of G. The following are equivalent:*

(i) *G has property (T);*

(ii) *there exists $\varepsilon > 0$ such that any representation of G with (ε, K)-invariant vectors has non-zero invariant vectors.*

Proof. We prove the non-trivial implication $(i) \Longrightarrow (ii)$. Suppose by contradiction that there is no ε with the indicated property, i.e. we can find a sequence $(\pi_n)_{n \geqslant 1}$ of representations of G without non-zero invariant vectors and with $(\frac{1}{n^2}, K)$-invariant vectors. Set $K_n = (K \cup \{1\} \cup K^{-1})^n$; because K is generating, $(K_n)_{n \geqslant 1}$ is an increasing sequence of compact subsets such that $G = \bigcup_{n=1}^{\infty} K_n$. Let $\xi_n \in \mathcal{H}_{\pi_n}^1$ be such that

$$\max_{k \in K} \|\pi_n(k)\xi_n - \xi_n\| < \frac{1}{n^2}.$$

Then

$$\max_{k \in K_n} \|\pi_n(k)\xi_n - \xi_n\| < \frac{1}{n},$$

276

so that $\pi = \bigoplus\limits_{n=1}^{\infty} \pi_n$ almost has invariant vectors. By property (T), π must have non-zero invariant vectors, which means that some π_n must have non-zero invariant vectors; this is a contradiction. $\qquad\square$

1.10. Definition. The *dual* of G, denoted by $\hat{G}$, is the space of (unitary equivalence classes of) irreducible unitary representations of G.

There is a topology on $\hat{G}$, called the *Fell topology* (sometimes Fell-Jacobson topology), and based on the notion of weak containment: for $S \subset \hat{G}$, we say that $\pi \in \hat{G}$ is in the closure of S if π is weakly contained in S, i.e. any *normalized positive definite function* of π

$$\varphi_\xi : G \to \mathbb{C} : g \mapsto < \pi(g)\xi | \xi > \qquad (\xi \in \mathcal{H}_\pi^1)$$

can be approximated uniformly on compact sets of G by linear combinations of normalized positive definite functions associated with representations in S.

The relevance of this topology for property (T) is given by the next result.

1.11. Proposition. G *has property* (T) *if and only if the trivial representation* 1_G *is isolated in* $\hat{G}$.

Proof. To prove the direct implication, we have to see that, if $S \subseteq \hat{G}$ is such that $1_G \in cl\, S$, then $1_G \in S$. But $1_G \in cl\, S$ means that 1_G is weakly contained in $\bigoplus\limits_{\pi \in S} \pi$; by property (T) this implies that 1_G coincides with some $\pi \in S$.

For the converse implication, assume that 1_G is isolated in $\hat{G}$. Let P be the convex hull of 0 and all normalized positive definite functions on G associated with all representations of G. Endowed with the topology of uniform convergence on compact subsets of G, the space P is compact ([Dix], 13.5.2). Let P_0 be the closed convex hull of all extremal points in P which are distinct from the constant function 1. Our assumption exactly means that 1 does not belong to P_0. Denote by K the union of all segments joining 1 to a point in P_0.

We claim that $P = K$. To see it, we first notice that K is convex compact (because P_0 is), and that K contains all extremal points of P. The desired equality then follows from the Krein-Milman theorem.

To prove that G has property (T), consider a representation π of G that almost has invariant vectors. Then select a normalized positive definite function φ of π which belongs to $P - P_0$. By the preceding claim, we may write

$$\varphi = \lambda + (1 - \lambda)\psi$$

where $\lambda \in]0, 1]$ and $\psi \in P_0$. By Proposition 9 in Chapter 5 of [HaV], this decomposition of φ implies that π has non-zero fixed vectors. (This proof is based on suggestions by M. Burger and G. Valette; for a different proof based on decomposition theory, see [DeK]; it has to be pointed out hat the proof of this result given in Proposition 14 of Chapter 1 of [HaV] is false for non-discrete groups). $\qquad\square$

1.12. Definitions. Let $\mathbb{F}$ be a local field, and let $G(\mathbb{F})$ be the group of $\mathbb{F}$-rational points of a semi-simple algebraic group G defined over $\mathbb{F}$.

a) A *split torus* in $G(\mathbb{F})$ is an algebraic subgroup T which is $\mathbb{F}$-isomorphic to $(\mathbb{F}^\times)^n$ for some n, called the dimension of T;

b) The $\mathbb{F}$-*rank* is the dimension of a maximal split torus in $G(\mathbb{F})$ (any two maximal split tori are conjugate in $G(\mathbb{F})$);

c) G has *higher rank* if the $\mathbb{F}$-rank of G is at least 2.

For example, a maximal split torus in $SL_n(\mathbb{F})$ is the subgroup of diagonal matrices, which is clearly $\mathbb{F}$-isomorphic to $(\mathbb{F}^\times)^{n-1}$; so the $\mathbb{F}$-rank of $SL_n(\mathbb{F})$ is $n-1$.

Here is now Kazhdan's basic result.

1.13. Theorem. *Let $\mathbb{F}$ be a local field, and let $G(\mathbb{F})$ be the group of $\mathbb{F}$-rational points of a simple algebraic group G defined over $\mathbb{F}$. If G has higher rank, then $G(\mathbb{F})$ has property (T).*

In particular, $SL_n(\mathbb{F})$ has property (T) for $n \geqslant 3$. Combining 1.8 and 1.13, we immediately get:

1.14. Corollary. *Let $\mathbb{F}, G$ be as in 1.13. If G has higher rank and Γ is a lattice in $G(\mathbb{F})$, then Γ has property (T).*

Thus, $SL_n(\mathbb{Z})$ has property (T) for $n \geqslant 3$, because it is a lattice in $SL_n(\mathbb{R})$. At this juncture, we recall a result of Borel [Bo1]: if G is a semi-simple algebraic group defined over $\mathbb{R}$ and of $\mathbb{R}$-rank at least 1, then $G(\mathbb{R})$ contains both uniform and non-uniform lattices.

Combining 1.14 with 1.5, we see that lattices in higher rank groups are finitely generated. This was one of the motivations of Kazhdan for [Kaz], as it solves a very important case of Selberg's conjecture: any lattice Γ in a real semi-simple Lie group G is finitely generated. (This conjecture is now fully solved in the affirmative: see the remark following 3.0). The geometric meaning of Selberg's conjecture is the following: if we let Γ act on the Riemannian symmetric space X of G, then Γ admits a fundamental domain which is a finite polyhedron (possibly with cusps) in X. Remark that the finite generation issue is of relevance only for *non-uniform* lattices. That a co-compact lattice is finitely generated follows from 3.1.

At the end of this chapter, we shall sketch a proof of Theorem 1.13 for $SL_n(\mathbb{R}), n \geqslant 3$. For proofs of the general case, see [DeK], [HaV], [Ma2], [Wan], [Zim]. It has to be remarked that these proofs do not use any deep result from representation theory; only some standard facts on the structure of simple algebraic groups are used (e.g. root theory).

Again, let G be a simple algebraic group defined over the local field $\mathbb{F}$. If G has $\mathbb{F}$-rank 0, then $G(\mathbb{F})$ is compact, and so is Kazhdan for trivial reasons. Theorem 1.13 leaves open the possibility that $G(\mathbb{F})$ has property (T) when G has $\mathbb{F}$-rank 1. We discuss that case now.

1.15. Proposition. *Let* $\mathbb{F}$ *be a non-archimedean local field, and let* G *be a simple algebraic group defined over* $\mathbb{F}$, *of* $\mathbb{F}$-*rank 1 (e.g.* $G = SL_2$). *Then* $G(\mathbb{F})$ *does not have property* (T).

Proof. It can be proved that any action of a Kazhdan group on an oriented tree must have a fixed vertex (see Proposition 4 in Chapitre 6 of [HaV]). Now there exists an oriented tree (the Bruhat-Tits building of $G(\mathbb{F})$, see [BrT]) on which $G(\mathbb{F})$ acts without fixed vertex. □

When $\mathbb{F}$ is archimedean, the situation is more subtle.

For $\mathbb{F} = \mathbb{R}$, simple real Lie groups of $\mathbb{R}$-rank 1 are classified: up to local isomorphism, they fall in three infinite families and one exceptional group:

$$SO_0(n,1)\,(n \geq 2);\ SU(n,1)\,(n \geq 2);\ Sp(n,1)\,(n \geq 2);\ F_{4(-20)}$$

(the adjoint groups of these groups are the connected components of identity in the full isometry groups of n-dimensional hyperbolic spaces respectively over $\mathbb{R}, \mathbb{C}, \mathbb{H}$ and the Cayley octonions $\mathbb{O}$; in the last case, $n = 2$). For $\mathbb{F} = \mathbb{C}$, the group $G(\mathbb{F})$ is locally isomorphic to $SL_2(\mathbb{C})$, itself locally isomorphic to $SO_0(3,1)$.

1.16. Theorem. *i) For* $n \geq 2, SO_0(n,1)$ *and* $SU(n,1)$ *do not have property* (T);
ii) $Sp(n,1)\,(n \geq 2)$ *and* $F_{4(-20)}$ *have property* (T).

Proof. i) It is known that a Kazhdan group acting isometrically on the n-dimensional real or complex hyperbolic space must have a fixed point (Corollaire 23 in Chapitre 6 of [HaV]).

ii) This is a hard result of Kostant [Kos]. We sketch here a proof suggested by M. Cowling in his review of [HaV] (Maths Review 90m22001). This proof solves a problem from the final section of [HaV], namely to find a proof that $Sp(n,1)$ and $F_{4(-20)}$ have property (T) without appealing to the explicit description of the dual of these groups (as available in [BaB]).

Write G for $Sp(n,1)$ or $F_{4(-20)}$, and let $G = KAN$ be an Iwasawa decomposition (for fine structure of semi-simple Lie groups, we refer to Helgason's book [Hel]). Let $\mathfrak{g}, \mathfrak{n}, \mathfrak{a}$ be the Lie algebras of G, N, A respectively. Then

$$\mathfrak{n} = \mathfrak{g}_\alpha + \mathfrak{g}_{2\alpha}$$

where $\mathfrak{g}_\alpha, \mathfrak{g}_{2\alpha}$ are the positive root subspaces associated with the adjoint representation of $\mathfrak{a}$ in $\mathfrak{g}$. Write $2p = \dim_\mathbb{R} \mathfrak{g}_\alpha$, $q = \dim_\mathbb{R} \mathfrak{g}_{2\alpha}$. Define a positive definite quadratic form on $\mathfrak{n}$ by

$$|X + Y|^2 = -(2p + 4q)^{-1} B\left(\frac{X}{2} + \frac{Y}{4}, \frac{X}{2} + \frac{Y}{4}\right) \qquad (X \in \mathfrak{g}_\alpha, Y \in \mathfrak{g}_{2\alpha})$$

where B is the Killing form on $\mathfrak{g}$.

We shall use $\mathrm{Exp}\left(X + \frac{Y}{4}\right)$ (with $X \in \mathfrak{g}_\alpha$, $Y \in \mathfrak{g}_{2\alpha}$) as coordinates on N. We shall have to express these coordinates in the Cartan decomposition $G = KAK$. So fix $H_\alpha \in \mathfrak{a}$ such that $\alpha(H_\alpha) = 1$. Then, for $X \in \mathfrak{g}_\alpha$, $Y \in \mathfrak{g}_{2\alpha}$, one has

$$\mathrm{Exp}\left(X + \frac{Y}{4}\right) = k_1 \cdot \mathrm{Exp}(a\, H_\alpha) \cdot k_2$$

where $k_1, k_2 \in K$ and

$$2\,\mathrm{sh}\, a = (4|X|^2 + |X|^4 + |Y|^2)^{1/2}$$

(see [CoH], Proposition 2.1). As a consequence of this, one checks easily that the space $C^\infty(G/\!/K)|_N$ of restrictions to N of K-bi-invariant C^∞-functions on G coincides with the space of functions of the form $\mathrm{Exp}\left(X + \frac{Y}{4}\right) \longrightarrow f(4|X|^2 + |X|^4 + |Y|^2)$ where f is a C^∞-function on $\mathbb{R}$.

Let $B(N)$ be the Fourier-Stieltjes algebra of N, i.e. the space of all matrix coefficients $\varphi : n \mapsto \langle \pi(n)\xi|\eta\rangle$ of all unitary representations π of N. Then $B(N)$ is an algebra under pointwise multiplication, and it is a Banach algebra for the norm

$$\|\varphi\|_B = \inf\{\|\xi\|\|\eta\| : \varphi(n) = \langle \pi(n)\xi|\eta\rangle\} \ .$$

(Note that $\|\varphi\|_\infty \leqslant \|\varphi\|_B$). Using harmonic analysis on groups of Heisenberg type (this is a special class of 2-step nilpotent Lie groups, containing the groups N considered above), Cowling and Haagerup proved the following result ([CoH], Theorem 5.4).

Theorem. *Let $(u_k)_{k\geqslant 1}$ be a uniformly bounded sequence of K-bi-invariant C^∞-functions on G that converges to 1 uniformly on compact subsets of G. Let f_k be a C^∞-function on $\mathbb{R}$ such that*

$$u_k\left(\mathrm{Exp}\left(X + \frac{Y}{4}\right)\right) = f_k(4|X|^2 + |X|^4 + |Y|^2)$$

for $X \in \mathfrak{g}_\alpha$, $Y \in \mathfrak{g}_{2\alpha}$. Suppose that the following decay conditions at infinity are satisfied for all $k \geqslant 1$:
(i) $f_k^{(i)}(t) = o(t^{-i})$ for $0 \leqslant i \leqslant \frac{p}{2}$;
(ii) $\int_1^\infty |f_k^{(p/2)}(t)|\, t^{(p/2)-1}\,dt < \infty$.

Then

$$\liminf_{k\to\infty} \|u_k|_N\|_B \geqslant G(p,q) =: \frac{\pi^{1/2}\Gamma\left(\frac{p+q}{2}\right)}{\Gamma\left(\frac{q}{2}\right)\Gamma\left(\frac{p+1}{2}\right)} \ .$$

(Note that the decay conditions are empty if u_k has compact support).

For $G = Sp(n, 1)$, one has $p = 2(n-1)$, $q = 3$ and $G(p, q) = 2n-1$ (which is bigger than 1 provided $n \geqslant 2$); for $G = F_{4(-20)}$, one has $p = 4$, $q = 7$, and $G(p, q) = 21$.

We shall also need the spherical functions φ_s on G ($s \in \mathbb{C}$). We parametrize them in such a way that φ_s corresponds to the spherical principal series of representations, for $s \in i\mathbb{R}$, and φ_1 corresponds to the trivial representation i.e. $\varphi_1 \equiv 1$. We have $\varphi_s = \varphi_t$ if and only if $s = \pm t$ ([Hel], Theorem 4.3 of Chapter IV), so that we may assume $\operatorname{Re} s \geqslant 0$.

A function φ on G is *hermitian* if $\varphi(g^{-1}) = \overline{\varphi(g)}$ for all $g \in G$. One checks that φ_s is hermitian if and only if $s \in \mathbb{R} \cup i\mathbb{R}$ ([Val], Corollary 11). For $\operatorname{Re} s \geqslant 0$, the function $a \mapsto \varphi_s(\operatorname{Exp} aH_\alpha)$ has the following asymptotic behaviour, for $a \mapsto \infty$:

$$\varphi_s(\operatorname{Exp} aH_\alpha) \sim c(s)e^{(s-1)(p+q)a}$$

(see [Hel], Theorem 6.14 of Chapter IV). In particular, φ_s is bounded if and only if $|\operatorname{Re} s| \leqslant 1$.

If f_s is a C^∞-function on $\mathbb{R}$ such that $\varphi_s\left(\operatorname{Exp}\left(X + \frac{Y}{4}\right)\right) = f_s(4|X|^2 + |X|^4 + |Y|^2)$, one has then the following asymptotic behaviour, for $t \to \infty$:

$$f_s(t) \sim c(s)t^{\frac{(s-1)(p+q)}{2}} .$$

Suppose by contradiction that G does not have property (T). Then we find a sequence $(\varphi_{s_k})_{k \geqslant 1}$ of *positive definite* spherical functions on G that converges to 1 uniformly on compact subsets of G. Since φ_{s_k} is both hermitian and bounded by 1, we may assume $s_k \in i\mathbb{R} \cup [0, 1[$ for all $k \geqslant 1$.

It follows from the asymptotic behaviour given above that f_{s_k} satisfies the two decay conditions in the theorem of Cowling-Haagerup. Hence

$$\liminf_{k \to \infty} \|\varphi_{s_k}|_N\|_B \geqslant G(p, q) > 1 .$$

But this is absurd because, φ_{s_k} being positive definite, one has $\|\varphi_{s_k}|_N\|_B = 1$ for all $k \geqslant 1$. $\qquad\square$

It may be interesting to note that the theorem of Cowling and Haagerup is also valid for $SU(n, 1)$; but in that case, one has $p = n-1$, $q = 1$ and $G(p, q) = 1$, so that the above proof (fortunately!) fails.

The fact that $Sp(n, 1)$ has property (T) will play a crucial role in Chapter 3, in connection with genericity properties of discrete groups.

Let now G be a connected, semi-simple real Lie group. The adjoint group $G/Z(G)$ is a direct product of simple algebraic real Lie groups; by 1.6 (ii) and 1.16, $G/Z(G)$ has property (T) if and only if each of its simple factors has, if and only if no simple factor is locally isomorphic either to $SO_0(n, 1)$ or $SU(n, 1)$.

If the center $Z(G)$ is finite, then by 1.6(i) G and $G/Z(G)$ are simultaneously Kazhdan or not, so we can tell if G is Kazhdan just by looking at its simple factors. If $Z(G)$ is infinite, the answer will be provided by the following result, obtained by S.P. Wang (lemma 1.7 in [Wan]; see also [HaV], Chapter 2, Theorem 12).

1.17. Theorem. *let G be a compactly generated locally compact group. Suppose that:*

a) there is a closed, central subgroup Z in G such that G/Z has property (T);
b) $G/cl\,G'$ is compact.
Then G has property (T).

We shall not repeat the proof here. If G is again a semi-simple real Lie group, we know that G is perfect ($G = G'$) so that 1.17 states that G has property (T) if and only if $G/Z(G)$ has; in other words, for semi-simple real Lie groups, property (T) is invariant under local isomorphism. The classification is as follows:

1.18. Corollary. *Let G be a semi-simple real Lie group. G has property (T) if and only if no simple factor of G is locally isomorphic to $SO_0(n,1)$ or $SU(n,1)$.*

1.c. The case of $SL_n(\mathbb{R})$

We finish this chapter by sketching a proof that $SL_n(\mathbb{R})$ is Kazhdan for $n \geqslant 3$. We shall outline the proof given in the recent book of Howe and Tan [HoT], based on a careful study of the decay at infinity of coefficient of representations of $SL_n(\mathbb{R})$ without non-zero fixed vector. It allows a quantitative proof of property (T) for $SL_n(\mathbb{R})$.

We first introduce some notation. First, let $K = SO(n)$ be the standard maximal compact subgroup of $SL_n(\mathbb{R})$, and let A be the diagonal subgroup, i.e.

$$A = \left\{ \begin{pmatrix} a_1 & & O \\ & \ddots & \\ O & & a_n \end{pmatrix} : a_i > 0 \text{ for } i = 1, \ldots, n, \ a_1 a_2 \ldots a_n = 1 \right\}.$$

Set

$$A^+ = \left\{ \begin{pmatrix} a_1 & & O \\ & \ddots & \\ O & & a_n \end{pmatrix} \in A : a_1 \geqslant a_2 \geqslant \ldots \geqslant a_n \right\}.$$

From the polar decomposition in $SL_n(\mathbb{R})$ and the fact that a symmetric matrix is diagonalizable, we deduce easily the Cartan, or KAK, decomposition:

$$SL_n(\mathbb{R}) = K A^+ K.$$

This means that, to study the behaviour at infinity of coefficients, we may restrict to A^+.

The first step is to collect information about the left regular representation of $SL_2(\mathbb{R})$. For $a \geqslant 1$, set

$$\tilde{a} = \begin{pmatrix} a & 0 \\ 0 & a^{-1} \end{pmatrix} \in A^+$$

The *Harish-Chandra* Ξ-*function* is the function

$$\Xi : A^+ \to \mathbb{R}_+ : \tilde{a} \mapsto \frac{1}{a} \int_0^{2\pi} \left(\frac{\cos^2 \theta}{a^4} + \sin^2 \theta \right)^{-\frac{1}{2}} d\theta$$

(Ξ is in fact the restriction to A^+ of the spherical function associated with the bottom of the spherical principal series).

1.19. Proposition. *Let λ be the left regular representation of $SL_2(\mathbb{R})$. For ξ, η unit K-finite vectors in $L^2(SL_2(\mathbb{R}))$, one has, for $\tilde{a} \in A^+$:*

$$|< \lambda(\tilde{a})\xi|\eta >| \leqslant (\dim < \lambda(K)\xi > \cdot \dim < \lambda(K)\eta >)^{1/2} \Xi(\tilde{a})$$

(where $< \lambda(K)\xi >$ denotes the linear span of the orbit $\lambda(K)\xi$, and dim denotes dimension of a complex vector space).

Proof. We follow the proof of Theorem 3.2.1 of Chapter 5 in [HoT], and proceed in several steps.

1) Since K is abelian, we way write ξ as a finite sum of K-eigenvectors:

$$\xi = \sum_{k \in I \subset \mathbb{Z}} \xi_k .$$

Suppose that we can prove the estimate for K-eigenvectors. Then from

$$\sum_{k \in I} \|\xi_k\|_2 \leqslant |I|^{\frac{1}{2}} \left(\sum_{k \in I} \|\xi_k\|_2^2 \right)^{1/2} = (\dim\langle\lambda(K)\xi\rangle)^{1/2} \|\xi\|_2$$

we also have the estimate for ξ. So we may assume that ξ, η are K-eigenvectors.

2) From

$$|\langle\lambda(\tilde{a})\xi|\eta\rangle| = |\int_G \xi(\tilde{a}^{-1}x)\overline{\eta(x)}dx| \leqslant \int_G |\xi|(\tilde{a}^{-1}x)|\eta|(x)dx$$

we may assume that ξ, η are non-negative and K-fixed.

3) Let N be the standard unipotent subgroup of $G = SL_2(\mathbb{R})$:

$$N = \left\{ \begin{pmatrix} 1 & n \\ 0 & 1 \end{pmatrix} : n \in \mathbb{R} \right\} .$$

We shall appeal to the Iwasawa decomposition $G = KAN$ and to the associated integral formula ([HoT], Proposition 1.1.11 in Chapter 5).

$$\langle\lambda(\tilde{a})\xi|\eta\rangle = \int_K \int_A \int_N \xi(\tilde{a}^{-1}\tilde{kbn})\eta(\tilde{kbn}) \, dk \, b \, db \, dn$$

$$\leqslant \int_K dk \left[\int_A \int_N \xi^2(\tilde{a}^{-1}\tilde{kbn})b \, db \, dn \right]^{1/2} \left[\int_A \int_N \eta^2(\tilde{kbn})b \, db \, dn \right]^{1/2} .$$

By K-invariance of η, the second factor is just $\|\eta\|_2 = 1$. Hence

$$\langle \lambda(\tilde{a})\xi|\eta \rangle \leqslant \int_K dk \left[\int_A \int_N \xi^2(a^{-1}k\,\tilde{b}n)\,b\,db\,dn \right]^{1/2} .$$

Write now $\tilde{a}^{-1}k = k_1\,\tilde{b}_1\,n_1$, so that

$$\tilde{a}^{-1}k\,\tilde{b}n = k_1\,\tilde{b}_1\,n_1\,\tilde{b}n = k\,\tilde{b}_1\,\tilde{b}(\tilde{b}^{-1}\,n_1\,\tilde{b}n)$$

and the term in braces belongs to N, since A normalizes N.

Then, using K-invariance of ξ:

$$\int_A \int_N \xi^2(a^{-1}k\,\tilde{b}n)bdb\,dn = \int_A \int_N \xi^2(\tilde{b}_1\,\tilde{b}(\tilde{b}^{-1}n_1\,\tilde{b})n)bdb\,dn$$

$$= \int_A \int_N \xi^2(\tilde{b}_1\,\tilde{b}n)bdb\,db = b_1^{-2}\int_A \int_N \xi^2(\tilde{b}n)bdb\,dn .$$

By K-invariance of ξ, the last double integral is just $\|\xi\|_2^2 = 1$.

Finally:

$$\langle \lambda(\tilde{a})\xi|\eta \rangle \leqslant \int_K b_1^{-1}\,dk .$$

An easy computation then shows that $\int_K b_1^{-1}\,dk = \Xi(\tilde{a})$. $\qquad\qquad \square$

Like in most other proofs of property (T) for $SL_n(\mathbb{R})$, we now have to say something about representations of the semi-direct product $\mathbb{R}^2 \rtimes SL_2(\mathbb{R})$.

1.20. Proposition. *Let π be a representation of $\mathbb{R}^2 \rtimes SL_2(\mathbb{R})$ without non-zero $\mathbb{R}^2$-invariant vectors. There exists a dense subspace $\mathcal{K}$ of $\mathcal{H}_\pi$ such that, for any $\xi, \eta \in \mathcal{K}, \tilde{a} \in A$ and $k_1, k_2 \in K$:*

$$| < \pi(k_1\tilde{a}k_2)\xi|\eta > | \leqslant c_{\xi,\eta}a^{-1}$$

where $c_{\xi,\eta}$ is a constant depending only on ξ and η.

Proof. This result is really the most crucial in the proof of property (T) for $SL_n(\mathbb{R})$, $n \geqslant 3$. We follow [HoT], Theorem 3.3.1 of Chapter 5.

We consider $\pi|_{\mathbb{R}^2}$, the restriction of π to $\mathbb{R}^2$; it has a *spectral measure*, i.e. a projection-valued measure

$$P : \{\text{Borel subsets of } \hat{\mathbb{R}}^2\} \ \rightarrow \ \{ \text{ Projections on } \mathcal{H}_\pi \} : X \mapsto P_X .$$

Since π has no non-zero $\mathbb{R}^2$-invariant vector, we have

$$P_{\{0\}} = 0 ;$$

284

since π is a representation of $\mathbb{R}^2 \rtimes SL_2(\mathbb{R})$, we have the equivariance relation

$$\pi(g) P_X \pi(g^{-1}) = P_{g(X)}$$

for any $g \in SL_2(\mathbb{R})$ and any Borel subset X in $\hat{\mathbb{R}}^2$. For $s > 0$, set

$$X_s = \left\{ x \in \hat{\mathbb{R}}^2 : \frac{1}{s} \leqslant |x| \leqslant s \right\} .$$

Since $\hat{\mathbb{R}}^2 - \{0\} = \bigcup_{s>0} X_s$, we see that P_{X_s} converges strongly to 1, for $s \to \infty$. Denote by $\mathcal{H}_{\pi,K}$ the space of K-finite vectors in $\mathcal{H}_\pi$, and set

$$\mathcal{K} = \left[\bigcup_{s>0} P_{X_s}(\mathcal{H}_\pi) \right] \cap \mathcal{H}_{\pi,K} .$$

Because X_s is K-invariant, so is each P_{X_s}, which implies that $\mathcal{K}$ is dense is $\mathcal{H}_\pi$. By bilinearity of coefficients, we may assume that ξ, η are K-eigenvectors of norm 1 in $P_{X_s}(\mathcal{H}_\pi)$. That is, if

$$R_\theta = \begin{pmatrix} \cos\theta & -\sin\theta \\ \sin\theta & \cos\theta \end{pmatrix}$$

is a rotation in K, we may assume

$$\pi(R_\theta)\xi = e^{ik\theta}\xi \qquad \text{for } k \in \mathbb{Z} .$$

Fix $\theta = \frac{2\pi}{n}$ ($n \in \mathbb{N} - \{0\}$). Let S_θ be a fundamental domain for the action of R_θ on $\hat{\mathbb{R}}^2$-$\{0\}$; we assume that S_θ is a sector centered on the x-axis. Because of the equivariance relation, we have for any $R_\psi \in K$:

$$\pi(R_\psi) P_{S_\theta} \xi = P_{R_\psi(S_\theta)} \pi(R_\psi)\xi = e^{ik\theta} P_{R_\psi(S_\theta)}\xi$$

so that

$$\|P_{S_\theta}\xi\| = \|P_{R_\psi(S_\theta)}\xi\| \qquad \text{for any } \psi .$$

Take now $\psi_j = \frac{2\pi j}{n}$ ($j = 0, 1, \ldots, n-1$). Since we have an orthogonal decomposition $1 = \sum_{j=0}^{n-1} P_{R_{\psi_j}(S_\theta)}$, we deduce

$$\|P_{S_\theta}\xi\|^2 = \frac{1}{n} .$$

Then

$$\langle \pi(\tilde{a})\xi|\eta \rangle = \langle P_{X_s} \pi(\tilde{a}) P_{X_s}\xi | P_{X_s}\eta \rangle = \langle P_{X_s} P_{\tilde{a}(X_s)} \pi(\tilde{a})\xi | P_{X_s}\eta \rangle$$
$$= \langle P_{X_s} \pi(\tilde{a})\xi | P_{\tilde{a}(X_s)} P_{X_s}\eta \rangle = \langle \pi(\tilde{a}) P_{\tilde{a}^{-1}(X_s)}\xi | P_{\tilde{a}(X_s) \cap X_s}\eta \rangle$$
$$= \langle \pi(\tilde{a}) P_{\tilde{a}^{-1}(X_s) \cap X_s}\xi | P_{\tilde{a}(X_s) \cap X_s}\eta \rangle ;$$

285

therefore

$$|\langle \pi(\tilde{a})\xi | \eta \rangle| \leqslant \|P_{\tilde{a}^{-1}(X_s) \cap X_s} \xi\| \, \|P_{\tilde{a}(X_s) \cap X_s} \eta\| \, .$$

Fix $a \geqslant 1$. Then $\tilde{a}(X_s)$ is contained in the horizontal strip

$$\left\{ (x, y) \in \hat{\mathbb{R}}^2 : |y| \leqslant \frac{s}{a} \right\}$$

and $\tilde{a}(X_s) \cap X_s$ is contained in the truncated strip

$$\left\{ (x, y) \in \hat{\mathbb{R}}^2 : x^2 + y^2 \geqslant \frac{1}{s^2} \, , \, |y| \leqslant \frac{s}{a} \right\} \, .$$

Define n as the integer part of $\dfrac{\pi}{\arcsin \frac{s^2}{a}}$, and $\theta = \frac{2\pi}{n}$. Then $\tilde{a}(X_s) \cap X_s$ is contained in $S_\theta \cup R_\pi(S_\theta)$, and $\tilde{a}^{-1}(X_s) \cap X_s$ is contained in $R_{\pi/2}(S_\theta) \cup R_{-(\pi/2)}(S_\theta)$ (the reader is strongly encouraged to make a drawing). For a big, the 4 sectors S_θ, $R_{\pi/2}(S_\theta)$, $R_\pi(S_\theta)$, $R_{-(\pi/2)}(S_\theta)$ are disjoint, so we have

$$|\langle \pi(\tilde{a})\xi | \eta \rangle| \leqslant 4 \|P_{S_\theta} \xi\| \, \|P_{S_\theta} \eta\| = \frac{4}{n} \, .$$

This means that, for a large, $|\langle \pi(\tilde{a})\xi | \eta \rangle|$ is dominated by $\left(\frac{4s^2}{\pi} \right) \cdot \frac{1}{a}$. $\qquad \square$

1.21. Corollary. *Let π be as in 1.20. Then, for any $\varepsilon > c$ and $\xi, \eta \in \mathcal{K}$:*

$$\int_{SL_2(\mathbb{R})} | < \pi(g)\xi | \eta > |^{2+\varepsilon} dg < \infty.$$

Proof. We appeal to the integral formula for the KAK decomposition ([HoT], 1.1.11 in Chapter 5).

$$\int_{SL_2(\mathbb{R})} | < \pi(g)\xi | \eta > |^{2+\varepsilon} dg =$$

$$= \int_K \int_{A+} \int_K | < \pi(k_1 \tilde{a} k_2)\xi | \eta > |^{2+\varepsilon} dk_1 \left(a^2 - \frac{1}{a^2} \right) \frac{da}{a} dk_2$$

$$\leqslant c_{\xi,\eta}^{2+\varepsilon} \int_1^\infty \frac{1}{a^{2+\varepsilon}} \left(a^2 - \frac{1}{a^2} \right) \frac{da}{a} \qquad \text{by 1.20}$$

$$\leqslant c_{\xi,\eta}^{2+\varepsilon} \int_1^\infty \frac{da}{a^{1+\varepsilon}} = \frac{c_{\xi,\eta}^{2+\varepsilon}}{\varepsilon}$$

$$\qquad \qquad \qquad \qquad \qquad \qquad \qquad \qquad \qquad \qquad \qquad \square$$

The next result exploits the fact that $SL_2(\mathbb{R})$ is *not* Kazhdan.

1.22. Corollary. *Let π be as in 1.20. Then, for $\xi, \eta \in \mathcal{H}_\pi$:*

$$| < \pi(\tilde{a})\xi|\eta > | \leqslant (\dim < \pi(K)\xi > \cdot < \pi(K)\eta >)^{1/2} \cdot \Xi(\tilde{a}) \ \text{for all } \tilde{a} \in A^+ .$$

Proof. (sketch) The basic trick, due to Cowling [Cow], is to tensor $\pi|_{SL_2(\mathbb{R})}$ with a complementary series π_s of $SL_2(\mathbb{R})$ $(0 < s < 1)$. It is known that coefficients of π_s decay on A like a^{s-1}; thus, by 1.20, coefficients of $\pi|_{SL_2(\mathbb{R})} \otimes \pi_s$ decay on A^+ like a^{s-2}; the same computation as in 1.21 shows that $\pi|_{SL_2(\mathbb{R})} \otimes \pi_s$ has L^2-coefficients, so is contained in a multiple of the left regular representation of $SL_2(\mathbb{R})$.

So the estimate of 1.19 applies to $\pi|_{SL_2(\mathbb{R})} \otimes \pi_s$; in particular if φ_s is the spherical function associated with π_s, we have

$$| < \pi(\tilde{a})\xi|\eta > | \cdot |\varphi_s(\tilde{a})| \leqslant (\dim < \pi(K)\xi > \cdot \dim < \pi(K)\eta >)^{1/2} \cdot \Xi(\tilde{a})$$

for $\tilde{a} \in A^+$. We now let s tend to 1, and use the fact that π_s converges to the trivial representation, i.e. φ_s converges to the constant 1, to get the result. For complete details of the proof, see [HoT], Theorem 3.3.10 in Chapter 5. □

The next step is to pass to $SL_n(\mathbb{R}), n \geqslant 3$. Set

$$\tilde{a} = \begin{pmatrix} a_1 & & 0 \\ & \ddots & \\ 0 & & a_n \end{pmatrix} \in A^+$$

and, for $k_1, k_2 \in K = SO(n)$:

$$\Psi(k_1 \tilde{a} k_2) = \min_{i \neq j} \Xi \left(\sqrt{\widetilde{\frac{a_i}{a_j}}} \right)$$

so that Ψ is a K-bi-invariant function on $SL_n(\mathbb{R})$. □

1.23. Proposition. *Let π be a representation of $SL_n(\mathbb{R}), n \geqslant 3$, without non-zero fixed vector. Then, for any $\xi, \eta \in \mathcal{H}_\pi^1, \tilde{a} \in A^+$:*

$$| < \pi(\tilde{a})\xi|\eta > | \leqslant (\dim < \pi(K)\xi > \cdot \dim < \pi(K)\eta >)^{1/2} . \Psi(\tilde{a})$$

Proof. (sketch) We embed $H = \mathbb{R}^2 \rtimes SL_2(\mathbb{R})$ in $SL_n(\mathbb{R})$ by

$$\left(\begin{pmatrix} x \\ y \end{pmatrix}, \begin{pmatrix} a & b \\ c & d \end{pmatrix} \right) \mapsto \begin{pmatrix} a & b & x & 0 \\ c & d & y & 0 \\ 0 & 0 & 1 & 0 \\ 0 & 0 & 0 & 1_{n-3} \end{pmatrix}.$$

By the Howe-Moore vanishing theorem ([HoM], Theorem 5.2), coefficients of π vanish at infinity on $SL_n(\mathbb{R})$, so the same holds for the restriction $\pi|_H$; in particular $\pi|_H$ has no non-zero $\mathbb{R}^2$-invariant vector. Now write

$$
\tilde{a} = \begin{pmatrix} \alpha & & & & 0 \\ & \alpha^{-1} & & & \\ & & 1 & & \\ & & & \ddots & \\ 0 & & & & 1 \end{pmatrix} \begin{pmatrix} \beta & & & & 0 \\ & \beta & & & \\ & & a_3 & & \\ & & & \ddots & \\ 0 & & & & a_n \end{pmatrix}
$$

where $\alpha = \sqrt{\frac{a_1}{a_2}}$ and $\beta = \sqrt{a_1 a_2}$. By 1.22, we have

$$| < \pi(\tilde{a})\xi | \eta > |$$

$$
\leqslant \left[\dim < \pi \begin{pmatrix} SO(2) & 0 \\ 0 & 1_{n-2} \end{pmatrix} \pi \begin{pmatrix} \beta & & & & 0 \\ & \beta & & & \\ & & a_3 & & \\ & & & \ddots & \\ 0 & & & & a_n \end{pmatrix} \xi > \cdot \dim < \pi \begin{pmatrix} SO(2) & 0 \\ 0 & 1_{n-2} \end{pmatrix} \eta > \right]^{\frac{1}{2}} \cdot \Xi(\tilde{\alpha})
$$

$$
= \left[\dim < \pi \begin{pmatrix} \beta & & & & 0 \\ & \beta & & & \\ & & a_3 & & \\ & & & \ddots & \\ 0 & & & & a_n \end{pmatrix} \pi \begin{pmatrix} SO(2) & 0 \\ 0 & 1_{n-2} \end{pmatrix} \xi > \cdot \dim < \pi \begin{pmatrix} SO(2) & 0 \\ 0 & 1_{n-2} \end{pmatrix} \eta > \right]^{\frac{1}{2}} \cdot \Xi(\tilde{\alpha})
$$

$$
= \left[\dim < \pi \begin{pmatrix} SO(2) & 0 \\ 0 & 1_{n-2} \end{pmatrix} \xi > \cdot \dim < \pi \begin{pmatrix} SO(2) & 0 \\ 0 & 1_{n-2} \end{pmatrix} \eta > \right]^{\frac{1}{2}} \cdot \Xi(\tilde{\alpha})
$$

$$
\leqslant [\dim < \pi(K)\xi > \cdot \dim < \pi(K)\eta >]^{\frac{1}{2}} \cdot \Xi(\tilde{\alpha}).
$$

Taking various embeddings of $\mathbb{R}^2 \rtimes SL_2(\mathbb{R})$ in $SL_n(\mathbb{R})$, one gets the result; see Theorem 3.3.12 in [HoT] for details. □

Notice now that, for $\delta \in]0,1[$, the set $\Psi^{-1}[\delta, 1]$ contains K and is a compact generating subset of $SL_n(\mathbb{R})$. Also, the function $[0,1] \to \mathbb{R}_+ : \varepsilon \mapsto 4 \sin \left(\frac{\arcsin \varepsilon}{2} \right) + \varepsilon$ is strictly increasing, so that the equation $4 \sin \left(\frac{\arcsin \varepsilon}{2} \right) + \varepsilon = \sqrt{2(1-\delta)}$ has a unique solution (necessarily $\varepsilon < 1$).

We are now reaching our goal.

1.24. Theorem. Set $G = SL_n(\mathbb{R}), n \geqslant 3$. Fix $\delta \in]0,1[$, and let ε be the unique solution of $4 \sin(\frac{\arcsin \varepsilon}{2}) + \varepsilon = \sqrt{2(1-\delta)}$. Set $C = \Psi^{-1}[\delta, 1]$. Any representation of G with (ε, C)-invariant vectors has non-zero invariant vectors. In particular G is Kazhdan.

Proof. (inspired by [HoT], Theorem 4.1.1). Let π be a representation of G that has an (ε, C)-invariant vector $\xi \in \mathcal{H}_\pi^1$. Set $\eta = \int_K \pi(k)\xi \, dk$. Since $K \subseteq C$, we have

$$\|\xi - \eta\| \leqslant \int_K \|\pi(k)\xi - \xi\| \, dk < \varepsilon.$$

Hence $\|\eta\| > 1 - \varepsilon > 0$, so that $\eta \neq 0$. It follows then from elementary trigonometry that

$$\left\|\xi - \frac{\eta}{\|\eta\|}\right\| \leqslant 2\sin\left(\frac{\arcsin\|\xi - \eta\|}{2}\right) < 2\sin\left(\frac{\arcsin\varepsilon}{2}\right).$$

Suppose by contradiction that π has no non-zero invariant vector. Then, by 1.23, we have for all $g \in G$:

$$Re\left\langle \pi(g)\frac{\eta}{\|\eta\|} \,\Big|\, \frac{\eta}{\|\eta\|} \right\rangle \leqslant \Psi(g)$$

because η is K-fixed. Then

$$\left\|\pi(g)\frac{\eta}{\|\eta\|} - \frac{\eta}{\|\eta\|}\right\| \geqslant \sqrt{2}(1 - \Psi(g))^{\frac{1}{2}}$$

Take now $g \in C$ such that $\Psi(g) = \delta$. Then

$$\varepsilon > \|\pi(g)\xi - \xi\| \geqslant \left\|\pi(g)\frac{\eta}{\|\eta\|} - \frac{\eta}{\|\eta\|}\right\| - 2\left\|\xi - \frac{\eta}{\|\eta\|}\right\|$$

$$> \sqrt{2(1 - \delta)} - 4\sin\left(\frac{\arcsin\varepsilon}{2}\right)$$

which contradicts the definition of ε. $\qquad\square$

2. A RESULT OF S. P. WANG
ON NON-SEMI-SIMPLE LIE GROUPS WITH PROPERTY (T).

In the first chapter, we have given the complete list of those semi-simple real Lie groups with property (T) (see Corollary 1.18).

Here, we shall deal with a general connected real Lie group G and ask when G is Kazhdan. The complete characterization of those $G's$ is a little-known result of S.P. Wang(1) [Wan].

Before stating the result, we recall some basic facts from structure theory of Lie groups (see e.g. [Hoc]). Indeed G has a decomposition

$$G = RS$$

where R is the *radical* of G, i.e. the unique maximal solvable normal subgroup of G, and S is a maximal analytic semi-simple subgroup (R is always closed, S not always). Now we have a factorization in commuting factors $S = S_c S_n$ where S_c is the maximal compact normal subgroup of S, and S_n is the maximal semi–simple normal subgroup of S without compact factors; $S_c \cap S_n$ is a finite central subgroup of S.

Here is now Wang's result ([Wan], Theorem 1.9).

---

(1) Maybe because Wang gives absolutely no example this result attracted the attention of none of the experts who wrote about property (T). This criticism especially points at the authors of [HaV]; had we known about Wang result, it is clear that our chapter 2 would have been different.

2.0. Theorem. *Let G be a connected, real Lie group, with*

$$G = RS_cS_n$$

as above. The following are equivalent:
  (i) G has property (T);
  (ii) RS_n has property (T);
  (iii) S_n has property (T) and $R \cap cl \, S_n[S_n, R]$ is co-compact in R (where $S_n[S_n, R]$ is the smallest subgroup of G containing $\{x \, y \, r \, y^{-1}r^{-1} : x, y \in S_n, \ r \in R\}$).

The implications (i) $\Rightarrow$ (ii) $\Rightarrow$ (iii) are farly easy, while (iii) $\Rightarrow$ (i) is hard: Wang's proof rests on some deep representation-theoretic results of Moore. We shall not give his proof here. Instead, we shall give a new, simplified proof in the case of groups with abelian radical. Our proof will exploit the relative property (T), that we now define.

2.1. Definition. Let G be a locally compact group, and let H be a closed subgroup. The pair (G, H) has *property (T)* if any representation of G that almost has invariant vectors has non-zero H-invariant vectors.

This relative property was introduced by Margulis in [Ma1]; he used it to solve an outstanding problem in the theory of graphs and communication networks; we shall come back on this in Chapter 5.

The following lemma makes Proposition 1.6(i) more precise:

2.2. Lemma. *Let $1 \to N \to G \to G/N \to 1$ be a short exact sequence of locally compact groups. The following are equivalent:*

  (i) G has property (T);
  (ii) G/N has property (T), and the pair (G, N) has property (T).

The proof of this lemma is easy, and left to the reader.

We now turn to semi-direct products of semi-simple real Lie groups with finite-dimensional real vector spaces. So let $S = S_cS_n$ be a *non-compact* semi-simple real Lie group; and let V be a finite-dimensional real vector space; fix a continuous homomorphism $\rho : S \to GL(V)$. We denote by V_n the space of $\rho(S_n)$-fixed points in V. We are interested in the semi-direct product $V \rtimes_\rho S$, especially when the pair $(V \rtimes_\rho S, V)$ has property (T).

2.3. Proposition. *We keep the above notations. The following are equivalent:*

  (i) the pair $(V \rtimes_\rho S, V)$ has property (T);
  (ii) there is no S-invariant probability measure on the projective space $\mathbb{P}(V^*)$ deduced from the vector space V^* dual of V.

(iii) there exists a finite family $(\xi_i)_{i \in I}$ in V such that each orbit $\rho(S)\xi_i$ is a punctured cone, and $\bigcup_{i \in I} \rho(S)\xi_i$ generates V;

(iv) $V_n = \{0\}$.

Proof. We prove the proposition following the scheme (i) $\Longrightarrow$ (iv) $\Longrightarrow$ (ii) $\Longrightarrow$ (i) and (iii) $\Longleftrightarrow$ (iv).

(i) $\Longrightarrow$ (iv). Assume that $(V \rtimes_\rho S, V)$ has property (T).

From the fact that S_n is normal in S, it follows that V_n is an S-invariant subspace. Because every finite dimensional representation of S is completely reducible, we find W, an S-invariant complement to V_n in V. Let then $P : V \to V_n$ be the projection on the first factor in the decomposition $V = V_n \oplus W$.

Since P commutes with $\rho(S)$, and the action of S on V_n factors through S/S_n, we have an homomorphism

$$\alpha : V \rtimes_\rho S \to V_n \rtimes (S/S_n) : (v, s) \longmapsto (P(v), sS_n)$$

which is onto. Composing the left regular representation of $V_n \rtimes (S/S_n)$ with α, we get a representation of $V \rtimes_\rho S$ that almost has invariant vectors, because $V_n \rtimes (S/S_n)$ is amenable. Now by the assumption, this representation must have non-zero V-invariant vectors, i.e. non-zero L^2-functions on $V_n \rtimes (S/S_n)$ that are constant on cosets of V_n. This forces $V_n = \{0\}$

(iv) $\Longrightarrow$ (ii). We endow V^* with the contragredient representation ρ^* of ρ.

We first claim that $V_n^* = \{0\}$. Indeed, if not, we find $f \in V_n^* - \{0\}$; then Ker f is a $\rho(S_n)$-invariant hyperplane in V. By complete reducibility of ρ, we find a 1-dimensional $\rho(S_n)$-invariant complement of Ker f; because there is no non-trivial homomorphism $S_n \to \mathbb{R}^\times$, any element in this complement must be in V_n, contradicting the assumption $V_n = \{0\}$.

The claim has the following consequence: if E^* is a non-zero $\rho^*(S)$-invariant subspace in V^*, then composing $\rho^*|_{E^*}$ with the canonical map $GL(E^*) \to PGL(E^*)$, we get a homomorphism $S \to PGL(E^*)$ with not relatively compact range.

We now assume by contradiction that there exists an S-invariant probability measure μ on $\mathbb{P}(V^*)$. Let E^* be a non-zero subspace of V^*, such that E^* is $\rho^*(S)$-invariant, $\mu(\mathbb{P}(E^*)) > 0$, and E^* is minimal for these 2 properties. The probability measure $\mu/\mu(\mathbb{P}(E^*))$ is then S-invariant on $\mathbb{P}(E^*)$; since S is not relatively compact in $PGL(E)$, a famous result of Furstenberg ([Zim], Cor. 3.2.2) ensures the existence of a proper, non-zero subspace F^* of E^*, such that $\mu(\mathbb{P}(F^*)) > 0$, and invariant under a finite index subgroup of $\rho^*|_{E^*}(S)$. Since $\rho^*|_{E^*}(S)$ has no proper subgroup of finite index, we see that F^* must be $\rho^*(S)$-invariant, and this contradicts minimality of E^*.

(ii) $\Longrightarrow$ (i). This is a result of Burger, valid under the assumption "S locally compact" (Proposition 7 in [Bu2]); see also the Proposition in n. 9 (V) of Chapter 2 in [HaV].

(iii) $\Longrightarrow$ (iv). Assume $V_n \neq \{0\}$. As above, let W be a $\rho(S)$-invariant complement of V_n in V. We observe that orbits of $\rho(S)$ in V_n are compact. So, if $\xi \in V$ is such that 0 is in the closure of $\rho(S)\xi$, then ξ lies in W. Thus, for any family $(\xi_i)_{i \in I}$ of vectors in V whose orbits are punctured cones, the subgroup generated by $\bigcup_{i \in I} \rho(S)\xi_i$ is contained in W.

(iv) $\Longrightarrow$ (iii). We assume $V_n = \{0\}$. By complete reducibility of finite-dimensional representations of S, we may assume that ρ is irreducible. Now, let $S = KAN$ be an Iwasawa decomposition of S. It is known (see [Mos], 2.6) that $\rho(A)$ is contained in a maximal split torus of $GL(V)$ (i.e. a conjugate of the subgroup of diagonal matrices). In particular, we find $\xi \in V - \{0\}$ and a non-trivial homomorphism $\alpha : A \to \mathbb{R}_+^\times$ such that

$$\rho(a)\xi = \alpha(a)\xi \quad \text{for any } a \in A.$$

We claim that the orbit $\rho(S)\xi$ is a punctured cone. Indeed, for $\lambda \in \mathbb{R}_+^\times$, we find $a \in A$ such that $\alpha(a) = \lambda$, hence for $s \in S$:

$$\lambda\rho(s)\xi = \rho(s)\alpha(a)\xi = \rho(sa)\xi.$$

Because ρ is irreducible, ξ is a cyclic vector, so that the orbit $\rho(S)\xi$ generates V linearly. $\qquad\square$

From that proposition, we immediately deduce the criterion for a semi-direct product $V \rtimes_\rho S$ to be Kazhdan.

2.4. Proposition. *We keep the above notations. The following are equivalent.*

(i) $V \rtimes_\rho S$ has property (T).
(ii) S has property (T) and $V_n = \{0\}$.

Proof. Combine Lemma 2.2 with Proposition 2.3. For a different proof of this result, see Theorem 0.1 in [Gui]. $\qquad\square$

2.5. Examples. Take $S = SL_n(\mathbb{R})$ with $n \geqslant 3$. Let $\rho : S \to GL_N(\mathbb{R})$ be a representation on $\mathbb{R}^N$; it follows from Proposition 2.4 that $\mathbb{R}^N \rtimes_\rho SL_n(\mathbb{R})$ has property (T) if and only if ρ has no non-zero fixed vector (compare with Proposition 10 in Chapitre 2 of [HaV], where this was only proved for ρ the direct sum of d copies of the standard representation, $1 \leqslant d < n$).

On the other hand, it follows from Corollaire 7.13 in [Bo2] that $\rho(SL_n(\mathbb{Z}))$ preserves at least one lattice in $\mathbb{R}^N$. So, up to a change of basis, we may assume

$$\rho(SL_n(\mathbb{Z})) \subseteq SL_N(\mathbb{Z})$$

and $\mathbb{Z}^N \rtimes_\rho SL_n(\mathbb{Z})$ is a lattice in $\mathbb{R}^N \rtimes_\rho SL_n(\mathbb{R})$. Thus $\mathbb{Z}^N \rtimes_\rho SL_n(\mathbb{Z})$ is Kazhdan if and only if ρ has no non-zero fixed vector. This is to say that Proposition 2.4 also provides many new examples of discrete groups with property (T).

We now treat the general case of connected Lie groups with abelian radical. It is essentially a matter of behaviour of property (T) under extensions, since any Lie group with abelian radical has a quotient of the form $V \rtimes_\rho S$, considered in 2.3 and 2.4.

Let us fix some notation. For $G = RS$ a connected Lie group with $S = S_c S_n$ as above, we denote by R_n the group of fixed points of S_n in R, i.e. $R_n = \{r \in R : s\,r\,s^{-1} = r$ for all $s \in S_n\}$. Now $R \cap S$ is a discrete subgroup of S, which is normal in S, hence central in S (by connectedness of S). As a consequence

$$R \cap S \subseteq R_n.$$

We recall from [Hoc] that R has a unique maximal compact torus T; it is thus necessarily normal in G, but more is true: since Aut T is a discrete group and S is connected, the map $S \to$ Aut $T : s \mapsto (t \to sts^{-1})$ is constant, i.e. T is central in G; in particular T is contained in R_n.

2.6. Proposition. *Let $G = RS$ be a connected Lie group with abelian radical. The following are equivalent:*
    (i) G has property (T);
    (ii) S has property (T) and $cl(R \cap S)$ is co-compact in R_n.

Proof. (i) $\Longrightarrow$ (ii). If G has property (T), then so has $G/R \simeq S/R \cap S$; since, for semi-simple Lie groups, property (T) is invariant under local isomorphism, S has property (T) as well. Let us prove that $cl(R \cap S)$ is co-compact in R_n. As in the proof of the implication (i) $\Longrightarrow$ (iv) in 2.3 we construct an S-equivariant homomorphism $P : R \to R_n$ which is onto (to construct P, we work at the Lie algebra level and exponentiate; here is used the fact that R is abelian).

It is an easy check that the map

$$\alpha : G \to G : rs \mapsto P(r)s$$

is a *well-defined* homomorphism with range $R_n S$; in particular, $R_n S$ is closed since

$$R_n S = \{g \in G : g = \alpha(g)\}$$

This implies that the closure $cl\,S_n$ is contained in $R_n S$. Since S_n centralizes R_n, we see that $cl\,S_n$ is normal in $R_n S$, so we have a quotient group $R_n S/cl\,S_n$ which is clearly amenable, and has property (T) as a quotient of G. Thus $R_n S/cl\,S_n$ is compact. But $R_n S/cl\,S_n = R_n S_c/cl\,S_n \cap R_n S_c$ so that $cl\,S_n \cap R_n S_c$ is co-compact in $R_n S_c$. This implies that $cl\,S_n \cap R_n$ is co-compact in R_n; a fortiori $cl(R \cap S)$ is co-compact in R_n.

(ii) $\Longrightarrow$ (i). We assume that S has property (T), and that $cl(R \cap S)$ is co-compact in R_n. Note that $cl(R \cap S)$ is central in G.

1st *step:* $G/cl(R \cap S)$ has property (T). Indeed, the radical of $G/cl(R \cap S)$ is $R/cl(R \cap S)$. Since $R_n/cl(R \cap S)$ is a compact torus, it follows from the remark preceding 2.6 that $R_n/cl(R \cap S)$ is the maximal compact torus of $R/cl(R \cap S)$. To prove that $G/cl(R \cap S)$ has property (T), it is therefore enough to prove that $G/cl(R \cap S) \big/ R_n/cl(R \cap S.) \simeq G/R_n$ has (T).

Now G/R_n is a semi-direct product of a vector space by a semi-simple Lie group; indeed, Ker P is a vector space and since Ker $P \cap S = \{1\}$, we have $G/R_n \simeq$ Ker $P \rtimes_\rho (S/R \cap S)$, where ρ denotes the restriction to Ker P of the action of S on R by conjugation. Since $S/R \cap S$ has no non-trivial fixed point in Ker P, it follows from 2.4 that G/R_n is Kazhdan.

2nd *step:* G/clG' is compact.

Indeed, on the one hand, G' contains S (semi-simple Lie groups are perfect); on the other hand, G' contains the subgroup H generated by the commutators $rsr^{-1}s^{-1}$ with $r \in$ Ker P and $s \in S$. Clearly $H \subseteq$ Ker P and actually $H =$ Ker P (otherwise, we could find a non-trivial $f \in ($Ker $P)^*$ taking the value 1 on H; such an f would be S-invariant, a fortiori S_n-invariant. But we know from the proof of the implication (iv) $\implies$ (ii) in 2.3 that S_n has no non-trivial fixed point on (Ker $P)^*$. Since $cl(R \cap S)$ is co-compact in R_n, the subgroup $cl($Ker $P \cdot S) = cl(HS)$ is co-compact in G; this subgroup is contained in clG', so that clG' is itself co-compact.

3rd *step:* Since $G/cl(R \cap S)$ has property (T) and G/clG' is compact, Theorem 1.17 applies to conclude that G has property (T). □

2.7. Examples.

a) Let H be a semi-simple Lie group with property (T) and centre isomorphic to $\mathbb{Z}$ (e.g. $H = \widetilde{SU(2,2)}$). Fix an isomorphism $\alpha : Z(H) \to \mathbb{Z}$. Consider now the group

$$G = H \times \mathbb{R}/Z$$

where $Z = \{(z, \alpha(z)) : z \in Z(H)\}$.

In this example, $R \cap S \simeq \mathbb{Z}$ is a lattice in $R_n = R \simeq \mathbb{R}$. It follows from 2.6 that G is Kazhdan; thus G is a reductive group with centre $\mathbb{R}$ and property (T). (These examples of Kazhdan groups with centre $\mathbb{R}$ are much easier than the ones presented in n.14 of Chapitre 2 in [HaV]).

b) Let H be a semi-simple Lie group with property (T) and centre isomorphic to $\mathbb{Z}^2$ (e.g. $S = \widetilde{SU(2,2)} \times \widetilde{SU(2,2)}$). Let θ be an irrational number, and let α be an isomorphism $Z(H) \to \mathbb{Z} + \theta\mathbb{Z}$. Consider the group

$$G = H \times \mathbb{R}/Z$$

where $Z = \{(z, \alpha(z)) : z \in Z(H)\}$.

294

In this example, $R \cap S \simeq \mathbb{Z} + \theta\mathbb{Z}$ is dense in $R_n = R \simeq \mathbb{R}$. Again, it follows from 2.6 that G is a reductive group with centre $\mathbb{R}$ and property (T), the basic difference between this example and the preceding one is that here S is not closed in G; actually S is dense.

We conclude this chapter by treating some Lie groups with non–abelian radical.

Next lemma arose in a conversation between M. Picardello and the author, on the subject of Proposition 2.3.

2.8. Lemma. *Let G be a locally compact group. Suppose that there is a decomposition $G = NH$ in closed subgroups N, H, with $N \lhd G$ and H Kazhdan. Suppose moreover that there exists a family $(\mathcal{O}_i)_{i \in I}$ of H-orbits in N such that:*
(i) for all $i \in I$, one has $1 \in cl\, \mathcal{O}_i$;
(ii) N is the closed subgroup generated by $\bigcup_{i \in I} \mathcal{O}_i$.
Then G is Kazhdan.

Proof. Let π be a representation of G that almost has invariant vectors. Since H has property (T), we find an H-fixed vector $\xi \in \mathcal{H}_\pi^1$. We want to show that ξ is N-fixed as well. But, for $h \in H$, $n \in N$, we have:

$$\langle \pi(n)\xi|\xi\rangle = \langle \pi(n)\pi(h)\xi|\pi(h)\xi\rangle = \langle \pi(h^{-1}nh)\xi|\xi\rangle \ ,$$

so that the coefficient function $g \mapsto \langle \pi(g)\xi|\xi\rangle$ is constant on each $\mathcal{O}_i$. By assumption (i), this coefficient is 1 on each $\mathcal{O}_i$, meaning that ξ is fixed by each $\mathcal{O}_i$. By assumption (ii), ξ is N-fixed. $\square$

Next result was obtained in a slightly less general form (and with a completely different proof) in Theorem 0.4 of [Gui].

2.9. Proposition. *Let $G = RS$ be a connected Lie group. Denote by $\mathfrak{R}$ the Lie algebra of R, and by $Ad : S \to GL(\mathfrak{R})$ the adjoint action of S on R. If the semi-direct product $\mathfrak{R} \rtimes_{Ad} S$ has property (T), then so does G.*

Proof. Suppose that $\mathfrak{R} \rtimes_{Ad} S$ has property (T). Then by 2.3 and 2.4, S has property (T) and there exists a family $X_1, \ldots, X_k$ in $\mathfrak{R}$ such that $\bigcup_{i=1}^{k} Ad(S)X_i$ generates $\mathfrak{R}$ linearly, and each orbit $Ad(S)X_i$ is a punctured cone. Let $\mathrm{Exp} : \mathfrak{R} \to R$ be the exponential map, and set $x_i = \mathrm{Exp}\, X_i$, for $i = 1, \ldots, k$. Denote by $\mathcal{O}_i$ the orbit of x_i in R under the adjoint action of $cl\, S$. Since Exp is equivariant with respect to adjoint actions of S, we see that $1 \in cl\, \mathcal{O}_i$ for each i. Finally, let $Y_1, \ldots, Y_n$ be a basis of $\mathfrak{R}$ contained in $\bigcup_{i=1}^{k} Ad(S)X_i$. By [Hoc], Proposition 4.1 in Chapter 7, there is a neighbourhood U of 1 in R such that any element of U may be written

$$\mathrm{Exp}\, t_1\, Y_1 \cdot \mathrm{Exp}\, t_2\, Y_2 \cdots \mathrm{Exp}\, t_n\, Y_n$$

with $|t_j|$ small enough. This means that the subgroup generated by $\bigcup_{i=1}^{k} \mathcal{O}_i$ contains U, hence coincides with R. We conclude by appealing to lemma 2.8, with $N = R$ and $H = cl\, S$. $\square$

2.10. Example. Let G be a simple non–compact Lie group with property (T). We denote by $\mathfrak{g}$ the Lie algebra of G, and endow $\mathfrak{g} \oplus \mathfrak{g}$ with the following multiplication rule:

$$(X,Y)(X',Y') = (X + X', Y + Y' + [X,X']) .$$

It is readily seen that, in this way, $\mathfrak{g} \oplus \mathfrak{g}$ becomes a 2-step nilpotent Lie group, which we denote by $N(\mathfrak{g})$. Clearly, G acts by automorphisms on $N(\mathfrak{g})$, so we may form the semi-direct product $N(\mathfrak{g}) \rtimes G$. We now apply 2.9 to show that $N(\mathfrak{g}) \rtimes G$ has property (T). Indeed, the Lie algebra of $N(\mathfrak{g})$ is just $\mathfrak{g} \oplus \mathfrak{g}$ with the Lie bracket

$$[(X,Y),(X',Y')] = (0,[X,X'])$$

and the action of G on $\mathfrak{g} \oplus \mathfrak{g}$ is given by two copies of the adjoint representation Ad of G on $\mathfrak{g}$. It follows from 2.4 that $(\mathfrak{g} \oplus \mathfrak{g}) \rtimes_{Ad} G$ has property (T).

Our final example shows that the converse of Proposition 2.9 is not true in general.

2.11. Definition. Let ω be a symplectic form on $\mathbb{R}^{2n}$ (i.e. a non-degenerate, alternating, bilinear form). The $(2n+1)$-*dimensional Heisenberg group* H_{2n+1} is the space $\mathbb{R}^{2n} \times \mathbb{R}$ endowed with the following group structure:

$$(x,s)(y,t) = (x + y, s + t + \omega(x,y)) \text{ for } x,y \in \mathbb{R}^{2n}; s,t \in \mathbb{R}.$$

H_{2n+1} is a 2-step nilpotent connected Lie group. We may take for instance the standard symplectic form

$$\omega((x_1,\cdots,x_{2n}),(y_1,\cdots,y_{2n})) = \sum_{i=1}^{n} x_i y_{i+n} - \sum_{i=1}^{n} x_{i+n} y_i$$

It is clear from the definition that the symplectic group $Sp_{2n}(\mathbb{R})$ acts on H_{2n+1} via automorphisms, so we may form the semi-direct products $H_{2n+1} \rtimes Sp_{2n}(\mathbb{R})$ and (Lie $H_{2n+1}) \rtimes Sp_{2n}(\mathbb{R})$. We observe that, by 2.4, the second one never has property (T), since $Sp_{2n}(\mathbb{R})$ fixes the centre of Lie H_{2n+1}.

2.12. Proposition. $H_{2n+1} \rtimes Sp_{2n}(\mathbb{R})$ *has property* (T) *if and only if* $n \geqslant 2$.

Proof. For $n = 1$, the group $H_3 \rtimes Sp_2(\mathbb{R})$ has $Sp_2(\mathbb{R}) \simeq SL_2(\mathbb{R})$ as a quotient, so it cannot have property (T).

Assume now $n \geqslant 2$. The centre $Z(H_{2n+1})$ is isomorphic to $\mathbb{R}$, and it is fixed pointwise under the action of $Sp_{2n}(\mathbb{R})$. So it remains central in $H_{2n+1} \rtimes Sp_{2n}(\mathbb{R})$, which appears thus as a central extension:

$$0 \to \mathbb{R} \xrightarrow{\cdot} H_{2n+1} \rtimes Sp_{2n}(\mathbb{R}) \to \mathbb{R}^{2n} \rtimes Sp_{2n}(\mathbb{R}) \to 1.$$

Now $\mathbb{R}^{2n} \rtimes Sp_{2n}(\mathbb{R})$ has property (T), by Proposition 2.4. Moreover, $\mathbb{R}^{2n} \rtimes Sp_{2n}(\mathbb{R})$ is perfect, and since $H'_{2n+1} = Z(H_{2n+1})$, we have that $H_{2n+1} \rtimes Sp_{2n}(\mathbb{R})$ is perfect as well. Again by 1.17, we see that $H_{2n+1} \rtimes Sp_{2n}(\mathbb{R})$ is Kazhdan. $\qquad \square$

Let us take for ω the standard symplectic form on $\mathbb{R}^{2n}$; it takes integer values on $\mathbb{Z}^{2n}$, so that
$$H_{2n+1}(\mathbb{Z}) = \big\{(x,s) \in H_{2n+1} : x \in \mathbb{Z}^{2n}, \ s \in \mathbb{Z}\big\}$$
is a subgroup of H_{2n+1}, called the *discrete Heisenberg group;* $H_{2n+1}(\mathbb{Z})$ is clearly invariant under the action of $Sp_{2n}(\mathbb{Z})$, and the semi-direct product $H_{2n+1}(\mathbb{Z}) \rtimes Sp_{2n}(\mathbb{Z})$ is a lattice in $H_{2n+1} \rtimes Sp_{2n}(\mathbb{R})$; so we see from 2.12 that $H_{2n+1}(\mathbb{Z}) \rtimes Sp_{2n}(\mathbb{Z})$ is Kazhdan if and only if $n \geqslant 2$.

3. Genericity of Discrete Kazhdan Groups

3.a. Results of Gromov, Delzant, Champetier

All the examples of infinite discrete Kazhdan groups that we have seen up to now were essentially lattices in products of semisimple Lie groups over local fields. That this yields only countably many examples (up to isomorphism) is a consequence of the following Lemma.

3.0. Lemma. *Let $G = G_1(k_1) \times \cdots \times G_n(k_n)$ be a finite product of simple isotropic algebraic groups over local fields. If G has property (T), then G contains countably many lattices, up to isomorphism.*

Proof. It is enough to see that G contains at most countably many irreducible lattices. Assume that either $n \geqslant 2$ or $n = 1$ and rank $G_1 \geqslant 2$. In either case, rank $G \geqslant 2$ so that Margulis' arithmeticity theorem applies to show that any irreducible lattice in G is arithmetic ([Ma2], Theorem 1.11 of Chapter IX); the result then follows from the fact that there are countably many arithmetic constructions.

Assume now that $n = 1$ and rank $G = 1$; it follows from 1.15 and 1.16 that we must have $k_1 = \mathbb{R}$ and either $G = Sp(n,1)$ or $G = F_{4(-20)}$. Now, by the Gromov-Schoen arithmeticity theorem [GrS], any lattice in G is arithmetic, hence the same argument as above applies. $\qquad \square$

Remark. If all fields k_i are archimedean, i.e. G is a semi-simple real Lie group, Lemma 3.1 is true without the property (T) assumption; indeed, in that case, any irreducible lattice in G is finitely presentable, and there are only countably many finitely presentable groups (for finite presentation of lattices in G, see [Rag], Corollary 13.20, in the case where at least one G_i is of rank 1, and [BGS], § 13 and Appendix 2, for the general case). A similar argument, based on finite presentability of lattices, probably works when the $k_i's$ are merely assumed to be of characteristic 0 (it certainly works when all $k_i's$ are non-archimedean, because by Proposition 3.7 in Chapter IX of [Ma2] any lattice in G is then co-compact, hence finitely presented). But the argument breaks down in the case where G is a simple split algebraic group over a local field k of positive characteristic; if G has rank 1 (e.g $G = SL_2(k)$), any non-uniform lattice in G is not finitely generated, a fortiori not finitely presented ([Lu1], Corollary 4); in the rank 2 case, $SL_3(\mathbb{F}_q[t])$ is a lattice in $SL_3(\mathbb{F}_q((t)))$, therefore $SL_3(\mathbb{F}_q[t])$ is finitely generated (e.g. because of property (T)) but it is known that $SL_3(\mathbb{F}_q[t])$ is not finitely presented ([Beh], main result); this was proved using the action of $SL_3(\mathbb{F}_q[t])$ on the associated $\tilde{A}_2$-building. Note that the existence of not finitely presented discrete Kazhdan group answers in the negative a question in (the russian version of) Kazhdan's original paper [Kaz]. In opposition, it is known [ReS] that, for $n \geqslant 4$, the group $SL_n(\mathbb{F}_q[t])$ is finitely presented.

Since we alluded several times to the fact that a co–compact lattice has to be finitely generated, we now present a proof of this fact.

Definitions. Let (E, d) be a metric space.
    1) E is *proper* if closed balls in E are compact.
    2) E is *geodesic* if, for any $c > 0$, any two points x, y at distance c in E may be joined by an isometric embedding of the interval $[0, c]$ in $\mathbb{R}$.

3.1. Proposition. *Let Γ be a discrete group. Assume that there exists a proper geodesic metric space (E, d) on which Γ acts isometrically, properly, and with compact quotient. Then Γ is finitely generated.*

Proof. (taken from [GhH], n. 19 in Chapter 3). The quotient $\Gamma \backslash E$ is a compact metric space, so it has finite diameter $R > 0$. Fix a base-point x_0 in E, and denote by B the closed ball with radius R centered at x_0. Set

$$S = \{\gamma \in \Gamma : \gamma \neq 1 \text{ and } \gamma B \cap B \text{ is non–empty}\} .$$

Clearly $S = S^{-1}$. Because E is proper and Γ acts properly, the set S is finite. We claim that S generates Γ. Set

$$r = \inf\{d(B, \gamma B) : \gamma \in \Gamma - (S \cup \{1\})\} .$$

We first show that $r > 0$. To see this, set

$$T = \{\gamma \in \Gamma - (S \cup \{1\}) : d(x_0, \gamma x_0) \leqslant 4R\} ;$$

since the action is proper, T is a finite set. Define then

$$r' = \min\{d(B, \gamma B) : \gamma \in T\} .$$

Clearly $r' > 0$. Suppose that $r < R$, and find a sequence $(\gamma_n)_{n \geqslant 1}$ in Γ such that $R > d(\gamma_n B, B)$ for all $n \geqslant 1$ and

$$\lim_{n \to 1} d(\gamma_n B, B) = r .$$

298

Then $d(x_0, \gamma_n x_0) < 3R$ for all $n \geqslant 1$, i.e. $\gamma_n \in T$. This means that $r = r' > 0$.
Now, we observe that r has the following property.

Fix $\sigma \in \Gamma$. If there exists $y \in B, z \in \sigma B$ such that $d(y, z) < r$,
then $\sigma \in S \cup \{1\}$. $\hspace{6cm}$ (*)

Finally, we prove that any element γ of Γ is a product of elements in S. So, let k be the smallest integer such that $d(x_0, \gamma x_0) < kr + R$. We then choose a geodesic from x_0 to γx_0 and subdivide it by points $x_0, x_1, x_2, \ldots, x_k, x_{k+1} = \gamma x_0$ such that $d(x_0, x_1) \leqslant R$ and $d(x_i, x_{i+1}) < r$ for $i = 1, 2, \ldots, k$. Since $E = \Gamma B$, we find, for $i = 1, 2, \ldots, k+1$, an element $\gamma_i \in \Gamma$ such that $x_i \in \gamma_i B$. We may take $\gamma_1 = 1$ and $\gamma_{k+1} = \gamma$. We then set $\sigma_i = \gamma_i^{-1} \gamma_{i+1}$ (for $(i = 1, 2, \ldots, k)$, so that $\gamma = \sigma_1 \sigma_2 \ldots \sigma_k$. Now we notice that $\gamma_i^{-1} x_i \in B$ and that $\gamma_i^{-1} x_{i+1} = \sigma_i \gamma_{i+1}^{-1} x_{i+1} \in \sigma_i B$; moreover $d(\gamma_i^{-1} x_i, \gamma_i^{-1} x_{i+1}) < r$. From property (*), this implies that $\sigma_i \in S \cup \{1\}$, and the proof is complete. $\hspace{5cm}$ $\square$

It came as a surprise when Gromov ([Gro], Corollary 4.4 E) constructed uncountably many discrete Kazhdan groups out of his theory of hyperbolic metric spaces and groups. Since we shall need part of this theory, we recall now the basic definitions and results.

3.2. Definition.
A geodesic metric space (X, d) is δ-hyperbolic, for $\delta \geqslant 0$, if for any geodesic triangle abc in X and any $x \in [a, b]$, the distance from x to $[b, c] \cup [c, a]$ is $\leqslant \delta$. The geodesic metric space (X, d) is hyperbolic if it is δ-hyperbolic for some $\delta \geqslant 0$.

Basic examples of hyperbolic metric spaces are trees on the one hand (here $\delta = 0$), and on the other hand the Riemannian symmetric spaces of non-compact type and rank 1, i.e. the classical hyperbolic spaces over $\mathbb{R}, \mathbb{C}, \mathbb{H}$ or the Cayley numbers (see [GhH], Chapitre 3, Corollaire 10).
If now Γ is a finitely generated group, with a finite symmetric generating subset $S = S^{-1}$, the Cayley graph $\mathcal{G}(\Gamma, S)$ becomes a geodesic metric space when replacing each edge by an isometric copy of the unit interval [0,1].

3.3. Definitions.
(i) The group Γ is hyperbolic if $\mathcal{G}(\Gamma, S)$ is hyperbolic, for some finite symmetric generating subset S of Γ.
(ii) An hyperbolic group Γ is non-elementary if it has no cyclic subgroup of finite index.
(iii) An hyperbolic group has cyclic centralizers if all abelian subgroups are cyclic.

It is an important fact that hyperbolicity is a genuine property of Γ, i.e. it is independent of the choice of S (the constant δ may depend on S however). A useful criterion for hyperbolicity is the following: if Γ acts properly, with compact quotient, on a hyperbolic metric space, then Γ is itself hyperbolic (see [GhH], Chapitre 5,

Theorem 12). Therefore, any group acting properly on a tree with finite quotient graph is hyperbolic; similarly, so is any co-compact lattice in the isometry group of a Riemannian symmetric space of non-compact type and rank 1. It is also interesting to notice that an hyperbolic group is non-elementary if and only if it contains a non-abelian free subgroup ([GhH], Chapitre 8, Théorème 37).

Now, here is Gromov's result.

3.4. Theorem. *Let Γ be a torsion-free non-elementary hyperbolic group. There exists a sequence $(m_j)_{j\geqslant 1}$ of positive integers having the property that, for any sequence $(n_j)_{j\geqslant 1}$ with $n_j \geqslant m_j$ for all $j \geqslant 1$, there exists an infinite quotient Δ of Γ and subgroups Δ_j of Δ such that:*

(a) Δ_j is cyclic of order n_j for $j \geqslant 1$;
(b) $\gamma\Delta_j\gamma^1 \cap \Delta_k = \{1\}$ for all $\gamma \in \Delta$ and $j \neq k$;
(c) any proper subgroup of Δ is conjugate within Δ to a subgroup of some Δ_j.

If, in the above theorem, we take for Γ a co-compact lattice in $Sp(n,1)$ or $F_{4(-20)}$ we get a hyperbolic group with property (T), which therefore has uncountably many pairwise non-isomorphic quotients which are infinite torsion groups with property (T); in particular, these groups are non-amenable and do not contain any non-abelian free subgroup, and so provide a very strong negative answer to the famous question of von Neumann [vNe] whether non-amenability of a group is always due to the presence of a non-abelian free subgroup. (Other examples of non-amenable torsion groups were previously given by Olshanskii [Ols]).

The question of von Neumann is still open for finitely presented groups. Note that there are only countably many finitely presented groups. So, by a trivial cardinality argument, 3.4 gives the existence of uncountably many Kazhdan groups which are not finitely presented.

Theorem 3.4 also exhibits an interesting difference between lattices in $Sp(n,1)$ or $F_{4(-20)}$ and lattices in higher rank groups. Indeed, for lattices Γ in higher rank, we have Margulis' normal subgroup theorem (see [Zim], [Ma2]) telling us that any normal subgroup of Γ is either of finite index, or contained in the (finite) centre of Γ; in particular, since there are only finitely many homomorphisms from a finitely generated group onto a finite group of given order, Γ has only countably many pairwise non-isomorphic quotients. (Note however that lattices in $Sp(n,1)$ or $F_{4(-20)}$ share many properties with higher rank lattices, like property (T), superrigidity, arithmeticity,...).

An interesting property of Gromov's groups is their simplicity; this was noticed in [Ch1], from which the following proof is taken.

3.5. Lemma. *Any infinite group Δ satisfying conditions (a), (b), (c) of Theorem 3.4 is simple.*

Proof. We first show that any proper normal subgroup N of Δ is central. Indeed, since N is finite, its centralizer $Z_\Delta(N) = \{\gamma \in \Delta : \gamma n \gamma^{-1} = n$ for all $n \in N\}$ is of finite index in Δ. (This is best seen by viewing $Z_\Delta(N)$ as the kernel of the homomorphism $\Delta \to Aut(N)$ obtained by letting Δ act on N by inner automorphisms). But since Δ has no proper subgroup of finite index, we must have $Z_\Delta(N) = \Delta$, i.e. N is central.

The proof will be finished by showing that Δ has trivial centre. So, suppose by contradiction that there exists a non-trivial $x \in Z(\Delta)$. Then the subgroup $< x >$ generated by x must be contained in a unique Δ_n. Take then a non-trivial g in Δ_m, with $m \neq n$. Because x is central, the subgroup $< x, g >$ generated by x and g is finite, so it is contained in some $\gamma \Delta_k \gamma^{-1}$; by uniqueness of Δ_n, we have $k = n$, but then $g \in \gamma \Delta_n \gamma^{-1} \cap \Delta_m$, and we have reached a contradiction. $\qquad \square$

The study of quotients of non-elementary hyperbolic groups, initiated by Gromov (see section 4.5 of [Gro]), was pursued by Delzant ([Del], Theoreme 3.5) who proved the following result, generalizing the result of Higman-Neumann-Neumann that any countable group can be embedded in a 2-generator group (see e.g. [LyS]).

3.6. Theorem. *Let Γ be a non-elementary hyperbolic group, and let A be a countable group. There exists a quotient group of Γ that contains A as a subgroup.*

In particular, any countable group may be embedded in a countable group with property (T); this answers positively question 4 in the final section of [HaV].

Delzant's result is a first sign that discrete Kazhdan groups might be "generic" in a suitable sense. We now turn to Champetier's results, where the last sentence is made precise, and genericity is meant in the sense of Baire category.

3.7. Definitions. (a) A *marked group* is a pair (Γ, S) where Γ is a finitely generated group and S is a (totally ordered) finite generating subset of Γ.

(b) If $(\Gamma_1, S_1), (\Gamma_2, S_2)$ are marked groups, a *marked isomorphism* between (Γ_1, S_1) and (Γ_2, S_2) is given by a group isomorphism $\alpha : \Gamma_1 \to \Gamma_2$ such that $\alpha_{|S_1}$ is an increasing bijection from S_1 to S_2.

If $\mathbb{F}_m$ denotes the free non-abelian group on the generators $a_1, \ldots, a_m$, we see that there is a canonical bijection between quotients of $\mathbb{F}_m$ and marked isomorphism classes of marked groups (Γ, S) with $|S| = m$. If N is a normal subgroup of $\mathbb{F}_m$, the quotient $\mathbb{F}_m/N$ is marked by the images of $a_1, \ldots, a_m$ in $\mathbb{F}_m/N$.

Fix a integer $m \geqslant 1$, and denote by $\mathcal{G}_m$ the set of marked isomorphism classes of marked groups (Γ, S) with $|S| = m$. Following Grigorchuk [Gri], we are going to equip $\mathcal{G}_m$ with the structure of a compact metric space, which will allow us to state genericity results in the sense of Baire category.

We first define a distance an $\mathcal{G}_m$. According to the preceding remark, it is enough to define it on the set of all normal subgroups of $\mathbb{F}_m$. For $n \in \mathbb{N}$, let $B(n)$ be the ball of radius n centered at the identity of $\mathbb{F}_m$, for the word metric associated with $a_1, \ldots, a_m$. Then, for N_1, N_2 normal subgroups of $\mathbb{F}_m$, we set:

$$v(N_1, N_2) = \sup\{n \in \mathbb{N} : N_1 \cap B(n) = N_2 \cap B(n)\}.$$

Note that $v(N_1, N_2) \in \mathbb{N} \cup \{\infty\}$, with $v(N_1, N_2) = \infty$ if and only if $N_1 = N_2$. Moreover, for normal subgroups N_1, N_2, N_3 one has:

$$v(N_1, N_3) \geqslant \min\{v(N_1, N_2), v(N_2, N_3)\}.$$

Setting now

$$d(N_1, N_2) = e^{-v(N_1, N_2)}$$

we obtain a distance on $\mathcal{G}_m$ that satisfies the ultrametric inequality

$$d(N_1, N_3) \leqslant \max\{d(N_1, N_2), d(N_2, N_3)\}.$$

An easy but important property of $\mathcal{G}_m$ is that, if (Γ, S) is a *finitely presented* marked group, there is a neighbourhood of (Γ, S) in $\mathcal{G}_m$ that consists of quotients of Γ.

3.8. Lemma. *Equipped with the metric d, the space $\mathcal{G}_m$ is compact.*

Proof. We show that any sequence $(N_k)_{k \geqslant 1}$ of normal subgroups in $\mathbb{F}_m$ has a cluster point. By repeated application of the pigeon-hole principle and repeated extraction of subsequences, we find an increasing family $(I_n)_{n \geqslant 1}$ of finite subsets of $\mathbb{F}_m$ such that $I_n = B(n) \cap N_k$ for infinitely many k's. Set $N = \bigcup_{n=1}^{\infty} I_n$; an easy check shows that N is a normal subgroup of $\mathbb{F}_n$, so that N is a cluster point of $(N_k)_{k \geqslant 1}$. $\qquad \square$

An important ingredient for what follows is the

Approximation Theorem. (Gromov, [Gro], 5.5.A; Champetier, [Cha], 4.19). *Let $\Gamma_1, \ldots, \Gamma_n$ be non-elementary hyperbolic groups with cyclic centralizers; let N be a positive integer. There exists a non-elementary hyperbolic group Γ with cyclic centralizers such that*

1) there exists homomorphisms $\Gamma_i \to \Gamma$ which are onto;

2) via these homomorphisms, the balls of radius N in Γ_i are mapped injectively in Γ.

Moreover Γ can be taken on two generators.

The meaning of this theorem for $\mathcal{G}_m$ is the following. Suppose that $\Gamma_1, \ldots, \Gamma_n$ belong to $\mathcal{G}_m$; let S_i be the generating subset of Γ_i, with $|S_i| = m$, appearing implicitly in the theorem; let $\overline{S_i}$ be the image of S_i in Γ. Then the approximation theorem says that the marked groups (Γ_i, S_i) and $(\Gamma, \overline{S_i})$ are e^{-N}–close in $\mathcal{G}_m$. It is then natural to introduce the equivalence relation $\mathcal{R}$ on $\mathcal{G}_m$ given by abstract isomorphism of groups. So the approximation theorem can be rephrased by saying that, if $\Gamma_1, \ldots, \Gamma_n$ are marked non–elementary hyperbolic groups with cyclic centralizers in $\mathcal{G}_m$, then for any $\varepsilon > 0$ there exists a marked non–elementary hyperbolic group Γ with cyclic centralizers in $\mathcal{G}_m$ such that $\Gamma_1, \ldots, \Gamma_n$ are ε–close to the equivalence class of Γ in $\mathcal{G}_m$.

Here is now Champetier's remarkable genericity result ([Cha], Théorème 1.3).

3.9. Theorem. *Fix $m \geqslant 2$. Denote by $\mathcal{H}_m$ the closure in $\mathcal{G}_m$ of the set of marked groups which are non-elementary hyperbolic, and with cyclic centralizers(1). Then $\mathcal{H}_m$ contains a dense G_δ of marked groups (Γ, S), each having the following properties:*

(a) the equivalence class of (Γ, S) is dense in $\mathcal{H}_m$;

(b) Γ is an infinite torsion group;

(c) Γ contains all finite groups with cyclic centralizers;

(d) Γ has property (T);

(e) Γ is a 2-generator group;

(f) Γ is perfect, i.e. Γ is equal to its commutator subgroup.

Sketch of proof. Since the intersection of a finite number of dense G_δ's is still a dense G_δ, it is enough to prove genericity for each of the properties appearing in 3.9.

a) This is really the core of the result. We say that an equivalence class of $\mathcal{R}$ is e^{-n}-*dense* in $\mathcal{H}_m$ if every point in $\mathcal{H}_m$ is within distance e^{-n} of this equivalence class. We denote by X_n the union of all e^{-n}-dense equivalence classes in $\mathcal{H}_m$, and proceed in several steps.

First, we check that X_n is non-empty. Using compactness of $\mathcal{H}_m$, we may cover $\mathcal{H}_m$ with a finite number k of balls of radius e^{-n}, whose centers $\Gamma_1, \ldots, \Gamma_k$ are marked non-elementary hyperbolic groups with cyclic centralizers. By the approximation theorem, we find $\Gamma \in \mathcal{H}_m$ whose equivalence class is e^{-n}-close to $\Gamma_1, \ldots, \Gamma_k$. Since the distance on $\mathcal{G}_m$ is ultrametric, this means that the equivalence class of Γ is e^{-n}-dense in $\mathcal{H}_m$.

Next, we show that X_n is dense in $\mathcal{H}_m$. So let Γ_0 be a marked non-elementary hyperbolic group with cyclic centralizers; fix $\varepsilon > 0$ with $\varepsilon < e^{-n}$. With $\Gamma_1, \ldots, \Gamma_k$ as above, the approximation theorem applied to $\Gamma_0, \Gamma_1, \ldots, \Gamma_k$, with N big enough to have $e^{-N} < \varepsilon$, provides a group Δ whose equivalence class in e^{-N}-close to $\Gamma_0, \Gamma_1, \ldots, \Gamma_k$.

Moreover, X_n is open in $\mathcal{H}_m$. To see this, first observe that there is a canonical embedding of $\mathcal{G}_m$ into $\mathcal{G}_{m+1}$ (any group on m generators can be viewed as a group on $m+1$ generators); then $\mathcal{G}_m$ is open and closed in $\mathcal{G}_{m+1}$. One can show ([Cha], Proposition 2.3) that the equivalence classes of $\mathcal{R}$ on $\mathcal{G}_m$ are restrictions to $\mathcal{G}_m$ of orbits of a group of homeomorphisms of $\mathcal{G}_{2m}$. This has the important consequence that, if Γ, Γ' are in the same equivalence class and if we fix $\varepsilon > 0$, then for any Δ which is close enough to Γ we can find Δ' in the equivalence class of Δ such that Δ' is ε-close to Γ'. Now, assume that Γ belongs to X_n. Then we can find $\Gamma = \Gamma^1$, $\Gamma^2, \ldots, \Gamma^k$ in the equivalence class of Γ such that any point in $\mathcal{H}_m$ is e^{-n}-close to some Γ^i. Then, for Δ close enough to Γ in $\mathcal{H}_m$, we find $\Delta^1, \Delta^2, \ldots, \Delta^k$ in the equivalence class of Δ such that Δ^i is e^{-n}-close to Γ^i, for $i = 1, 2, \ldots, k$. This means that the equivalence class of Δ is e^{-n}-dense in $\mathcal{H}_m$.

---

(1) It is somewhat unfortunate that marked groups in $\mathcal{H}_m$ seem to be hard to characterize!

Finally, by Baire's theorem, $\bigcap_{n \geqslant 1} X_n$ is a dense G_δ of groups, the equivalence class of which is dense in $\mathcal{H}_m$.

b) See [Cha], lemma 4.13 and § 5.

c) We first note that the fact that $\mathcal{R}$ is given by the restriction to $\mathcal{G}_m$ of a group of homeomorphisms of $\mathcal{G}_{2m}$ also implies that $\mathcal{R}$ *saturates open sets:* i.e., if U is open in $\mathcal{G}_m$, then the union of equivalence classes meeting U is still open.

As a consequence, because of the existence (proved in (a)) of groups with dense orbits in $\mathcal{H}_m$, a property of groups will be satisfied on an open dense subset of $\mathcal{H}_m$ as soon as it is satisfied on some open subset in $\mathcal{H}_m$.

Fix then a finite group F with cyclic centralizers. The set

$$X_F = \{\Gamma \in \mathcal{H}_m : F \text{ embeds in } \Gamma\}$$

is non–empty: indeed, the approximation theorem applied to $\mathbb{F}_2 * F$ and $\mathbb{F}_m$, with $N \geqslant |F|$, provides a non–elementary hyperbolic group Γ with finite centralizers, that belongs to X_F. A sufficiently small neighbourhood of Γ will consist of quotients of Γ that contain F. So the property "F embeds in Γ" is satisfied on an open subset, meaning that X_F is open and dense in $\mathcal{H}_m$.

By Baire's theorem, $\bigcap_F X_F$ is dense in $\mathcal{H}_m$ (where the intersection runs over all finite groups with cyclic centralizers). In passing, we notice that there are other finite groups with cyclic centralizers than the finite cyclic groups; e.g., there is the dihedral group D_n, with n odd, or the group $PSL_2(p)$, for prime p (see [NaS], § 5 of Chapter 2, for this last example).

d) By the remark made at the beginning of (c), we just have to notice that there are finitely presented groups with property (T) in $\mathcal{H}_m$ (for example a torsion-free co-compact lattice in $Sp(n,1)$; if, by bad luck, Γ requires more than m generators, we replace it by a common quotient of Γ and $\mathbb{F}_m$, as given by the approximation theorem).

e) $\mathbb{F}_2$, viewed as an element of $\mathcal{H}_m$, has a neighbourhood consisting of 2–generators groups.

f) Consider the triangle group

$$\Gamma = \Gamma(2,3,7) = \langle r, s, t : r^2 = s^2 = t^2 = (rs)^2 = (rt)^3 = (st)^7 = 1 \rangle \ .$$

It is known that Γ is non–elementary hyperbolic, with cyclic centralizers, and perfect; therefore Γ has a neighbourhood in $\mathbb{F}_m$ consisting of perfect groups.

The "generic" groups appearing in 3.9 are different from Gromov's groups from 3.4; indeed, the latter have the property that every proper subgroup is cyclic.

3.b. A C^*-algebraic application

For G a locally compact group, $L^1(G)$ denotes as usual the convolution algebra of integrable functions on G. Any strongly continuous unitary representation π of G extends to a *-representation of $L^1(G)$, still denoted by π, and defined by

$$\pi(f) = \int_G f(g)\pi(g)dg.$$

Important representations of G are, on the one hand, the universal representation π_{un}, defined as the direct sum of all cyclic representations of G; on the other hand, the left and right regular representations λ and ρ of G on $L^2(G)$, defined respectively by:

$$(\lambda(g)\xi)(h) = \xi(g^{-1}h)$$
$$(\rho(g)\xi)(h) = \xi(hg)\Delta(g)^{1/2} \qquad (\xi \in L^2(G); g, h \in G).$$

(Here $\Delta : G \to \mathbb{R}_+^\times$ is the modular homomorphism of G). Note that λ and ρ are unitarily equivalent via $\xi \to (s \to \Delta(s)^{-1/2}\xi(x^{-1}))$.

3.10. Definition. (a) The *full C^*-algebra* of G, denoted by $C^*(G)$, is the norm closure of $\pi_{un}(L^1(G))$;

(b) the *reduced C^*-algebra* of G, denoted by $C_{red}^*(G)$, is the norm closure of $\lambda(L^1(G))$ equivalently of $\rho(L^1(G)))$.

We shall also need some basic definitions from the theory of C^*-tensor products (see e.g. [Lan], [Tak]). For A, B two C^*-algebras, we denote by $A \otimes B$ their algebraic tensor product, and view it as a *-algebra.

3.11. Definitions.

(a) The *maximal tensor product* of A and B, denoted by $A \otimes_{\max} B$, is the completion of $A \otimes B$ for the norm $||x||_{\max} = \sup\{||\pi(x)|| : \pi$ is a *-representation of $A \otimes B\}$;

(b) the *minimal (or spatial) tensor product* of A and B, denoted by $A \otimes_{min} B$, is the completion of $A \otimes B$ for the norm $||x||_{\min} = \sup\{||(\pi \otimes \sigma)(x)|| : \pi$ is a *-representation of A, σ is a *-representation of $B\}$;

(c) the C^*-algebra A is *nuclear* if, for any C^*-algebra B and any $x \in A \otimes B$, one has $||x||_{\max} = ||x||_{\min}$.

Now, let Γ be a discrete group. Define a representation Θ of $\Gamma \times \Gamma$ on $l^2(\Gamma)$ by $\Theta(g, h) = \lambda(g)\rho(h)$ for $(g, h) \in \Gamma \times \Gamma$. We denote by the same letter Θ the associated *-representation of $C^*(\Gamma \times \Gamma) = C^*(\Gamma) \otimes_{\max} C^*(\Gamma)$ on $l^2(\Gamma)$. Clearly this *-representation always factorizes through $C_{red}^*(G) \otimes_{\max} C_{red}^*(G)$.

3.12. Definition. Γ has the *factorization property* if the *-homomorphism $\Theta : C^*(\Gamma) \otimes_{\max} C^*(\Gamma) \to \mathcal{L}(l^2(\Gamma))$ also factorizes through $C^*(\Gamma) \otimes_{min} C^*(\Gamma)$.

The factorization property was introduced by Kirchberg in [Ki1]; in [Ki2], it was proved that a wide class of discrete groups satisfies this property, namely those groups Γ that embed (not necessarily discretely!) into a locally compact group G such that $C^*(G)$ is nuclear. (It follows from [Con] that any almost connected G fulfills that condition). Because of the abundance of examples, it was asked whether any discrete group has the factorization property. The following result of Kirchberg, giving a negative answer ([Ki3], Theorem 1.1) is therefore quite striking.

3.13. Theorem. *Let Γ be a discrete Kazhdan group. The following are equivalent:*

(i) Γ is maximally almost periodic, i.e. the finite-dimensional unitary representations separate points of Γ;

(ii) Γ is residually finite, i.e. the homomorphisms from Γ onto finite groups separate points of Γ;

(iii) Γ has the factorization property.

Since Gromov's examples presented in 3.4 are simple infinite Kazhdan groups, they cannot be residually finite, hence they do not have the factorization property.

Proof of Theorem 3.13.

(i) $\Rightarrow$ (ii) This implication is true for any finitely generated group, and follows from Malcev theorem that any finitely generated subgroup of a compact Lie group is residually finite [Mal].

(ii) $\Rightarrow$ (iii) Let G be the product of all finite quotients of Γ. Since Γ is residually finite, Γ embeds into G; moreover, G being compact, $C^*(G)$ is nuclear. By Kirchberg's result quoted above, Γ has the factorization property. (See also Proposition 3 of [HRV2] for a direct proof of (ii) $\Rightarrow$ (iii)).

(iii) $\Rightarrow$ (i) We present a proof, based on an idea of M.E.B. Bekka, which is simpler than the original one from [Ki3] (still for a different approach, see [Rob]). For any representation σ of Γ, we denote by $\bar{\sigma}$ the conjugate representation of σ, i.e. the representation σ but viewed on the conjugate Hilbert space of $\mathcal{H}_\sigma$. If π, σ are representations of Γ, it is well-known that the representation $\pi \otimes \bar{\sigma}$ of $\Gamma \times \Gamma$ is unitarily equivalent to the representation of $\Gamma \times \Gamma$ on the space $HS(\mathcal{H}_\sigma, \mathcal{H}_\pi)$ of Hilbert-Schmidt operators from $\mathcal{H}_\sigma$ to $\mathcal{H}_\pi$, given by:

$$(g, h) \rightarrow [X \rightarrow \pi(g)X\sigma(h^{-1})] \qquad (g, h \in \Gamma, X \in HS(\mathcal{H}_\sigma, \mathcal{H}_\pi)).$$

Now we begin the proof. By the factorization property, the representation Θ of $\Gamma \times \Gamma$ is weakly contained in the family $\{\pi \otimes \bar{\sigma} : \pi, \sigma \text{ are cyclic representations of } \Gamma\}$.

1st *step:* Θ is weakly contained in $T = \{\pi \otimes \bar{\sigma} : \pi, \sigma \in \hat{\Gamma}\}$.

This follows from decomposition theory (every cyclic representation τ of Γ is weakly contained in $\hat{\Gamma}$), the fact that weak containment is inherited by tensor products, and transitivity of weak containment.

Let F be the set of elements $\pi \otimes \bar{\sigma}$ in T such that $\pi = \sigma$ and $\dim \pi < \infty$.

2nd *step:* Θ is weakly contained in F.

Θ is a cyclic representation whose coefficient on the cyclic vector δ_1 is given by:

$$< \Theta(g, h)\delta_1 | \delta_1 > = \delta_{g,h} \qquad (g, h \in \Gamma).$$

By the first step, we may approximate pointwise this coefficient by a sequence $(\phi_k)_{k \geqslant 1}$ of normalized positive definite functions on $\Gamma \times \Gamma$ associated with direct sums of elements in T. We may split any such ϕ_k as a sum $\phi_k = \phi'_k + \phi''_k$ where ϕ'_k, ϕ''_k are positive definite functions associated respectively with F and $T - F$. Now, the set of positive definite functions ψ on $\Gamma \times \Gamma$ such that $\psi(1,1) \leqslant 1$ is compact for the topology of pointwise convergence. Therefore, passing if necessary to a subsequence, we may assume that the sequences $(\phi'_k)_{k \geqslant 1}, (\phi''_k)_{k \geqslant 1}$ converge respectively to positive definite functions ϕ', ϕ'' such that $\phi'(g, h) + \phi''(g, h) = \delta_{g,h}$ for all $g, h \in \Gamma$.

What we are left to prove is $\phi'' = 0$.

To see this, we assume by contradiction that $\phi''(1,1) > 0$. We now restrict to the diagonal subgroup Δ of $\Gamma \times \Gamma$, which is canonically identified with Γ. We then have:

$$\phi'(g, g) + \phi''(g, g) = 1 \text{ for all } g \in \Gamma.$$

Since the constant 1 is a pure positive definite function on Δ, we see that $\phi'|_\Delta$ and $\phi''|_\Delta$ must be constant. In particular $\phi''(g, g) = \phi''(1, 1)$ for any $g \in \Gamma$. So $\left(\frac{\phi''_k|_\Delta}{\phi''_k(1,1)}\right)_{k \geqslant 1}$ is a sequence of normalized positive definite functions on Δ converging pointwise to 1, meaning that the trivial representation 1_Δ is weakly contained in the set of restrictions of $T - F$ to Δ. By property $(T), 1_\Delta$ must be contained in that set, i.e. 1_Δ is a subrepresentation of $(\pi \otimes \overline{\sigma})_{|\Delta}$ for some $\pi \otimes \overline{\sigma}$ in $T - F$. But this is a contradiction because it is readily seen, using irreducibility of π, σ and the description of $\pi \otimes \overline{\sigma}$ as a representation on $HS(\mathcal{H}_\sigma, \mathcal{H}_\pi)$ that 1_Δ is a subrepresentation of $(\pi \otimes \overline{\sigma})_{|\Delta}$ if and only if $\pi = \sigma$ and $\dim \pi < \infty$.

3rd step: coda. By the second step, we may approximate $\delta_{g,h}$ pointwise on $\Gamma \times \Gamma$ by a sequence of normalized positive definite functions of the form

$$(g, h) \to \sum_k \operatorname{Tr} \pi_k(g) X_k \pi_k(h^{-1}) X_k^* \,,$$

where π_k is an irreducible, finite-dimensional unitary representation of Γ. Now, fix $g \in \Gamma - \{1\}$ and set $h = 1$; eventually we will have $\sum_k \operatorname{Tr} \pi_k(g) X_k X_k^* \neq 1$, meaning that $\pi_k(g) \neq 1$ for at least one k. This concludes the proof. $\qquad \square$

Notice that the implication (iii) $\Rightarrow$ (i) is false without the property (T) assumption; for an example of a finitely generated non-Kazhdan group, with the factorization property but not residually finite, see §5 in [Ki3].

3.c. A von Neumann algebraic application

We finish this chapter with an application of the existence of discrete Kazhdan groups with infinite centre, obtained by combining 1.14 and 1.18.

3.14. Definitions. Let Γ be a countable group;

(a) Γ is an *infinite conjugacy classes* group (ICC) if any nontrivial conjugacy class of Γ is infinite;

(b) The *von Neumann algebra* of Γ, denoted by $vN(\Gamma)$, is the commutant of the left regular representation $\lambda(\Gamma)$.

It is well-known that $vN(\Gamma)$ is a *factor*, i.e. has centre reduced to $\mathbb{C} \cdot 1$, if and only if Γ is ICC (see e.g. Proposition 7.9 in [Tak]).

Connes and Jones introduced in [CoJ] property (T) for a von Neumann algebra M. We shall not give here the formal definition (see Chapter 10 in [HaV]), but just mention that it is defined in terms of *correspondences* of M, i.e. normal Hilbert bi-modules over M; there is a *trivial correspondence of M*, and M has property (T) if any correspondence of M that weakly contains the trivial correspondence of M, contains it as a sub-correspondence. A basic result of Connes and Jones (Theorem 2 in [CoJ]) is the following.

3.15. Theorem. *Let Γ be a countable ICC group. The following are equivalent:*
(i) $vN(\Gamma)$ has property (T);
(ii) Γ has property (T).

Jolissaint investigated what this result becomes when Γ is not necessarily ICC, and proved ([Jol], Theorem A):

3.16. Theorem. *Let Γ a countable group. The following are equivalent:*
(i) $vN(\Gamma)$ has property (T);
(ii) Γ has property (T) and the subgroup of elements with finite conjugacy class is finite.

Since there are discrete Kazhdan groups with infinite centre, this yields the unexpected result that there exists countable groups Γ with property (T) such that $vN(\Gamma)$ does not have property (T).

4. KAZHDAN CONSTANTS

4.a. General results

Kazhdan constants provide a quantitative approach to property (T). If G is a Kazhdan group and K is a compact generating subset of G, the constant $\mathcal{K}(G, K)$ that we are going to define can be intuitively thought of as the "distance" between the trivial representation 1_G and the rest of the dual of G.

We recall from 1.9 that, if G is a locally compact group with property (T) and K is a compact generating subset, there exists $\varepsilon > 0$ such that, for any unitary representation π of G without non-zero fixed vector, and for any $\xi \in \mathcal{H}_\pi^1$, there exists $k \in K$ such that $||\pi(k)\xi - \xi|| \geqslant \varepsilon$.

4.1. Definition. The *Kazhdan constant* $\mathcal{K}(G, K)$ is the supremum of all $\varepsilon's$ satisfying the above property, i.e.

$$\mathcal{K}(G, K) = \inf_\pi \inf_{\xi \in \mathcal{H}_\pi^1} \max_{k \in K} ||\pi(k)\xi - \xi||$$

where the first infimum is taken over all (equivalence classes of) unitary representations of G without non-zero fixed vector.

If $\mathcal{R}$ is a set of unitary representations of G without non-zero fixed vector, it is convenient to define a constant

$$\mathcal{K}(\mathcal{R}, G, K) = \inf_{\pi \in \mathcal{R}} \inf_{\xi \in \mathcal{H}_\pi^1} \max_{k \in K} ||\pi(k)\xi - \xi||.$$

If $\mathcal{R} = \{\pi\}$, we write $\mathcal{K}(\pi, G, K)$ instead of $\mathcal{K}(\mathcal{R}, G, K)$. Although our primary interest is in the constants associated with a compact generating subset of G, we notice that the definitions of the constants make sense for any compact subset of G.

Remark. The infimum over representations appearing in 4.1 is actually a minimum. Indeed, if $(\pi_n)_{n \geqslant 1}$ is a sequence of representations of G without fixed vector such that $\lim_{n \to \infty} \mathcal{K}(\pi_n, G, K) = \mathcal{K}(G, K)$, set $\pi = \oplus_{n \geqslant 1} \pi_n$; then $\mathcal{K}(\pi, G, K) = \mathcal{K}(G, K)$.

Clearly $\mathcal{K}(G, K)$ is less or equal to 2. However, for non-compact groups, there is a better upper bound.

4.2. Lemma. *Let G be a non-compact group. For any compact subset K of G, one has $\mathcal{K}(G, K) \leqslant \mathcal{K}(\lambda, G, K) \leqslant \sqrt{2}$.*

Proof. Let C be a compact subset of G with non-empty interior. Let $\mu(C) > 0$ denote the Haar measure of C. Let ξ_C be the normalized characteristic function of C, i.e. $\xi_C(g) = \begin{cases} \mu(C)^{-1/2} & \text{if } g \in C \\ 0 & \text{otherwise.} \end{cases}$

Then $||\xi_C|| = 1$ and $< \lambda(g)\xi_C | \xi_C > \geqslant 0$ for any $g \in G$. Hence $||\lambda(g)\xi_C - \xi_C|| \leqslant \sqrt{2}$ for any $g \in G$. $\qquad\square$

If we care about computations of Kazhdan constants in explicit examples, the first case to look at is the one of a compact group G, with $K = G$. The answer in this case is given by the following result, taken from [DeV].

4.3. Proposition. *Let G be a compact group; denote by λ^0 the restriction of the left regular representation λ to the orthogonal of constants in $L^2(G)$. Then:*

(i) if G is finite of order n, then $\mathcal{K}(G, G) = \mathcal{K}(\lambda^0, G, G) = \sqrt{\frac{2n}{n-1}}$;

(ii) if G is infinite, then $\mathcal{K}(G, G) = \mathcal{K}(\lambda^0, G, G) = \sqrt{2}$.

For the circle group $\mathbb{T} = \mathbb{R}/\mathbb{Z}$, the dependence of the constants on the compact generating set was studied in [De2]; there it was shown that, for $I_\delta = [0, \delta]$, one has $\lim_{\delta \to 0} \mathcal{K}(\mathbb{T}, I_\delta) = 0$ and $\lim_{\delta \to 1} \mathcal{K}(\mathbb{T}, I_\delta) = \sqrt{2}$.

For finite groups, the few exact computations that I am aware of are summarized in the following proposition.

4.4. Proposition.

(i) Let C_n be the cyclic group of order n; if g generates C_n, then $\mathcal{K}(C_n, \{g\}) = 2\sin\pi/n$ (Deutsch, [De1], 2.3);

(ii) For $n \geqslant 3$, let D_n be the dihedral group of order $2n$, presented as $D_n = \langle s, t | s^2 = t^2 = (st)^n = 1 \rangle$; then $\mathcal{K}(D_n, \{s, t\}) = 2\sin 2\pi/n$ (Deutsch, [De1], 2.9);

(iii) Let D_n be now presented as $D_n = < r, s | r^n = s^2 = (sr)^2 = 1 >$; then $\mathcal{K}(D_n, \{r, s\}) = \frac{2\sin\pi/n}{\sqrt{1+\sin^2\pi/n}}$ (Cherix, [Ch2]).

(iv) Let $\mathcal{S}_n$ be the symmetric group on $\{1, 2, \ldots, n\}$ with generating set $S = \{(1,2), (2,3), \ldots, (n-1, n)\}$. Let π be the permutation representation of $\mathcal{S}_n$ on $l^2\{1, 2, \ldots, n\}$, and π^0 be the restriction of π to the orthogonal of constants in $l^2\{1, 2, \ldots, n\}$. Then $\mathcal{K}(\mathcal{S}_n, S) = \mathcal{K}(\pi^0, \mathcal{S}_n, S) = \sqrt{\frac{24}{n^3-n}}$ (Bacher and de la Harpe [BaH]).

From (ii) and (iii) above, we see that Kazhdan constants, as expected, highly depend on the choice of the compact generating subset. About (iii), notice that Cherix actually proved a somewhat more general result: let G be a compact abelian group and let the group $\mathbb{Z}/2\mathbb{Z}$ act on G by $\alpha(g) = g^{-1}$ (here α denotes the non-trivial element of $\mathbb{Z}/2\mathbb{Z}$); let K be any compact generating subset of G; then by taking $K \cup \{\alpha\}$ as generating subset of the semi-direct product $G \ltimes \mathbb{Z}/2\mathbb{Z}$, one has

$$\mathcal{K}(G \ltimes \mathbb{Z}/2\mathbb{Z}, K \cup \{\alpha\}) = \frac{2\mathcal{K}(G, K)}{\sqrt{4 + \mathcal{K}(G, K)^2}} .$$

At this juncture, we may mention two results (due to the author) on the behaviour of Kazhdan constants under short exact sequences and direct products. Next lemma gives a quantitative proof for Proposition 1.6 (i).

4.5. Lemma. Let $1 \to G_1 \to G \xrightarrow{p} G_2 \to 1$ be a short exact sequence of locally compact groups, with G/G_1 isomorphic to G_2. Let K_1, K_2 be compact generating subsets of G_1, G_2 respectively. Let K be a compact generating subset of G such that $K \cap G_1 \supseteq K_1$ and $p(K) \supseteq K_2$. Then

$$\mathcal{K}(G, K) \geqslant \frac{\mathcal{K}(G_1, K_1)\mathcal{K}(G_2, K_2)}{\sqrt{\mathcal{K}(G_1, K_1)^2 + \mathcal{K}(G_2, K_2)^2}} .$$

Proof. Let π be a representation of G without non-zero fixed vector. Denote by $\mathcal{H}^1$ the subspace of $\pi(G_1)$-fixed vectors in $\mathcal{H}_\pi$, and by $\mathcal{H}^\perp$ the orthogonal of $\mathcal{H}^1$; notice that both subspaces are invariant under $\pi(G)$, since G_1 is normal in G. Any vector $\xi \in \mathcal{H}_\pi$ has a decomposition $\xi = \xi^1 \oplus \xi^\perp$ in the direct sum decomposition $\mathcal{H}_\pi = \mathcal{H}^1 \oplus \mathcal{H}^\perp$. We set $\alpha = \frac{\mathcal{K}(G_2, K_2)}{\sqrt{\mathcal{K}(G_1, K_1)^2 + \mathcal{K}(G_2, K_2)^2}}$ and fix ξ in the unit sphere of $\mathcal{H}_\pi$. We then consider two cases:

310

(i) $||\xi^\perp|| \geqslant \alpha$. Then

$$\max_{k \in K} ||\pi(k)\xi - \xi|| \geqslant \max_{k_1 \in K_1} ||\pi(k_1)\xi - \xi|| =$$

$$= \max_{k_1 \in K_1} ||\pi(k_1)\xi^\perp - \xi^\perp|| \geqslant \alpha \cdot \mathcal{K}(G_1, K_1),$$

because by definition $\pi(G_1)$ has no non-zero fixed vector in $\mathcal{H}^\perp$.

(ii) $||\xi^\perp|| \leqslant \alpha, i.e. ||\xi^1|| \geqslant \sqrt{1 - \alpha^2}$. Then, because the action of G on $\mathcal{H}^1$ factorizes through G_2, we have:

$$\max_{k \in K} ||\pi(k)\xi - \xi|| = \max_{k \in K}(||\pi(k)\xi^1 - \xi^1||^2 + ||\pi(k)\xi^\perp - \xi^\perp||^2)^{1/2}$$

$$\geqslant \max_{k \in K} ||\pi(k)\xi^1 - \xi^1|| \geqslant \max_{k_2 \in K_2} ||\pi(k_2)\xi^1 - \xi^1||$$

$$\geqslant \sqrt{1 - \alpha^2} \mathcal{K}(G_2, K_2).$$

(The last inequality follows from the fact that G has no non-zero fixed vector on $\mathcal{H}_\pi$, hence G_2 has no non-zero fixed vector on $\mathcal{H}^1$).

Thus, in both cases, we have:

$$\max_{k \in K} ||\pi(k)\xi - \xi|| \geqslant \frac{\mathcal{K}(G_1, K_1)\mathcal{K}(G_2, K_2)}{\sqrt{\mathcal{K}(G_1, K_1)^2 + \mathcal{K}(G_2, K_2)^2}} .$$

The result follows by taking the infimum first over all unit vectors in $\mathcal{H}_\pi$, then over all π's without non-zero fixed vectors. □

4.6. Proposition. *Let G_1, G_2 be Kazhdan groups, with compact generating subsets $K_1 \subseteq G_1, K_2 \subseteq G_2$. In the direct product $G_1 \times G_2$, consider the compact generating subset $K = (K_1 \times \{1\}) \cup (\{1\} \times K_2)$. Then:*

$$\mathcal{K}(G_1 \times G_2, K) = \frac{\mathcal{K}(G_1, K_1)\mathcal{K}(G_2, K_2)}{\sqrt{\mathcal{K}(G_1, K_1)^2 + \mathcal{K}(G_2, K_2)^2}} .$$

Proof. Fix $\varepsilon > 0$. For $i = 1, 2$, let π_i be a representation of G_i, without non-zero fixed vector, such that $\mathcal{K}(G_i, K_i) = \mathcal{K}(\pi_i, G_i, K_i)$; find a unit vector η_i in $\mathcal{H}_{\pi_i}$ such that

$$\max_{k_i \in K_i} ||\pi_i(k_i)\eta_i - \eta_i|| \leqslant \mathcal{K}(\pi_i, G_i, K_i) + \varepsilon.$$

Let π be the representation of $G_1 \times G_2$ on the Hilbert space $\mathcal{H}_{\pi_1} \oplus \mathcal{H}_{\pi_2}$ given by $\pi(g_1, g_2) = \pi_1(g_1) \oplus \pi_2(g_2)$; with α as in the preceding proof, set

$$\xi = \alpha \eta_1 \oplus \sqrt{1 - \alpha^2}\, \eta_2$$

so that $||\xi|| = 1$. Then

$$\max_{k_1 \in K_1} ||\pi(k_1, 1)\xi - \xi|| = \alpha \max_{k_1 \in K_1} ||\pi_1(k_1)\eta_1 - \eta_1||$$
$$\leqslant \alpha(\mathcal{K}(\pi_1, G_1, K_1) + \varepsilon);$$

$$\max_{k_2 \in K_2} ||\pi(1, k_2)\xi - \xi|| = \sqrt{1 - \alpha^2} \max_{k_2 \in K_2} ||\pi_2(k_2)\eta_2 - \eta_2||$$
$$\leqslant \sqrt{1 - \alpha^2}(\mathcal{K}(\pi_2, G_2, K_2) + \varepsilon).$$

Hence

$$\mathcal{K}(\pi, G_1 \times G_2, K) \leqslant \max_{k \in K} ||\pi(k)\xi - \xi||$$
$$\leqslant \max\{\alpha(\mathcal{K}(\pi_1, G_1, K_1) + \varepsilon); \sqrt{1 - \alpha^2}(\mathcal{K}(\pi_2, G_2, K_2) + \varepsilon)\}.$$

Letting ε tend to 0, we get

$$\mathcal{K}(\pi, G_1 \times G_2, K) \leqslant \frac{\mathcal{K}(G_1, K_1)\mathcal{K}(G_2, K_2)}{\sqrt{\mathcal{K}(G_1, K_1)^2 + \mathcal{K}(G_2, K_2)^2}}.$$

The converse inequality follows from 4.5. $\qquad\square$

4.b. The case of SL_n

We now turn to explicit computations -or estimations- of Kazhdan constants for non-compact groups. Here is a result of Burger [Bu2] concerning $SL_3(\mathbb{Z})$; this result is impressive in that the proof is a *tour de force* of ingenuity and clever computation. Before stating it, we need to know that any finite-dimensional unitary irreducible representation π of $SL_3(\mathbb{Z})$ factorizes through some principal congruence subgroup

$$\Gamma(N) = \{g \in SL_3(\mathbb{Z}) : g \equiv 1 \ (\text{mod. } N)\}$$

(see e.g. [Ste] for this). Thus, there exists a smaller integer N_π such that π has this property.

4.7. Theorem. *Let S be the 12-element generating subset of $SL_3(\mathbb{Z})$ consisting of the matrices*

$$S = \begin{pmatrix} 1 & 2 & j \\ 0 & 1 & 0 \\ 0 & 0 & 1 \end{pmatrix}, \quad \begin{pmatrix} 1 & 0 & 0 \\ 2 & 1 & j \\ 0 & 0 & 1 \end{pmatrix},$$

for $j = -1, 0, 1$, and of the transposed inverses of these 6 matrices.

(i) Let π be a non-trivial finite-dimensional unitary irreducible representation of $SL_3(\mathbb{Z})$. Let n_π be the product of all distinct prime factors of N_π. Then

$$\mathcal{K}(\pi, SL_3(\mathbb{Z}), S) \geqslant \left[2 - 2\sqrt{1 - \left(\frac{1 - n_\pi^{-1/2}}{4}\right)^2}\right]^{1/2} \geqslant 0.0732725\ldots$$

(ii) Let Γ be a subgroup of infinite index in $SL_3(\mathbb{Z})$, and let π be the permutation representation of $SL_3(\mathbb{Z})$ on $l^2(SL_3(\mathbb{Z})/\Gamma)$. Then

$$\mathcal{K}(\pi, SL_3(\mathbb{Z}), S) \geqslant [8 + 2\sqrt{15}]^{-1/2} \approx 0.2520086\ldots$$

Burger gives the following non-trivial example where (ii) applies. Let ρ be an irreducible non-trivial representation of $SL_3(\mathbb{R})$ on $\mathbb{R}^n$ (ρ is necessarily non-unitary!). By [Bor], Corollaire 7.13, $\rho(SL_3(\mathbb{Z}))$ preserves a lattice in $\mathbb{R}^n$, so up to a change of basis we may assume $\rho(SL_3(\mathbb{Z})) \subseteq SL_n(\mathbb{Z})$. This gives us a measure-preserving action of $SL_3(\mathbb{Z})$ on the n-dimensional torus $\mathbb{T}^n = \mathbb{R}^n/\mathbb{Z}^n$, hence a permutation representation σ on $L^2(\mathbb{T}^n)$. We claim that the restriction σ^0 of σ to the orthogonal of constants in $L^2(\mathbb{T}^n)$ is a direct sum of representations of the type appearing in 4.7(ii); indeed, by Fourier transform, it is enough to see that $\rho(SL_3(\mathbb{Z}))$ has infinite orbits on $\mathbb{Z}^n - \{0\}$; so let v be a non-zero vector in $\mathbb{Z}^n$ and let Γ be its stabilizer in $SL_3(\mathbb{Z})$; clearly $\rho(\Gamma)$ cannot act irreducibly on $\mathbb{R}^n$; on the other hand, by Borel's density theorem ([Zim], 3.2.5), the restriction of an irreducible representation of $SL_3(\mathbb{R})$ to any lattice remains irreducible; so Γ cannot be a lattice, i.e. Γ must be of infinite index in $SL_3(\mathbb{Z})$. Therefore, we may conclude from 4.7(ii):

$$\mathcal{K}(\sigma^0, SL_3(\mathbb{Z}), S) \geqslant (8 + 2\sqrt{15})^{-1/2}.$$

Concerning $SL_n(\mathbb{R})$ ($n \geqslant 3$), we now rephrase Theorem 1.24 in terms of Kazhdan constants.

4.8. Theorem. *Let G be $SL_n(\mathbb{R})$, $n \geqslant 3$. Let $\Psi : G \rightarrow \mathbb{R}$ be the function defined before 1.23. Fix $\delta \in]0,1[$, and let ε be the unique solution of*

$$4 \sin\left(\frac{\arcsin \varepsilon}{2}\right) + \varepsilon = \sqrt{2(1 - \delta)}.$$

Then $\mathcal{K}(G, \Psi^{-1}[\delta, 1]) \geqslant \varepsilon$.

It seems an interesting question to try to use this result to estimate Kazhdan constants for $SL_n(\mathbb{Z})$ (e.g. by means of induction of representations - see the proof of Proposition 1 in [Bu2]), and to compare the result for $SL_3(\mathbb{Z})$ with the one obtained in Theorem 4.7.

Before we move to our final example, let us introduce a useful element in the full C^*-algebra $C^*(\Gamma)$ of a finitely generated group Γ. If S is a finite generating subset of Γ, set:

$$h = \frac{1}{|S|} \sum_{s \in S} \pi_{un}(s) \in C^*(\Gamma).$$

We shall denote by $Sp(h)$ the spectrum of h_S in $C^*(\Gamma)$, i.e.

$$Sp(h) = \{z \in \mathbb{C} : h - z \text{ is not invertible in } C^*(\Gamma)\}.$$

We recall from the general theory that $Sp(h)$ is a non-empty compact subset of $\mathbb{C}$, actually contained in the unit disk because $\|h\| \leqslant 1$. An extra property is that $1 \in Sp(h)$, because the trivial representation 1_Γ maps h to 1. The following result is essentially taken from Proposition III of [HRV1].

4.9. Proposition. *Let Γ be a finitely generated group, and let S be a finite generating subset.*

(i) If Γ has property (T), then 1 is an isolated point of $Sp(h)$; more precisely,
$$Sp(h) \cap \{z \in \mathbb{C} : |z - 1| < \tfrac{\mathcal{K}(\Gamma,S)^2}{2|S|}\} = \{1\};$$

(ii) Suppose that h is normal, i.e. h commutes with h^*; if $\{1\}$ is isolated in $Sp(h)$, then Γ has property (T); moreover, if $Sp(h) - \{1\} \subseteq \{z \in \mathbb{C} : \operatorname{Re} z \leqslant 1 - \varepsilon\}$ for some positive ε, then $\mathcal{K}(\Gamma, S) \geqslant \sqrt{2\varepsilon}$.

Notice that h is certainly normal if it is self-adjoint, which occurs exactly when S is symmetric $(S = S^{-1})$. In [HRV1], (ii) was stated only in the symmetric case, by lack of non-trivial examples with h normal but non-self-adjoint. But since these examples are now available (see below), and since the proof of (ii) goes over to the normal case, it is worthwhile to mention it that way.

It is not known whether (ii) remains valid without the normality assumption; see §1 in [HRV3] for partial results in that direction.

Proof of Proposition 4.9.

(i) Set $\delta = \tfrac{\mathcal{K}(\Gamma,S)^2}{2|S|}$. Let π denote the restriction of the universal representation of Γ to the orthogonal of fixed vectors. It is clearly enough to prove that $Sp(\pi(h))$ does not meet the open disk $\{z \in \mathbb{C} : |z - 1| < \delta\}$. Now, fix $\xi \in \mathcal{H}_\pi^1$; by definition of Kazhdan constants, we find $s \in S$ such that

$$\|\pi(s)\xi - \xi\| \geqslant \mathcal{K}(\Gamma, S) = \sqrt{2|S|\delta}.$$

It follows from a simple computation on averages of unitary operators (see e.g. Lemma 3 in [HRV1]) that $\|\pi(h)\xi - \xi\| \geqslant \delta$. Choose $w \in \mathbb{C}$ such that $|w| < \delta$. Then

$$\|\pi(h)\xi - (1 + w)\xi\| \geqslant \delta - |w|$$

and similarly, using $\mathcal{K}(\Gamma, S) = \mathcal{K}(\Gamma, S^{-1})$:

$$\|\pi(h^*)\xi - (1 + \overline{w})\xi\| \geqslant \delta - |w|.$$

Since this holds for any $\xi \in \mathcal{H}_\pi^1$, the operator $\pi(h) - (1 + w)$ is invertible, which is what we wanted.

(ii) Let π be a representation of Γ without non-zero fixed vector. We first claim that $1 \notin Sp(\pi(h))$; indeed, suppose by contradiction that 1 is in $Sp(\pi(h))$; since 1 is isolated in $Sp(\pi(h))$, it must be an eigenvalue (by the spectral theorem for normal operators); if ξ is a corresponding eigenvector of norm 1, we have:

$$1 = \frac{1}{|S|} \sum_{s \in S} < \pi(s)\xi|\xi >$$

so that 1 is an average of complex numbers of modulus $\leqslant 1$; by strict convexity of the unit disk:

$$< \pi(s)\xi|\xi >= 1 \text{ for any } s \in S, \text{ i.e. } \pi(s)\xi = \xi \text{ for any } s \in S;$$

since S generates Γ, this means that ξ is a non-zero fixed vector of Γ, a contradiction. Thus $Sp(\pi(h)) \subseteq \{z \in \mathbb{C} : Re\ z \leqslant 1 - \varepsilon\}$; by spectral theory, we have

$$Re < \pi(h)\eta|\eta > \leqslant 1 - \varepsilon \text{ for any } \eta \in \mathcal{H}_\pi^1;$$

so, for any such η, there is at least one $s \in S$ such that $Re \langle \pi(s)\eta|\eta \rangle \leqslant 1 - \varepsilon$. This implies

$$||\pi(s)\eta - \eta||^2 \geqslant 2\varepsilon,$$

hence $\mathcal{K}(\pi, \Gamma, S) \geqslant \sqrt{2\varepsilon}$. The result follows by taking the infimum over π. $\square$

The following consequence of 4.9, due to Deutsch and Robertson (Corollary 5 in [DeR]) ought to be compared to the bound $\mathcal{K}(\Gamma, S) \leqslant \sqrt{2}$ from 4.2.

4.10. Corollary. *For Γ a discrete Kazhdan group:*

$$\sup_F \mathcal{K}(\Gamma, F) \geqslant 1$$

where the supremum is taken over all finite subsets F of Γ.

Proof. Fix S a finite symmetric generating subset of Γ, containing the unit of Γ. Let π be a representation of Γ without non-zero fixed vector. Let F be any finite subset of Γ, and let f be a non-negative function supported on F, such that

$$\sum_{x \in F} f(x) = 1.$$

Then, for any $\xi \in \mathcal{H}_\pi^1$:

$$1 = \sum_{x \in F} [f(x) < \xi - \pi(x)\xi|\xi >] + < \pi(f)\xi|\xi >$$

$$\leqslant \sum_{x \in F} [f(x)||\xi - \pi(x)\xi||] + ||\pi(f)|| \leqslant \max_{x \in F} ||\xi - \pi(x)\xi|| + ||\pi(f)||.$$

Taking the infimum over ξ :

$$1 \leqslant \mathcal{K}(\pi, \Gamma, F) + ||\pi(f)||.$$

We now apply this with $f = h_S^{2n}$, a function supported in S^{2n}. We then have:

$$1 - \mathcal{K}(\pi, \Gamma, S^{2n}) \leqslant ||\pi(h_S^2)||^n.$$

Since π has no non-zero fixed vector and $\pi(h_S^2)$ is a positive operator, it follows from 4.9 (applied to h_S^2) that $||\pi(h_S^2)|| < 1$. Because $S^{2n} \subseteq S^{2(n+1)}$, the sequence $[1 - \mathcal{K}(\pi, \Gamma, S^{2n})]_{n \geqslant 1}$ is decreasing, and bounded below by -1. Letting n tend to ∞, we have $\lim_{n \to \infty}[1 - \mathcal{K}(\pi, \Gamma, S^{2n})] \leqslant 0$, hence

$$1 \leqslant \sup_F \mathcal{K}(\pi, \Gamma, F).$$

The result follows by taking the infimum over π. $\qquad\square$

4.c. Groups acting on $\tilde{A}_2$-buildings

We now describe the only known family of infinite discrete Kazhdan groups where an exact computation of Kazhdan constants was possible. These groups have been introduced by Cartwright, Mantero, Steger and Zappa [CMSZ1], and the computation of Kazhdan constants is due to Cartwright, Młotkowski and Steger [CMS]. This class of groups is especially appealing because of its nice geometric content. Indeed, these groups have a simply transitive action on the vertex set of some Bruhat-Tits buildings (also called affine or euclidean buildings). We shall just give what we need from the theory of buildings.

4.11. Definition. A *projective plane* is a set $\mathcal{P}$ endowed with a family $\mathcal{L}$ of subsets, called *lines*, such that:

(i) any two distinct points determine a unique line;
(ii) any two distinct lines meet in a unique point;
(iii) there are at least two lines.

To any projective plane $(\mathcal{P}, \mathcal{L})$ we associate its *incidence graph*, a bipartite graph whose set of vertices is the disjoint union $\mathcal{P} \amalg \mathcal{L}$, and $x \in \mathcal{P}, L \in \mathcal{L}$ determine an edge if and only if x and L are incident, i.e. $x \in L$.

4.12. Definition.(2) An $\tilde{A}_2$-*building* is a 2-dimensional, connected, contractible simplicial complex X such that the link of any vertex x is the incidence graph of

---

(2) The reader who knows about buildings will find this definition quite different from the traditional one via apartments. They are however equivalent; for definitions of buildings in both global and local terms, we refer to [Ti1].

some projective plane $(\mathcal{P}, \mathcal{L})_x$. A *chamber* in X is a 2-dimensional simplex, i.e. a triangle. The vertex set of X will be denoted by X^0.

For X an $\tilde{A}_2$-building, a crucial feature that follows from simple connectedness is that there exists a well-defined *type function* $t : X^0 \to \{0, 1, 2\}$ such that t restricted to the vertex set of any chamber is a bijection. An automorphism g of X is *type-rotating* if there exists an integer c_g such that for any $v \in X^0 : t(g(v)) \equiv t(v) + c_g$ (mod 3).

Let k be a non-archimedean local field with residual field $\mathbb{F}_q$, the finite field with q elements (q a prime power). It is known (see e.g. [BrT]) that $PGL_3(k)$ has a type-rotating transitive action on some $\tilde{A}_2$-building such that the link of any vertex is the incidence graph of the projective plane $\mathbb{P}_2(\mathbb{F}_q)$ over $\mathbb{F}_q$.

The questions that motivated the authors of [CMSZ1] were:

– Is it possible to construct, and possibly characterize, the groups with a simply transitive type-rotating action on some $\tilde{A}_2$-building?

– Can one exhibit some of them as lattices in $PGL_3(k)$ for a suitable k?

Suppose that Γ has a simply transitive type-rotating action on the $\tilde{A}_2$-building X. Then the link of any vertex of X is isomorphic to the incidence graph of a projective plane $(\mathcal{P}, \mathcal{L})$. Fix a vertex $v_0 \in X^0$, say of type 0; in the identification between the incidence graph of $(\mathcal{P}, \mathcal{L})$ and the link of v_0, $\mathcal{P}$ will go to (say) the neighbours of type 1 of v_0, and $\mathcal{L}$ will go to the neighbours of type 2.

For any neighbour u of v_0, there exists a unique $g_u \in \Gamma$ such that $g_u(v_0) = u$. Then $g_u^{-1}(v_0)$ is still a neighbour of v_0, and we may write

$$g_u^{-1}(v_0) = g_{\lambda(u)}(v_0)$$

for a unique $g_{\lambda(u)} \in \Gamma$. The map $\lambda : \mathcal{P} \amalg \mathcal{L} \to \mathcal{P} \amalg \mathcal{L}$ so defined is an involutive bijection. Moreover, by the type-rotating assumption, $t(u) \neq t(\lambda(u))$, i.e. λ exchanges $\mathcal{P}$ and $\mathcal{L}$; we then view λ as a bijection between $\mathcal{P}$ and $\mathcal{L}$.

Now, let u, v be neighbours of v_0 such that $\lambda(u) = g_u^{-1}(v_0)$ and $v = g_v(v_0)$ are incident (i.e. $v_0, \lambda(u), v$ are the 3 vertices of a chamber). Applying g_u, we see that $g_u g_v(v_0)$ is a neighbour of v_0, which we call $\lambda(w)$; then $g_u g_v(v_0) = g_w^{-1}(v_0)$ which, by simple transitivity, implies:

$$g_u g_v g_w = 1.$$

Note that, in the above construction, the points u, v, w have the same type.

Retracing the steps in the construction, we see that, if $g_u g_v g_w = 1$, then u, v, w must have the same type and $\lambda(u)$ and v must be incident (which, in the projective plane $\mathcal{P}, \mathcal{L}$), translates into $\lambda(u) \in v$ or $\lambda(u) \ni v$). Set

$$\mathcal{I} = \{(u, v, w) \in \mathcal{P}^3 : g_u g_v g_w = 1\};$$

The set $\mathcal{I}$ has two basic properties:

a) $\mathcal{I}$ is invariant under cyclic permutations;

b) For $u, v, \in \mathcal{P}$: there exists $w \in \mathcal{P}$ such that $(u, v, w) \in \mathcal{I}$ if and only if $v \in \lambda(u)$; moreover, if such an element w does exist, it must be unique.

Here is now the first important result ([CMSZ1], Theorem 3.1); it follows from simple connectedness of the building.

4.13. Proposition. Γ *is generated by the* $g'_u s$ *with* $u \in \mathcal{P}$; *moreover*

$$< g_u \ (u \in \mathcal{P}) | g_u g_v g_w = 1 \ \text{for any} \ (u, v, w) \in \mathcal{I} >$$

is a presentation of Γ.

A remarkable thing is that there is a converse ([CMSZ1], Theorem 3.4).

4.14. Theorem. *Let* $(\mathcal{P}, \mathcal{L})$ *be a projective plane. Assume we are given a bijection* $\lambda : \mathcal{P} \to \mathcal{L}$ *and a subset* $\mathcal{I}$ *of* $\mathcal{P}^3$ *satisfying condition a), b) above. Form the group* $\Gamma_{\mathcal{I}}$ *with presentation*

$$\Gamma_{\mathcal{I}} = < g_u(u \in \mathcal{P}) | g_u g_v g_w = 1 \ \text{for any} \ (u, v, w) \in \mathcal{I} > .$$

The Cayley graph of $\Gamma_{\mathcal{I}}$ *with respect to the* $g'_u s$ *and their inverses is the 1-skeleton of an* $\tilde{A}_2$-*building on which* $\Gamma_{\mathcal{I}}$ *acts simply transitively in a type-rotating way.*

A group presentation of the form $< g_u(u \in \mathcal{P}) | g_u g_v g_w = 1$ for any $(u, v, w) \in \mathcal{I} >$ with $(\mathcal{P}, \mathcal{L})$, λ and $\mathcal{I}$ as above is a *triangle presentation*. Theorem 4.14 gives us a complete algebraic characterization of those groups admitting a simply transitive type-rotating action on an $\tilde{A}_2$-building. A basic lemma in the proof of 4.14 is the following one (Proposition 3.2 in [CMSZ1]), giving the normal form of elements in $\Gamma_{\mathcal{I}}$.

4.15. Lemma. *Any element in* $\Gamma_{\mathcal{I}}$ *can be written uniquely as*

$$g_{x_1} g_{x_2} \cdots g_{x_m} g_{y_1}^{-1} g_{y_2}^{-1} \cdots g_{y_n}^{-1}$$

where m, n *are non-negative integers and*

(a) $x_{i+1} \notin \lambda(x_i)$ *for* $1 \leqslant i < m$;

(b) $y_j \notin \lambda(y_{j+1})$ *for* $1 \leqslant j < n$;

(c) $x_m \neq y_1$ *if* $m, n \geqslant 1$.

Of course, the question now arises of the existence of triangle presentations. This is settled in Theorem 5.1 of [CMSZ1], where it is proved that if we start from the projective plane $\mathbb{P}_2(\mathbb{F}_q)$, there exists at least $2^{q/3}$ triangle presentations associated with $\mathbb{P}_2(\mathbb{F}_q)$. It is known that some of them may be embedded as arithmetic cocompact lattices in $PGL_3(k)$, for suitable non-archimedean local fields k. For $q = 2$ (resp. 3), the projective plane $\mathbb{P}_2(\mathbb{F}_q)$ has order 7 (resp. 13), so the symmetric group of $\mathbb{P}_2(\mathbb{F}_q)$ has a "reasonable" order, in the sense that it is possible to run a computer search that will find all possible triangle presentations associated with all possible bijections $\mathcal{P} \rightarrow \mathcal{L}$, and will determine when distinct presentations give isomorphic groups. This is done in Theorems 1, 2 of [CSMZ2]; the neat result is as follows.

4.16. Proposition.

a) For $q = 2$, there are 8 different groups that admit a triangle presentation associated with $\mathbb{P}_2(\mathbb{F}_q)$; they can be realized as arithmetic co-compact lattices either in $PGL_3(\mathbb{Q}_2)$ or $PGL_3(\mathbb{F}_2((X)))$.

b) For $q = 3$, there are 89 different groups that admit a triangle presentation associated with $\mathbb{P}_2(\mathbb{F}_q)$; 16 of them can be realized as arithmetic co-compact lattices in $PGL_3(\mathbb{F}_3((X)))$; 8 of them can be realized as arithmetic co-compact lattices in $PGL_3(\mathbb{Q}_3)$; the remaining 65 groups do not embed in $PGL_3(k)$, for any non-archimedean local k with residual field $\mathbb{F}_3$.

Part b) of 4.16 is especially interesting, for it points at the existence of non-classical $\tilde{A}_2$-buildings, i.e. buildings that do not come from an algebraic group over a local field. Such $\tilde{A}_2$-buildings were already studied by Tits [Ti3], where an invariant allowing one to distinguish between classical and "exotic" buildings was introduced. This invariant involves the geometry of spheres of radius 2 in the building. A somewhat similar invariant allowed the authors of [CMSZ2] to prove that 65 of the buildings they constructed for $q = 3$ were indeed exotic.(3)

Fix now a triangle presentation $\Gamma_{\mathcal{I}} =< g_u \ (u \in \mathcal{P})|g_u g_v g_w = 1$ for any $(u, v, w) \in \mathcal{I} >$, and consider the operator

$$h = \frac{1}{|\mathcal{P}|} \sum_{u \in \mathcal{P}} g_u$$

in $C^*(\Gamma_{\mathcal{I}})$, as in 4.9.

4.17. Lemma. The operator h is normal in $C^*(\Gamma_{\mathcal{I}})$.

Proof. It is enough to show that, for any pair (x, y) of distinct elements in $\mathcal{P}$, there exists a unique pair (x', y') of distinct elements such that:

$$g_x^{-1} g_y = g_{x'} g_{y'}^{-1}.$$

Uniqueness will follow from the uniqueness of the normal form in 4.15. For the existence, x and y determine a unique line $L \in \mathcal{L}$; write $L = \lambda(z)$ for some $z \in \mathcal{P}$.

---

(3) Here is a remark for the reader who knows a bit about buildings. What makes these exotic affine buildings of dimension 2 even more striking is the fact that, in dimension 3 or more, there is no exotic building; more precisely, there is a result of Tits [Ti2] that any affine thick building of dimension $\geqslant 3$ comes from a simple algebraic group over a non-archimedean local field.

By property b) of a triangle presentation, we find x', y' in $\mathcal{P}$ such that the triples (z, x, x') and (z, y, y') belong to $\mathcal{I}$; this means

$$g_z g_x g_{x'} = 1 = g_z g_y g_{y'}$$

from which $g_x^{-1} g_y = g_{x'} g_{y'}^{-1}$ follows immediately. □

In a remarkable piece of work, Cartwright, Mlotkowski and Steger determine explicitly the spectrum of h, when the triangle presentation $\mathcal{I}$ is associated with a finite projective plane $\mathbb{P}_2(\mathbb{F}_q)$ (see Theorem of [CMS]); they observe that 1 is isolated in $Sp(h)$ which, by 4.9.(ii), implies that $\Gamma_\mathcal{I}$ has property (T), and gives a lower bound for the Kazhdan constant $\mathcal{K}(\Gamma_\mathcal{I}, \mathcal{P})$. It turns out that in this case the bound is sharp, as is shown by exhibiting some special representations of $\Gamma_\mathcal{I}$.

4.18. Theorem. *Suppose that $\mathcal{I}$ is a triangle presentation associated with $\mathbb{P}_2(\mathbb{F}_q)$. Then $Sp(h)$ is the subset of $\mathbb{C}$ consisting of all 3 cubic roots of 1 and the region bounded by the curve*

$$\gamma(t) = \frac{q}{q^2 + q + 1} \left[(q^{1/2} + q^{-1/2})e^{it} + e^{-2it} \right] \quad (0 \leqslant t \leqslant 2\pi).$$

Set $\varepsilon_q = 1 - \gamma(0)$; then $\mathcal{K}(\Gamma_\mathcal{I}, \mathcal{P}) = \sqrt{2\varepsilon_q}$; in particular $\Gamma_\mathcal{I}$ has property (T).

About the proof, we shall just say that the abelian C^*-subalgebra $C^*(h)$ generated by 1 and h in $C^*(\Gamma_\mathcal{I})$ plays the role of (a discrete analogue of) the convolution algebra of K-bi-invariant functions with respect to a maximal compact subgroup K of an algebraic group G; in other words, morally we are in a Gelfand pair. (When $\Gamma_\mathcal{I}$ embeds as a lattice into $G = PGL_3(k)$, one could quite probably use the Gelfand pair of G with respect to a maximal compact subgroup K to get the result).

We shall also notice that the representation of $\Gamma_\mathcal{I}$ used to find the exact value $\mathcal{K}(\Gamma_\mathcal{I}, \mathcal{P})$ belongs to a family of representations realized on the building at infinity associated with the building of $\Gamma_\mathcal{I}$. Again, when $\Gamma_\mathcal{I}$ embeds into $G = PGL_3(k)$, this family of representations would be just the restriction to $\Gamma_\mathcal{I}$ of the spherical complementary series of G, realized on the Furstenberg boundary. We conclude by mentioning that 4.18 solves in one stroke two problems from the final section of [HaV], namely the one of computing explicity Kazhdan constants for non-compact groups, and the one of proving that some infinite discrete groups are Kazhdan without appealing to the theory of lattices in semi–simple algebraic groups. In particular, for those exotic groups with a triangle presentation, there is no other known proof of property (T) than the one via computation of $Sp(h)$.

5. APPLICATIONS OF KAZHDAN CONSTANTS

5.a. Finite graphs

We begin with historically the first application, by Margulis [Ma1], to solve a problem in the theory of graphs and communication circuits.

We are going to describe here the approach of Alon-Milman [AlM] to Margulis' idea. For background, the reader should consult the excellent book [Lu2].

First, here are some graph-theoretic definitions (we follow the Appendix of [BaH]).

5.1. Definitions. *Let X be a finite connected graph, with vertex set X^0 and edge set X^1. Set $n = |X^0|$.*

a) For $A \subset X^0$, the boundary ∂A of A is $\partial A = \{x \in X^0 - A : x$ is connected to at least one vertex in $A\}$

b) For $A, \ B \subset X^0$, we set $E(A,B) = \{e \in X^1 : e$ connects a vertex in A with a vertex in $B\}$;

c) The expanding constant of X is

$$c_{\max}(X) = \max\{c > 0 : \frac{|\partial A|}{|A|} \geqslant c\,(1 - \frac{|A|}{n}) \quad \text{for all } A \subset X^0, A \neq \emptyset\};$$

d) The Cheeger constant of X is

$$h(X) = \min_A \frac{|E(A, X^0 - A)|}{\min\{|A|, |X^0 - A|\}}$$

where the minimum is taken over all subsets A of X^0 with $\emptyset \neq A \neq X^0$;

e) The combinatorial Laplacian of X is the operator $\Delta : l^2(X^0) \to l^2(X^0)$ defined by

$$\Delta\xi(x) = d(x)\xi(x) - \sum_{y \text{ neighbour of } x} \xi(y) \qquad (x \in X^0, \xi \in l^2(X^0)).$$

Here $d(x)$ is the degree of the vertex x, i.e. the number of neighbours of x. The operator Δ is a positive operator, so it has a non-negative spectrum:

$$\lambda_0 = 0 < \lambda_1 \leqslant \lambda_2 \leqslant \ldots \leqslant \lambda_{n-1} \ .$$

f) We say that X is k-regular if $d(x) = k$ for all $x \in X^0$.

Here are the relations between $c_{\max}, h$ and λ_1.

5.2. Proposition. *Let X be a finite, connected, k-regular graph. Then:*

$$c_{\max}(X) \geqslant \frac{h(X)}{k};$$

$$h(X) \geqslant \frac{1}{2}c_{\max}(X);$$

$$h(X) \geqslant \frac{1}{2}\lambda_1(X);$$

$$\lambda_1(X) \geqslant \frac{1}{2k}h(X)^2.$$

Proof. See 1.1.4, 4.2.4 and 4.2.5 in [Lu2]. □

A problem important for communication networks (see 1.1 in [Lu2]) is the one of constructing a family $(X_i)_{i \geqslant 1}$ of finite k-regular graphs (k fixed), with $|X_1^0| \to \infty$ for $i \to \infty$, and $c_{max}(X_i)$ bounded below by a positive constant. Such families of graphs are called *infinite families of expanders*. It is not terribly difficult to prove, by counting arguments, that infinite families of expanders do exist (see e.g. 1.2.1 in [Lu2]). It is much harder to construct explicitly such families. Here comes in Margulis' remarkable idea of appealing to property (T).

5.3. Definition. Let Γ be a finitely generated group, with a finite symmetric generating subset S. Let H be a subgroup of Γ. The *Schreier graph* $\mathcal{G}(\Gamma/H, S)$ is the graph with vertex set Γ/H, two vertices aH and bH being connected by an edge if there exists $s \in S$ such that $saH = bH$. (For $H = \{1\}$, this is nothing but the Cayley graph $\mathcal{G}(\Gamma, S)$).

Here is now the Alon-Milman explicit construction of families of expanders [AlM].

5.4. Theorem. *Let Γ be an infinite discrete Kazhdan group. Fix a finite, symmetric generating subset S. Assume that Γ contains a family $(H_i)_{i \geqslant 1}$ of finite index subgroups with $\lim_{i \to \infty} [\Gamma : H_i] = \infty$ (this is the case if Γ is residually finite).*

Then the family of Schreier graphs $X_i = \mathcal{G}(\Gamma/H_i, S)$ is an infinite family of expanders with, for all i:

$$c_{max}(X_i) \geqslant \frac{\mathcal{K}(\Gamma, S)^2}{2};$$

$$\lambda_1(X_i) \geqslant \frac{\mathcal{K}(\Gamma, S)^2}{2};$$

$$h(X_i) \geqslant \frac{\mathcal{K}(\Gamma, S)^2}{4}.$$

Proof. Consider the permutation representation π_i of Γ on $l^2(\Gamma/H_i)$, and the operator

$$h = \frac{1}{|S|} \sum_{s \in S} \pi_{un}(s) \in C^*(\Gamma)$$

appearing before 4.9. Since X_i is $|S|$-regular, the combinatorial Laplacian on X_i is given by

$$\Delta_i = |S|(1 - \pi_i(h));$$

Since, by 4.9 (i):

$$Sp(\pi_i(h)) \subseteq [-1, 1 - \frac{\mathcal{K}(\Gamma, S)^2}{2|S|}] \cup \{1\},$$

we have $\lambda_1(X_i) \geqslant \mathcal{K}(\Gamma, S)^2/2$.

The inequality $h(X_i) \geqslant \mathcal{K}(\Gamma, S)^2/4$ follows from 5.2.

From 5.2, it would also follow

$$c_{\max}(X_i) \geqslant \frac{\mathcal{K}(\Gamma, S)^2}{4|S|};$$

for the proof of the better inequality

$$c_{\max}(X_i) \geqslant \frac{\mathcal{K}(\Gamma, S)^2}{2}$$

we refer to 3.3.1 in [Lu2] or the proof of Theorem 4 of Chapter 8 in [HaV]. □

5.b. Infinite graphs

5.5. Definitions. Let X be an infinite, connected, k-regular graph.

a) The *combinatorial Laplacian* of X is the operator $\Delta : l^2(X^0) \to l^2(X^0)$ defined as for finite graphs. It is a bounded, positive operator, so it has non-negative spectrum and we define

$$\lambda_0(X) = \min Sp(\Delta)$$

b) The *isoperimetric constant* of X is defined as $i(X) = \inf\{|\partial A|/|A|, a \subset X^0, A$ finite non-empty $\}$.

c) For $x_0 \in X^0$, the ball of radius n centered at x_0 is denoted by $B(x_0, n)$; the *growth rate* of X is

$$\tau(X) = \liminf_{n \to \infty} |B(x_0, n)|^{1/n}.$$

(it is easy to see, using connectedness, that this is independent of x_0); we say that X has *exponential growth* if $\tau(X) > 1$.

The relations between $\lambda_0(X), i(X)$ and $\tau(X)$ are as follows.

5.6. Lemma. *Let X be as in 5.5. Then*

a) $i(X) \geqslant \lambda_0(X)/k$

b) $\tau(X) \geqslant i(X) + 1$.

Proof. a) For $A \subset X^0$, denote by χ_A the characteristic function of A. Choose an auxiliary orientation on X; this defines, for any edge e, its origin e^- and extremity e^+. From this, we get the coboundary operator

$$d : l^2(X^0) \to l^2(X^1) : \xi \mapsto (e \mapsto \xi(e^+) - \xi(e^-)).$$

It is easy to check that, for any $\xi \in l^2(\Delta^0)$:

$$< \Delta\xi|\xi >= ||d\xi||^2 = \sum_{e \in X^1} |\xi(e^+) - \xi(e^-)|^2 .$$

Now:

$$\lambda_0(X) = \inf_{||\xi||=1} < \Delta\xi|\xi >= \inf_{||\xi||=1} ||d\xi||^2$$

$$\leqslant \inf_{A \subset X^0, A \text{ finite}, A \neq \emptyset} \frac{||d\chi_A||^2}{||\chi_A||^2}$$

$$= \inf_A \frac{1}{|A|} \sum_{e \in X^1} |\chi_A(e^+) - \chi_A(e^-)|^2$$

$$= \inf_A \frac{|E(A, X^0 - A)|}{|A|} \leqslant \inf_A \frac{k \cdot |\partial A|}{|A|} = k \cdot i(X)$$

b) For any $n \geqslant 0$ we have

$$|B(x_0, n+1)| - |B(x_0, n)| = |\partial B(x_0, n)| \geqslant i(X)|B(x_0, n)|$$

hence

$$|B(x_0, n+1)| \geqslant (i(X) + 1)|B(x_0, n)|,$$

and by an obvious induction:

$$|B(x_0, n+1)| \geqslant (i(X) + 1)^{n+1},$$

which gives the result. $\qquad\qquad\square$

The following result was observed by Mohar ([Moh], corollary 5.2). It shows that the growth rate of Schreier graphs associated with a Kazhdan group is uniformly bounded away from 1.

5.7. Theorem. *Let Γ be an infinite discrete Kazhdan group. Fix a finite, symmetric, generating subset S. For any subgroup H of infinite index in Γ, the Schreier group $X = \mathcal{G}(\Gamma/H, S)$ satisfies:*

i) $\lambda_0(X) \geqslant \mathcal{K}(\Gamma, S)^2/2$;

ii) $i(X) \geqslant \mathcal{K}(\Gamma, S)^2/2|S|$;

iii) $\tau(X) \geqslant 1 + \mathcal{K}(\Gamma, S)^2/2|S|$.

Proof. The first inequality is proved as in 5.4; the second and third inequalities then follow from 5.6. $\qquad\qquad\square$

5.c. Riemannian manifolds and coverings

Now we consider applications of Kazhdan constants to Riemannian geometry. We first recall some basic notions.

5.8. Definitions. Let (M, g) be a compact, connected, Riemannian manifold.

a) The *Laplace operator* on M is the second order differential operator on $C^\infty(M)$ defined by

$$\Delta = -\text{div grad}.$$

Δ is a positive unbounded operator on $L^2(M)$; it has non-negative spectrum consisting of a sequence of eigenvalues

$$\lambda_0 = 0 < \lambda_1 \leqslant \lambda_2 \leqslant \dots$$

increasing to $+\infty$.

b) The *Cheeger constant* of M is defined by

$$h(M) = \inf_S \frac{\text{area}(S)}{\min\{\text{vol}(A), \text{vol}(B)\}}$$

where S runs over hypersurfaces of M that divide M into two pieces A and B.

The link between $h(M)$ and $\lambda_1(M)$ is provided by the following two inequalities; the first one in due to Cheeger [Che]; the second to Buser [Bus].

5.9. Theorem. (i) $\lambda_1(M) \geqslant h(M)^2/4$;

(ii) There exists a constant $A(M) > 0$ (depending on the dimension and the Ricci curvature of M) such that:

$$\lambda_1(M) \leqslant Ah(M) + 10h(M)^2.$$

Let $\Gamma = \pi_1(M)$ be the fundamental group of M. Compactness of M ensures that Γ is finitely generated (see 3.1), so we fix a finite, symmetric generating subset S of Γ.

Let $M' \to M$ be a finite-sheeted covering of M; then $H = \pi_1(M')$ is a finite index subgroup of Γ, and we may form the Schreier graph $\mathcal{G}(\Gamma/H, S)$. We endow M' with the Riemannian structure pulled back from M via the covering map. The following result is due to Burger (Corollaire 1.a in [Bu1]).

5.10. Proposition. (Notations being as above). There exists a constant $c(M) > 0$ such that, for any finite-sheeted covering $M' \to M$,

$$\lambda_1(M') \geqslant c(M)\lambda_1(X)$$

where $X = \mathcal{G}(\Gamma/H, S)$.

From this, 5.4 and 5.9, we immediately deduce the following result, due to Brooks ([Bro], Theorem 3).

5.11. Theorem. *Let* (M, g) *be a Riemannian manifold. Suppose that* $\Gamma = \pi_1(M)$ *has property* (T). *Let* $M' \to M$ *be a finite sheeted covering of* M. *Then*

(a) $\lambda_1(M') \geqslant \frac{c(M)\mathcal{K}(\Gamma, S)^2}{2}$ *where* $c(M)$ *is the constant appearing in 5.10;*

(b) $h(M') \geqslant \frac{-A(M) + \sqrt{A(M)^2 + 20c(M)\mathcal{K}(\Gamma, S)^2}}{20}$ *where* $A(M)$ *is the constant appearing in 5.9 (ii).*

(Concerning (b), it has to be remarked that $A(M)$, depending only on the dimension and Ricci curvature of M, is invariant under finite-sheeted coverings, i.e. $A(M') = A(M)$).

For results similar to 5.11 for infinite-sheeted coverings, we refer to [Bro] and [Bu1]; these results are analogous to 5.7.

Proposition 5.10 makes it clear that there is a strong relation between the first eigenvalue of the Laplace operator on a finite-sheeted covering of M, and the first eigenvalue of the combinatorial Laplacian on the corresponding Schreier graph. Actually one has the following result, advertised in [Lu2] 4.3.1 and [LuZ], Proposition 1.2.

5.12. Theorem-Definition. *Let* Γ *be a finitely generated group, and let* S *be a finite, symmetric, generating subset. The following are equivalent.*

(i) *The trivial representation* 1_Γ *is isolated in the subset of* $\hat{\Gamma}$ *consisting of all representations that factor through a finite quotient of* Γ;

(ii) *there exists* $\varepsilon_1 > O$ *such that, for any representation* π *that factors through a finite quotient of* Γ *and without non-zero fixed vector:*

$$\mathcal{K}(\pi, \Gamma, S) = \inf_{\xi \in \mathcal{H}_\pi^1} \max_{s \in S} \|\pi(s)\xi - \xi\| \geqslant \varepsilon_1;$$

(iii) *there exists* $\varepsilon_2 > O$ *such that, for any normal subgroup* N *of finite index in* Γ:

$$c_{\max}(\mathcal{G}(\Gamma/N, S)) \geqslant \varepsilon_2;$$

(iv) *there exists* $\varepsilon_3 > O$ *such that, for any normal subgroup* N *of finite index in* Γ:

$$\lambda_1(\mathcal{G}(\Gamma/N, S)) \geqslant \varepsilon_3;$$

(v) *there exists* $\varepsilon_4 > O$ *such that, for any normal subgroup* N *of finite index in* Γ:

$$h(\mathcal{G}(\Gamma/N, S)) \geqslant \varepsilon_4.$$

Γ *is said to have property* (τ) *if it satisfies the equivalent properties* $(i) \leftrightarrow (v)$.

Suppose moreover that $\Gamma = \pi_1(M)$ for some compact Riemannian manifold (M, g). Then $(i) \leftrightarrow (v)$ are still equivalent to:

(vi) there exists $\varepsilon_5 > O$ such that, for any finite-sheeted Galois covering $M' \to M$:

$$\lambda_1(M') \geqslant \varepsilon_5;$$

(vii) there exists $\varepsilon_6 > O$ such that, for any finite-sheeted Galois covering $M' \to M$:

$$h(M') \geqslant \varepsilon_6.$$

Sketch of proof.

(i) $\Rightarrow$ (ii) Denote by $\hat{\Gamma}_f$ the set of (classes of) irreducible representations of Γ that factor through some finite quotient of Γ. We suppose by contradiction that 1_Γ is isolated in $\hat{\Gamma}_f$, but that there is no ε_1 as in (ii). This means that we can find a sequence $(\pi_n)_{n\geqslant 1}$ of representations factoring through finite quotients of Γ, such that $\pi = \bigoplus_{n\geqslant 1} \pi_n$ almost has invariant vectors but no non-zero fixed vector. Decomposing π into irreducibles, we have $\pi = \bigoplus_{k\geqslant 1} \rho_k$, where each summand ρ_k is in $\hat{\Gamma}_f$. Let then $V(S,\varepsilon)$ be a neighbourhood of 1_Γ in $\hat{\Gamma}$ (see 1.10) such that $V(S,\varepsilon) \cap \hat{\Gamma}_f = \{1_\Gamma\}$. Fix $\delta > 0$; since π almost has invariant vectors, we find $\xi \in \mathcal{H}_\pi^1$ such that $\delta \geqslant \sum_{s \in S} \|\pi(s)\xi - \xi\|^2$. We write $\xi = (\xi_k)_{k\geqslant 1}$ in the decomposition $\mathcal{H}_\pi = \bigoplus_{k\geqslant 1} \mathcal{H}_{\rho_k}$; we may clearly assume that $\xi_k \neq 0$ for any k. Then

$$\delta \geqslant \sum_{s \in S} \sum_{k \geqslant 1} \left\| \rho_k(s) \frac{\xi_k}{\|\xi_k\|} - \frac{\xi_k}{\|\xi_k\|} \right\|^2 \cdot \|\xi_k\|^2 .$$

For at least one k, we then have

$$\delta \geqslant \sum_{s \in S} \left\| \rho_k(s) \frac{\xi_k}{\|\xi_k\|} - \frac{\xi_k}{\|\xi_k\|} \right\|^2 .$$

By taking δ small enough, we certainly have

$$\varepsilon > \max_{s \in S} \left| \langle \rho_k(s) \frac{\xi_k}{\|\xi_k\|} \Big| \frac{\xi_k}{\|\xi_k\|} \rangle - 1 \right|$$

i.e. $\rho_k \in V(S,\varepsilon) \cap \hat{\Gamma}_f$. Thus π contains non-zero fixed vectors, which is a contradiction.

(ii) $\Rightarrow$ (iv) This is proved as in 5.4.

(iv) $\Rightarrow$ (i) If $\rho \in \hat{\Gamma}_f - \{1_\Gamma\}$ factors through Γ/N, then ρ appears in the restriction of the left regular representation $\lambda_{\Gamma/N}$ to the orthogonal $l_0^2(\Gamma/N)$ of constants in $l^2(\Gamma/N)$. Viewing ρ as a subrepresentation of $\lambda_{\Gamma/N}$, we then have, for $\xi \in \mathcal{H}_\rho^1$:

$$\varepsilon_3 \leqslant \langle \Delta\xi | \xi \rangle = \sum_{s \in S} \langle \xi - \lambda_{\Gamma/N}(s)\xi | \xi \rangle$$

$$= \sum_{s \in S} \langle \xi - \rho(s)\xi | \xi \rangle \leqslant \sum_{s \in S} \| \xi - \rho(s)\xi \| .$$

Therefore $\max_{s \in S} \|\xi - \rho(s)\xi\| \geqslant \varepsilon_3/|S|$. From this it clearly follows that 1_Γ is isolated in $\hat{\Gamma}_f$. (This proof is taken from [Lu2], 4.3.2).

The equivalences (iii) $\Leftrightarrow$ (iv) $\Leftrightarrow$ (v) follow from 5.2, while (vi) $\Leftrightarrow$ (vii) follows from 5.9, and (iv) $\Rightarrow$ (vi) follows from 5.10.

(vii) $\Rightarrow$ (v) We give an idea of the proof of this implication, according to Theorem 1 of [Bro]. We claim that there exists a constant $C > 0$ such that, for any finite-sheeted Galois covering M' of M:

$$h(M') \leqslant C\, h(\mathcal{G}(\Gamma/N, S)) \qquad (*)$$

where $N = \pi_1(M')$. Let F be a fundamental domain for the action of Γ on the universal covering $\tilde{M}$. Then $M' = \tilde{M}/N$ is tesselated by finitely many copies of F, indexed by Γ/N. (This is to say that we may think of $\mathcal{G}(\Gamma/N, S)$ as being drawn on M'.) Any decomposition of $\mathcal{G}(\Gamma/N, S)$ then gives rise to a decomposition of M', implying the inequality $(*)$. $\qquad\square$

Property (τ) was investigated in [LuZ]. Clearly, a discrete Kazhdan group has property (τ), but the converse is not true. Indeed, $\Gamma = SL_2(\mathbb{Z}[\frac{1}{p}])$, where p is a prime, is certainly not Kazhdan, since it is dense in $SL_2(\mathbb{R})$ (see 1.4.(i)).

Let us check that Γ has property (τ). So, let N be a normal subgroup of finite index. By Serre's solution to the congruence subgroup problem [Ser], we find n, not divisible by p, such that N contains the *congruence subgroup* $\Gamma(n)$, i.e. the kernel of the reduction modulo n:

$$SL_2\left(\mathbb{Z}\left[\frac{1}{p}\right]\right) \longrightarrow SL_2\left(\frac{\mathbb{Z}\left[\frac{1}{p}\right]}{n\,\mathbb{Z}\left[\frac{1}{p}\right]}\right).$$

Now $\Gamma/N \simeq \Gamma/\Gamma(n)\big/N/\Gamma(n)$, which means that the Schreier graph $\mathcal{G}(\Gamma/\Gamma(n), S)$ is a covering of $\mathcal{G}(\Gamma/N, S)$; this implies

$$\lambda_1(\mathcal{G}(\Gamma/\Gamma(n), S)) \leqslant \lambda_1(\mathcal{G}(\Gamma/N, S)) \, .$$

On the other hand $\mathbb{Z}\left[\frac{1}{p}\right]/n\,\mathbb{Z}\left[\frac{1}{p}\right] \simeq \mathbb{Z}/n\,\mathbb{Z}$, so that

$$\Gamma/\Gamma(n) \simeq SL_2(\mathbb{Z})/(SL_2(\mathbb{Z}) \cap \Gamma(n)).$$

Then we appeal to Selberg's theorem $\lambda_1 \geqslant \frac{3}{16}$ for arithmetic surfaces [Sel]. More precisely, if M' is the quotient of the Poincaré upper half-plane by any congruence subgroup of $SL_2(\mathbb{Z})$, one has:

$$\lambda_1(M') \geqslant \frac{3}{16} \, ,$$

328

i.e., Γ has property (τ).

By way of contrast with Selberg's theorem quoted above, we have *Selberg's conjecture*: let M be the quotient of the Poincaré upper half-plane by an *arithmetic lattice*; then $\lambda_1(M) \geqslant \frac{1}{4}$.

In opposition with $SL_2\left(\mathbb{Z}\left[\frac{1}{p}\right]\right)$, the group $SL_2(\mathbb{Z})$ does *not* have property (τ); this follows from

5.13. Proposition. *Let Γ be a lattice in $PSL_2(\mathbb{R})$. Then Γ does not have property (τ).*

Proof. Let H be a finite index subgroup of Γ. It is easy to see that Γ has property (τ) if and only if H has. So, by passing to a torsion-free finite index subgroup of Γ, we may assume that Γ is torsion-free. We appeal then to a result of Randol [Ran]: if M is a Riemann surface with constant curvature-1 and finite area, there exists a sequence $(M_n)_{n \geqslant 1}$ of finite-sheeted Galois coverings of M such that

$$\lim_{n \to \infty} \lambda_1(M_n) = 0 \ .$$

Alternatively, we may use the fact that Γ is either a surface group (in the uniform case) or a free group (in the non–uniform case). In both cases Γ has a quotient isomorphic to $\mathbb{Z}$, which clearly does not have property (τ). The result follows by observing that property (τ) is inherited by quotients. $\qquad\square$

5.d. Miscellany

Here is a final application of Kazhdan constants to discrete groups. Let Γ be a discrete Kazhdan group. It is known that, for any $d \geqslant 1$, the group Γ has only finitely many unitary irreducible representations in degree d. The number of such representations was estimated in [HRV1], Proposition IV.

5.14. Proposition. *Let Γ be a discrete Kazhdan group, with a finite generating subset S. For $d \geqslant 1$, denote by Irrep (d) the number of unitary irreducible representation of Γ of degree $\leqslant d$. Then, for any*

$$C > -2|S|\log\left(\frac{\mathcal{K}(\Gamma, S)^2}{4|S|}\right)$$

one has:

$$\text{Irrep } (d) = O(e^{Cd^2}) \ .$$

We conclude this Chapter with a few words about applications of Kazhdan constants to non-discrete groups. So let G be a locally compact group, with a compact generating subset K. Recall that we endowed the dual $\hat{G}$ with the Fell topology (usually a bad topology!). Kaniuth and Taylor studied in [KaT] continuity properties of the map

$$\hat{G} \to \mathbb{R} : \pi \mapsto \mathcal{K}(\pi, G, K).$$

It turns out that, at least for connected amenable groups, continuity of this map imposes stringent conditions on G.

5.15. Proposition. *([KaT], Theorem 1): Let G be an almost connected amenable group. The following are equivalent.*

(i) $\mathcal{K}(\cdot, G, K)$ *is continuous on $\hat{G}$;*

(ii) *there exists a compact normal subgroup N in G such that G/N is abelian.*

For specific groups, $\mathcal{K}(\cdot, G, K)$ may be continuous on "big" parts of $\hat{G}$.

5.16. Proposition. *([KaT], Corollary 1 and Theorem 5).*

(i) *Let G be the motion group $\mathbb{R}^n \rtimes SO(n)$ ($n \geqslant 2$). Then $\mathcal{K}(\cdot, G, K)$ is continuous on the set of infinite-dimensional representations in $\hat{G}$.*

(ii) *Let G be a semi-simple real Lie group with finite centre. Then $\mathcal{K}(\cdot, G, K)$ is continuous on the set of principal series representations of G.*

REFERENCES

[AlM]   N. Alon, V. D. Milman, λ_1, *Isoperimetric inequalities for graphs, and superconcentrators*, J. Combin. Theory, Ser. B **38** (1985), 73-88.

[BaH]   R. Bacher, P. de la Harpe, *Exact values of Kazhdan constants for some finite groups*, Journ. of Algebra **163** (1994), 495-515.

[BaB]   M.W. Baldoni–Silva, D. Barbasch, *The unitary spectrum for real rank one groups*, Invent. Math. **72** (1983), 27-55.

[BGS]   W. Ballman, M. Gromov, V. Schroeder, *Manifolds of nonpositive curvature*, Birkhäuser, 1985.

[Beh]   H. Behr, $SL_3(F_q[t])$ *is not finitely presentable*, in Homological Group Theory, C.T.C. Wall ed., Cambridge Univ., London, 1979, pp. 213-224.

[Bo1]   A. Borel, *Compact Clifford-Klein forms of symmetric spaces*, Topology, vol. 2, 1963, pp. 111-122.

[Bo2]   A. Borel, *Introduction aux groupes arithmétiques*, Actualités Sci. et Indus., vol. 1341, Hermann, 1969.

[Bro]   R. Brooks, *The spectral geometry of a tower of coverings*, J. Differential Geometry **23** (1986), 97-107.

[BrT]   F. Bruhat, J. Tits, *Groupes réductifs sur un corps local (données radicielles valuées)*, Publ. Math. IHES **41** (1972), 5-252.

[Bu1]   M. Burger, *Spectre du Laplacien, graphes et topologie de Fell*, Comment. Math. Helvetici **63** (1988), 226-252.

[Bu2]   M. Burger, *Kazhdan constants for $SL_3(\mathbb{Z})$*, J. Reine Angew. Math. **43** (1991), 36-67.

[CMSZ1] D.I. Cartwright, A. Mantero, T. Steger, A. Zappa, *Groups acting simply transitively on the vertices of a building of type $\tilde{A}_2$, I*, Geom. Ded. **47** (1993), 143-166.

[CMSZ2] D.I. Cartwright, A.M. Mantero, T. Steger, A. Zappa, *Groups acting simply transitively on the vertices of a building of type $\tilde{A}_2$, II: The cases $q = 2$ and $q = 3$*, Geom. Ded. **47** (1993), 167-226.

[CMS]   D.I. Cartwright, W. Młotkowski, T. Steger, *Property (T) and $\tilde{A}_2$-groups*, Ann. Inst. Fourier (Grenoble) **44**, 1 (1993), 213-248.

[Cha]   C. Champetier, *L'espace des groupes de type fini*, Preprint, Grenoble (1993).

[Ch1]   P.-A. Cherix, *Propriétés de groupes héritées par commensurabilité*, Travail de Licence, Neuchâtel (1991).

[Ch2]   P.-A. Cherix, *Property (T) and expanding constants for semi–direct products* (to appear in Linear & Multilin. Algebra).

[Con]   A. Connes, *Classification of injective factors*, Ann. Math. **103** (1976), 73-115.

[CoJ]   A. Connes, V. Jones, *Property T for von Neumann algebras*, Bull. London Math. Soc. **17** (1985), 57-62.

[Cow]   M. Cowling, *The Kunze-Stein phenemenon*, Ann. Math. **107** (1978), 209-234.

[CoH]   M. Cowling, U. Haagerup, *Completely bounded multipliers of the Fourier algebra of a simple Lie group of real rank one*, Invent. Math **96** (1989), 507-549.

[DeK]   C. Delaroche, A. Kirillov, *Sur les relations entre l'espace dual d'un groupe et la structure de ses sous-groupes fermés*, Sém. Bourbaki, Exposé 343, 20ème année (1967-1968).

[Del]   T. Delzant, *Sous-groupes distingués et quotients des groupes hyperboliques*, Preprint, Univ. de Strasbourg (1991).

[De1]   A.J. Deutsch, *Kazhdan's property (T) and related properties of locally compact and discrete groups*, Ph.D. Thesis, Univ. of Edinburgh (1992).

[De2]   A.J. Deutsch, *Kazhdan constants for the circle* (to appear in Bull. London Math. Soc.).

[DeR]   A.J. Deutsch, A.G. Robertson, *Functions of conditionally negative type on Kazhdan groups* (to appear in Proc. Amer. Math. Soc.).

331

[DeV]   A.J. Deutsch, A. Valette, *On diameters of orbits of compact groups in unitary representations* (to appear in J. Austral. Math. Soc.).

[Dix]   J. Dixmier, C^*-algebras, North–Holland, 1982.

[Fel]   J.M.G. Fell, *Weak containment and induced representations of groups*, Canad. J. Math. **14** (1962), 237-268.

[GhH]   E. Ghys, de la Harpe (eds.), *Sur les groupes hyperboliques d'après M. Gromov*, Birkhäuser, 1990.

[Gri]   R. I. Grigorchuk, *Degrees of growth of finitely generated groups and the theory of invariant means*, Math. USSR Izvestiya **25** (1985), 259-300.

[Gro]   M. Gromov, *Hyperbolic groups*, in "Essays in Group Theory", S.M. Gersten ed., Springer, 1987, pp. 75-263.

[GrP]   M. Gromov, P. Pansu, *Rigidity of lattices: an introduction*, in "Geometric topology: recent developments", Lect. Notes in Math., vol. 1504, Springer, 1992, pp. 39-137.

[GrS]   M. Gromov, R. Schoen, *Harmonic maps into singular spaces and p-adic superrigidity for lattices in groups of rank one*, Preprint (1991).

[Gui]   A. Guichardet, *Etude de la 1-cohomologie et de la topologie du dual pour les groupes de Lie à radical abélien*, Math. Ann. **228** (1977), 215-232.

[HRV1]   P. de la Harpe, A.G. Robertson, A. Valette, *On the spectrum of the sum of the generators of a finitely generated group*, Israel J. Math. **81** (1993), 65-96.

[HRV2]   P. de la Harpe, A.G. Robertson, A. Valette, *On exactness of group C^*-algebras*, to appear in Quart. J. Math.

[HRV3]   P. de la Harpe, A.G. Robertson, A. Valette, *On the spectrum of the sum of the generators of a finitely generated group, II*, Colloquium Math. **LXV** (1993), 87-102.

[HaV]   P. de la Harpe, A. Valette, *La propriété (T) de Kazhdan pour les groupes localement compacts*, Astérisque, vol. 175, Soc. Math. France, Paris, 1989.

[Hoc]   G. Hochschild, *The structure of Lie groups*, Holden-Day, 1965.

[HoM]   R.E. Howe, C.C. Moore, *Asymptotic properties of unitary representations*, J. Funct. Anal. **32** (1979), 72-96.

[HoT]   R.E. Howe, E. Tan, *Nonabelian harmonic analysis (applications of $SL_2(\mathbb{R})$)*, Springer, 1992.

[Jol]   P. Jolissaint, *Property T for discrete groups in terms of their regular representation*, Math. Ann. **297** (1993), 539-551.

[KaT]   E. Kaniuth, K.F. Taylor, *Kazhdan constants and the dual space topology*, Math. Ann. **293** (1992), 495-508.

[Kaz]   D. Kazhdan, *Connection of the dual space of a group with the structure of its closed subgroups*, Funct. Anal. and Appl. **1** (1967), 63-65.

[Ki1]   E. Kirchberg, *Positive maps and C^*-nuclear algebras*, Proc. Intern. Conf. on Operator Algebras, Ideals and their Applications in Theoretical Physics (1978), Teubner (Leipzig), 327-328.

[Ki2]   E. Kirchberg, *On non-semisplit extensions, tensor products and exactness of group C^*-algebras*, Invent. Math. **112** (1993), 449-489.

[Ki3]   E. Kirchberg, *Discrete groups with Kazhdan's property T and factorization property are residually finite* (to appear in Math. Ann.).

[Kos]   B. Kostant, *On the existence and irreducibility of certain series of representations*, Bull. Amer. Math. Soc. **75** (1969), 627-642.

[Lan]   E.C. Lance, *Tensor products and nuclear C^*-algebras*, Proc. Symp. Pure Math. **38-1**, Amer. Math. Soc. (1982), 379-399.

[Lu1]   A. Lubotzky, *Trees and discrete subgroups of Lie groups over local fields*, Bull. (New series) Amer. Math. Soc. **20** (1989), 27-30.

[Lu2]   A. Lubotzky, *Discrete groups, expanding graphs and invariant measures*, Birkhäuser, (to appear).

332

[LuZ]   A. Lubotzky, R.J. Zimmer, *Variants of Kazhdan's property for subgroups of semisimple groups*, Israel J. Math. **66** (1989), 289-299.

[LyS]   R.C. Lyndon, P.E. Schupp, *Combinatorial group theory*, Springer, 1977.

[Mal]   A.I. Mal'cev, *On the faithful representations of infinite groups by matrices*, Amer. Math. Soc. Transl. **45** (1965), 1-18.

[Ma1]   G.A. Margulis, *Explicit constructions of concentrators*, Problems Inform. Transmission 9-4 (1973), 325-332.

[Ma2]   G.A. Margulis, *Discrete subgroups of semisimple Lie groups*, Springer, 1991.

[Moh]   B. Mohar, *Some relations between analytic and geometric properties of infinite graphs*, Discrete Math. **95** (1991), 193-219.

[Mos]   G.D. Mostow, *Strong rigidity of locally symmetric spaces*, Princeton Univ. Press, 1973.

[NaS]   M.A. Naimark, A.I. Štern, *Theory of group representations*, Grundlehren der Math. Wiss. **246**, Springer, 1982.

[vNe]   J. von Neumann, *Zur allgemeinen Theorie des Masses*, Fund. Math. **13** (1929), 73-116.

[Ols]   A.Y. Olshanski, *On a geometric method in the combinatorial group theory*, Proc. ICM, Warszawa **1** (1984), 415-424.

[Rag]   M.S. Raghunathan, *Discrete subgroups of Lie groups*, Springer, 1979.

[Ran]   B. Randol, *Small eigenvalues of the Laplace operator on compact Riemann surfaces*, Bull. Amer. Math. Soc. **80** (1974), 996–1000.

[ReS]   U. Rehmann, C. Soulé, *Finitely presented groups of matrices*, in "Algebraic K-theory", Evanston 1976, Lect. Notes in Math. **551**, Springer, 1976, 164–169.

[Rob]   A.G. Robertson, *Property (T) for II_1-factors and unitary representations of Kazhdan groups*, Math. Ann. **296** (1993), 547–555.

[Sel]   A. Selberg, *On the estimation of Fourier coefficients of modular forms*, Proc. Sympos. Pure Math. **8**, Amer. Math. Soc. (1965), 1–15.

[Ser]   J.P. Serre, *Le problème des groupes de congruence pour SL_2*, Ann. Math. **92** (1970), 489–527.

[Ste]   R. Steinberg, *Some consequences of the elementary relations in SL_n*, Contemporary Math. **45** (1985), 335-350.

[Tak]   M. Takesaki, *Theory of operator algebras I*, Springer, 1979.

[Ti1]   J. Tits, *A local approach to buildings*, in The Geometric Vein, The Coxeter Festschrift, Springer, 1981, pp. 519-547.

[Ti2]   J. Tits, *Immeubles de type affine*, in Buildings and the Geometry of Diagrams (Como 1984), Springer, Lect. Notes in Math., vol 1181, 1986, pp. 159-190.

[Ti3]   J. Tits, *Spheres of radius 2 in triangle buildings*, in Finite Geometries, Buildings and Related Topics, W.M. Kantor et al. eds., Clarendon Press, 1990, pp. 17-28.

[Val]   A. Valette, *A global approach to spherical functions on rank 1 symmetric spaces*, Nieuw Archief voor Wiskunde **5** (1987), 33–52.

[Wan]   S.P. Wang, *On the Mautner phenomenon and groups with property (T)*, Amer. J. Math. **104** (1982), 1191-1210.

[Zim]   R.J. Zimmer, *Ergodic theory and semisimple groups*, Birkhäuser, 1984.

Université de Neuchâtel, Rue Emile Argand, 11, CH-2007 Switzerland

e–mail: `valette@maths.unine.ch`